Self-Tests

	Chapter–Test Number	Title of Self-Test
Part I	2-1	Significant figures
	2-2	The arithmetic mean and the median
	2-3	Average deviation, standard deviation, and the range
	2-4	The Q test
	2-5	Reporting laboratory results
	4-1	Preparation for using the single-pan balance
	5-1	Calculation of gravimetric factors
	5-2	Calculation of solubility in pure water
	6-1	Calculation of percentage composition and titrant concentration
	7-1	Deciding whether a reaction is quantitative
	8-1	Strong and weak acids and bases
	9-1	Choosing indicators
	9-2	Calculation of acid-base titration results
	10-1	Calculation of precipitation titration results
	10-2	Calculation of EDTA titration results
	11-1	The ion-electron method
	11-2	Equivalent weights and other calculations
Part II	13-1	Absorbance, transmittance, and radiant power
	13-2	Beer-Lambert law calculations
	14-1	Electronic absorption
	15-1	Fluorescence-concentration relation
	15-2	Quantitative fluorometric analysis problems

Quantitative Analytical Chemistry

This book is part of
THE ALLYN AND BACON CHEMISTRY SERIES
and was developed under the co-consulting editorship of
DARYLE H. BUSCH AND HARRISON SHULL

Quantitative Analytical Chemistry

Principles and Life Science Applications

George H. Schenk
Wayne State University

Richard B. Hahn
Wayne State University

Arleigh V. Hartkopf
Mobil Chemical Company

Allyn and Bacon, Inc.   Boston, London, Sydney, Toronto

Copyright © 1977 by Allyn and Bacon, Inc.,
470 Atlantic Avenue, Boston, Massachusetts 02210

All rights reserved. Printed in the United States of America. No part of the material protected by this copyright notice may be reproduced or utilized in any form or by any means, electronic or mechanical, including photocopying, recording, or by any information storage and retrieval system, without written permission from the copyright owner.

Library of Congress Cataloging in Publication Data

Schenk, George H
  Quantitative analytical chemistry.

  Includes bibliographical references and index.
  1. Chemistry, Analytic—Quantitative. I. Hahn, Richard B., joint author. II. Hartkopf, Arleigh V., joint author. III. Title.
QD101.2.S33   543   76-50555
ISBN 0-205-05700-4

To the memory of CLAIRE M. SCHENK, *who joined the glory of our risen Lord on January 26, 1976*

Contents

Preface   *xi*

PART I   FUNDAMENTALS   1

1. Introduction to Chemical Analysis   3
2. Data Handling   17
3. Sampling and Preparation of the Sample   43
4. Fundamentals of Weighing and Related Measurements   54
5. Gravimetric Analysis   73
6. Fundamentals of Volumetric Methods of Analysis   99
7. Introduction to Titration Equilibria and End Point Detection   121
8. Elementary Acid-Base Equilibria   144
9. Acid-Base Titrations   174
10. Precipitation and Complexation Titrations   211
11. Oxidation-Reduction Methods   236

PART II   SOME INSTRUMENTAL METHODS OF ANALYSIS   269

12. Introduction to Instrumental Methods   271
13. Introduction to Spectrophotometry and Spectrophotometric Instruments   284
14. Applications of Spectrophotometry   310
15. Fluorometry, Refractometry, and Light Scattering   339
16. Emission and Absorption of Radiation by Gaseous Atoms   363
17. Direct Potentiometric and pH Measurements   384
18. Electrochemical Methods of Analysis   399

PART III SEPARATIONS AND CLINICAL ANALYSIS 421

19. Introduction to Separations: Liquid-Liquid Extraction   423
20. Theory of Chromatography   444
21. Applications of Chromatography   461
22. Clinical Analysis   479

PART IV SELECTED EXPERIMENTS 493

A. Treatment of Data from Weighing (Sec 4–6)   68
B. Use of the Volumetric Pipet (Sec 6–3)   115

1. Five Methods for the Determination of Chloride   495
   - 1A. Determination of Chloride by Density Measurement   495
   - 1B. Gravimetric Determination of Chloride   498
   - 1C. Volumetric Determination of Chloride by the Mohr and Fajans Methods   502
   - 1D. Determination of Chloride by Refractive Index Measurement   504
   - 1E. Determination of Chloride by Specific Ion Electrode Measurement   506
2. Preparation and Standardization of NaOH   508
3. Preparation and Standardization of HCl   509
4. Analysis of Milk of Magnesia   511
5. Analysis of Impure Sodium Carbonate   512
6. Acid-Neutralizing Capacity of an Antacid Tablet   513
7. Nonaqueous Titration of Amines and Acid Salts   514
8. Determination of Water Hardness   516
9. Preparation and Standardization of Iodine   517
10. Analysis of Vitamin C Tablets Using Iodine   519
11. Iodometric Determination of Arsenic   521
12. Using a Spectrophotometer to Obtain an Absorption Spectrum   522
13. Colorimetric Determination of Glucose   524
14. Spectrophotometric Determination of Manganese   525
15. Fluorometric Determination of Quinine in Quinine Water or Bitter Lemon   527
16. Separation of Amino Acids by Paper Chromatography   529

Appendix 1. Equilibrium Constants   532
Appendix 2. Standard Electrode Potentials   537
Appendix 3. Using Logarithms on the Electronic Calculator to Find Roots   538
Appendix 4. Four-place Logarithms   541
Appendix 5. Answers to Even-numbered Problems   543

Index   553

Preface

This text emphasizes *principles* rather than the theoretical development of chemical equilibrium and kinetics which appeals mainly to chemistry majors in a department emphasizing theory. Our purpose is to *broaden* the chemistry major's perception of analytical chemistry as well as to educate the nonchemist. We have therefore chosen to apply the principles to, and illustrate them with, relevant health-science analytical problems and analytical chemistry problems of various types.

We have included a number of special features in this text. In particular, we have included a strong discussion of how spectrophotometry and fluorometry are used for *all kinds* of analysis, rather than just quantitative analysis. We have chosen to place these chapters in Part II, rather than earlier in the book, but they may be read earlier if the instructor chooses. We have also treated chromatography somewhat differently by trying to discuss all of the theory together in a modern sense, and then all of the applications. Finally, we have *organized* the problems (see below) and included some *challenging problems* at the end of each problem set of the first fourteen chapters in hopes of challenging the better students.

Many texts stress the constant calculation of titration curves, but we feel they have sometimes missed the more important question of whether a reaction is quantitative or not. Therefore we have established the principle of a quantitative reaction (Chs. 5 and 7) and have illustrated that principle in discussing different kinds of analytical reactions. At the same time, we have introduced and discussed the titration curve (Ch. 7) without overemphasizing it.

In the above sense, the material in Part I is at a slightly "lower" level than many texts; we prefer to say that students of this text are liberated from equilibrium calculations to focus on principles, including the principle of quantitative reaction.

Because we have written a text for chemists and nonchemists, we have tried to discuss methods and techniques of broad interest to a number of health science–related fields, not just to analytical chemistry. Because instrumental analysis is extensively used by those in the health science fields, as well as by chemists, we have included a number of chapters in Parts II and III devoted to this area.

Since we anticipate that many nonchemists will use this text, we have included a number of helpful features for them as well as for chemistry students. Some of these are:

(1) Self-tests over topics that might need extra work (see list on inside front cover).
(2) Important principles or rules are boxed off in the text for easy access.
(3) There are numerous *numbered* examples for better understanding and easy referencing, especially between chapters.

(4) The equations and reactions are for the most part *numbered* for easy referencing between chapters.
(5) Problems are divided into logical groups with a heading for each group. Answers to most even-numbered problems are given in Appendix 5.

The self-tests are important because they help the student to delve a little more deeply into a topic, as well as helping him review the principles in the preceding material.

We have been very selective in including experiments because we feel that most instructors have accumulated experiments of their own. We have tried to include mostly experiments related to health science, mixed with a few standard experiments.

Quantitative Analytical Chemistry

PART I

Fundamentals

This part of the book is a treatment of the fundamentals of analytical chemistry. In Chapter 1 we attempt to introduce the student to chemical analysis from both qualitative and quantitative viewpoints. In Chapter 2 we present certain fundamentals of data handling that are necessary in all of the succeeding chapters. Sampling and preparation of the sample are discussed in Chapter 3. Chapter 4 covers the use of the analytical balance and density measurements. Gravimetric analysis is described in Chapter 5.

Chapters 6–11 are concerned with various aspects of volumetric, or titrimetric, analysis, thus concluding the discussion of fundamentals.

1. Introduction to Chemical Analysis

"There is our result—and a very workmanlike little bit of analysis it was,"
said Holmes.

ARTHUR CONAN DOYLE
The Valley of Fear, Chapter 1

Chemical Analysis and Detective Work

This chapter and following chapters begin with a quotation from the cases of Sherlock Holmes for a good reason. Chemical analysis *is* essentially detective work, and the analytical chemist is surely a scientific detective. If you have ever read any of Sherlock Holmes' cases, you know that detective work consists of observations, testing, interpretation, and the final glorious deduction. The same is true for chemical analysis and the analytical chemist. The analyst is usually faced with the question of Why? or What is wrong? To answer such a question, he first makes certain observations to see what tests or analyses he must conduct. Then he judiciously chooses the proper sample and tests it. Finally, he interprets his results to someone, and may even get to make the final deduction, although it perhaps may not be as glamorous as those made by Mr. Holmes.

In this book we are concerned with all of the above four steps, but observation is difficult to teach; it must be learned in the laboratory or in the real world. We will therefore be more concerned with understanding how to use data in our analysis (Ch. 2), choosing a sample (Ch. 3), testing the sample (Chs. 4–22), and interpreting the results (Ch. 2) for that final deduction.

1-1 INTRODUCTION TO CHEMICAL ANALYSIS

What Kinds of Chemical Analysis Are There?

Chemical analysis encompasses numerous types of measurements. We might identify four unique areas of chemical analysis as follows:

1. Measurements of physical and chemical properties
2. Qualitative analysis
3. Quantitative analysis
4. Diagnostic analysis based on a combination of qualitative and quantitative analyses

We will attempt to discuss each of these areas to give you a feeling for chemical analysis in its broadest sense.

Measurement of Physical and Chemical Properties. New compounds, crystals, polymers, and natural substances are being discovered and synthesized all of the time. One of the most important things the chemical analyst is concerned with is the characterization of any of these substances by measuring their physical and chemical properties. If possible, he will try to measure the melting point, boiling point, density, index of refraction, and so on. If the substance is an acid or a base, he will measure the ionization constant. If the substance is insoluble in water, he may try to measure its solubility and/or evaluate a solubility product constant. In a similar manner, he may try to evaluate other chemical properties.

Qualitative Analysis. Qualitative analysis involves the *identification* of a compound whose identity is unknown or only partially known, as well as *screening* on a routine basis for the presence or absence of significant amounts of a compound whose presence is only suspected. One of the most powerful tools for qualitative analysis is spectroscopy; both Chapters 14 and 15 give examples of the application of spectroscopy to qualitative analysis. Chapter 14 discusses the application of ultraviolet, visible, and infrared spectrophotometry to qualitative analysis, and Chapter 15 discusses the use of fluorescence in qualitative analysis. It is important to employ several different measurements for rigorous proof in qualitative analysis. These measurements may consist of a combination of spectroscopic measurements with measurement of physical properties such as melting point, or just a number of spectroscopic measurements.

Quantitative Analysis. Quantitative analysis involves the determination of the *amount* of a substance present, either its percentage or its concentration in a solution. Quantitative analysis may be performed just once, or it may be performed continuously to monitor changes in percentage purity or concentration of a sample with time. Quantitative analysis may be performed by the classical gravimetric (Ch. 5) or titration methods (Chs. 6–11), as well as by spectroscopic (Chs. 13–16) and electrical means (Chs. 17 and 18). Different types of quantitative measurements give different kinds of information, depending on the information needed. When speed is important, it is imperative that we measure only what is necessary and transmit the results quickly. Some of the different types of quantitative analysis are described below.

A *complete analysis* is an analysis in which each compound in a sample is identified and the percentage or concentration measured. For example, the complete analysis of an aspirin tablet would include the percentage of acetylsalicylic acid (ASA), the percentage of impurities such as salicylic acid (and possibly any acetic acid which has not volatilized), and the amount of inert material used as binder to hold the ASA particles together in tablet form.

An *elemental analysis* is an analysis in which a mixture or pure compound is analyzed for the percentage or concentration of each element present. This type of analysis may be used together with a molecular weight determination to establish the identity of a compound, or it may be used to characterize a complex sample. Some samples such as petroleum products are such complex mixtures that it is not worth running a complete analysis. It is just as useful to determine the percentage of certain key constituents, such as sulfur, oxygen, and phosphorus, to characterize the petroleum.

A *partial analysis* is an analysis in which the concentration or percentage of one or more impurities is all that must be determined. For example, the main impurity in aspirin tablets is salicylic acid, which arises from the hydrolysis of acetylsalicylic acid (ASA):

$$[C_6H_4(OH)]C\text{—}O\text{—}\underset{\underset{O}{\|}}{C}\text{—}CH_3 + H_2O \rightarrow [C_6H_4(OH)]C\text{—}OH + CH_3\underset{\underset{O}{\|}}{C}\text{—}OH$$

(ASA)   (Salicylic acid)   (Acetic acid)

Acetic acid is also produced in the reaction, but it gradually evaporates from the aspirin tablets. Therefore, to evaluate the purity of the aspirin, it is simpler to determine only the salicylic acid rather than measure the ASA present. In fact, the U.S. Food and Drug Administration (FDA) does just that. It has established a *tolerance* of 0.15% salicylic acid for aspirin tablets. If a company's aspirin exceeds this tolerance, the FDA issues a warning to them to improve the product.

Diagnostic Analysis. This may involve a combination of qualitative and quantitative tests. It may also involve quantitative measurements that establish primarily whether a certain concentration is within or outside a certain normal *range* of values. The best example of this is of course the standard clinical tests that are run on blood and/or urine samples.

Since most clinical measurements are performed on automated instrumentation, the results are frequently plotted by a recorder across a standard graph paper displaying the range of values possible for each test. A section of such a graph paper for blood analysis is shown in Figure 1-1. The shaded area in each vertical column of values represents the normal range of concentration for each chemical species. For example, for calcium(II) ion, a normal blood sample should contain 8.5 to 10.5 mg per cent of Ca^{+2}.

An example of the usefulness of a combination of diagnostic tests follows. Suppose that a certain blood sample is found to be low in calcium(II) ion from a protein deficiency. Since the calcium(II) ion concentration controls the phosphate ion, then we would expect the phosphate ion concentration to be raised above normal. If instead the phosphate concentration is lower than normal (Fig. 1-1),

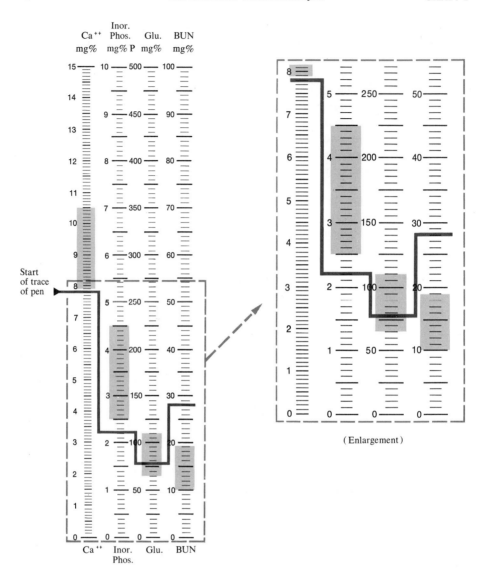

FIGURE 1-1. A section of an automated clinical analysis graph paper and a typical result plotted by a recorder pen.

this may indicate a kidney dysfunction resulting in retention of phosphate in the kidney. The increase in blood urea nitrogen (BUN) (Fig. 1-1) also indicates a kidney dysfunction.

Sample Size

Analytical methods can be classified according to a number of criteria, one of which is sample size. Table 1-1 contains a summary of the classification according to the weight and the concentration of the sample.

TABLE 1-1. Sample Size Classification of Analytical Methods

Name of method	Approximate sample weight	Approximate molarity: sample of 100 ml
Macro method	100–1000 mg	0.01–$0.1M$
Semimicro method	10–100 mg	10^{-3}–$10^{-2}M$
Micro method	1–10 mg	10^{-4}–$10^{-3}M$
Ultramicro method	0.001–1 mg	10^{-7}–$10^{-4}M$
Submicrogram method	10^{-5}–10^{-3} mg	10^{-9}–$10^{-7}M$

As we will discuss below, most analytical techniques are not suitable for measuring more than one or two sample sizes, unless the sample can be diluted or concentrated. It is important for an analytical chemist to become experienced with analytical techniques so that he can use the proper technique to obtain an accurate answer.

Analytical Measurement Techniques

Quantitative analysis, and to some extent qualitative analysis, is performed by using one of the following measurement techniques:

1. Gravimetric (precipitation)
2. Volumetric (titrimetric)
3. Optical (spectroscopy)
4. Electrical
5. Gasometric
6. Separation

Gravimetric Analysis. Gravimetric analysis involves isolating and weighing a pure compound which is or which contains the desired constituent. It is mainly a macro or semimicro method since the solubility of most of the insoluble compounds determined by gravimetry is not small enough for micro determination. Micro samples would have to be concentrated to be analyzed gravimetrically.

Volumetric Analysis. Volumetric analysis is primarily performed by titrating the desired constituent with a solution of known concentration. It is mainly used as a macro method, but can be adapted to the semimicro or micro level for reactions with large equilibrium constants. Such reactions include most acid-base reactions, most complex-forming reactions, a few precipitation reactions, and most oxidation-reduction reactions.

Optical Analysis (Spectroscopy). Optical analysis involves the measurement of light or other radiation after it has interacted, or failed to interact, with ions or molecules in solution, or with atoms in the vapor phase. It is usually very sensitive to small concentrations and is used as a micro method (absorption spectroscopy), as an ultramicro method (emission or absorption by gaseous atoms), or as a submicrogram method (fluorescence in solutions). Macro samples have to be diluted to be analyzed by optical methods.

Electrical Analysis. Electrical analysis involves the measurement of electrical potential, or the electrical output from an electrochemical reaction. It is usually used as a semimicro or a micro method. However, potential measurements can be used over a wide range of concentration from ultramicro to the macro sample level. A few electrical methods of analysis can be used as ultramicro methods also.

Gasometric Analysis. We will arbitrarily define a gasometric method as one in which a *molecular* sample is measured in the gaseous form. This could include measurement of the volume of gas in a gas buret, but mainly refers to the gas chromatographic method in which gases are separated and measured at the same time on the same instrument. This latter method covers a wide range of sample sizes and no dilution or concentration is needed for most samples.

Separations. This operation primarily involves isolating the various constituents in the sample so that the amount of each constituent can be measured. Separation methods include solvent extraction, liquid-liquid chromatography, thin-layer chromatography, gas chromatography, dialysis of small, water-soluble molecules, and so on. All such separations make qualitative and quantitative analysis much easier to perform by removing possible interferences from the constituent to be identified or determined.

1-2 REVIEW OF SOME USEFUL CONCEPTS

In this section we hope to review some useful concepts that you will need through the rest of the book. We will begin with a survey of the nomenclature of inorganic and organic compounds. It is taken for granted in the book that you understand the naming of these compounds.

SEC. 1-2   Review of Some Useful Concepts

Inorganic Nomenclature

This will be a short review of the naming of metal ions, anions, molecules, compounds, acids, and complex ions.

Metal Ions. The Stock nomenclature system of naming cations involves the use of the oxidation state of the ion in parentheses immediately following the name of the element. For example, Fe^{+2} and Fe^{+3} are named

$$Fe^{+2}: iron(II) \text{ ion} \qquad Fe^{+3}: iron(III) \text{ ion}$$

We prefer this system, although occasionally we will use the common or trivial names like ferrous ion or ferric ion where it is convenient. Even in the case of the Hg_2^{+2} ion the name used is simply mercury(I) ion. It is understood that this name implies a dimeric cation.

In the case of the alkali and alkaline earth metal ions it is usually not necessary to write the oxidation state after the element's name. However, we will occasionally use it to be sure that non-chemistry majors are aware of the oxidation state. Thus the Na^+ ion may be named either way,

$$Na^+: \text{sodium ion or sodium(I) ion}$$

Anions and Molecules. Anions and molecules are named differently when they exist alone or in simple compounds than when they are bonded to metal ions in complex ions. A list of a few representative anions and molecules and their names follows:

Formula	Alone (or in simple compound)	Complex ion
H_2O	Water	Aquo
NH_3	Ammonia	Ammine
$NH_2(CH_2)_2NH_2$	Ethylenediamine	(Ethylenediamine)
Cl^-	Chloride	Chloro
OH^-	Hydroxide	Hydroxo
CN^-	Cyanide	Cyano
SO_4^{-2}	Sulfate	Sulfato
PO_4^{-3}	Phosphate	Phosphato

When the molecule attached to a metal is complex, such as ethylenediamine, its name is used unchanged and parentheses are placed around its name in the name of the complex ion.

Compounds. When compounds are named, the cation is named first using the Stock system, and the anion is named next. No prefix is necessary to indicate the number of anions since this is assumed to be understood from the oxidation state in the name. Some examples:

Formula	Name
$FeSO_4$	Iron(II) sulfate
$Fe_2(SO_4)_3$	Iron(III) sulfate
Hg_2Cl_2	Mercury(I) chloride
Na_2CO_3	Sodium carbonate

Acids. Acids are referred to by their common names. Some examples:

Formula	Name
H_2CO_3	Carbonic acid
HCl	Hydrochloric acid
HNO_3	Nitric acid
$HClO_4$	Perchloric acid
H_3PO_4	Phosphoric acid

Compounds that are acid salts, such as $NaHCO_3$, are formally named using *hydrogen* with the proper prefix for the number in the formula after the name of the metal ion. Thus $NaHCO_3$ is sodium hydrogen carbonate, although it is also known as sodium bicarbonate. The compound NaH_2PO_4 would be named sodium dihydrogen phosphate.

Complex ions. In *positively*-charged complex ions the ligand is named first, preceded by a prefix (di-, tri-, tetra-, penta-, or hexa- for simple ligands; bis-, tris-, etc. for complex ligands) to indicate the number of ligands. The metal ion is named next. Any anions not complexing the metal ion are named last. Some examples:

Formula	Name
$Ag(NH_3)_2^+$	Diamminesilver(I) ion
$FeCl_2^+$	Dichloroiron(III) ion
$Cu(NH_2C_2H_4NH_2)_2^{+2}$	Bis(ethylenediamine)copper(II) ion
$[Fe(SO_4)^+]Cl^-$	Sulfatoiron(III) chloride

In *negatively*-charged complex ions, the name of the metal ion is followed by the suffix *-ate*; elements whose names end in *y* drop the *y* when the name of the metal ion is formed. In addition, certain metal ions are named according to their Latin names. The important metal ions for which this is done are listed below:

Formula	Name
$AgCl_2^-$	Dichloroargentate(I) ion
$Fe(CN)_6^{-3}$	Hexacyanoferrate(III) ion; also, ferricyanide
$CuBr_4^{-2}$	Tetrabromocuprate(II) ion
$PbCl_4^{-2}$	Tetrachloroplumbate(II) ion
$SbCl_6^{-3}$	Hexachlorostibnate(III) ion

Organic Nomenclature

Rather than discuss this nomenclature in a complete, formal manner, we will simply list each important class of compounds and name one or two examples of that class that may be encountered in later chapters. You should consult an organic text for a complete review if needed.

Alkanes. These are straight chain or branched chain aliphatic hydrocarbons, such as butane, $CH_3CH_2CH_2CH_3$.

Alkenes. These are olefins with one or more double bonds, such as ethylene, $H_2C{=}CH_2$, or 1,3-butadiene, $H_2C{=}CH{-}CH{=}CH_2$.

Alkynes. These are triply bonded hydrocarbons having one or more triple bonds. An example is acetylene, $HC{\equiv}CH$.

Aromatic Hydrocarbons. These are cyclic conjugated unsaturated hydrocarbons, such as benzene (C_6H_6) or naphthalene ($C_{10}H_8$). Their structures are

Halogenated Hydrocarbons. These include halo-substituted alkanes, alkenes, aromatic hydrocarbons, etc. Examples are chloroform, $CHCl_3$, and chlorobenzene, C_6H_5Cl.

Hydroxyl Compounds. These are hydroxyl-substituted compounds. They include alcohols such as ethyl alcohol (ethanol), CH_3CH_2OH, and phenols such as phenol itself, C_6H_5OH.

Carbonyl Compounds. These are compounds containing a carbon-oxygen double bond, or carbonyl group, bonded to other carbon or hydrogen atoms. Examples include aldehydes such as formaldehyde, $H_2C{=}O$, and ketones such as acetone,
$$CH_3{-}\underset{\underset{O}{\|}}{C}{-}CH_3.$$

Carboxylic Acids and Esters. Carboxylic acids are aliphatic acids such as acetic acid, $CH_3\underset{\underset{O}{\|}}{C}{-}OH$, or aromatic acids such as phthalic acid $C_6H_4(COOH)_2$. Phthalic acid consists of two carboxylic groups bonded to a benzene ring. It is half neutralized with potassium hydroxide to make potassium acid phthalate, $C_6H_4(COOH)COO^-K^+$, an important primary standard for basic titrants.

Esters are compounds made from carboxylic acids and alcohols. A common example is ethyl acetate, $CH_3\underset{\underset{O}{\|}}{C}-O-CH_2CH_3$.

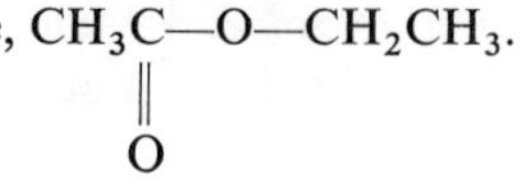

Sugars. This class of compounds contains hydroxyl groups and an aldehyde and/or ketone group. Examples are glucose

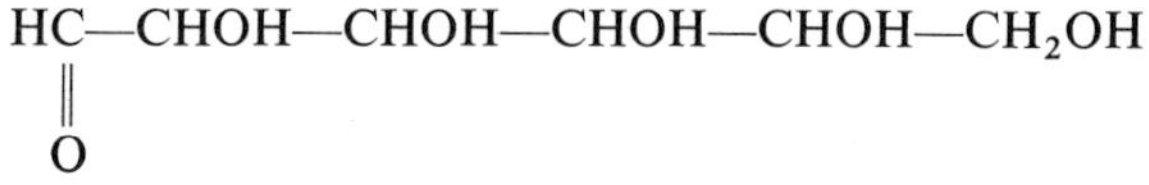

$$H\underset{\underset{O}{\|}}{C}-CHOH-CHOH-CHOH-CHOH-CH_2OH$$

and fructose

$$CH_2OH-\underset{\underset{O}{\|}}{C}-CHOH-CHOH-CHOH-CH_2OH$$

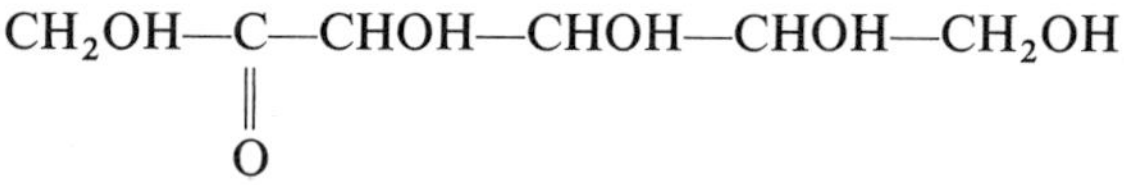

These compounds usually exist in cyclic forms as well as the straight chains shown above, but for our purposes the straight chain is sufficient.

Atomic Weights, Moles, and Concentration

This discussion is a useful review for the material in Chapters 5 and 6 where the calculation of results is introduced.

Atomic Weights and Moles. Although chemists use atomic weights as absolute quantities, atomic weights are really the weights of atoms relative to the weight of the carbon isotope of mass number 12. Thus the atomic weight of 24.312 for magnesium implies that it is a little over twice the weight of the carbon 12 isotope. A table of atomic weights is included on the inside back cover of the book.

The terms *molecular weight* and *formula weight* are sometimes used interchangeably, although a molecular weight strictly refers to the sum of the atomic weights of the atoms in a molecule and a formula weight refers to the sum of the atomic weight(s) of the atom(s) in an element, an ionic compound, or a molecule.

To simplify working with these terms, we can use the *mole* as a fundamental quantity to specify amounts of any species. A mole is defined as 6.023×10^{23} atoms or ions of an element or 6.023×10^{23} molecules. The working definition of a mole is that it is a formula weight of any species expressed in grams. Thus a mole of chloride ions is 35.45 g, and a mole of sodium chloride is 58.44 g. In addition to the gram and the mole, the milligram (mg) or microgram (μg) and the millimole (mmole) or micromole (μmole) are used. (1 μg $= 10^{-6}$ g; 1 μmole $= 10^{-6}$ mole.) Thus a millimole of Cl^- is 35.45 mg, and a micromole of sodium chloride is 58.44 μg.

The number of moles or millimoles is calculated as follows:

$$\text{moles Cl}^- = \frac{\text{g Cl}^-}{35.45 \text{ g/mole Cl}^- \text{ (formula weight)}}$$

$$\mu\text{moles NaCl} = \frac{\mu\text{g NaCl}}{58.44 \text{ }\mu\text{g/}\mu\text{mole NaCl (formula weight)}}$$

$$\text{mmoles NaCl} = \frac{\text{mg NaCl}}{58.44 \text{ mg/mmole NaCl (formula weight)}}$$

Note in the above calculations that the formula weight can be expressed in grams per mole, milligrams per millimole, or micrograms per micromole.

Concentration. Concentration will be discussed fully in Section 6-2; we will provide only a brief introduction here. The most general unit of concentration used in this text is molarity, M. The molarity of a solution can be expressed using moles and liters or millimoles and milliliters (ml).

$$M = \frac{\text{moles}}{\text{liter}} = \frac{\text{mmoles}}{\text{ml}}$$

Thus a $0.10M$ solution of sodium chloride implies that the solution contains 0.10 moles of sodium chloride per liter, or 0.10 mmoles of sodium chloride per ml.

An alternate method of expressing concentration is to use the normality, N. Normality depends on the type of reaction involved and will be fully defined in Section 6-1.

Clinical chemists frequently use the concentration term of milligram per cent (mg%). This is defined as milligrams of a substance per 100 g of solvent, or in the case of water, per 100 ml of water. The term can also be used to describe the milligrams of a substance present in 100 g of a solid.

Another concentration term that is frequently used is parts per million (ppm) which can be defined as

$$\text{ppm} = \frac{\text{mg of solute}}{\text{kg of solution or solid}} = \frac{\mu\text{g of solute}}{\text{g solution}}$$

Since a kilogram is almost the same as a liter for a dilute aqueous solution, this term can also be defined for aqueous solutions as

$$\text{ppm} = \frac{\text{mg of solute}}{\text{liters of aq soln}} = \frac{\mu\text{g of solute}}{\text{ml of aq soln}}$$

An Optional Look at Activity

It is necessary for an understanding of Chapter 17 and helpful for reading Chapter 8 to understand the concept of *activity* as a measure of concentration. Activity is a measure of concentration *not calculated directly* from the number of moles of a solute weighed into a certain number of liters of solution. It is really a measure of the effective molarity of an ion, corrected for interionic attraction.

Activity may be calculated from molarity as follows:

$$a_i \cong f_i [I] \tag{1-1}$$

where a_i is the activity of an individual ion, f_i is the activity coefficient of the ion, and [I] is the molarity of the ion I. The activity coefficient is generally a fraction except at infinite dilution where it approaches unity. Thus in very dilute solution, the activity of an ion is essentially equal to the molarity of an ion.

Ionic Strength. Activity coefficients can be estimated by a calculation involving the ionic strength, μ. The magnitude of the ionic strength depends on the concentrations and charges of the ions in the solution. For compounds of $+1$ cations and -1 anions, the ionic strength is the same as the molarity of the compound. Thus the ionic strength of a solution containing 0.01 mole per liter of potassium nitrate and 0.030 mole per liter of sodium perchlorate ($NaClO_4$) is 0.040. For compounds with ions of higher charges than one, the ionic strength must be *calculated using this equation*:

$$\mu = \frac{1}{2} \sum C_i (Z_i)^2 \tag{1-2}$$

where C_i is the concentration of an ion with charge Z_i.

EXAMPLE 1-1. Calculate the ionic strength of a solution containing $0.030M$ magnesium(II) nitrate and $1 \times 10^{-5} M$ hydrochloric acid.

Solution: First, we note that the molarity of the hydrochloric acid is not *significant* compared to the molarity of the magnesium(II) nitrate (see Sec. 2-1). Since the magnesium nitrate concentration is known only to the third digit to the right of the decimal, the total concentration of the ions must be limited to that digit also, making the amount of HCl negligible. Second, we calculate the molarity of each ion:

$$[Mg^{+2}] = M_{Mg(NO_3)_2} = 0.030M$$

$$[NO_3^-] = 2M_{Mg(NO_3)_2} = 0.060M$$

Third, we use Equation **1-2** to calculate the ionic strength:

$$\mu = \frac{1}{2} \sum 0.030M \; Mg^{+2}(+2)^2 + 0.060M \; NO_3^-(-1)^2$$

$$\mu = 0.090$$

At ionic strengths below 0.10, activity coefficients for univalent ions may be estimated using the abbreviated form of the DeBye-Huckel equation:

$$-\log f = 0.509 (Z_i)^2 \sqrt{\mu} \tag{1-3}$$

Equation **1-3** can also be used for divalent ions at ionic strengths below about 0.04.

Equation **1-3** yields a value for a *mean* activity coefficient for both cations and anions of the same charge magnitude. More accurate values for individual ions are given in Table 1-2 for selected ionic strengths.

TABLE 1-2. Individual Ion Activity Coefficients as a Function of Ionic Strength

Ion	Activity coefficient at				
	$\mu = 0.001$	$\mu = 0.005$	$\mu = 0.01$	$\mu = 0.05$	$\mu = 0.10$
H^+	0.967	0.933	0.914	0.86	0.83
Li^+	0.965	0.929	0.907	0.835	0.80
Na^+, IO_3^-, HSO_4^-	0.964	0.927	0.901	0.815	0.77
OH^-, F^-, ClO_4^-	0.964	0.926	0.900	0.81	0.76
K^+, Cl^-, Br^-, I^-	0.964	0.925	0.899	0.805	0.755
NH_4^+, Ag^+	0.964	0.924	0.898	0.80	0.75
Mg^{+2}, Be^{+2}	0.872	0.755	0.69	0.52	0.45
$Ca^{+2}, Cu^{+2}, Zn^{+2}, Mn^{+2}, Ni^{+2}, Co^{+2}$	0.870	0.749	0.675	0.485	0.405
Ba^{+2}, Cd^{+2}	0.868	0.744	0.67	0.465	0.38
Pb^{+2}	0.867	0.742	0.665	0.455	0.37
SO_4^{-2}, HPO_4^{-2}	0.867	0.740	0.660	0.445	0.355
$Al^{+3}, Fe^{+3}, Cr^{+3}$	0.738	0.54	0.445	0.245	0.18
PO_4^{-3}	0.725	0.505	0.395	0.16	0.095
$Th^{+4}, Zr^{+4}, Ce^{+4}$	0.588	0.35	0.255	0.10	0.065

From J. Kielland, *J. Am. Chem. Soc.* **59**, 1675 (1937).

Once the activity coefficient is known, it can be substituted into **1-1** to calculate the activity of an ion or ions.

EXAMPLE 1-2. Calculate the activity of both the hydrogen ion and the chloride ion of $1.0 \times 10^{-5} M$ HCl in (a) $0.030 M$ magnesium(II) nitrate and (b) $9.0 \times 10^{-5} M$ sodium(I) chloride.

Solution to a: As we have noted in Example 1-1 the contribution of the hydrochloric acid to the ionic strength is negligible compared to that of the magnesium(II) nitrate. The ionic strength (as calculated in Example 1-1) is 0.090. Using **1-3**, the activity coefficient for the hydrogen ion and for the chloride ion is estimated.

$$-\log f = 0.509 \sqrt{0.090} = 0.1527$$

$$\log f = -0.15_{27} = +0.84_{73} - 1$$

$$f = 0.70 \text{ (estimate)}$$

Note that the more accurate values in Table 1-2 are significantly different; for hydrogen ion f is about 0.83, and for chloride ion f is about 0.76. We will therefore use these values to calculate activities.

$$a_{H^+} = 0.83[1.0 \times 10^{-5} M] = 8.3 \times 10^{-6}$$

$$a_{Cl^-} = 0.76[1.0 \times 10^{-5} M] = 7.6 \times 10^{-6}$$

Solution to b: The ionic strength here is the sum of that of the hydrochloric acid and the sodium chloride, or 1.0×10^{-4}. Using **1-3**, the mean activity coefficient for both ions is

$$-\log f = 0.509\sqrt{1.0 \times 10^{-4}} = 0.00509$$

$$\log f = -0.00509 = +0.9949 - 1$$

$$f = 0.99 \text{ (estimate)}$$

This is so close to unity that the activity and the molarity of each ion can be said to be equal ($1.0 \times 10^{-5} M$).

QUESTIONS AND PROBLEMS

(Answers to most even-numbered problems are in Appendix 5.)

Definitions and Concepts
1. Define the following terms:
   a. Qualitative analysis
   b. Quantitative analysis
   c. Diagnostic analysis
   d. Macro method
   e. Micro method
2. Contrast the following pairs of methods as to the size of sample involved:
   a. Volumetric and gravimetric analysis
   b. Optical analysis and electrical analysis
   c. Optical analysis and gas chromatographic analysis
3. Name the following compounds using the Stock nomenclature system:
   a. Hg_2Cl_2
   b. $HgCl_2$
   c. $[Cu(NH_3)_4^{+2}]SO_4^{-2}$
   d. $K_4^+[Fe(CN)_6^{-4}]$

Concentration Calculations
4. Calculate the number of millimoles and micromoles in 50 mg of sodium chloride. Also calculate the molarity of a solution containing 500 μg of sodium chloride in 10 ml of solution.
5. A solution contains 100 mg of potassium(I) ion and 100 mg of chloride ion in 50 ml of water. Are the molarities of each ion the same? Show by calculation.
6. A common level of water hardness is 100 ppm calcium(II) carbonate. Calculate the mg and the μg of $CaCO_3$ in 0.01 liter. Also calculate its molarity.

Activity Calculations
7. Calculate the activity of both H^+ and NO_3^- in a $1.0 \times 10^{-6} M$ HNO_3 solution in $0.030 M$ potassium(I) sulfate and in $9.0 \times 10^{-6} M$ potassium(I) chloride.
8. Calculate the activity of both Mg^{+2} and SO_4^- in a $0.01 M$ $MgSO_4$ solution.

2. Data Handling

*"What is the meaning of it all,
Mr Holmes?" "Ah. I have no data.
I cannot tell," he said.*

ARTHUR CONAN DOYLE
The Adventure of the Copper Beeches, 1892

The main objective of a chemical analysis is to obtain meaningful data. Without it you will be as helpless as Sherlock Holmes was in the above story. You must also learn how to handle and report numbers before you make chemical measurements to obtain data. You will be using both *exact* and *measured* numbers in your work, and you should know the difference between them.

Exact numbers include factors such as the factor of 100 used to calculate a percentage, exponents such as 10^{-5}, and small counted numbers such as 3 or 4 in the formula H_3PO_4.

Measured numbers are uncertain numbers, in contrast to exact numbers, because any measurement involves some uncertainty. For example, a balance capable of weighing only to the nearest gram was used to measure a person's daily intake of protein. One such weighing gave a weight of 141 g of protein. The last digit in the weight is uncertain, however, since the balance can only be read to ± 1 g. Hence we should say that the weight of the protein is 141 $\pm$ 1 g.

The above example is fairly typical; measured numbers are usually assumed to be uncertain by ± 1 in the last digit. This is true whether the last digit is to the left or to the right of the decimal point. Thus the number 1.83 is assumed to imply 1.83 ± 0.01 unless otherwise specified. We will describe the treatment of measured numbers throughout this chapter.

A single number or set of numbers may be characterized by its *accuracy*, or closeness to the true value, and also may be characterized by its *precision*, or agreement, among the numbers in the set. For a single number, there can be no such thing as its precision, but we may refer to its *uncertainty*. All three of the above concepts—accuracy, precision, and uncertainty—can be expressed in mathematical terms only by calculating the error (for accuracy) or the deviation. Both error and deviation can be expressed in *absolute* and in *relative* terms. Each will be discussed separately.

Absolute and Relative Error

The absolute error, A.E., of any measurement, X, is the difference between that measurement and the true value, μ. Stated mathematically,

$$\text{A.E.} = X - \mu$$

Absolute errors cannot be compared with one another; instead the *relative error*, R.E., of each measurement must be calculated and used for comparison. The mathematical definition of relative error in parts per hundred (pph) is

$$\text{R.E.} = \frac{(X-\mu)}{\mu}(100) = \frac{\text{A.E.}}{\mu}(100)$$

Note that relative error is simply the absolute error divided by the true value; this is then converted to a pph basis by multiplying by one hundred.

For example, suppose an erroneous analysis of vitamin C in a urine sample gives a value of 402 mg for 24 hours when the true value is 400 mg for 24 hours. The absolute error is calculated as follows:

$$\text{A.E.} = 402 - 400 = 2$$

The relative error of the analysis is

$$\text{R.E.} = \frac{402 - 400}{400}(100) = 0.5 \text{ pph}$$

Absolute and Relative Deviation

The deviation of a number or a set of results may be used to describe the precision of the measurement(s) mathematically. In general, deviation is defined as the difference between a measurement and the mean, $\bar{X}$. Later on, you will be introduced to the standard deviation of a set of results.

Deviation and the standard deviation can be expressed in absolute and relative terms just like error. Absolute deviation like absolute error is just an *absolute difference* (A.D.) between two numbers; for absolute deviation this is the absolute difference between the measurement and the mean. Relative deviation (R.D.) is the absolute deviation divided by the mean and multiplied by one hundred:

$$\text{R.D.} = \frac{\text{A.D.}}{\bar{X}}(100)$$

This is a mathematical definition of relative deviation in parts per hundred (pph).

For example, suppose the first weighing of an aspirin tablet gives a value of 0.5011 g. Further weighings are made and the mean of all the weighings is 0.5010 g. The absolute deviation of the first weighing is

$$\text{A.D.} = 0.5011 \text{ g} - 0.5010 \text{ g} = 0.0001 \text{ g}$$

SEC. 2-1   Significant Figures   19

The relative deviation of the first weighing is

$$\text{R.D.} = \frac{0.5011 \text{ g} - 0.5010 \text{ g}}{0.5010 \text{ g}} (100) = 0.02 \text{ pph}$$

Uncertainty: Absolute and Relative

When something must be said about the deviation of a single result, all that can be done is to calculate its uncertainty. In the absence of qualifying information, it may be assumed that the last digit is uncertain by ± 1. Another way of saying this is to state that the absolute uncertainty is ± 1. The relative uncertainty, or relative deviation, of a single measurement is calculated as above. As an example, consider counting the number of aspirin tablets in a bottle supposedly containing 200 tablets. The relative uncertainty (R.U.) of counting this number assuming an absolute uncertainty of ± 1 is

$$\text{R.U.} = \frac{1}{200}(100) = 0.5 \text{ pph}$$

This raises the question of whether all such counting operations are uncertain by ± 1. The answer is certainly no. Where the sample size is small enough to preclude an error, such as a sample of 10, there the absolute uncertainty is usually zero. In an important situation such as counting the number of capsules of a potentially dangerous prescription the pharmacist will check the count so that the uncertainty is also zero.

2-1 SIGNIFICANT FIGURES

Before you master the handling of data, you should understand all the aspects of significant figures discussed below. This is necessary because most of the data you will handle are measured numbers, not exact numbers. (The concepts of significant figures should not be applied to exact numbers, only to measured numbers.) After reading this section, take the self-test at the end to see if you have mastered significant figures.

Concept of Significance

Significant figures may be defined as those digits in a number that are known with certainty, *plus* the first uncertain digit. All other digits are termed *nonsignificant figures*. In scientific work it is desirable to report data showing which digits are significant and which are not, or to report data using only significant figures. For scientific work it will be useful to follow the following simple rules.

1. Usually only the final result, or answer, to a calculation need be expressed with the correct number of significant figures. It is best to write such data in semi-exponential form, that is, write a result such as 0.0123 as 1.23×10^{-2} to show that three significant figures are justified.
2. Use appropriate symbolism to designate which digits are nonsignificant figures if it is desired to report digits which are not significant. Throughout this book nonsignificant figures will be written as subscripts; for example, if there are only two significant figures in a number such as 1.544, it will be written as 1.5_{44}, or 1.5_4.
3. If the last digit of a result is uncertain by more than ± 1, write the uncertainty after the number, for example, 3.1 ± 0.2. All results written without such qualifying information should be assumed to have a last digit that is uncertain by ± 1.

Zeroes

Special attention should be given to whether zeroes are significant or not. The following rules are helpful.

1. Zeroes which appear between other digits, as in 10.04, are significant.
2. Initial zeroes, such as in 0.104 or 0.0014, are usually not significant since they serve only to locate the decimal point and are not "certain" digits. Logs are an exception—initial zeroes to the right of the decimal in a log are significant; for example, the log of 1.010 = 0.0043; both zeroes to the right of the decimal point in 0.0043 are just as significant as those in 1.0043, which is the log of 10.10.
3. Terminal zeroes are generally considered to be significant. If a terminal zero is not significant but is only used to fix the decimal point, it should be eliminated and the number written using powers of ten. For example, 10,100 as so written is considered to have five significant figures. If only three digits are meant to be significant, then it should be written as 1.01×10^4.

Rounding Off

Use the following rules when rounding off is needed.

1. In general eliminate digits which are not significant by rounding off. Increase the last retained figure by 1 if the next discarded digit is 5 or greater. If it is less than 5, do not change the last retained figure.
2. To avoid rounding off report nonsignificant figures as subscripts. Where your results justify only one significant figure report an additional nonsignificant figure result as $1._4$ rather than 1; this indicates a *trend* towards a number higher than 1.0 than a number between 0.5 and 1.0.
3. Avoid rounding off when there is more than one calculation sequence or when data are reentered on a calculator. Carry along at least two nonsignificant figures to avoid errors in the last significant figure in the result.

Addition and Subtraction

The absolute uncertainty of each number controls the number of significant figures in the sum or the difference of the numbers. To apply this concept you should first write all the numbers in semiexponential form and *adjust all numbers so that they have the same exponent*. (For numbers close to one, an all-digital form is recommended, for example, $9.0 - 0.8 = 8.2$.)

To express a sum or difference to the correct number of significant figures use the following rule.

> RULE 1. Use the same number of significant figures to the right of the decimal in a sum or difference as the *number with the fewest number of significant figures to the right of the decimal*.

Do not round off any of the numbers before adding or subtracting; round off afterwards.

EXAMPLE 2-1. The saliva contains about 4×10^{-7} moles/liter of H^+. Find its $[H^+]$ after ingestion of a food acid adding 5.55×10^{-6} moles/liter of H^+ to the mouth.

Solution: Adjust the exponents of each number to be the same, then add.

5.55×10^{-6} moles H^+/liter $= 5.55 \times 10^{-6} M$ H^+ from food acid
4×10^{-7} moles H^+/liter $= 0.4 \times 10^{-6} M$ H^+ from saliva
$[H^+] = \overline{5.9_5 \times 10^{-6} M}$, or $6.0 \times 10^{-6} M$ (rounded off)

Multiplication and Division

The relative uncertainty of each number controls the number of significant figures in the product or the quotient of two or more numbers. Comparing relative uncertainties is time-consuming, and a shorter, but sometimes less accurate, method is recommended instead. To apply this method you should first write all the numbers in semiexponential form, using only one digit to the left of the decimal point. (This is not necessary for numbers between one and ten.)

To express a product or quotient to the correct number of significant figures or to take a root such as a square root use the following rule.

> RULE 2. Use the *same number* of significant figures in a product, quotient, or root as the number with the fewest number of significant figures. (Note that this is different from Rule 1 above.)

EXAMPLE 2-2. Find the $[H^+]$ of 0.020 liter of gastric juice containing 4.00×10^{-4} moles H^+.

Solution: Divide as follows:

$$[H^+] = \frac{4.00 \times 10^{-4} \text{ moles}}{2.0 \times 10^{-2} \text{ liters}} = 2.0 \times 10^{-2} M \quad \text{(two sig. figs.)}$$

Log Terms

Log terms such as pH should be expressed with the same number of significant figures to the right of the decimal as the total number of significant figures in the semiexponential number from which they are calculated [1]. This is because the digits to the left of the decimal come from the exponent which is an *exact* number, not a *measured* number. The following examples will illustrate this rule.

EXAMPLE 2-3. The equilibrium constant, K, for an oxidation-reduction reaction is 2.0×10^4. Calculate log K.

Solution: Take the log of the number and the exponent separately and add.

$$\log(2.0 \times 10^4) = \log 2.0 + \log 10^4 = 0.3010 + 4 = 4.30$$

Note that the log of 2.0 was expressed to four digits initially because it was found in the four-place log table in Appendix 4. After adding, it was rounded off to the proper number of significant figures, two to the right of the decimal.

EXAMPLE 2-4. Calculate the pH of gastric juice having $[H^+] = 2.0 \times 10^{-2} M$.

Solution: Take the negative log of the number and the exponent separately and combine algebraically (see Appendix 4 for four-place logs).

$$\text{pH} = -\log 2.0 + (-\log 10^{-2}) = -0.3010 + 2 = 1.69_9 \text{ or } 1.70 \quad \text{(two sig. figs. to right)}$$

Note that the *1* is really an *exact* number rather than a *measured* number because it is the result of a subtraction involving a digit that is the log of the 10^{-2} term.

Since the absorbance, A, is defined as the negative log of transmittance, T, on a spectrophotometer (see Ch. 13), it should be calculated from transmittance in the same manner as pH is. See the example below.

EXAMPLE 2-5. The transmittance of a solution is 0.10. Calculate the absorbance, A.

Solution: Write transmittance, T, in semiexponential form. Then take the negative log.

$$A = -\log T = -\log(0.10) = -\log(1.0 \times 10^{-1}) = -\log(1.0) + -\log(10^{-1}) = 1.00$$

Note that absorbance has three significant figures compared to two for transmittance, only because the first digit comes from the exponent.

Self-Test 2-1. Significant Figures

Directions: After reading the material on significant figures you should work each problem and write the answer in the space provided. It is important to understand each problem before proceeding further. After completing each

[1] D. E. Jones, *J. Chem. Ed.* **49**, 753 (1972).

SEC. 2-1   Significant Figures

problem you should consult the correct answer given at the end of the self-test before proceeding to the next problem. If an answer is wrong, you should reread the appropriate section before going farther.

A. Express the following numbers in scientific notation using the correct number of significant figures.
    a. 0.00202   b. 20.00   c. 1.10

B. Using a five-digit calculator the calculation of per cent chloride from three analyses of a single sample gives these results: 2.0510%, 2.0481%, and 2.0440%. Only two significant figures are justified for each calculated result.
    a. Round off each result without using a subscript.
    2.0510   2.0441   2.0480
    b. Round off each result using a subscript to avoid ambiguities in part *a*.
    2.0510   2.0441   2.0480

C. Add each of the following pairs of numbers using the correct number of significant figures.
    a. $1 \times 10^{-9} M$ OH^- + $1.0 \times 10^{-7} M$ OH^- from water
    b. $1 \times 10^{-8} M$ H^+ + $1.0 \times 10^{-7} M$ H^+ from water
    c. $5.0 \times 10^{-5} M$ H^+Cl^- + $1.0 \times 10^{-7} M$ H^+ from water

D. Using $K_w = 1.0 \times 10^{-14} = [H^+][OH^-]$ calculate each concentration using that given.
    a. If $[OH^-] = 2.0 \times 10^{-9} M$, $[H^+] = ?$
    b. If $[H^+] = 3.00 \times 10^{-7} M$, $[OH^-] = ?$
    c. If $[H^+] = 4 \times 10^{-4} M$, $[OH^-] = ?$

E. Calculate the pH of each solution below using the correct number of significant figures. Subtract a four-digit mantissa (0.3010, etc.) and round off after subtraction.
    a. $[H^+] = 3.00 \times 10^{-11} M$ b. $[H^+] = 3 \times 10^{-11} M$
    c. $[H^+] = 3.0 \times 10^{-10} M$ d. $[H^+] = 2.000 \times 10^{-10} M$

Answers to Self-Test 2-1

A. a. 2.02×10^{-3} b. 2.000×10^1 c. 1.10 (or 1.10×10^0)

B. a. 2.1, 2.0, 2.0 (Rounding off makes the differences appear larger than they are.)   b 2.0_5, 2.0_4, 2.0_{48} (Rounding to 2.0_5 is wrong; it implies further rounding to 2.1.)

C. a. $1.0 \times 10^{-7} M$ (1×10^{-9} isn't significant.)   b. $1.1 \times 10^{-7} M$
   c. $5.0 \times 10^{-5} M$ (H^+ from water is not significant.)

D. a. $[H^+] = 5.0 \times 10^{-6} M$ b. $[OH^-] = 3.3 \times 10^{-8} M$
   c. $[OH^-] = 2._5 \times 10^{-11} M$

E. a. pH = 11 − 0.4771 = 10.523 or 10.522_9
   b. pH = 11 − 0.4771 = 10.5 or 10.5_2
   c. pH = 10 − 0.4771 = 9.52 or 9.52_3
   d. pH = 10 − 0.3010 = 9.6990

2-2 THE ARITHMETIC MEAN AND THE MEDIAN

We have already discussed the true value, μ. This is also the arithmetic mean of an infinite number of valid results and is a reliable value. Unfortunately, time limits the number of analytical results that can be obtained so that only an *estimate* of μ can be obtained. In many cases this can be done by calculating the mean, $\overline{X}$, of the results. In other cases it has been shown that the median, M, is more effective and is to be preferred. The advantages and disadvantages of both will be discussed, but it should be stressed that both the mean and the median are only *estimates* of the true value; there is no "law" that either must be used all the time.

The Mean

The arithmetic mean, simply called the mean, is an estimate of μ that is calculated from *all* the results in a sample. The formula for calculating the mean from a set of n results arranged in increasing order $(X_1, X_2, ..., X_n)$ is

$$\overline{X} = \frac{X_1 + X_2 + \cdots + X_n}{n} \tag{2-1}$$

It can be seen that each analytical result contributes *equally* to the value of the mean. If the results are symmetrically (or normally) distributed about the mean, the mean is the most effective, or efficient, measure of the true value. The mean is not always more effective than the median if one or more of the results has a gross (large) error. As n decreases from 10 to 3, the effect of a gross error on the mean becomes larger and larger so that it is desirable to use the median in this area to obtain a better estimate of μ. In fact, it has been shown [2] that for three results from a symmetrical population it is more effective to use the median than to use the mean of the two closest results!

As the value of n increases above 10, the mean becomes a better and better estimate of μ, and there is no point in using the median beyond $n = 10$. This reflects the fact that a gross error is not as serious when ten or more "good" results are present.

The Median

The median, M, is the middle result of a set of results arranged in increasing order. If n is even, M is calculated by averaging the two middle results. In a *large* sample where the results occur in a symmetrical pattern about the mean the median is the same as the mean. In a *small* sample this pattern is usually not symmetrical, especially where n is 10 or less, so that M is not the same as $\overline{X}$.

[2] W. J. Blaedel, V. W. Meloche, and J. A. Ramsey, *J. Chem. Ed.* **28**, 643 (1951).

Except for $n = 2$, the median is less effective than the mean for symmetrical samples where no gross error is present [3]. It is interesting to compare the effectiveness of the median to the mean in a symmetrical sample. This is shown in Table 2-1. Note that the median is much more effective when n is even than when

TABLE 2-1. Effectiveness of the Median Compared to the Mean for a Symmetrical Sample (Normal Distribution)

Number of results, n	Effectiveness of median for even n	Effectiveness of median for odd n
2	1 (= same effectiveness as mean)	
3		0.74[a]
4	0.84	
5		0.69
6	0.78	
7		0.67
8	0.74	

[a] A fraction of 0.74 indicates that a median calculated from three results gives as much information about the true value as a mean calculated from 0.74×3, or 2.2 results.

it is odd (this is explored in detail in the following self-test). This is important when deciding when to use the median, which we will discuss next.

Choosing Between the Mean and the Median

If you are a beginning student, you are advised to use the mean in the laboratory except when the instructor advises otherwise. However, if the precision of a method is known, it may be used to judge whether a gross error may be present. If a method gives poorer precision than past experience indicates is possible, then the median should be used. (When using M it is advantageous to have four results rather than three. See Table 2-1.) It should be stressed that if the precision is poor or if one result appears to have a gross error, it is always preferable to run more analyses. The following example will illustrate all of the above points.

EXAMPLE 2-5. An FDA analyst obtains results of 0.13, 0.47, 0.57, and 0.59 ppm of mercury after analyzing a large catch of fish. The precision obtainable from past experience (a range of 0.30 ppm) indicates poor precision and the possible presence of a gross error in the 0.13 ppm result. The FDA tolerance for mercury in fish is 0.5 ppm. Should the fish be certified as meeting FDA standards or not?

Solution: Normally the mean is calculated first ($\bar{X} = 0.44$ ppm), but this is useless since the poor precision indicates the presence of a gross error. Calculation of the median

[3] R. B. Dean and W. J. Dixon, *Anal. Chem.* **23**, 636 (1951).

gives an M of 0.52 ppm, indicating that the fish barely fail to pass FDA standards. Since it is important to avoid wasting a large amount of fish as well as to avoid poisoning people, more analyses should definitely be obtained. Suppose, however, that this is not possible and that a decision must be made. What should be done? A comparison of the precision of the four results (range = 0.46 ppm) with that from past experience (range = 0.30 ppm) indicates that the precision definitely exceeds that obtained in the past. Assuming that the analysis is good to two significant figures, it is valid to use the median and discard the fish. Since the median is calculated from the two middle results it has an attractively large effectiveness in this case (see Table 2-1).

(Note: The typical FDA lab generally has *two* analysts analyze the same batch of fish for mercury. The mean and/or median of *both* analysts must exceed 0.5 ppm before the fish are rejected for certification. Thus in the above example the results of a second analyst must also exceed 0.5 ppm before the fish are rejected.)

Self-Test 2-2. The Arithmetic Mean and the Median

Directions: Work problems one at a time and check with the answers given at the end. The problems are meant to help you compare the usefulness of the mean and the median for increasingly larger samples from $n = 2$ through $n = 5$.

A. Two measurements of the pH of gastric juice give values of 1.1 and 1.50.
   a. Calculate the mean using the correct number of significant figures.
   b. Calculate the median using the correct number of significant figures.
   c. Is there any difference between the effectiveness of the mean and median for $n = 2$? Why?
B. Three measurements of the pH of saliva give values of 6.4, 6.6., and 4.4.
   a. Calculate the mean using the correct number of significant figures.
   b. Calculate the median using the correct number of significant figures.
   c. Is there any difference between the effectiveness of the mean and median for $n = 3$? Why?
   What is the effectiveness *in general* of the median relative to the mean for $n = 3$? Why?
   d. *In the case above*, is the mean or the median more effective? Why?
C. Four measurements yield values of 1.1, 1.2, 1.2, and 3.30.
   a. Calculate the mean using the correct number of significant figures.
   b. Calculate the median using the correct number of significant figures.
   c. What is the effectiveness *in general* of the median relative to the mean for $n = 4$? Why is it higher than for $n = 3$?
   d. *In the case above*, is the median or the mean more effective? Why?
D. Five measurements yield values of 1.10, 1.10, 1.20, 1.20, and 1.30.
   a. Calculate the mean using the correct number of significant figures; show additional uncertain digits using a subscript, if necessary.
   b. Calculate the median using the correct number of significant figures.

c. What is the effectiveness *in general* of the median relative to the mean for $n = 5$?
d. *In the above case*, is the median or the mean more effective? Why?
E. A symmetrical distribution of 101 results yields a mean of 1.50. What should the median of these results be? Why?

Answers to Self-Test 2-2

A. a. $\bar{X} = 1.3$ b. $M = 1.3$ (median is the average of both results).
   c. No. Both are calculated using the entire sample.
B. a. $\bar{X} = 5.8$ b. $M = 6.4$ c. Yes. The mean is generally more effective because all of the results are used. 0.74. It is calculated from only one-third of the results.   d. The median. It is not affected by the gross error in 4.4.
C. a. $\bar{X} = 1.7$ b. $M = 1.2$ c. 0.84. It is calculated from two-fourths of the results instead of one-third for $n = 3$.   d. The median. It is not affected by the gross error in 3.30.
D. a. $\bar{X} = 1.18$ b. $M = 1.20$ c. 0.69   d. The mean. There are no obvious gross errors.
E. Median should be 1.50. Mean and median should be equal for a symmetrical sample.

2-3 PRECISION

The precision, or agreement, of a set of results is governed by two types of errors that occur in the measurements. These are *systematic errors* and *random errors*. Systematic errors are those that cause the results to be consistently high or consistently low, but not both. Once systematic errors are detected they can usually be eliminated. An example of a systematic error is a titration with a dirty buret. This usually means some of the titrant remains in the buret and the actual volume used is smaller than the measured volume. The net effect is that all the results will be altered by a consistently high measured volume.

The other type of error is random error. Random errors are those that cause the results to be symmetrically scattered on either side of the true value. Random errors cannot be eliminated; hence they are always present and limit the precision of any analysis. An example of a random error is the dropwise approach to a titration end point. One time a drop too little of titrant will be added, the next time a drop too much, and so on. All of the results of the titration will have errors scattered symmetrically on either side of the true value.

Now, how do these errors affect precision? Recall that precision is the agreement among a set of measurements, or the ability to reproduce a measurement. If there are large systematic errors in an analysis, the results will tend to be scattered to one side or the other of the true value. If there is little or no systematic error, then only random error will limit precision, that is, the results will be symmetrically scattered on either side of the true value.

Although precision is not *defined* as the deviation of a set of measurements from the mean, it may be measured by calculating one of the three measures of deviation: standard deviation, range, or average deviation. First, let us discuss the standard deviation of a *population*.

Standard Deviation of a Population

If an infinite number of results could be obtained by an analysis, then the standard deviation (σ) of this infinite number, or population, would yield a 100% reliable measure of the precision of the population. The standard deviation would be calculated in this case by taking the square root of the average of the squares of the individual deviations, as follows:

$$\sigma = \left[\frac{\sum (X_i - \mu)^2}{n} \right]^{1/2} \tag{2-2}$$

where X_i represents all values from X_1 to X_n and where $\sum (X_i - \mu)^2$ is the sum of the squares of all the deviations from the mean. Now consider Figure 2-1. This is the

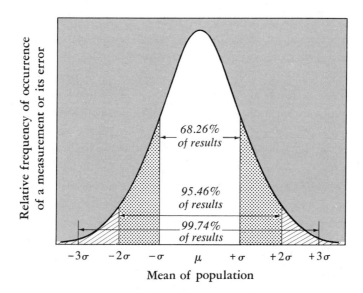

FIGURE 2-1. *The normal distribution curve.*

so-called normal distribution curve. It is formed by plotting an infinite number of measurements on the graph. If there are no systematic errors present, the ever-present random error would cause the results to be distributed in a random fashion about the true value. Also note that 99.74% of the results would occur within $\pm 3\sigma$ units of the true value, or population mean.

Now, the population standard deviation is a true measure of the precision of an analysis. Unfortunately, time and economics force the analyst to obtain a finite number of analyses, and he can only estimate the precision by calculating a sample standard deviation, a sample range, or a sample average deviation. The important concept of sample standard deviation will be discussed first.

Standard Deviation of a Sample

The standard deviation of a sample is calculated using an equation similar to **2-2**, except that the average is found by dividing by $(n-1)$, not n. The formula for the calculation is

$$s = \left[\frac{\sum(X_i - \bar{X})^2}{(n-1)}\right]^{1/2} \tag{2-3}$$

The reason $(n-1)$ is used instead of n is essentially that a degree of bias is introduced by using an estimate of the true value calculated from the very data to be used to find s. Because many analysts use s, it is important to be able to calculate it correctly. The following example will illustrate this.

EXAMPLE 2-6. Babies are tested for cystic fibrosis by measuring the chloride in their sweat. One such analysis gives molarities of 2.15×10^{-2}, 2.35×10^{-2}, and 1.50×10^{-2} chloride. Calculate the mean and standard deviation and evaluate the results.

Solution: First, calculate the mean of the three results. $\bar{X} = 2.00 \times 10^{-2} M$.
Second, calculate the deviations (omitting the 10^{-2} factor) and square them.

$$\left.\begin{array}{l}(2.15 - 2.00)^2 = 0.0225 \\ (2.35 - 2.00)^2 = 0.1225 \\ (1.50 - 2.00)^2 = 0.2500\end{array}\right\} \text{(Retain at least two nonsig. figs.)}$$

Third, sum the squares. $0.0225 + 0.1225 + 0.2500 = 0.39_{50}$
Fourth, divide by $(n-1)$.

$$\frac{0.39_{50}}{(3-1)} = 0.19_{75} \quad \text{(Retain two nonsig. figs.)}$$

Fifth, take the square root.

$$\sqrt{0.19_{75}} = 0.44$$

The *absolute* standard deviation is thus $0.44 \times 10^{-2} M$ chloride. The *relative* standard deviation in pph is calculated as follows:

$$\frac{0.44 \times 10^{-2} M}{2.00 \times 10^{-2} M} (100) = 22 \text{ pph}$$

Since the mean chloride level for normal children is $2.0 \times 10^{-2} M$, the analysis indicates that the child does not have cystic fibrosis (Sec. 17-3).

Because the relative standard deviation is expressed in pph, it may be compared with values of relative standard deviation from analyses which have a different mean. The relative standard deviation is independent of the mean by the very nature of the way it is calculated, making it more useful for comparison than the absolute value.

Generally the relative standard deviation for a highly precise method of analysis is less than 1 pph. This implies that the data in Example 2-6 are not very precise. Part of the problem is that sweat chloride is not always easy to measure in a baby (Sec. 17-3), but another part of the problem is that *only three analyses were performed*. As can be seen in Figure 2-2, the standard deviation is a very

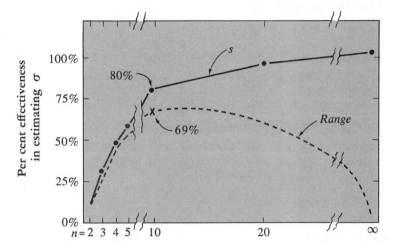

FIGURE 2-2. *The per cent effectiveness in the estimation of σ by using the s and the range of a sample.*

poor estimate of precision when $n = 3$. This is because there are only three values used to sample an infinite number of values in the population. For example, the only way of evaluating the large deviation between the results of 1.50 and 2.15 in Example 2-6 is the other large deviation between the results of 2.15 and 2.35. This deviation itself is fairly large; without more measurements it is difficult to conclude that the precision is poor or that these measurements happen to have a large deviation, possibly because of two gross errors.

The Range of a Sample

The range (also called the *spread*) of a sample of results is the difference between the largest value and the smallest value, $X_n - X_1$. It is also an estimate of σ, the standard deviation of a population, and is the simplest and most rapidly calculated measure of precision. Just as for any of the three measures, both absolute and relative values of the range may be calculated. The relative value of the range is calculated by dividing the absolute value by the mean and multiplying by 100 to express it in pph. The example below will illustrate the use and calculation of the range.

EXAMPLE 2-7. It is difficult to determine mercury in biological samples because there are so few stable materials containing known amounts of mercury. A photographic gelatin containing 2.0 ppm of mercury was developed to meet this need [4]. The following six results were obtained by flameless atomic absorption (Sec. 16-3): 1.95, 2.10, 1.90, 2.10, 2.05, and 2.00 ppm mercury. Calculate the mean, the range, and comment on the number of significant figures that are justified for the mean.

Solution: First, calculate the absolute value of the range. This is $2.10 - 1.90$, or 0.20 ppm.
Second, calculate the mean. $\bar{X} = 2.00$ ppm.
Third, calculate the relative value of the range.

$$\frac{0.20 \text{ ppm}}{2.02 \text{ ppm}} (100) = 10 \text{ pph}$$

Fourth, consider the number of significant figures that are justified for the mean. If a mean of 2.02 ppm is reported, this implies an absolute uncertainty of ± 0.01 ppm in the last significant figure. The relative uncertainty would be

$$\frac{0.01 \text{ ppm Hg}}{2.02 \text{ ppm Hg}} (100) = 0.5 \text{ pph}$$

Since the range has a relative value of 10 pph, the use of three significant figures for the mean is clearly not justified. The precision of the measurements is too poor to allow the analyst to make a definite statement about the third digit. The next question is, Can two significant figures be used to report the mean? The relative uncertainty in 2.0 ppm is easily seen to be 5 pph. Although this is still smaller than the relative value of the range, it is better to report the mean as 2.0 ppm rather than as 2 ppm. In the latter case, the relative uncertainty would be 50 pph, which would be much larger than the relative value of the range.

In general, the range can be used as an estimate of precision in preference to the standard deviation of a sample when $n = 3$ to 10. This is not to say that the range is a better estimate of precision than s. As can be seen from Figure 2-2, the range is in general less effective than s. However, there is very little difference between the two when $n = 3$ to 10 so that the *convenience* of using the range is the reason for using it. Only as n approaches 10 does s become appreciably more effective (see Fig. 2-2).

Average Deviation of a Sample

The average deviation of a sample is the average of the *absolute* deviations of the individual results from the mean. It is calculated as follows:

$$\text{Av. dev.} = \frac{\sum |X_i - \bar{X}|}{n} \tag{2-4}$$

[4] D. H. Anderson, *Anal. Chem.* **44**, 2099 (1972).

The vertical lines on either side of the numerator indicate that the absolute value of the deviation is used, that is, all negative signs are dropped. Note that the average deviation is found by dividing by n, not $(n-1)$ as is done for the standard deviation. It does not reflect values that deviate widely from the others as much as the standard deviation does and is therefore not as useful. Like s and the range it may be given as an absolute or a relative value.

Self-Test 2-3. Average Deviation, Standard Deviation, and the Range

A. Analysis of a blood sample for inorganic phosphate gives 4.00, 4.20, 3.60, and 4.20 mg% phosphate.
   a. Calculate the average deviation.
   b. Calculate the standard deviation.
   c. Calculate the range.
   d. Are the results in the normal range? (See Fig. 1-1.)
B. Using the *absolute* values in question A, calculate
   a. The *relative* average deviation in pph.
   b. The *relative* standard deviation in pph.
   c. The *relative* range in pph.
C. An analyst obtains the following values: 3.00, 4.20, 4.50, and 4.30 mg% phosphate.
   a. Calculate the average deviation and its relative value.
   b. Calculate the standard deviation and its relative value.
   c. Calculate the range and its relative value.
D. Make the following comparisons between the data in question A and that in question C.
   a. Compare the mean of each set of data.
   b. In which set of data is there a larger deviation of one value from the other three?
   c. Which set of data has the larger average deviation? The larger standard deviation? The larger range?
   d. What does the larger standard deviation reflect or indicate?
E. What measure of precision was the quickest indication of the larger deviation of the data in question C? Is it as effective, more effective, or almost as effective as the standard deviation?

Answers to Self-Test 2-3

A. a. 0.20 mg%   b. 0.28_3 mg%   c. 0.60 mg%   d. Yes
B. a. 5.0 pph   b. 7.0_8 pph   c. 15 pph
C. a. 0.50 mg%, $12._5$ pph   b. 0.67_8 mg%, 17 pph   c. 1.50 mg%, $37._5$ pph
D. a. The means are both 4.00 mg% phosphate.   b. 3.00 deviates more than 3.60 mg% does.   c. The data in question C.   d. The larger deviation of 3.00 mg%.
E. The range. Almost as effective.

2-4 USE OF PRECISION IN TESTING FOR GROSS ERRORS

In Section 2-2 we stated that the median could be used instead of the mean if the precision of the method indicated that a gross error was present in one of the results. This should be done only if we can compare the precision with the precision known from past experience. However, the latter may not always be known; it is therefore better to employ a general testing procedure to detect gross errors based on the precision obtained in the actual analysis. Some tests are based on s or the average deviation, but the most statistically valid test when n is 3 to 10 is based on the range. This test is the Q test.

The Q Test

The Q test is a convenient method for testing for a gross error in one result when n is 3 to 10. It involves identifying the result with a possible gross error, followed by dividing the *absolute value* of the difference, d, between the result and its nearest neighbor by the range. The resulting quotient is symbolized as Q.

$$Q = \frac{d}{(X_n - X_1)} \qquad (2\text{-}5)$$

The calculated Q is compared with a rejection quotient, $Q_{0.90}$ (Table 2-2). If the

TABLE 2-2. Rejection Quotients [3] for the Q Test[a]

n	$Q_{0.90}$
3	0.941
4	0.765
5	0.642
6	0.560
7	0.507
8	0.479
9	0.441
10	0.409

[a] If the calculated $Q \geq Q_{0.90}$, reject the result being tested.

calculated Q is larger than or the same as the rejection quotient, then the result is rejected [3]. This can be expressed as

> If $Q \geq Q_{0.90}$, reject the result being tested.

Note this specifically indicates that if the calculated Q equals $Q_{0.90}$, the result should also be rejected. When this equality occurs it means the result is rejected with *exactly* 90% statistical confidence that it is significantly different from the other results. (If Q is larger, the result is rejected with somewhat more than 90% confidence that it is different.) When the rejection quotients are given to only two significant figures it often happens that the calculated $Q = Q_{0.90}$. To avoid this situation the $Q_{0.90}$ values are tabulated to three significant figures.

EXAMPLE 2-8. Three high-precision pH measurements of the pH of gastric fluid give pH values of 1.701, 1.607, and 1.601. *Assuming* there is no instrumental (or experimental) error, decide whether any result should be rejected. (Because the first value deviates greatly, we might suspect an instrumental error from improper startup, warmup, etc.)

Solution: First identify the result with a possible gross error. In general this is the largest or the smallest value so both should be inspected. In this case the largest result of 1.701 is obviously the result with a possible gross error.

Next, use **2-5** to calculate Q.

$$Q = \frac{(1.701 - 1.607)}{(1.701 - 1.601)} = 0.94$$

Finally, decide whether or not to reject 1.701. Since the Q of 0.94 is not greater than the $Q_{0.90}$ of 0.941, we must retain the result. However, note that the pH readings make it impossible to calculate a third significant figure for Q; the fact that it is almost equal to $Q_{0.90}$ makes it highly desirable to obtain more measurements if possible. If not, the precision should be compared with past precision prior to deciding whether to report the median.

(If we had not assumed that an instrumental error was absent, the first thing we might realistically have done was to check this possibility. Most likely, we could have rejected 1.701 on the basis of an instrumental error or operator error.)

In general the possibility of an instrumental or experimental error should be evaluated *before using* the Q test. If the Q test rejects a result, then the mean of the remaining data should be calculated. If the Q test fails to reject a result, the mean of *all the data* should be reported unless the precision is poorer than that obtained in the past (Sec. 2-2). In that case the median should be reported. (If such information is not available, one can be sure that the precision is poor *only if a result differs from all others in the first digit.*)

SEC. 2-5   Reporting Laboratory Results; Lab Notebook   35

Self-Test 2-4. The Q Test

Directions: Compare your answers with those given at the end.

A. The following results are obtained: 9.00% N, 11.00% N, and 9.10% N.
   a. Identify the value with a possible gross error and test it using the Q test.
   b. Calculate the mean after applying the Q test.
B. The following analyses are obtained for the weight of vitamin C in tablets: 100.0 mg, 110.0 mg, and 100.8 mg.
   a. Identify the value with a possible gross error and test it using the Q test.
   b. Should the mean or median be reported? Why? What numerical value should you report?
C. The following results are obtained: 10.000% Cl, 11.000% Cl, 10.10% Cl, and 10.235% Cl.
   a. Identify the value with a possible gross error and test it using the Q test.
   b. Calculate the mean after applying the Q test.
D. The following results are obtained: 1.00 ppm, 3.00 ppm, 1.10 ppm, and 1.56 ppm Hg in fish.
   a. Identify the value with a possible gross error and apply the Q test.
   b. Can the median be reported? Why? What numerical value should you report?

Answers to Self-Test 2-4

A. a. Calculated Q of 0.95 is greater than 0.941 and 11.00% N is rejected.
    b. The mean = 9.05% N.
B. a. Calculated Q of 0.92 is less than 0.941, and 110.0 mg is retained.
    b. Mean. No information on past precision. $\bar{X}$ = 103.6 mg.
C. a. Calculated Q of 0.765 is the same as 0.765 and 11.000% Cl is rejected.
    b. The mean = 10.11_2% Cl. (Only four sig. figs.)
D. a. Calculated Q of 0.72 is smaller than 0.765 and 3.00 ppm Hg is retained.
    b. Yes. The 3.00 ppm result differs in the first digit. M = 1.33 ppm Hg.

2-5 REPORTING LABORATORY RESULTS; LAB NOTEBOOK

This section will summarize the important ideas of the preceding sections as they apply to reporting laboratory results in a lab notebook. First, the suggested format of a lab notebook and a report page in the notebook will be discussed.

The Laboratory Notebook

The lab notebook has two purposes—to record raw data and to report complete results for an analysis. Left-hand pages should be reserved for raw data. These data need not be neat; the idea is to avoid writing raw data on odd pieces of paper which can easily be lost. Always make all entries, no matter how messy, in the notebook.

All reports, in contrast, should be written neatly on right-hand pages. A vertical tabular form is recommended. Each sample number should be entered at the top and all data, in the order needed, should be entered below. See the example in Table 2-3. In general the weight of a dried sample (or the volume of the

TABLE 2-3. *A Report Form for a Gravimetric Analysis*

Title: Determination of Per Cent Chloride
Reaction: $Cl^- + AgNO_3 = AgCl(s) + NO_3^-$

	I	II	III
Weight Cl sample, g	0.5000	0.5001	0.5001
Weight crucible + AgCl	31.8002	32.8014	33.8250
Weight crucible, g	31.0001	32.0004	33.0100
Weight AgCl, g	0.8001	0.8010	0.8150
Gravimetric factor = Cl/AgCl	0.24737		
Weight Cl, g	0.1980	0.1984	0.2022
% Cl	39.60%	39.68%	40.44%
Application of Q test	—	—	$Q = 0.91$ (Retain)
Mean		39.91%	
Median		39.68%	

(The range of 0.84% above is greater than an expected range of 0.01 (1 pph) × 39.91% or 0.4%. Reporting the median is justified.)
The better estimate of μ: Median of 39.68%

sample) is recorded first since this is usually the first piece of data obtained. Then the actual analytical data are recorded: the volume of titrant or, in the case of gravimetric analysis, the weights of the crucible plus precipitate and the weights of the empty crucibles (Table 2-3). Other data may be added below the actual analytical data, for example, the actual weight of the precipitate in a gravimetric analysis and, if desired, the weight of the constituent sought (see *Weight Cl, g* in Table 2-3).

The last entry under each sample heading should always be the desired result; this might be a percentage or a concentration such as normality or g/liter. The application of the Q test to the result with the largest deviation should be shown, if necessary. If the instructor indicates the expected range, then this can be used to decide whether to report the mean or the median. (In Table 2-3, the expected range is calculated from an expected relative range of 1 pph and the mean.) Let us discuss this in more detail.

Deciding What Value to Report in the Lab

You should review Sections 2-2 and 2-4 when deciding what value to report after obtaining three or more results. If the results all appear close together and the range of the results is not large, then in general you should report the mean. If you have obtained a result in the lab that appears to have a gross error, you should follow these steps:

1. If an experimental error, such as overshooting the end point in the first titration, is suspected, reject the result and obtain one or two more results.
2. If no experimental errors are suspected, apply the Q test. If the suspected result is not rejected and there is time, obtain one or two more results.
3. If the result is not rejected by the Q test and there is no time to obtain additional results, then use the precision from past experience to decide whether or not the median can be reported (see Ex. 2-5). The relative range for many methods should be no larger than 1 pph; thus multiplying 0.01 times the mean will give a good estimate of the absolute value of the range (see Table 2-3). If the actual range is significantly larger than this estimated value, then there is good reason to use the median, even if the Q test does not reject a suspicious result.

Occasionally the Q test will reject a reasonable result because of the agreement of the other results. This important exception to the usually absolute reliability of the Q test will now be discussed.

Two "Lucky" Results Out of Three

It occasionally happens that two results out of the usual three are closer together than would be expected from the precision of the measurement used. For example, suppose results of 60.00%, 60.01%, and 61.00% are obtained. The Q test would reject the 61.00% because the first two results are so close together. (Note that 61.00% would not be rejected if the middle result were 60.10%.) The *relative* deviation of the first two results in fact is 0.02 pph. This is less than the relative uncertainty of 0.1 pph for buret reading error and less than the relative uncertainty of 0.04 pph for a gravimetric analysis (where the weighing error for a 0.500 g sample weighed by difference is assumed to be the least precise step). The first two results are "lucky" results because they are closer together than just the measuring error in any method would normally allow.

In a case such as this, the Q test should not be used to reject a value because it is not clear that the value has a gross error. It would be better to obtain additional results before running the Q test. If there is no time available to do this, then the only alternative is to report the mean. Reporting the median would be incorrect because, in effect, a possibly valid result would be rejected.

Self-Test 2-5. Reporting Laboratory Results

Directions: Work the problems one at a time and compare answers with those given at the end of the test.

A. You have just completed three gravimetric analyses for chloride in the laboratory and you are ready to enter your calculations in your notebook.
  a. What is the last entry you should make under each sample number heading?
  b. Before deciding whether to report the mean, what should you do?
  c. After applying the Q test, should you report the mean or median?
  d. Assuming that a range is too large if it is greater than 1 pph, what would the absolute value of an undesirable range be if the mean were 50.00%?
  e. Suppose you had a mean of 50.00% and had a value that could not be rejected by the Q test. If the range of your results were 0.60%, what should be reported instead of the mean? Why?
B. A student obtains gravimetric chloride results of 30.00%, 30.10%, and 31.00% Cl.
  a. In a case such as this where the results are not satisfactory, what should be done to improve the data?
  b. Even if the Q test were applied to the data, what would the calculated Q imply?
  c. Assume that the student obtains a fourth result of 30.20%. What would be the result of the Q test? Should the mean or median be calculated?
  d. Assume instead that the student obtains a fourth result of 30.30%. What would be the result of the Q test? Assuming that a range is undesirable in this case if it exceeds an absolute value of 1 pph × 30%, or 0.30%, what should be reported, the mean or median? Why?
C. A student obtains titrimetric chloride results of 50.00%, 50.01%, and 51.00% Cl under conditions where the relative uncertainty of the titration is no more than 0.1 pph.
  a. What is the result of the application of the Q test to the data?
  b. Is the result of the Q test justified in light of the uncertainty of the titration and the deviation between 50.00 and 50.01%?
  c. What could be done to check the results if you are not satisfied with them?

Answers to Self-Test 2-5

A. a. The % Cl in each sample   b. Identify the result with possible gross error and test it using the Q test.   c. If the Q test rejects a value, report the mean. If not, compare the range of results with the range from past experience and report the median if justified   d. $0.01 \times 50.00\% = 0.50\%$
  e. The median. The range of 0.60 % is greater than 0.50% (part d).
B. a. Obtain more results.   b. The Q of 0.90 implies that 31.00% should be retained.   c. The Q of 0.80 implies that 31.00% should be rejected. The mean.   d. The Q of 0.70 implies that 31.00% should be retained. The median. The range is greater than the range from past experience.
C. a. The Q of 0.99 implies that 51.00% should be rejected.   b. No, the relative deviation between 50.00 and 50.01% is only 0.02 pph; they are two "lucky" results.   c. Obtain more results.

2-6 CONFIDENCE LIMITS FOR THE MEAN

It is possible to use statistical theory to predict within what limits around the sample mean the true value will be found. These limits are called *confidence limits*, and they are calculated using values of *t*, or *Student's t*, and *s*, the sample standard deviation.

$$\text{confidence limits} = \overline{X} \pm \frac{ts}{\sqrt{n}} \qquad (2\text{-}6)$$

Unfortunately, statistical theory will not permit the limits for the true value to be calculated with 100% probability. There is always some fraction of risk, α, or percentage probability $(100 - 100\alpha)$ involved in such a prediction. The value of *t* varies with the fraction of risk, or percentage probability, involved and the number of results. Such values are listed in Table 2-4. (Some scientists prefer the degrees of freedom $(n-1)$ rather than *n*, and this is listed in Table 2-4 along with *n*.)

When confidence limits for the true value are calculated using **2-6** it may be said that the fraction of risk that the true value lies *outside these limits* is α. Since

TABLE 2-4. Values of t for Calculating Confidence Limits

Number of measurements, n	Degrees of freedom, $n-1$	Risk and probability level		
		0.10	0.05	0.01
		90%	95%	99%
2	1	6.314	12.706	63.657
3	2	2.920	4.303	9.925
4	3	2.353	3.182	5.841
5	4	2.132	2.776	4.604
6	5	2.015	2.571	4.032
7	6	1.943	2.447	3.707
8	7	1.895	2.365	3.499
9	8	1.860	2.306	3.355
10	9	1.833	2.262	3.250
11	10	1.812	2.228	3.169
12	11	1.796	2.201	3.106
13	12	1.782	2.179	3.055
14	13	1.771	2.160	3.012
15	14	1.761	2.145	2.977
16	15	1.753	2.131	2.947
21	20	1.725	2.086	2.845
26	25	1.708	2.060	2.787
31	30	1.697	2.042	2.750
41	40	1.684	2.021	2.704
61	60	1.671	2.000	2.660
$\infty + 1$	∞	1.645	1.960	2.576

the true value may lie above or below these limits, $\alpha/2$ is the fraction of risk that the true value is *either above or below* these limits. It may also be said that the probability that the true value lies *inside the limits* is $100 - 100\alpha$.

EXAMPLE 2-8. Suppose that ten results for the mercury content in fish have been obtained with a mean of 0.44 ppm Hg and an absolute standard deviation of 0.10 ppm Hg. Calculate the confidence limits for a fraction of risk of 0.10 and interpret them.

Solution: For $\alpha = 0.10$, $t = 1.833$ at $n = 10$. The limits are

$$0.44 \text{ ppm} \pm \frac{(1.833)(0.10)}{\sqrt{10}} = 0.44 \text{ ppm} \pm 0.057_9 = 0.38 \text{ to } 0.50 \text{ ppm}$$

The interpretation is that there is a 0.10 fraction of risk that the true value lies outside the range 0.38 to 0.50 ppm and a 90% probability that it lies inside this range. There is a 0.05 fraction of risk that the true value lies below 0.38 ppm or above 0.50 ppm (Fig. 2-3).

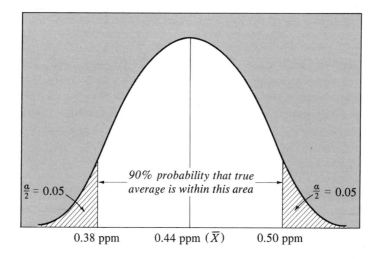

FIGURE 2-3. Graphical representation of the t function for a mean of 0.44 ppm mercury, a standard deviation of 0.10 ppm mercury for ten results and confidence limits of 90%.

QUESTIONS AND PROBLEMS

(Answers to most even-numbered problems are in Appendix 5.)

Concepts and Definitions
1. Define the following terms using "the true value" in your definitions.
   a. Absolute error
   b. Relative error
   c. Accuracy (Do not use error in your definition.)

2. Define the following terms without using mathematical formulas or symbols.
   a. Precision (Do not use deviation in your definition.)
   b. Standard deviation
   c. Range
3. Explain the advantage of *relative* standard deviation over *absolute* standard deviation.
4. Explain why the median for $n = 4$ has a higher effectiveness than
   a. The median for $n = 3$.
   b. The median for $n = 6$.
5. Explain why the median and the mean have the same effectiveness for $n = 2$.
6. Explain why the range is recommended over the standard deviation for $n = 3$ to 10 when the range is less effective than s.

Significant Figures
7. Add each of the following pairs of numbers and express the $[H^+]$ to the correct number of significant figures.
   a. $1 \times 10^{-10} M$ of HCl (a strong acid) $+ 1.0 \times 10^{-7} M$ H^+ from pure water
   b. $5.0 \times 10^{-5} M$ of HCl $+ 1.0 \times 10^{-7} M$ H^+ from pure water
   c. $4.0 \times 10^{-5} M$ of NaOH (a strong base) $+ 1.0 \times 10^{-7} M$ OH^- from pure water
8. Calculate the correct $[H^+]$ of pure water in each case below, assuming that K_w is known only to the number of significant figures given.
   a. For $K_w = 1.0 \times 10^{-14}$, $[H^+] = ?$
   b. For $K_w = 1.00 \times 10^{-14}$, $[H^+] = ?$
   c. For $K_w = 1.008 \times 10^{-14}$, $[H^+] = ?$
9. Calculate the pH of each solution below using the correct number of significant figures.
   a. $[H^+] = 1 \times 10^{-10}$
   b. $[H^+] = 1.40 M$
   c. $[OH^-] = 2.0 \times 10^{-11}$
10. Calculate the semiexponential value of K from the log of K for each case, using the correct number of significant figures.
    a. $\log K = 2.40$
    b. $\log K = 2.30$
    c. $\log K = 10._3$

Mean and Median
11. The absolute value of the range for a titrimetric chloride analysis is usually 0.50% for the % Cl values obtained below. Decide whether the mean or the median should be used in each case and calculate it.
    a. 50.1%, 50.60%, 50.3%
    b. 50.0%, 50.3%, 50.8%
    c. 50.1%, 50.2%, 51.4%, 51.0%
12. In each of the cases below decide, if possible, whether the mean or the median is to be preferred and report its value. (Assume the Q test does not reject any values.)
    a. 9.1%, 9.6%, 9.80%, and 11.0%
    b. 9.60% and 9.8%
13. Explain why the median is more effective for $n = 4$ than
    a. The median where $n = 6$.
    b. The median where $n = 3$.
    c. The median where $n = 5$.

Precision

14. Calculate the standard deviation and the range for each of the following samples of results from the determination of lead in glazed pottery.
  a. 0.100 ppm, 0.110 ppm, 0.120 ppm, and 0.150 ppm of Pb
  b. 1.00 ppm, 1.10 ppm, 1.20 ppm, and 1.50 ppm of Pb
15. Calculate the relative values of the standard deviation and the range for each of the samples in problem 14.
16. Calculate the standard deviation and the range for the following values of percentage chloride in a drug: 40.02, 40.11, 40.16, 40.18, 40.18, and 40.19%.

Testing for Gross Errors

17. The following three results for percentage carbon are obtained for a sample: 80.00%, 80.12%, and 82.00%.
  a. If there is time, what is the first thing that should be done to obtain a better estimate of the true value?
  b. Assuming that there is no time and that only the three results above can be considered, apply the Q test to the data and report either the median or the mean, whichever is more appropriate.
18. After obtaining chloride analyses of 10.00%, 10.10%, and 11.00% Cl an analyst desires to reject the 11.00% result.
  a. Can he do so using the Q test?
  b. Next he obtains a fourth result of 10.20% Cl. What is the result of the Q test now, and what value should he report for the mean?
  c. Assume that the fourth result is 10.30% Cl instead of 10.20% Cl. What would be the result of the Q test? Assuming that his precision is poorer than usual, should he use the median or the mean? Calculate the appropriate value.

Challenging Problems

19. The following results for percentage HCl were obtained by gravimetric analysis: 70.00%, 70.01%, and 71.00%. Assuming that the maximum deviation occurs from weighing 1.0000 g of silver chloride by difference, calculate the mean of the results after considering possible application of the Q test.
20. The calculation of the rejection quotient, $Q_{0.90}$, for $n = 3$ is done using the general formula of Dixon [*Annal. of Math. Statistics* **22**, 68 (1951)] for R_α:

$$R_\alpha = 0.500 + [(\sqrt{3})/2] \tan[\pi/3 (0.500 - \alpha)]$$

where α is the fractional probability that a suspicious value is *larger* than R_α; for $Q_{0.90}$, α is thus 0.05.
  a. Calculate $Q_{0.90}$ for $n = 3$ to three significant figures.
  b. Calculate $Q_{0.99}$ for $n = 3$ to three significant figures.
21. After analyzing the same sample of fish for mercury two FDA analysts find 0.495 and 0.497 ppm of mercury using the same method of analysis and the same equipment. Recall that the FDA tolerance for mercury is 0.5 ppm Hg. Apply the concepts of significant figures and decide whether the fish should be rejected as unsafe, whether the fish should be declared safe to eat, or whether more analyses should be done.
22. Calculate 95% confidence limits for both sets of data in problem 14.

3. Sampling and Preparation of the Sample

"What are you going to do, then?"
I asked. "To smoke," Holmes answered.
"It is quite a three pipe problem...."

Arthur Conan Doyle
The Red-Headed League

Sampling, as well as preparation of the sample, is indeed a "three pipe problem." The difficulty is that there is no one sampling procedure that is suitable for all samples. One reason for this is of course that a sample may have to be taken from a gas, a liquid, or a solid, but surely another reason that is equally important is the immense variety of samples and the incredible amount of information that may be required. Finally, there is the discouraging maxim that if the sample is not chosen properly and prepared properly for analysis, all of the results may be meaningless because they will not represent the true nature of the material sampled.

In this chapter we will try to describe things in the order the analyst does them: first, sampling from beginning to end and, second, the preparation of the sample for analysis.

3-1 SAMPLING

Since the sampling method chosen and the size of sample taken depend on the information required, we begin by discussing the latter.

Sampling and the Information Requested

Several different kinds of analytical information might be requested for a sample submitted for analysis. Recall from Section 1-1 that there are at least four unique areas of chemical measurement: measurement of physical and chemical properties, qualitative analysis, quantitative analysis, and diagnostic analysis. If a substance's physical or chemical properties are to be measured, then the sampling method is not as important as the purity of the sample. A sample of the substance that appears as pure as possible is chosen, and the sample is then purified by a technique such as distillation or recrystallization. Only a small amount of the purified sample is necessary unless a large number of measurements are to be made.

Suppose on the other hand that only a qualitative analysis is to be performed. In general a sample that is as representative as possible of the average composition of the substance to be analyzed is taken. This is done so that the minor constituents ($<1\%$) as well as the major constituent ($>1\%$) or constituents in the sample can be identified. In many cases the sample will not be purified because the minor constituents are just as important as the major constituent(s). In some cases, however, only the main constituent must be identified. In these cases the sample may be purified to prevent interference by the minor constituents in the identification of the main constituent. The size of the sample of course will depend on the number of qualitative tests applied to the sample. If mainly optical methods of analysis (Sec. 1-1) are to be used, the sample can generally be quite small.

If a quantitative analysis or diagnostic analysis is to be performed, then a sample as representative as possible of the substance to be analyzed is taken. As for qualitative analysis, this is done so that the amounts of the major constituent(s) and minor constituents found in the sample can be stated to be representative of the substance to be analyzed. The size of the sample of course will depend on the number of each type of quantitative measurement performed on the sample and on the type of measurement (Table 1-1).

Sampling of Solids

The sampling of solids involves the largest number of difficult sampling problems. Solid substances exist in so many shapes and sizes that it is difficult to decide how best to obtain a representative sample. To illustrate some of the problems involved, we will discuss the sampling of solids of large nonuniform particle size, solids of small uniform particle size, tablets, and contaminants from the surface of a solid.

Sampling of Large Nonuniform Particles. We will assume at this point that a large portion of the substance to be analyzed has been taken in a random fashion and brought to the laboratory for further treatment and that we are to select our sample from this portion. Solids consisting of large nonuniform particles first must be reduced to a small uniform particle size. This is accomplished by passing the substance through a disk pulverizer, grinding it in a ball mill, or grinding it with a mortar and pestle. A sieve may be used to insure that the final sample is selected from a solid of uniform particle size. Care must be taken during the grinding and sieving that the substance to be analyzed is not contaminated by dust, water, or other particles in the laboratory. Once the solid substance has been reduced to particles small enough to dissolve for analysis, the problem is to reduce the amount of the solid to a small enough size for an analytical sample. This will be discussed under the sampling of small particles.

Sampling of Small Particles. When a solid substance is judged to be composed of particles small enough to be dissolved for analysis without further grinding, the first step is to make sure that the particles are of uniform size. This is important because including too many particles of a larger or smaller size may yield a sample that is not homogeneous and therefore not representative.

To achieve uniform particle size the particles are sieved, and only the particles passing through a certain size sieve are retained for selection of the sample. If there is too much of the substance to be analyzed left at this point, the amount may be reduced by forming a pile, dividing the pile into quadrants, and keeping opposite quadrants for the final sample.

The above method is of course a recommended *general* approach for all samples; however, many bottled powders, bottled crystals, and student laboratory samples are already homogeneous and may be sampled directly without any treatment. Still other bottled products may only need thorough mixing in the bottle to insure homogeneity and a representative sample.

Sampling of Tablets. The sampling of pharmaceutical tablets, capsules, and so on, poses different problems. One problem common to tablets is that decomposition or other changes in the tablets at the top of the container may not have occurred to the same extent in the tablets near the bottom of the container. It is therefore not correct to take one or two tablets from only one area in the bottle.

A good illustration of the way to sample some tablets is provided by the method used by the FDA to sample aspirin tablets. The main decomposition reaction of aspirin tablets in bottles that have been opened to moist air (or bottled improperly by the manufacturer) is the hydrolysis of acetylsalicylic acid (ASA):

$$\text{Acetylsalicylic acid} + H_2O \rightarrow \text{Salicylic acid} + \text{Acetic acid}$$

(The balanced reaction has been given in Sec. 1-1.)

Naturally tablets at the top of an aspirin bottle will react to a greater extent with water than those at the bottom. The final sample should reflect the average composition of the aspirin since the manufacturer cannot be held responsible for the handling of the aspirin once he has bottled it. The FDA method calls for taking a representative sample of 20 tablets for analysis. The tablets are ground together to form a powder of uniform particle size and a sample weight equivalent to the average weight of one tablet is taken for analysis.

The sample taken also influences the constituent determined. Since there is so much ASA present in an aspirin tablet, the most accurate indication of decomposition is the amount of salicylic acid or acetic acid formed. Since acetic acid is volatile it will evaporate to varying degrees from tablets—certainly to a larger degree from tablets at the top of the bottle. Therefore the determination of non-volatile salicylic acid will be a more accurate indication of the degree of purity.

Sampling for Contaminants From the Surface of a Solid. Frequently there is the need to test certain solids which release contaminants into a liquid or solid stored in them by leaching from the surface. A good example is contamination of foods and liquids by lead- and cadmium-based pigments used on pottery and ceramic ware. Analysis of the entire solid will be useless since the problem is not how much lead or cadmium is coated on the pottery, but how much is leached off into the food or liquid stored therein.

The FDA solved this problem by adopting standard conditions for simulating the leaching of lead(II) and cadmium(II) into a standard liquid. The liquid selected was a dilute (4%) solution of acetic acid, similar in composition to table vinegar. This solution represents perhaps the most common degree of high food acidity so that the sampling represents the most possible contamination. The 4% acetic acid solution was then allowed to remain in the pottery to be tested for a period of 24 hours. The 24-hour period of testing of course represents a practical compromise between the possible longer term use by a consumer and the need for the FDA laboratory to evaluate pottery at a reasonable rate.

Sampling of Liquids

For liquids in general, the most difficult problem is to decide *how* to sample the liquid. For body fluids the most difficult problem is *when* to sample the fluid.

Liquids in General. If a liquid is pure, or at least homogeneous, it can be sampled with any appropriate volumetric device capable of opening and withdrawing a certain volume from the liquid to be analyzed. In the simplest case of a solution in the laboratory a volumetric pipet (Sec. 6-3) can be inserted into the solution and suction exerted on the mouth of the pipet to withdraw a specific volume of liquid. Naturally the pipet should be clean and dry so as not to contaminate the liquid. Liquids in commercial containers or tanks may be sampled by more elaborate devices such as a pipe inserted deep into the liquid.

If a liquid is not homogeneous or is a suspension, the sampling is more difficult. If the substance to be sampled can conveniently be shaken, then it may be that shaking is all that is necessary to obtain a homogeneous sample. For example, milk of magnesia may be sampled after thorough shaking to insure that withdrawing a pipetful of liquid will contain a representative amount of magnesium hydroxide. If the substance to be sampled cannot be mixed or shaken, then samples may have to be withdrawn from different levels of the liquid and a plot made of composition against depth.

Body Fluids. The sampling of blood and other body fluids must be done when the sample is representative of the state of the body to be tested. The ingestion of food, in particular, can disturb the levels of chemicals in the blood so that analysis is not meaningful.

It is interesting to consider the sampling of blood when testing for the body's response to glucose load. This is done in a *glucose tolerance test* in which the glucose level may be measured over a period as long as six hours. To establish a normal level of glucose, the blood is sampled before the patient has eaten any food in the morning. Then the patient drinks a flavored solution of glucose. Blood samples are withdrawn after the solution has been swallowed at exact time intervals, such as every hour for three hours or more. The glucose level at these times is plotted and compared to the normal glucose tolerance curve (the lower curve in Fig. 3-1).

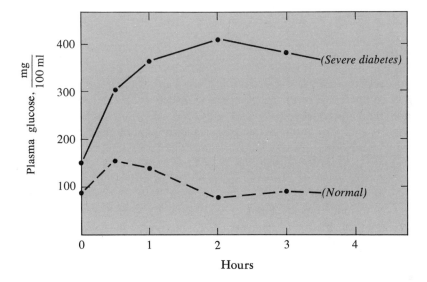

FIGURE 3-1. *Glucose tolerance curves for normal individual (lower curve) and diabetic individual (upper curve).*

For example, a rising trend in the glucose level after one hour instead of a gradual lowering and leveling off indicates that the body is not being stimulated to consume the glucose (upper curve, Fig. 3-1). Such a symptom is a possible indication of diabetes.

Sampling of Air and Other Gases

The sampling of gases is a specialized area so we will discuss just one aspect of gas sampling—the general problems of sampling polluted air in cities. This has been described in detail by Warner [1].

[1] P. O. Warner, *Analysis of Air Pollutants*, Wiley-Interscience, New York, 1976, pp. 196–249.

The general requirements of a sampling method for polluted air are:

1. Use of an accurate flow-measurement device to measure the total volume of air sampled.
2. Use of a sample collector to trap the desired pollutants. The collector is usually a filter and/or absorbing solution whose efficiency must be determined under actual operating conditions since air cannot usually be sampled with 100% efficiency.
3. Use of a pump that will insure a constant flow of air through the collector.

The problems of sampling polluted air are quite complex and only a few of them can be alluded to in this chapter. For example, there are the problems of how to sample, where to sample, how long to sample, how many sampling sites should be used in a large city, and what to sample.

Sampling of Gases. The problem of what to sample is of greatest interest. The most obvious constituents are the toxic gases which become a hazard when their concentrations reach certain levels. Some of these gases are carbon monoxide, sulfur dioxide, sulfur trioxide, nitrogen oxides (NO) such as NO and NO_2, ozone (O_3), and certain organic aldehydes.

The analysis for carbon monoxide is particularly important because it can affect driving efficiency on freeways; the government has set a level of 9 ppm as a maximum safe level. (Death occurs at 750 ppm when carbon monoxide prevents sufficient oxygen from reaching the brain via hemoglobin in the blood.) Carbon monoxide may be sampled by absorbing it in a solution of copper(I) chloride, with which it reacts to form a copper(I) complex ion. It may also be determined on a continuous basis by channeling it directly into an analyzer from the sample collector, omitting the absorption step.

The analysis for sulfur dioxide is even more important because of its toxic nature and because high-sulfur coal burning has for many years caused serious lung problems and even deaths. In the United States a monitoring system has been established for sulfur dioxide and other such gases in large cities, such as New York, Los Angeles, and Chicago, to sample, determine, and record the levels of such gases as a function of time. Keyed in with the analysis data are multistage alert plans that go into effect if the sulfur dioxide levels or other gas levels rise beyond safe limits.

Sampling of Particulates. In addition to gases, particulates (solids) suspended in the air are potentially dangerous and must also be sampled. Since there are many particles that are constantly settling because of gravity there must be some way of separating particulates that are truly suspended from those that are not.

The air sampler that is used to obtain a representative sample of truly suspended particulates is shown schematically in Figure 3-2. It is based on the vacuum cleaner principle; air is sucked under the eaves of the roof causing the air to change direction and exclude particles settling by gravity. Usually a standard 8 × 10-inch glass-fiber *filter mat* is employed to trap suspended particulate (aerosol).

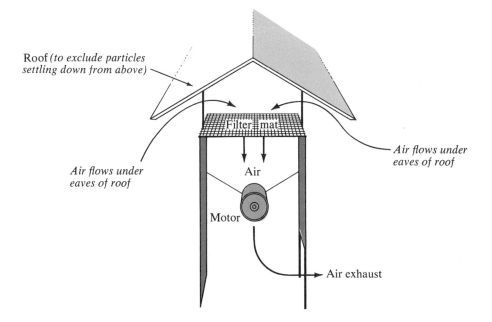

FIGURE 3-2. A schematic diagram of a high volume air sampler showing filtration of suspended particulate matter from polluted air.

A normal sampling period of 24 hours is used resulting in a typical particulate sample of about one-half gram [1]. Such particulates may consist of carbon soot, carcinogenic aromatic hydrocarbons, fly ash from power plant emission, iron oxide, silica (sand), and limestone from kilning.

3-2 PREPARATION OF THE SAMPLE FOR MEASUREMENT

Once a sample has been properly chosen, it must be prepared for analysis. Some of the steps that may be necessary are drying, measuring, dissolving, and separating interfering substances.

Drying Samples

Samples may contain water, and it must be decided whether it is necessary to dry the sample to obtain the proper analytical information. This decision will depend somewhat on the type of measurement to be made. In Section 1-1 four different areas were described: measurement of chemical and physical properties, qualitative analysis, quantitative analysis, and diagnostic analysis.

If a sample's physical or chemical properties are to be measured, it may have been purified and automatically dried during purification. Since most substances should be as pure as possible for this kind of measurement, removal of water is essential. Some samples may be solutions containing a constant amount of water and therefore no water will be removed.

If a qualitative analysis is to be performed, it may not be necessary to remove any water as long as it does not interfere with the qualitative tests to be done. In fact, the properties of hydrates (compounds containing stoichiometric amounts of water) and other such samples might be altered by removing water.

If a quantitative analysis is desired, then drying of most inorganic samples is preferred because the results can then be expressed as a percentage based on the dry sample rather than on a wet sample whose amount of water may vary. The drying of many organic and biological samples may be undesirable because of decomposition of the sample or the fact that the sample is largely water. Inorganic and organic samples that need drying are usually dried in an oven at 100° to 110°C for one to two hours. Some temperature-sensitive samples are dried in a desiccator containing a chemical desiccant to absorb water without heating. In either case the percentage of water loss may be measured by weighing the sample before and after drying and dividing the weight loss by the weight of the sample before drying.

If a diagnostic analysis is to be performed, no drying is usually done because the sample is a body fluid, and the objective is to determine the concentration of the constituent in a certain volume of the sample.

3-3 MEASURING THE AMOUNT OF SAMPLE TO BE USED

Weighing the Sample

Solids and some liquid samples are usually measured by weighing on the analytical balance (Ch. 4) to obtain a known amount for analysis. Most analysts weigh out three or four samples of approximately the same weight so that three or four results can be obtained to calculate an arithmetic mean (Sec. 2-2) which will then be reported. The results are usually calculated in terms of a weight percentage.

$$\frac{\text{Weight of constituent determined}}{\text{Weight of sample}} (100) = \text{Wt \% of constituent}$$

This is the type of calculation used in gravimetric analysis (Ch. 5) and in many volumetric analyses. Because the analytical balance is so accurate and precise, this is the most accurate way of measuring the amount of sample to be taken.

Measuring the Volume of a Sample

Many liquid samples and some gas samples are measured by obtaining a known volume of the sample in a pipet (Ch. 6) or a container whose volume is known.

Liquids and solutions containing a known weight of a sample can be measured by using a pipet to take several *aliquots* (exact volume) for analysis, so that three or four results can be obtained. The results are calculated in terms of concentration, or on a weight/volume basis.

$$\frac{\text{Weight of constituent determined}}{\text{Volume of sample}} (100) = \text{wt/vol \% of constituent}$$

If the weight is expressed in milligrams and the volume in liters, concentration in terms of ppm (Sec. 1-2) can be calculated by omitting the 100 multiplier. In general, measuring the volume of a sample is less accurate than weighing so that the results using this type of measurement are less accurate than when the sample is weighed.

Some gases are measured in terms of a flow rate and time of flow rather than by volume. This is usually adequate for measuring the total volume of polluted air sampled, for example.

3-4 DISSOLVING OR DILUTING SAMPLES

Dissolving Solid Samples

Since most analytical methods depend on measurements made on liquids or on solutions, a solid sample must be dissolved in an appropriate solvent. Solid samples are treated differently depending on whether they yield to dissolution by aqueous or nonaqueous type solvents.

Dissolution in Aqueous Type Solvents. Most inorganic compounds and some organic compounds can be dissolved in water or a mineral acid. If possible, water should be used as the solvent to avoid introducing anions of mineral acids that might cause interference. Many metals, oxides, carbonates, and even halides of silver(I), mercury(I), and lead(II) will not dissolve in water, however. Therefore it is necessary to employ an acid or mixture of acids to dissolve such compounds. Dilute nitric acid is a good general solvent for many metals and insoluble compounds because it forms soluble metal nitrates. It will not dissolve gold, platinum, antimony, certain sulfide compounds, and the halides of silver(I), mercury(I), and lead(II). Aqua regia, a mixture of nitric and hydrochloric acids, will dissolve gold, platinum, antimony, some sulfides, and the halides of lead(II) and mercury(I). Samples that fail to dissolve in the above acids can frequently be converted to a soluble form by high-temperature fusion using a flux of sodium carbonate, sodium peroxide (Na_2O_2), or a mixture thereof. The sample is ground together with the flux, and the mixture is heated to melting in a crucible to dissolve the sample. The melted material is allowed to cool and is then dissolved in water or an aqueous dilute acid to bring the sample into aqueous solution.

Dissolution in Nonaqueous Type Solvents. Most organic compounds and biological compounds can be dissolved in organic solvents. It is difficult to make too many generalities regarding the appropriate solvent for each class of compound. In general, however, like compounds dissolve like compounds. For example, ketones will dissolve in acetone, alcohols dissolve in ethyl alcohol, and so on. If a compound has a polar functional group such as the carbonyl group, it should dissolve in a hydrogen-bonding solvent such as methyl alcohol or ethyl alcohol.

3-5 SEPARATION OF INTERFERENCES

Once a sample has been dissolved in an appropriate solvent, the desired constituent may have to be separated from other constituents that may interfere in the measurement step. The nature of the interference will depend on which of the four general types of chemical measurements (Sec. 1-1) is to be made and on the specific method to be employed.

For example, suppose that a quantitative analysis is to be performed by precipitating the chloride ion from a solution. If silver(I) nitrate is to be added to precipitate the chloride as silver(I) chloride and if the bromide ion is also present, the bromide ion will interfere by also precipitating silver(I). In that case the bromide ion will have to be separated by some separation technique. Some interferences will not have to be separated if an adjustment in the composition of the solution is possible. For example, if ammonia is present when silver(I) ion is added to precipitate chloride ion, the acidity of the solution can be increased to convert the ammonia to the ammonium ion, thus preventing it from complexing the silver(I) ion as the diamminesilver(I) ion.

Various separation techniques can be employed: precipitation, electrodeposition, solvent extraction, chromatography, and so on. These techniques are introduced in Chapter 19 and elaborated on in Chapters 20 and 21. In discussing the various methods of quantitative and qualitative analysis in Parts I and II we are assuming that a separation is not necessary. In many cases, however, separations are absolutely essential for an accurate analysis.

QUESTIONS

(Answers to most even-numbered questions appear in Appendix 5.)

Sampling
1. What special problems are involved in samping
    a. Large nonuniform solid particles?
    b. Small solid particles?
    c. A nonhomogeneous liquid?
    d. Polluted air for particulate matter?
2. What differences might there be in the sampling difficulties with aspirin tablets as opposed to aspirin capsules?

3. What difference is there in the treatment of a sample to be used for qualitative analysis and the treatment of a sample to be used for the measurement of physical properties?
4. How can a liquid which is not homogeneous be sampled if
    a. It is contained in a one pint bottle?
    b. It is contained in a large tank?
5. Explain the two different sampling requirements involved in the glucose tolerance test. One requirement should cover the first measurement and the other requirement should cover all succeeding measurements.
6. What difficulties are involved in sampling city air for air pollution as opposed to sampling a laboratory for natural gas pollution?

Preparation of the Sample
7. Explain the different requirements for drying samples for qualitative analysis as opposed to drying samples for measurement of physical properties.
8. Why does drying a sample enable a more accurate determination of the weight per cent to be made?
9. Explain how measuring the flow rate of a gas and the time it flows can give an estimate of the volume of the gas taken for a sample.
10. Why is nitric acid a better solvent for lead and silver than hydrochloric acid is?

Challenging Question
11. Suggest a suitable solvent for each substance:
    a. Cholesterol, $C_{27}H_{45}OH$
    b. Glucose
    c. Glycine
    d. Vitamin A

4. Fundamentals of Weighing and Related Measurements

"You will also find beside it the revolver, string and weight *with which this vindictive woman attempted to disguise her own crime and to fasten a charge of murder upon an innocent victim,"* Holmes said.

ARTHUR CONAN DOYLE
The Problem of Thor Bridge

As we see from the above Sherlock Holmes quotation, weights had a certain "nonscientific" utility before they came to be used on the analytical balance. Nevertheless, the most important use of weights is in connection with accurate and precise measurements of sample weights on the analytical balance. In addition the analytical balance is also used for gravimetric analysis (Ch. 5), for density and specific gravity measurements, and so on.

From the 1800s until around the 1950s weighing was slow and tedious. Then the rapid-weighing single-pan balance appeared, making weighing the quickest part of most analyses. Weighing is so simple and so rapid that it is easy to take weight measurements for granted. Therefore before discussing the principles of weighing, the accuracy and precision of weighing will be discussed.

4-1 THE ACCURACY AND PRECISION OF WEIGHING

As stated in Chapter 2 the accuracy of a measurement such as a balance measurement can be expressed in terms of *error*. Since objects are generally measured only once, the precision of a balance measurement has to be expressed as the *uncertainty* of a single weighing. Since some objects are weighed directly using one weighing, and some objects are weighed by difference (two weighings), the accuracy and precision of these two methods will be discussed separately.

A Single Weighing

The absolute error of a single weighing using a standard analytical balance is of the order of ± 0.0001 g; the absolute uncertainty of a single weighing is also ± 0.0001 g. Both of these quantities will be smaller for semimicro and micro balances; for the latter, an absolute uncertainty of ± 0.001 to 0.003 mg is common.

The relative error and relative uncertainty of a single weighing both are found by dividing the corresponding absolute quantity by the weight; the quotient is then multiplied by 100 to express the quantity in parts per hundred (pph). For example, the relative uncertainty of weighing a 1.0000 g object is calculated as follows:

$$\text{relative uncertainty} = \frac{0.0001 \text{ g}}{1.0000 \text{ g}} \times 100 = 0.01 \text{ pph}$$

Since the desirable relative deviation or relative uncertainty in analytical work should be no more than 0.1 to 0.2 pph, the relative uncertainty of weighing 1.0000 g is well within the desirable precision.

You should be able to calculate the minimum size sample that can be weighed for analysis given the desirable precision in terms of relative uncertainty, or relative deviation. This can be done for a single weighing by rearranging the above equation to

$$\text{min wt, g} = \frac{\text{absolute uncertainty, g}}{\text{relative uncertainty, pph}} \times 100$$

For an absolute uncertainty of 0.0001 g and a relative uncertainty of 0.1 pph the minimum size sample is

$$\text{min wt, g} = \frac{0.0001 \text{ g}}{0.1 \text{ pph}} \times 100 = 0.1 \text{ g}$$

The implication of this calculation is that you should generally weigh samples of this size or larger for methods where a precision of 0.1 pph is desired. Remember that this applies only for a *single weighing*, not weighing by difference.

Weighing by Difference

When a sample is weighed by difference the weight of the container plus sample is found first. Then some of the sample is withdrawn from the container, and the container plus the remaining sample is then weighed again, and so on. In this manner only four weighings are necessary for three samples, rather than weighing each sample on a sheet of preweighed paper, or watch glass, and using six weighings for three samples.

When weighings are made in this manner the total *possible* absolute uncertainty for a given sample is 0.0002 g. This arises from the possibility that one weighing may be high by +0.0001 g and the other weighing may be low by −0.0001 g. To be sure, the total uncertainty may be less than this, but we must proceed on the assumption that any given weighing-by-difference operation could have the total possible uncertainty of 0.0002 g.

When a sample is weighed by difference, the *relative* uncertainty of that weighing should be double that of weighing it by a single weighing. The minimum size sample necessary to avoid exceeding a relative uncertainty of 0.1 pph should also be doubled. It is calculated using the above equation.

$$\min \text{wt, g} = \frac{0.0002 \text{ g}}{0.1 \text{ pph}} (100) = 0.2 \text{ g}$$

This increase in the minimum sample size is not usually a problem, considering the fact that weighing by difference reduces the number of weighings needed.

4-2 THE BALANCE AS A LEVER

Although most modern balances are now single-pan balances, the double-pan balance provides a better illustration of how the balance operates as a lever. In Figure 4-1 the pans and knife-edge of a double-pan balance are pictured. Since this

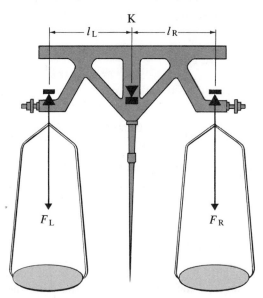

FIGURE 4-1. *The pans and knife-edge as part of the lever system of the analytical balance. The fulcrum is at K, the points of application of the forces F_L and F_R, and the hypothetical lever arms are l_L and l_R.*

is an unloaded balance, you should assume that the pointer (between the pans) is at a balanced position, or rest point, equidistant from both pans. This is the *original rest point*. The discussion that follows is concerned with forces exerted by putting weights on both pans of the analytical balance.

In principle the analytical balance is a lever of the first class; its fulcrum lies between the points of application of the forces. As shown in Figure 4-1 the fulcrum is the point at which the knife-edge, K, rests on the agate plate. The fulcrum lies halfway between the points of application of the forces, F_L and F_R, exerted by weights on the pans of the balance.

Suppose that the force F_L just balances the force F_R (this is true if the balance comes to rest at its original rest point). Then,

$$(F_L)(l_L) = (F_R)(l_R)$$

where l_L and l_R are the lengths of the arms of the lever from the knife-edges to the fulcrum.

Now, the forces F_L and F_R are proportional to the mass on each pan.

The origin of force is the attraction of the earth's gravity on an object of a given mass. Thus, mass is the quantity of matter in a body, and weight is the force exerted by gravity on that body. The relationship between force and mass is $F = (m)(g)$ where m is the mass and g is the acceleration due to gravity. Although the force of gravity varies somwehat throughout the world, it is constant for any given locality. Therefore the force or weight of an object is proportional to its mass.

As it is customary to speak of the mass of an object as its weight, the following equation holds true for the analytical balance:

$$(W_L)(l_L) = (W_R)(l_R)$$

Since the balance is so constructed that l_L and l_R are equal within an uncertainty of one part in 10^5, then $W_L = W_R$. The weight of the object on the left pan is then known directly from the sum of the weights on the right pan.

4-3 SENSITIVITY AND CAPACITY

Sensitivity

The sensitivity of a balance is defined as the amount of deflection of the beam produced by a given unit of weight. For the standard analytical balance the unit is the milligram (mg). For a micro analytical balance it may be a microgram (0.001 mg).

Sensitivity is affected most significantly by the mass of the beam, the pans, and the weights on the pan(s). A common relationship between sensitivity and the weight on the pan(s) is that as the weight increases, the sensitivity decreases. This reduces the response of the balance needed to reach a rest point for large weights. In effect, this also slightly increases the uncertainty of weighings when samples are weighed by difference on a double-pan balance.

With the modern single-pan, direct-reading balance, weighing operations are always conducted at a constant load (Sec. 4-4) so that the sensitivity of the balance is constant no matter what the weight.

Capacity

The capacity of many modern single-pan balances is of the order of 150 to 200 g. It is neither desirable nor necessary to weigh anything over 100 g on an analytical balance since the relative uncertainty of weighing 100 g on a trip balance accurate to ± 0.01 g is usually satisfactory. Objects weighing less than 10 mg should be weighed on a balance having a smaller capacity—a semimicro or micro type balance.

4-4 CONSTRUCTION OF SINGLE-PAN BALANCES

We will now discuss essentially the construction of the single-pan, direct-reading balance. However, there are certain features common to both the single-pan and double-pan balances that will be described first. Refer to Figure 4-1 when reading about these. The first feature is the *beam* which supports the pan(s) and weights. It must be rugged in construction, yet lightweight to allow for high sensitivity.

The *knife-edges* support the beam and the pan(s) and permit the balance to reach a rest point rapidly. These must be made of extremely hard material such as agate or sapphire. Although this insures that the knife-edges will remain sharp for a long time, it also means that the knife-edges can be easily damaged by the shock of releasing an unbalanced pan(s). Damage can be avoided by following the habit of never releasing the beam until the object and weights are almost equally balanced.

The *pan* (or pans) is made to hold an object or sample to be weighed. Pans are made of lightweight alloys which are resistant to air oxidation, but which can undergo chemical attack by reagents used in quantitative analysis. It is recommended that any reagent be weighed on a waxed paper or in a weighing bottle.

General Features of Single-Pan Balances

In general most single-pan balances have the same characteristic construction. As shown in Figure 4-2, such a balance is enclosed in a casing so that only the single pan is visible, the beam and weights completely enclosed. The weights are controlled by means of three or four knobs on the front of the balance case. Other knobs on the front or side of the case control the beam and zero adjustment.

SEC. 4-4   Construction of Single-Pan Balances   59

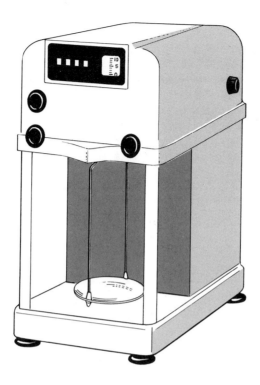

FIGURE 4-2. Front view of the Mettler H-15 balance. (Courtesy of the Mettler Instrument Corporation.)

Instead of weighing by *direct comparison* as on the double-pan balance, you find an object's weight by the *substitution method* on the single-pan balance. When the pan is empty the weight of the pan plus the weight of a set of weights suspended from the front of the beam is balanced by the weight of a large *counterweight* suspended from the rear of the beam. When an object is placed on the pan, the increase in weight is compensated for by *removing* an equal weight from the set of weights suspended from the front of the beam. Since no change occurs in the weight of the counterweight suspended from the rear of the beam, the weight of any object is found at a constant weight, or load.

It is important enough to reemphasize that the operation of the single-pan balance at constant load assures a constant sensitivity. A constant sensitivity insures a uniform precison for all weighings, particularly for weighing by difference (Sec. 4-3).

A side view of a typical single-pan balance is shown in Figure 4-3. The beam arrangement is not symmetrical as it is in the double-pan balance (Fig. 4-1). The removable weights are located *above* the pan, just behind the weight control knobs on the front of the balance. The counterweight is located at the back, much farther away from the central knife-edge than the pan. Since the weights and the

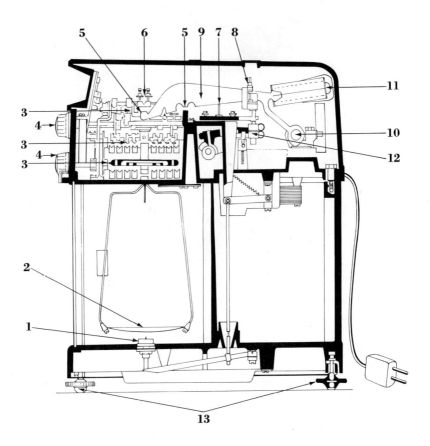

FIGURE 4-3. *Cutaway (side) view of the Mettler H-15 balance. The following parts are shown: (1) pan brake; (2) pan; (3) set of weights; (4) weight-control knob; (5) sapphire knife-edge; (6) stirrup for hangers and weights; (7) lifting device; (8) movable weights for adjustment of sensitivity; (9) beam; (10) engraved optical scale; (11) air damper; (12) counterweight; (13) foot screws. (Not shown are the arrest knob (upper left side) and the movable weight for zero adjustment (left side of beam).) (Courtesy of the Mettler Instrument Corporation.)*

knife-edges are completely enclosed, they are protected much better against atmospheric corrosion than they would be in an open double-pan balance.

Weight Readout

In many of the earlier single-pan balances, the weight readout was divided into three parts. The number of grams plus tenths of a gram were read from a digital scale. Hundredths and thousandths (mg) of a gram were read directly from a lighted optical scale. The fourth decimal (.0001 g) was estimated from a vernier on the optical scale.

The modern weight readout is a complete digital readout, as shown in Figure 4-4. All of the weights are read from a single scale, without need for a vernier

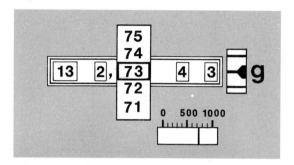

FIGURE 4-4. Optical scale with full range weight readout found on the newer model single-pan, direct-reading balances. The weight readout is 132.7343 g. Note that the first two decimals are displayed on the vertical scale and that the number (73) in the middle of the vertical scale gives the correct first two decimals. (Courtesy of the Mettler Instrument Corporation.)

to estimate the last decimal place. This prevents errors from incorrect vernier estimations as well as incorrect reading of hundredths and thousandths of a gram from an optical scale.

Special Features

Before 1975 balances were not protected from the mechanical shock caused by releasing the balance before the weights were adjusted to balance the weight of an object. In 1975 a number of balances, such as that shown in Figure 4-5, were introduced with a special "air-release" device. This is essentially a piston which compresses air when the balance is released so that complete release occurs gradually over a five-second period rather than instantly.

Other special features found on many balances include automatic preweighing, a scale-type weight-filling guide, weight locking, automatic weight printout, and automatic weight recording. The automatic preweighing features give an instantaneous indication of the approximate weight of the sample on the pan. The weight-filling guide shows the approximate sample weight between 0 and 100 g on a scale reading in 10 g units. The weight-locking feature prevents operation of the knobs controlling the heavier weights when the beam is completely released. Some balances also provide automatic printout of weights on paper tape or provide electronic signals which can be used to record weights.

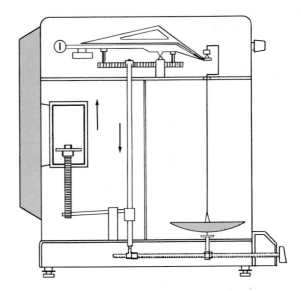

FIGURE 4-5. The "air-release" device for gentle release of the beam on balances such as the Mettler H78AR. As the beam moves down (right arrow), the piston moves up (left arrow), compressing air and cushioning the release.

4-5 USE OF THE SINGLE-PAN BALANCE

After you finish this section, you will have read all the information needed to operate a single-pan balance. You should then take the self-test at the end of this section to review this information.

Handling of Objects

It is important to remember that your fingertips are coated with perspiration and perhaps some dirt; handling objects with your fingers will always increase the weight of an object to some degree because of this. When handling heavy objects to be weighed you should use a pair of tongs or finger gloves. When handling light objects, such as a weighing bottle, you may use finger gloves or a paper loop (see Sec. 4-6).

Chemicals can corrode and change the weight of the balance pan. You may weigh nonhygroscopic chemicals by difference on a waxed paper or in a weighing bottle. Always weigh hygroscopic chemicals by difference in a weighing bottle which should be kept covered as much as possible.

A more detailed discussion of the weighing operations needed for various types of samples is given in the next section.

Rules for Operation of a Single-Pan Balance

Read the rules below carefully before using a single-pan balance for the first time. You should take the self-test at the end of this section as preparation also.

1. Treat the balance as carefully as you would any other piece of expensive equipment. It's better to appear ignorant and ask the instructor about something than to take the chance of damaging the balance.
2. Before you operate the balance or adjust its zero point, you should *always* check the following:
   a. The balance pan should be empty and free of dust and chemicals.
   b. All the weight knobs should be adjusted to the zero setting.
   c. The glass doors of the balance should be closed, thus preventing air currents from disturbing the equilibrium position of the pan when the balance is in use.
3. Check the zero point of the balance before you perform any weighings. This needs to be done only once for a series of weighings, but if someone uses the balance after you have set the zero point, you should clean the pan if necessary and check to see that the setting is still at zero. Adjust the zero point with the balance beam in the full release position, not in the semirelease position.
4. To weigh an object, follow the steps below:
   a. Put the object on the pan only when the balance beam is in the *arrest*, or *secured*, position. If this is not done, the shock of the sudden change in weight may cause misalignment of or damage to the balance.
   b. Close the doors and adjust the beam to the semirelease position.
   c. To locate the weight of an object quickly, develop a systematic approach of varying the weights. First, try a small weight, such as 1 to 5 g, and see if the beam moves. If not, try a larger weight, such as 90 g, and see if the beam moves. Repeat, increasing the smaller weights and decreasing the larger weights.
   d. When the weights have been adjusted to the nearest weight that is just less than the object's weight, adjust the beam to the *full-release* position and record the weight from the lighted scale of the balance.
5. Always record the weights of the object to *four* decimal places (0.0001 g) even if the last digit is zero. You should record the weight in a notebook, not on a piece of loose paper which may be easily lost.
6. Immediately adjust the beam to the arrest, or secure, position after finishing a weighing. Then remove the object from the pan and clean the pan if necessary. Return all weight knobs to the zero setting and close the glass doors.

Self-Test 4-1. Preparation for Using the Single-Pan Balance

Directions: After reading over the material on using the direct-reading single-pan balance, answer each of the questions below. Check each answer below before going on to the next question.

A. Before adjusting the zero point of the balance, you should automatically check at least three things:
   a. The weights on the optical scale should all read _____.
   b. Both glass doors should be _____.
   c. The pan of the balance should be _____ and _____.
B. Before any weighings are performed, the zero point of the balance should be checked.
   a. To avoid air currents that can cause a mistake in the zero point the check should be performed only after the glass side doors are _____.
   b. When the zero point is checked the beam of the balance should be in the _____ position.
   c. If someone else uses the balance after you have checked the zero point, you should _____ the pan and _____ the zero point.
C. To adjust or change weights while trying to find the weight of an object, you should follow the steps below:
   a. Before the object is put on the pan, the beam of the balance should be in the _____ position.
   b. Weights may be adjusted after the beam of the balance has been turned to the _____ position.
   c. To locate the weight of an object quickly, first try a small weight such as __.0 g, and then a heavy weight such as __.0 g. Repeat until the weight is between two consecutive readings such as 11 and 12 g.
D. To obtain the exact weight of the balance,
   a. Adjust the beam to the _____ position and read the optical scale.
   b. Since the balance is accurate to four decimal places, a weight such as 12.1100 g should be recorded as __ g.
E. After the weighing is finished and the weight has been read,
   a. The object cannot be removed from the pan until the beam _____.
   b. All weights are then returned to the _____ positions.
   c. The pan is _____ if necessary and the glass side doors are _____.

Answers to Self-Test 4-1

A. a. zero   b. closed completely   c. empty and clean
B. a. closed   b. full-release   c. clean; recheck
C. a. secured   b. semirelease   c. 1; 90
D. a. full-release   b. 12.1100 (both zeroes are significant)
E. a. is secured   b. zero   c. cleaned; closed

4-6 WEIGHING OPERATIONS

Drying the Sample

The time and temperature for drying depends on the nature of the sample. Samples such as organic and biological materials usually are not dried because drying would

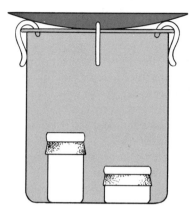

FIGURE 4-6. *Beaker covered with a watch glass and weighing hooks.*

partially decompose them. Inorganic samples such as sodium carbonate should properly be heated at high temperatures (300°C) to change the sodium bicarbonate impurity back to sodium carbonate. Still other samples, such as primary-standard arsenic(III) oxide which has not been appreciably exposed to air, are not hydroscopic and need not be dried.

If drying is needed, place the samples in a drying oven for one to two hours at 110°C. Place the sample in an open weighing bottle in a beaker. Cover the beaker with a watch glass on glass hooks, as shown in Figure 4-6. If the sample decomposes, oxidizes, or sublimes at 110°C, dry it by placing it in a desiccator containing a drying agent efficient enough to absorb water vaporized from the sample.

Desiccators. After the sample has been oven-dried, allow it to cool somewhat and then place it in a desiccator (Fig. 4-7). Leave the lid ajar for about 10 minutes

FIGURE 4-7. *A desiccator.*

so that a partial vacuum will not form in the desiccator after the air inside has cooled. Allow the sample to cool in the desiccator for at least 30 minutes before weighing.

To insure a tight seal and efficient drying, grease the surfaces of the desiccator and the lid lightly with petroleum jelly. To open or close the desiccator slide the lid sideways with a steady pressure. Never jerk or lift the lid directly upwards. Keep one hand on the lid when carrying the desiccator to prevent a loose lid from sliding off.

Before using the desiccator charge it with fresh desiccant. Although calcium chloride is often used, Drierite ($CaSO_4 \cdot \frac{1}{2}H_2O$) or anhydrous magnesium perchlorate absorbs water more efficiently and either is recommended. A colored indicating form of Drierite is also available.

A vacuum desiccator is a special type of desiccator that may be used for samples that cannot be dried by heating. Simply place the sample in the vacuum desiccator and connect a vacuum pump. Evacuate to the desired vacuum and close the port of the desiccator to retain the vacuum.

The freeze-drying technique may be used to dry samples that decompose even at room temperature. The sample is cooled to or below the freezing point, and water is removed by continuously evacuating with a vacuum pump.

Weighing the Sample

Three methods are given below for weighing solid samples and one method for weighing liquids.

Method 1. Solid samples in weighing bottles. After the sample is cooled, weigh it by difference from the weighing bottle. This is especially recommended for deliquescent samples. Throughout the weighing process handle the weighing bottle only with a paper loop, as shown in Figure 4-8; moisture from the fingers

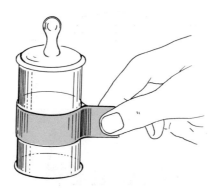

FIGURE 4-8. *Handling a weighing bottle with a paper loop.*

will change the weight of the bottle and cause errors. Another technique in handling glassware is to use finger gloves on the thumb and first two fingers.

To weigh by difference first weigh the bottle plus sample and record the weight in your notebook. Then carefully remove the estimated amount of sample from the bottle with a clean spatula and place in a marked beaker or flask. Weigh the bottle plus sample to the nearest 50 mg to see whether the amount of sample withdrawn is within the desired weight range. If necessary, withdraw more sample and again weigh the bottle approximately. When enough sample has been transferred, weigh the bottle plus remaining sample to the nearest 0.0001 g and record the weight. Subtract the final weight from the initial weight to obtain the exact weight of the sample.

As an alternative to using a spatula, insert a small aluminum scoop into the weighing bottle and weigh the scoop along with the bottle and sample for both the initial and final weighings. Handle the scoop with paper or finger gloves and use the scoop to transfer the sample.

Method 2. Solid sample in a scoop. Weigh accurately a larger, lightweight scoop, such as shown in Figure 4-9. With a clean spatula add sample to the scoop until

FIGURE 4-9. Lightweight scoop for handling solid samples.

a weighing to the nearest 50 mg indicates that the sample weight is within the desired range. Then weigh the scoop plus sample to the nearest 0.0001 g and record the weight. Carefully transfer the sample from the scoop to a marked beaker, handling the scoop with a piece of paper. Brush the last traces of sample into the beaker with a small camel hair brush.

Method 3. Waxed paper or polyethylene coated weighing paper. Weigh the weighing paper alone on the balance. Then add the sample to the weighing paper using a clean sptaula. Weigh the sample plus weighing paper. Subtract the weights to obtain the weight of the sample. This method is only valid for samples which do not absorb water from the air on standing.

Method 4. Liquid samples. Most liquid samples can be weighed out with a small bottle fitted with a medicine dropper and a ground-glass stopper. Weigh the bottle, sample, and medicine dropper accurately and record the weight. Use the dropper to transfer part of the liquid from the bottle to a marked flask or beaker. Replace the stopper, being careful not to get any liquid on the ground-glass surface of the bottle. When the proper amount of sample has been removed, as indicated by a weighing to the nearest 50 mg, weigh the bottle, dropper, and remaining sample to the nearest 0.0001 g and record the weight. The difference is the weight of the liquid sample.

Experiment A. Treatment of Data from Weighing

This experiment illustrates the statistical treatment of data from the weighing of ten nearly identical objects, such as ten coins, ten pencils. You will assume that all ten objects have been made by the same machine at the same time and that they constitute a representative sample. Review Chapter 2 before making the calculations.

PROCEDURE

1. Obtain ten identical objects to weigh from your instructor and record the sample number.
2. Check and, if necessary, adjust the zero point of the balance before making any weighings. This is necessary because the weighings are absolute.
3. Weigh each object to the nearest 0.1 mg and record each weight in your notebook.
4. Calculate the range of your weighings and use the Q test (Ch. 2) to see whether any result should be discarded because of a gross error.
5. Calculate the mean, the absolute and relative values of the average deviation, and the absolute and relative values of the standard deviation.
6. Report all data in tabular form and indicate the result of the application of the Q test to the most suspicious weighing of the ten weighings.

4-7 MEASUREMENT OF DENSITY AND SPECIFIC GRAVITY

Definitions

The density, or specific gravity, of a liquid is accurately and easily measured by weighing a known volume on the analytical balance. The density of an object is defined as its weight per unit volume at a specified temperature. In scientific work density is expressed in units of g/ml or g/cm^3, but in other areas it may be expressed in units such as lb/gal. Thus the density of water at 20°C is 0.99823 g/ml and also 8.330 lb/gal. (The density of water is at its maximum of 1.0000 g/ml at 4°C.)

The specific gravity of a liquid is the ratio of its weight to the weight of an equal volume of water. Specific gravity thus has no units, but it may be expressed as the density at 20°C relative to the density of water at 4°C $\left(\frac{20°}{4°}\right)$ or as the density at 20°C relative to the density of water at 20°C. Thus carbon tetrachloride has a specific gravity of 1.594 $\left(\frac{20°}{4°}\right)$ as well as a specific gravity of 1.606 $\left(\frac{20°}{20°}\right)$.

The density of a solvent is increased by dissolving a solute in it; that is, the solution resulting from adding a solute to a solvent has a greater density than the pure solvent. For example, the density of sea water is 1.025 g/ml at 20°C; the dissolved sodium chloride increases the density above that of pure water. As the concentration of the salt increases, the density also increases. The measurement of the density of a solution will therefore give an estimate of the concentration of the salt in the solution.

Measurement of Density

The density of a liquid is readily measured by carefully weighing a known volume of the liquid at a constant temperature. Depending on the accuracy required for measuring the volume, the volumetric apparatus used may be either a volumetric flask or a pycnometer (Fig. 4-10).

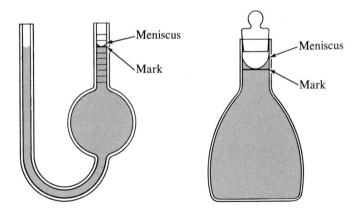

FIGURE 4-10. Two volumetric containers for the determination of density. Left: the pycnometer with a narrow capillary for a highly accurate volume measurement. Right: a volumetric flask for moderately accurate volume measurement. Note that a small discrepancy between the meniscus and the mark will cause a larger error in the volumetric flask because of its larger diameter.

Volumetric Flask Method. The measurement of density with a volumetric flask is the less accurate method of the two, but it is more convenient and slightly faster. There are two factors that limit the accuracy of the method. One is how closely the meniscus is adjusted to correspond to the mark on the neck of the flask (Fig. 4-9). This cannot be evaluated because it will vary according to who is making the measurement. Because the neck of the flask is relatively wide even a slight error may limit the accuracy so that it is only moderately good. The other factor that limits the accuracy of this method is the tolerance, or maximum allowable error of the volume of the volumetric flask. This can be evaluated. For example, a 25 ml volumetric flask has tolerance of ± 0.030 ml. This is the absolute value; the relative maximum allowable error is

$$\frac{0.030 \text{ ml}}{25.000 \text{ ml}} (100) = 0.12 \text{ pph}$$

A relative error of the order of 0.1 pph is typical for most volumetric flasks. This implies that the density will also have a relative error of ± 0.1 pph. For a liquid like water with a density near 1 g/ml the density may only be reported with an absolute uncertainty of ± 0.001 g/ml. This means that water may be reported as 1.000 g/ml or 0.999 g/ml, but not 1.0000 g/ml. The density of liquids in the range of 1.5 to 2.0 g/ml must be reported with an uncertainty of ± 0.002 g/ml. This means that they may be reported as 1.500 ± 0.002 g/ml or 1.50 g/ml, but not as 1.500 g/ml. To report the latter would imply that the last significant figure is only uncertain by ± 0.001 g/ml, which is not true.

Pycnometer Method. The measurement of density with a pycnometer is more accurate than using a volumetric flask, but it is slightly slower because the adjustment of the volume is more time-consuming. The pycnometer has a capillary device (Fig. 4-10) for accurate measurement of the volume; it requires careful adjustment to match the meniscus to the mark. The narrow diameter of the capillary compared to the wide diameter of the volumetric flask reduces the error arising from an inexact matching of the meniscus and the mark.

Because the pycnometer is more accurate, densities measured using it can be reported to four (or five) decimal places; for example, water can be reported to 1.0000 g/ml or 0.99823 g/ml, instead of 1.000 g/ml or 0.998 g/ml. The same is true for specific gravities. Most data reported in handbooks contain four significant figures to the right of the decimal point; the data for water are expressed to five significant figures to the right of the decimal point. Apparently a more accurate type of pycnometer was used for water.

Estimation of Concentration

Measurement of the density of a solution can be used to estimate the concentration of a solute. For example, consider the data in Table 4-1. As the percentage of

TABLE 4-1. The Density of Aqueous Sodium Chloride Solutions

Per Cent NaCl	Density, g/ml at 20°C
1.0%	1.0053
2.0%	1.0125
4.0%	1.0268
6.0%	1.0413
10.0%	1.0707
12.0%	1.0857
14.0%	1.1009
16.0%	1.1162
18.0%	1.1319
20.0%	1.1478

sodium chloride increases the density of the sodium chloride solution increases from 1.0053 to 1.1478 g/ml. Thus a sample may be analyzed for its percentage of sodium chloride by measuring its density (Exp. 1A). One approach would be to estimate the density to the nearest 0.1% by interpolation using the data in Table 4-1. Since smaller differences cannot be estimated so easily a better approach would be to plot the density of the sodium chloride solutions against the concentration, as shown in Experiment 1A. The concentration of an unknown solution can then be easily read by entering its density on the plot.

QUESTIONS AND PROBLEMS

(Answers to most even-numbered problems appear in Appendix 5.)

Definitions and Concepts

1. What is the absolute uncertainty on the usual analytical balance for
   a. A single weighing?
   b. Weighing by difference?
2. Explain weighing by difference.
3. Explain the difference between the operation of a double-pan balance and a single-pan balance.
4. Explain the advantage of the substitution method of weighing used on single-pan balances.
5. What is the purpose of the air-release device on single-pan balances?
6. What are some ways to handle samples without touching them during the weighing process?
7. Define:
   a. Density
   b. Specific gravity
   c. Pycnometer
8. Summarize the important rules for operation of a single-pan balance.

Balance Problems

9. Calculate the relative uncertainty of weighing a 0.150 g object on the single-pan balance using
   a. A single weighing.
   b. Weighing by difference.
10. Suppose that a semimicro balance has an absolute uncertainty of 0.01 mg. Calculate the minimum weight for a sample using a relative uncertainty of 0.1 pph for
    a. A single weighing.
    b. Weighing by difference.
11. Suppose that the absolute uncertainty of a single weighing on a balance were ± 0.0002 g. What would be the minimum weight required for a sample to be weighed by difference if the relative uncertainty could be no larger than 0.1 pph?
12. During World War I a clerk in a French intelligence bureau became curious about a suspected German agent who carried many pencils. He suspected that something was being carried inside the pencils. How could he have checked each pencil without breaking it open? (See A. A. Hoehling, *Women Who Spied*, Dodd, Mead & Co., New York, 1967, p. 58.)
13. Discover what the maximum weight is that can be weighed on the analytical balance in your laboratory and calculate the relative uncertainty (pph) of weighing that weight once.

Density-Specific Gravity Problems

14. A 5.000 ml volume of carbon tetrachloride weighs 7.970 g at 20°C. Calculate its density in g/ml.
15. Calculate the specific gravity $\left(\dfrac{20°}{4°}\right)$ of the carbon tetrachloride in the previous problem. Also calculate the specific gravity $\left(\dfrac{20°}{20°}\right)$.
16. A solution of unknown percentage sodium chloride has a density of 1.056 g/ml at 20°C. Determine its percentage sodium chloride to the nearest 0.5% by plotting the data in Table 4-1.

Challenging Problems

17. One of a set of five 1-g weights is either too heavy or too light. Devise a method that will identify that weight using a sixth correct weight and only three weighings on a double-pan analytical balance.
18. A combination of seven weights excluding the 50-g weight in a set of weights will give any weight up to 60 g. Devise a scheme to give any weight up to 63 g with the use of only six weights of any denomination and a double-pan balance.

5. Gravimetric Analysis

> "... his (Holmes') face showed rather the quiet and interested composure of the chemist who sees the crystals falling into position from his oversaturated solution."
>
> ARTHUR CONAN DOYLE
> The Valley of Fear, Chapter 2

In this chapter we encounter our first quantitative method of analysis—gravimetric analysis. It is such an old method that it could have been used by Sherlock Holmes after all of his crystals had precipitated from "his oversaturated solution!" Certainly, there is a close relationship between the precipitation of crystals and gravimetric analysis, which we will describe below.

5-1 INTRODUCTION

Gravimetric analysis is a quantitative analysis method in which the amount of the desired constituent is determined by isolating it in a known pure form and then weighing it. A simple example is the precipitation of cholesterol, $C_{27}H_{45}OH$, by digitonin, $C_{55}H_{90}O_{29}$, as an insoluble complex:

$$C_{27}H_{45}OH + C_{55}H_{90}O_{29} + H_2O \rightarrow C_{27}H_{45}OH(C_{55}H_{90}O_{29})H_2O(s)$$

In the above reaction the cholesterol is precipitated *unchanged*.
  Most gravimetric analysis methods involve a change in that an *ion* is precipitated as a *compound*:

$$Ag^+ + Cl^- \rightarrow AgCl(s)$$

In the above reaction either the silver(I) ion or the chloride ion may be determined by adding an excess of the other ion to precipitate silver(I) chloride. The precipitate is then filtered, dried, weighed, and the amount of the silver(I) ion (or the amount of chloride) calculated from the weight of the silver(I) chloride.

Gravimetric analysis is one of the oldest yet most accurate methods of analysis. It has been and still is used widely by the analytical chemist. It has certain advantages and disadvantages when compared with other methods of analysis, and these will be discussed in the following paragraphs.

Major Advantages

The major advantages of gravimetric analysis lie in its simplicity and its accuracy. Because it is simple and straightforward the fundamental chemistry of gravimetric analysis is easily understood. For example, the sulfur content of an organic compound can be determined gravimetrically by oxidizing the sulfur to the sulfate anion and then precipitating and weighing the sulfate as barium sulfate. In carrying out this determination no accurately known standard solution (titrant) needs to be prepared as in volumetric analysis, nor is there any need to prepare a *series* of known standards as in instrumental analysis.

In terms of accuracy a substance can be weighed on an ordinary analytical balance with an uncertainty of only ± 0.0001 g (Ch. 4), or for a 1 g sample an uncertainty of *one part in 10,000*. Comparing this with volumetric analysis we find that an ordinary buret or pipet can be read with an uncertainty of ± 0.01–0.02 ml, or one to two parts in a hundred for a 1 ml sample. On that basis a gravimetric method is one hundred times more accurate than a volumetric method.

Major Disadvantages

Gravimetric methods do suffer from certain disadvantages; because of them most chemists prefer to use other analytical methods. There are several disadvantages worth discussing.

Although simple in theory, a gravimetric analysis becomes very exacting and lengthy when carried out in the laboratory. The desired constituent is usually isolated by precipitating it from a solution. The precipitation process suffers from several inherent errors. Some of the desired constituent may be lost owing to the solubility of the precipitate, thus giving low results. On the other hand, if any impurities are present during the precipitation process, they may be carried down along with the desired constituent, thus giving high results. This phenomenon is known as *coprecipitation*.

Even when the precipitate is pure and no solubility losses occur, it must be quantitatively transferred and filtered to separate it from the solution associated with it, and then it must be washed to free it from impurities dissolved in the solution. After this it must be carefully dried and finally weighed. All of these manipulations are tedious and time-consuming, and the loss of any of the precipitate at any point will result in errors.

Applications

Gravimetric methods are seldom used in the analysis of biological substances since most of these compounds cannot be dried for weighing without decomposition. However, gravimetric methods can be used for many pharmaceuticals and many inorganic substances in biological samples.

Many pharmaceutical preparations can be analyzed by separating, purifying, and weighing an organic medicinal compound without using a chemical reaction at all. Preparations containing the sodium salt of a weak acid may be isolated by adding strong acid to the salt. For example, sodium phenobarbital, $C_{12}H_{11}N_2O_3{}^-Na^+$, is converted to phenobarbital, $C_{12}H_{11}N_2O_3H$, by the addition of acid.

$$C_{12}H_{11}N_2O_3{}^-Na^+ + H^+ \rightarrow C_{12}H_{11}N_2O_3H(s) + Na^+ \tag{5-1}$$

This will be discussed in detail in Section 5-12.

A good example of an inorganic substance determined in organic and biological samples is the chloride ion. Water is also frequently determined in inorganic as well as in organic samples. The sample is simply weighed, placed in a drying oven at 110°C for a few hours, and then removed and weighed. The loss in weight represents the amount of water.

The mineral content of biological samples is determined by the so-called *ashing procedure*. The weighed sample is heated to red heat in a crucible using a burner. The organic matter burns away and forms volatile gases. The residue of nonvolatile inorganic matter is then weighed to determine the "mineral" content. This is called a nonspecific analysis since the residue is a complex mixture of sodium, potassium, calcium, and phosphorus compounds with traces of iron, magnesium, and so on. A detailed analysis can then be made by dissolving the residue in dilute acid and analyzing the solution for the above constituents.

5-2 THE GRAVIMETRIC ANALYSIS METHOD

A typical gravimetric analysis method involves a number of steps which may be summarized as follows:

1. A representative sample must be obtained.
2. The sample must be weighed accurately and then dissolved; if it is a liquid, a known volume must be measured accurately using a pipet.
3. The conditions for quantitative and accurate precipitation must be achieved before precipitation is carried out. This may include some adjustment of the pH of the solution, heating it, and possibly removing some substance which may interfere with the precipitation by coprecipitation with the constituent to be determined.
4. The appropriate precipitating reagent must be added to achieve quantitative precipitation. This may involve addition of an excess of the precipitating reagent, settling of the precipitate, and testing for complete precipitation by dropwise addition of more reagent.

5. The solution may be heated (digestion) to achieve filterable crystals and to cause impurities to dissolve.
6. The precipitate must be filtered, washed, and dried in the oven to achieve a precipitate of "constant weight."
7. The weight of the pure precipitate must be converted to the weight of the substance to be determined by multiplying it by the *gravimetric factor*. The per cent of the substance to be determined in the impure sample is then calculated.

These steps will be discussed in detail in the following paragraphs.

5-3 OBTAINING A REPRESENTATIVE SAMPLE

In order to achieve a meaningful analysis you must choose a sample which represents the composition of the material which is being analyzed. If you do not, the results may not give any idea at all of the true composition and will thus mislead you. The problems of obtaining a truly representative sample are discussed in Chapter 3 and will not be described any further here.

5-4 WEIGHING AND DISSOLUTION OF THE SAMPLE

Unless the exact weight (or volume) of the original sample is accurately known, the results of the analysis are meaningless. The original weight (or volume) is always used in calculating the final result. If there is any doubt at all of this measurement, the sample should be discarded and a new one taken.

Most gravimetric procedures involve the formation of a precipitate from a solution; hence if the sample to be analyzed is a solid, it first must be dissolved in an appropriate solvent. In most instances the solvent is water or a dilute acid solution, but other solvents such as alcohol also may be used. The procedure itself usually gives exact instructions as to which solvent to use and how to dissolve the sample. If the sample is a liquid instead of a solid, it may be diluted with water or some other solvent before analysis.

5-5 PREPARATION OF THE SOLUTION FOR PRECIPITATION

After dissolving the sample, certain adjustments must be made before or during addition of the precipitating reagent. In most cases the pH of the solution is important. For example, in the precipitation of phenobarbital (**5-1**) by the addition of acid to sodium phenobarbital sufficient acid must be added to convert all of the sodium salt to phenobarbital. Thus the pH *during* the precipitation must be slightly on the acid side. In most cases the pH must be adjusted *before* the precipitating reagent is added. For example, in the precipitation of chloride ion by the addition of silver(I) nitrate the solution must first be made slightly acid by the addition of nitric acid. If the solution is not acidic, some silver oxide may form and coprecipitate with the silver chloride, causing incorrect results.

In certain gravimetric methods, the pH must be controlled *very exactly*. For example, the precipitation of barium chromate is quantitative only at pH 5.7. At lower pH values losses occur from increased solubility of the precipitate; at higher pHs strontium chromate is coprecipitated with barium chromate.

In the precipitation of pharmaceuticals a preliminary separation of binders and/or lubricants is often necessary. For example, in the analysis of amobarbital tablets or similar substances (Sec. 5-12) the sample of the tablets is frequently washed with petroleum ether to remove any binder or lubricant in the tablets. The amobarbital or other medicinal agent is then isolated by extraction into chloroform or ether and ultimately weighed in its pure form.

Frequently, inorganic substances require a preliminary oxidation or reduction. In the determination of iron by precipitation of iron(III) hydroxide the solution is treated with bromine water or hydrogen peroxide to insure that all the iron is oxidized to the $+3$ state.

In general laboratory procedures give detailed instructions for preparing the solution for precipitation. Special samples may require special treatment not given in the procedure, however.

5-6 PRECIPITATION OF THE DESIRED CONSTITUENT

Precipitation of the desired constituent is accomplished by adding a dilute solution of a reagent which reacts with the desired constituent to form an insoluble precipitate. A slight excess of this reagent is always added. This insures that the desired ion is completely precipitated and the solubility of the precipitate is decreased by the common ion effect. In the determination of the chloride ion, for example, silver nitrate solution is added to the unknown solution, which reacts to form a precipitate of insoluble silver chloride. In order to understand the conditions necessary for a successful gravimetric determination let us consider the precipitation process in some detail. This was first explained successfully in 1925 by a German chemist, P. P. von Weimarn. According to von Weimarn the following steps are involved in the production of a precipitate:

1. Formation of a supersaturated solution.
2. Combination of ions to form minute, invisible particles to serve as nuclei for the insoluble precipitate.
3. Spontaneous growth of these nuclei progressing through a colloidal state and finally reaching the final stage of large, visible, insoluble particles.

These steps take place very rapidly, usually within a fraction of a second, and hence are not noticeable to an observer. When you add silver nitrate solution to a soluble chloride, the precipitate seems to appear the instant that the two solutions are mixed. Studies made with instruments in which observations can be made in time intervals of 10^{-9} second (a nanosecond), however, confirm the postulates made by von Weimarn.

Let us now consider some practical implications derived from von Weimarn's theory. In carrying out a gravimetric determination it is important to have a precipitate which can be readily separated from a solution by filtration and which can easily be washed free of impurities. The particles must be large enough not to pass through the pores of the filter. Precipitates may be categorized into the following different types:

1. **Crystalline precipitates.** These are recognized by the presence of many regularly shaped, discrete particles having smooth, shiny surfaces. A crystalline precipitate looks like dry sugar or salt. It is the most desirable of all precipitates since it settles rapidly and is easy to filter and wash. The individual particles are large and compact. Errors caused by *occlusion* (entrapment within the crystal) and *adsorption* of lattice ions on the surface of the crystal can be minimized in most cases.
2. **Granular precipitates.** These look like coffee grounds. They consist of small, irregularly shaped, individual particles but the smooth surfaces and regular shape of a crystalline precipitate are absent. A granular precipitate, like a crystalline precipitate, is easily filtered and washed, but the particles tend to be porous which increases errors caused by adsorption and occlusion.
3. **Finely divided precipitates.** As the name implies these precipitates are composed of very small particles. The individual particles are invisible to the naked eye, but the microscope shows these particles to be extremely small, individual crystals. A finely divided precipitate has a flourlike appearance. This type of precipitate is not desirable since the small particles tend to pass partially through the pores of the filter. Because of the large surface area of the precipitate errors due to adsorption are increased.
4. **Gelatinous precipitates.** These form a sticky, jellylike mass that looks like jam or jelly, having no discrete particles and forming amorphous clumps. A gelatinous precipitate is not desirable since it is difficult to filter and it entraps impurities which are impossible to wash out. Gelatinous precipitates are formed, for example, when the precipitate is highly insoluble or when concentrated solutions are mixed together.

In carrying out a gravimetric determination the chemist tries to arrange conditions so that a crystalline or granular precipitate is obtained rather than a finely divided or gelatinous one. To achieve this let us first consider what occurs at the molecular or ionic level using the precipitation of silver chloride as an example. In an *unsaturated* solution chloride ions and silver ions may unite momentarily to form silver chloride molecules, but these soon redissolve to form silver and chloride ions again.

In a *saturated* solution the silver chloride molecules grow more rapidly than they dissolve and quickly grow large enough to form stable nuclei. These nuclei continue to grow in size, forming colloidal particles and, finally, large, visible particles in the form of a precipitate. If the solution is only slightly supersaturated, only a few nuclei will form, and these will then proceed to grow into fairly large particles. If the solution is highly supersaturated in the beginning many nuclei will form which then grow into a multitude of finely divided particles as a precipitate. In extreme conditions of supersaturation one nucleus forms on top of the other resulting in long molecular strands which form a gelatinous precipitate. In a crystalline precipitate the molecules are arranged orderly, but in a gelatinous precipitate they are arranged in a helter-skelter fashion. One might use the following analogy:

Suppose several truckloads of bricks are being delivered to a construction site (the bricks representing the molecules within a precipitate). If the trucks quickly dump their loads into one large heap, we have the equivalent of a gelatinous or amorphous precipitate. The molecules have no definite arrangement and there are large voids between the molecules where impurities are entrapped.

Now suppose that the same load of bricks is slowly unloaded by hand into neat piles. Here we have the equivalent of a crystalline precipitate where the molecules are stacked in a compact, orderly fashion and the spaces between the molecules are very small thus resulting in a pure, crystalline precipitate.

In light of these considerations we now can make some recommendations for the most favorable conditions for a quantitative precipitation.

1. Dilute solutions should be used and the reagent should be added slowly with constant stirring. Dilute solutions are used because they become only slightly supersaturated and relatively few nuclei form. As more reagent is slowly added these nuclei grow larger and the molecules arrange themselves in an orderly fashion resulting in large particles of a crystalline precipitate. The solution should be constantly stirred to avoid local regions of high supersaturation. In dilute solution the impurities are also diluted, thus causing less contamination.
2. Precipitation should be made from hot solutions, if possible. Keeping the solution hot increases the solubility and lowers the degree of supersaturation resulting in fewer nuclei and larger particles. In certain instances it is impossible to use hot solutions since the precipitate may decompose or may become too soluble. In some determinations the precipitation is made in hot solution, but the mixture is allowed to cool before filtration reducing the solubility of the precipitate. If possible, however, it is best to filter the solution while hot in order to increase the solubility of the impurities. Furthermore, hot solutions filter more rapidly through filter paper.

5-7 DIGESTION OF THE PRECIPITATE

In many procedures the precipitate along with the supernatant solution, sometimes called the "mother liquor," is allowed to "digest." Digesting a solution means to heat it below the boiling point for a period of time and to stir it occasionally. The digestion process accomplishes several things.

Small particles are more soluble than large particles, hence the finely divided particles of the precipitate dissolve and redeposit on the larger particles.

Hence the old axiom that the large grow larger and the small grow smaller is upheld. Impurities entrapped within the precipitate are leached out, thus improving the quality of the precipitate.

The student is warned, however, against allowing very long periods of digestion or allowing a precipitate to stand too long in contact with the mother liquor. Certain impurities do not coprecipitate immediately with the desired precipitate, but slowly precipitate out as the solution stands. This phenomenon is called *postprecipitation*. Certain precipitates are more prone to contamination by postprecipitation than others. Barium sulfate, for example, is easily contaminated, while silver chloride is only slightly affected.

5-8 FILTRATION

Filtration is required to separate the precipitate quantitatively from its mother liquor and impurities. The media commonly employed for filtration are:

1. The sintered glass filtering crucible
2. The Gooch crucible with an asbestos mat
3. Filter paper

Each of these will be discussed in detail.

Sintered Glass Filtering Crucibles

The sintered glass filtering crucible is made entirely from glass (Fig. 5-1), having

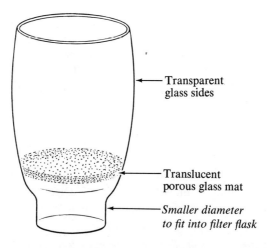

***FIGURE 5-1.** A schematic diagram of a sintered glass filter crucible.*

glass sides and a porous glass mat. The glass mats are made in different porosities usually designated as coarse, medium, or fine. Suction must be employed when using these filters. The filters of course must be cleaned and then dried to constant weight in a drying oven before use.

These crucibles should be used only with a crystalline or granular type of precipitate. Gelatinous precipitates tend to seep through the filter when suction is applied; finely divided precipitates run through or plug up the pores in the filter.

After filtration and washing, the crucible containing the precipitate is dried in an oven at temperatures ranging from 110° to 250°C, depending on the nature of the precipitate. Sintered glass crucibles should never be heated with a burner nor be heated to high temperatures as they may crack or melt.

Gooch Crucibles

The Gooch crucible is made of porcelain and has small round holes about the diameter of a pin in its base. To prepare it for use a suspension of asbestos fibers in water (asbestos soup) is poured into it and suction is applied. This forms a porous mat of asbestos 2–3 mm in thickness over the base. After washing the mat with distilled water, the crucible is dried and ready for use. Although not as convenient as the sintered glass crucible, the Gooch crucible must be used when the precipitate has to be heated to high temperatures such as 800°–1000°C.

Filter Paper

Filter paper is the most widely used filtering medium. It is made in a variety of sizes and porosities. For quantitative analysis a so-called *ashless* filter paper is used. The minerals, or ash, are leached out from the paper by washing in a mixture of hydrochloric and hydrofluoric acids during the manufacturing process. The paper is then essentially pure cellulose and leaves a residue of less than 0.1 mg after it is ignited in a crucible.

The size of the filter paper to be employed is determined by the bulk of the precipitate, not by the volume of the solution to be filtered. The entire precipitate at the end of the filtration should fill the filter paper no more than half full, otherwise losses may occur by overfilling the filter.

In quantitative filtrations with filter paper the filter paper is folded and placed in a long-stem funnel. (Techniques of filtration are discussed in the introduction to Exp. 1B, Part IV.) Suction usually is not employed in quantitative filtrations with filter paper since the paper may burst or the precipitate may be pulled through the pores of the paper.

After filtering and washing the precipitate the paper containing it is folded and placed in a porcelain crucible. The paper and precipitate are carefully dried, and then the paper is slowly burned away by heating the crucible and paper in a gas flame or in an electric muffle furnace.

After the filter paper is completely burned away the crucible and precipitate are cooled in a desiccator and then weighed. Not all precipitates may be treated in this fashion since some chemical compounds are decomposed on heating to a high temperature with filter paper. Silver(I) chloride is one such precipitate. When heated to high temperatures with filter paper it is partially reduced to free silver metal, leaving a precipitate of unknown composition. Barium(II) sulfate, on the other hand, is completely stable and is usually filtered using filter paper, followed by ignition and weighing the residue of pure barium(II) sulfate.

5-9 WASHING THE PRECIPITATE

During and after the filtration process the precipitate must be washed to free it from impurities. Washing is most effectively accomplished by stirring the precipitate in a beaker with the wash solution, then decanting the supernatant liquid through the filter. The wash solution also is used to quantitatively transfer the precipitate from the beaker to the filter. After the transfer is complete the precipitate and the filter is also washed. A wash solution must accomplish several things. It must wash out all impurities from the precipitate but leave it unchanged in composition. It should introduce no insoluble product into the precipitate, and the wash solution should be completely volatile at the drying temperature of the precipitate.

The composition of a wash solution is determined by the chemical composition of the precipitate and the nature of the impurities associated with it. Pure water usually is not used since it may cause peptization (form colloids) of the precipitate. If silver chloride, for example, is washed with pure water, it is partially transformed to colloidal particles which run through the filter. A dilute solution of an electrolyte is usually employed in washing inorganic precipitates. Dilute volatile acids, such as hydrochloric acid or nitric acid, or some ammonium salts, such as ammonium nitrate, are commonly used. All of these substances are volatile and are eliminated from the precipitate during the drying process. An electrolyte containing an ion common with the precipitate is desirable since this decreases the solubility of the precipitate. Let us use the silver chloride precipitate again as an example. If this precipitate is formed during the determination of silver precipitated by adding an excess of a soluble chloride to a solution containing silver ions, it would be desirable to use dilute hydrochloric acid as a wash solution. Hydrochloric acid would reduce the solubility of the silver chloride owing to the presence of the chloride ion. If silver chloride is precipitated to determine the amount of chloride in a solution, hydrochloric acid could not be used as wash solution since it would introduce an ion which is being determined. In this situation one should theoretically wash the silver chloride precipitate with a dilute solution of silver nitrate. Although silver nitrate would decrease the solubility of the silver ion, this wash solution would be undesirable since a residue of nonvolatile silver nitrate would remain and cause errors. Hence in the chloride determination the silver chloride precipitate is washed with dilute nitric acid.

Organic solvents such as ethyl alcohol or acetone are sometimes used as a final wash to eliminate any water associated with the precipitate and thus hasten the drying process. Organic liquids are not effective in washing away inorganic impurities but may be very useful in washing out organic impurities.

5-10 DRYING THE FINAL PRECIPITATE

The final gravimetric precipitate must be in a pure, dry form of known composition for weighing. Most inorganic compounds are stable at high temperatures and these precipitates are easily dried by placing the crucible containing them in an oven heated to 110°–250°C. They are kept in the oven until all moisture has evaporated, are placed in a desiccator to cool, and finally are weighed. Certain organic compounds, however, are decomposed even at temperatures of 100°C; hence they cannot be heated to hasten the drying process. These are usually placed in a desiccator, perhaps even a vacuum desiccator, to remove the last traces of water.

Precipitates filtered through filter paper must be stable at temperatures as high as 1000°C. Filter paper cannot be dried to constant weight since it begins to decompose at temperatures as low as 100°C; hence it is eliminated completely from the precipitate by burning it away at a high temperature. This process is known as *igniting* the precipitate. Barium sulfate, used in the determination of sulfate ion, is filtered through ashless filter paper, then ignited, and the remaining precipitate of barium sulfate is then weighed. Precipitates such as ferric hydroxide frequently are ignited and finally weighed as oxides.

5-11 CALCULATION OF RESULTS USING THE GRAVIMETRIC FACTOR

The Gravimetric Factor

A gravimetric analysis gives the weight of a pure precipitated compound, not the weight of the ion or compound sought. To convert the weight of the precipitated compound to the weight of the species sought the *gravimetric factor* is used. This may be defined as the stoichiometric ratio of the formula weight (form wt) of the species sought to the formula weight of the compound precipitated and weighed. Because *this ratio must reflect the stoichiometry of the reaction*, one or both of the formula weights may have to be multiplied by a number, R.

The multiplication of the weight in grams of the pure precipitate by the gravimetric factor gives the weight in grams of the species sought.

$$\text{g precipitate} \times (R) \frac{\text{form wt species sought}}{\text{form wt precipitate}} = \text{g species sought}$$

Note that the *precipitate* term falls out of the expression making it routine to check whether the gravimetric factor is correct or must be inverted.

As an example, consider the analysis of N.F. pharmaceutical grade sodium phosphate by precipitation of the phosphate ion to form magnesium ammonium phosphate, $MgNH_4PO_4$, which is heated to form magnesium pyrophosphate, $Mg_2P_2O_7$, the compound actually weighed. The gravimetric factor for the calculation of per cent sodium phosphate (actually $\%Na_2HPO_4$) is

$$\frac{2(\text{form wt } Na_2HPO_4)}{\text{form wt } Mg_2P_2O_7}$$

Note first that the formula weight of $Mg_2P_2O_7$ is in the denominator because it is the formula weight of the precipitate. Second, note that it is necessary to multiply the formula weight of Na_2HPO_4 by two; this reflects the stoichiometry of the reaction since two phosphates are equivalent to one pyrophosphate.

The precipitation of chloride ion from various chloride salts using silver nitrate will serve to illustrate further the calculation of the gravimetric factor. One such reaction is

$$AlCl_3 + 3Ag^+(\text{excess}) = 3AgCl(s) + Al^{+3}$$

Note that the solid state of the silver chloride precipitate is indicated by the small s in parenthesis. Also note that for every one chloride ion *one* silver chloride is precipitated, but for every one aluminum ion *three* silver chloride molecules are precipitated. The examples below will illustrate these two relationships.

EXAMPLE 5-1. Calculate the gravimetric factors for (a) finding the grams of chloride (form wt 35.45 g/mole) in impure aluminum chloride (form wt 133.3 g/mole) by precipitating the chloride as silver chloride (form wt 143.3 g/mole), and (b) finding the grams of aluminum chloride in impure aluminum chloride by precipitating the chloride as silver chloride.

Solution to a: For finding the grams of chloride from silver chloride the stoichiometric ratio of the formula weight of the species sought (Cl) to that of the compound precipitated is 1:1. Thus the gravimetric factor is simply the formula weight of the chloride divided by that of the silver chloride. The calculation of the grams of chloride is

$$\text{g AgCl precipitate} \times \frac{35.45 \text{ g/mole Cl}}{143.3 \text{ g/mole AgCl precipitate}} = \text{g Cl}$$

Note that the *AgCl precipitate* term falls out of the expression as a check on the correct form of the gravimetric factor.

Solution to b: For finding the grams of aluminum chloride from the silver chloride the stoichiometric ratio of the formula weight of the species sought ($AlCl_3$) to that of the compound precipitated is 1:3. This ratio equalizes the numbers of the *key atom*, the chloride, in the gravimetric factor, and it reflects the stoichiometry. The calculation of the grams of aluminum chloride is

$$\text{g AgCl precipitate} \times \frac{133.3 \text{ g/mole AlCl}_3}{(3) \, 143.3 \text{ g/mole AgCl precipitate}} = \text{g AlCl}_3$$

Calculation of Results Using the Gravimetric Factor

Stoichiometric Ratios. Note that the ratios of the formula weights in the two gravimetric factors in Example 5-1 varied and depended on the formulas of the species sought. At first, it may seem contradictory that the gravimetric factors are different for the same analysis. However, note that in both cases the number of chloride atoms in the numerator and denominator of each gravimetric factor are equal. The chloride atom is the *key atom*, the common atom, in this case. If the numbers of the key atom are the same in the numerator and denominator, then the ratio of the formula weights will be stoichiometric and therefore correct.

Unfortunately, not all gravimetric factors involve a key atom. Consider the following reaction:

$$C_2O_4^{-2} + Ca^{+2}(\text{excess}) = CaC_2O_4(s) \xrightarrow{\text{heat}} CaCO_3(s) + CO(g)$$

The gravimetric factor for the calculation of grams of oxalate is

$$\frac{88.0 \text{ g/mole } C_2O_4^{-2}}{100.1 \text{ g/mole } CaCO_3}$$

In this reaction there is no key atom because one of the carbon atoms and one of the oxygen atoms are lost during heating. Therefore the gravimetric factor must be based on the 1:1 stoichiometric ratio of the oxalate ion and the calcium ion in the precipitation reaction.

In most cases the key atom is a metal or nonmetal such as Cl, Br, or S (but not O) which is common to both the species sought and the compound precipitated. Some typical examples are tabulated below:

Species sought	Compound precipitated	Gravimetric factor
S	$BaSO_4$	form wt S/form wt $BaSO_4$
P	$Mg_2P_2O_7$	2(form wt P)/form wt $Mg_2P_2O_7$
C	CO_2 (abs'd in NaOH)	form wt C/form wt CO_2
Fe	Fe_2O_3	2(form wt Fe)/form wt Fe_2O_3
$HgNH_2Cl$	HgS	form wt $HgNH_2Cl$/form wt HgS

Work the self-test below for more examples of gravimetric factor calculations.

Self-Test 5-1. Calculation of Gravimetric Factors

Directions: Calculate the gravimetric factor for each problem to four significant figures. Check your answers with those given below.

A. The chloride ion in the antihistamine diphenhydramine hydrochloride, $C_{17}H_{21}ONH^+Cl^-$, formula weight 291.6, can be analyzed by precipitating it as silver chloride, formula weight 143.3. Calculate the gravimetric factor for

a. The per cent chloride (atomic weight of Cl = 35.45).
b. The per cent diphenhydramine hydrochloride.

B. The chloride ion in calcium chloride, formula weight 110.99, can be analyzed by precipitating it as silver chloride. Calculate the gravimetric factor for
   a. The per cent chloride.
   b. The per cent calcium chloride, $CaCl_2$.

C. The purity of magnesium citrate, $MgHC_6H_5O_7 \cdot 5H_2O$, can be found by precipitating the magnesium as $MgNH_4PO_4$, formula weight 137.4, and weighing it as $Mg_2P_2O_7$, formula weight 222.6. Calculate the gravimetric factor for finding the per cent purity of the magnesium citrate, formula weight 304.5.

D. Cholesterol, $C_{27}H_{45}OH$, with a formula weight of 386.6 is precipitated by digitonin, $C_{55}H_{90}O_{29}$, with a formula weight of 1214 to give the compound $C_{27}H_{45}OH \cdot C_{55}H_{90}O_{29} \cdot H_2O$. Calculate the gravimetric factor for finding the per cent cholesterol from the weight of the latter compound.

Answers to Self-Test 5-1

A. a. 0.2473_8 b. 2.034_9
B. a. 0.2473_8 b. 0.3872_6
C. 2.735_8
D. 0.2388_{48} (use 1618.6 as form wt)

Calculation of Results

Gravimetric analysis results may be expressed in terms of per cent composition, molarity, the number of halogen atoms present in a known molecule, or empirical formula.

Calculation of Per Cent Composition. Once the gravimetric factor has been used to find the weight of the species sought the percentage purity of the species sought can then be calculated. This is calculated as follows:

$$\frac{\text{g species sought}}{\text{g impure sample}}(100) = \% \text{ species sought}$$

Values for percentage may vary from zero to 100 or less, depending on whether the species sought is a compound or not. For example, if impure sodium chloride is analyzed and per cent sodium chloride is sought, the percentage of sodium chloride may vary from 0% to close to 100%. However, if per cent chloride is sought, the percentage chloride may vary from 0% to 60.66%, the percentage of chloride in pure sodium chloride.

Gravimetric analyses are capable of high accuracy and precision for percentage composition. For example, the relative deviation between two chloride analyses from precipitation of silver chloride may be as low as 0.02 pph. That means that results of 50.10% and 50.12% chloride can be obtained by an experienced analyst. Other gravimetric methods of analysis are not quite as accurate or precise as the silver chloride precipitation. For example, the precision of the barium sulfate precipitation is not nearly as good as that of the silver chloride precipitation.

Calculation of Molarity. The molarity of a solution can be determined if the cation or the anion of the solute can be precipitated quantitatively. For example, hydrochloric acid titrant can be standardized to a high degree of accuracy and precision by precipitation.

$$H^+Cl^- + Ag^+(\text{excess}) = AgCl(s) + H^+$$

Since the moles of the silver chloride precipitate equals the moles of the acid, the molarity of the acid can be calculated by finding the moles of silver chloride precipitate obtained per measured volume of the acid.

$$M \text{ HCl} = \frac{\text{moles HCl}}{\text{liters HCl}} = \frac{\text{g AgCl precipitate}/143.3 \text{ g/mole AgCl}}{\text{liters HCl}}$$

Since molarity is by definition also equal to mmoles/ml it is also calculated as follows:

$$M \text{ HCl} = \frac{\text{mmoles HCl}}{\text{ml HCl}} = \frac{\text{mg AgCl precipitate}/143.3 \text{ mg/mmole AgCl}}{\text{ml HCl}}$$

Calculation of the Number of Halogens in a Molecule. In several organic reactions molecules are reacted with halogens to give products substituted with an unknown number of halogen atoms. Many times the number of carbons, hydrogens, and oxygens in the product are known and only the number of halogens is unknown. Other times the number of hydrogens and halogens vary; the number of hydrogens decreases as the number of halogens increases.

Consider the following reaction.

$$C_4H_{12}N + yBr_2 \rightarrow C_4H_{12}NBr_x$$

The number of bromines, x, substituted on the product is unknown, but a gravimetric analysis of the percentage bromine will allow x to be calculated, as the following example demonstrates.

EXAMPLE 5-2. The gravimetric analysis of 0.0962 g of *pure* $C_4H_{12}NBr_x$ yields 0.1730 g of silver bromide, formula weight 187.78. Find x.

Solution: Calculate the percentage of bromine, formula weight 79.90, in the pure compound and compare it with the percentage of bromine in all of the possible combinations of $C_4H_{12}NBr_x$.

The percentage bromine in the pure compound is

$$\frac{0.1730 \text{ g AgBr ppt} \times \dfrac{79.90 \text{ g/mole Br}}{187.78 \text{ g/mole AgBr}} (100)}{0.0962 \text{ g pure sample}} = 76.5_2\% \text{ Br}$$

Calculation of the percentage bromine for the various possible values of x gives: $C_4H_{12}NBr$, 51.8% Br; $C_4H_{12}NBr_2$, 68.4% Br; and $C_4H_{12}NBr_3$, 76.4% Br. It is obvious that the *theoretical* percentage of bromine in the latter compound matches quite closely the *experimental* percentage of bromine found in the analysis and that $x = 3$.

5-12 BIOLOGICAL AND PHARMACEUTICAL GRAVIMETRIC ANALYSIS

There are several important gravimetric methods that are important in biological and pharmaceutical chemistry that deserve special attention. As an example of a biological gravimetric method we will discuss the determination of cholesterol. Thereafter, we will discuss some useful pharmaceutical gravimetric methods.

Determination of Cholesterol

Cholesterol is a steroid alcohol, or 3-sterol, with the formula $C_{27}H_{45}OH$. It can be precipitated by a high molecular weight organic saponin called digitonin. Digitonin has the formula $C_{55}H_{90}O_{29}$ and a formula weight of 1214. Cholesterol and digitonin react in a 1:1 molecular ratio to form an insoluble complex which apparently contains a molecule of water.

$$C_{27}H_{45}OH + C_{55}H_{90}O_{29} + H_2O \rightarrow C_{27}H_{45}OH(C_{55}H_{90}O_{29})H_2O(s)$$

The precipitation may be used to isolate cholesterol for other measurements or for the gravimetric determination of cholesterol. The theoretical gravimetric factor for calculating the per cent cholesterol is 0.2388. If esters of cholesterol must be converted to cholesterol before the precipitation, there is a small loss of cholesterol which is corrected for by increasing the gravimetric factor to 0.243 [1]. Several investigators have found that an accurate analysis depends on the purity of the digitonin that is used [2]; a correction curve must be established for each sample of digitonin to obtain high accuracy.

The digitonin precipitation is quite specific for cholesterol among the steroids having a hydroxyl group. Even cholesterol esters are not precipitated, nor are sterols with a 3-hydroxyl group of a different steric arrangement than cholesterol. Unfortunately the precipitation is slow so that it is recommended that it take place overnight [2].

Determination of Pharmaceuticals

As mentioned previously, some pharmaceuticals can be determined gravimetrically by isolating the pure form of the organic medicinal agent without any chemical

[1] R. Caminade, *Bull. Soc. Chim. Biol.* **4**, 601 (1922).
[2] J. J. Kabara in D. Glick, *Methods of Biochemical Analysis*, Vol. 10, Wiley-Interscience, New York, 1962, p. 270.

reaction. Others can be determined by conversion of the sodium salt to the acid form.

A partial list of pharmaceuticals determined without chemical reaction is given in Table 5-1. One such pharmaceutical is amobarbital, $C_{11}H_{17}N_2O_3H$, a weak acid. As long as it is the only pharmaceutical in the preparation it can be isolated by extraction. However, before the amobarbital is extracted from solid preparations the sample must be washed with petroleum ether to remove any inert binder or lubricant that might be extracted with the amobarbital. After washing, the amobarbital is extracted with chloroform or ether. The extracting solvent is then evaporated and the pure dry residue is weighed. The weight gives the amount of amobarbital present in whatever amount of the sample was taken for analysis.

A partial list of pharmaceuticals determined by conversion to the acid form is given in Table 5-2. One typical pharmaceutical is sodium phenobarbital,

TABLE 5-1. Pharmaceuticals Determined Without Chemical Reaction

U.S.P. Pharmaceuticals	N.F. Pharmaceuticals
Amobarbital tablets	Amobarbital elixir
Aurothioglucose injection	Citrated caffeine
Caffeine and sodium benzoate	Ephedrine sulfate
Phenacetin tablets	Mephobarbital tablets
Sodium lauryl sulfate	Progesterone tablets

$C_{12}H_{11}N_2O_3^-Na^+$. It is converted to phenobarbital (acid form) by adding a strong acid such as hydrochloric acid to it (**5-1**). After washing to remove any binder or lubricant it may also be extracted into ether or chloroform. (The sodium phenobarbital would not be soluble in an organic solvent such as chloroform.) Evaporation of the ether or chloroform gives a pure dry solid which can then be weighed to give an amount of phenobarbital equivalent to the sodium phenobarbital originally present in the pharmaceutical preparation. A gravimetric factor would have to be used to calculate the amount of sodium phenobarbital equivalent to the phenobarbital [3].

TABLE 5-2. Pharmaceuticals Determined by Conversion to the Acid Form

Pharmaceutical	Weighed as (Acid Form)
Sodium amobarbital, U.S.P.	$C_{11}H_{17}N_2O_3H$
Sodium butabarbital, N.F.	$C_{18}H_{19}O_2H$
Sodium fluorescein	$C_{20}H_{11}O_5H$
Sodium pentobarbital	$C_{11}H_{17}N_2O_3H$
Sodium phenobarbital	$C_{12}H_{11}N_2O_3H$

[3] G. L. Jenkins, A. M. Kenevel, and F. E. DiGangi, *Quantitative Pharmaceutical Chemistry*, McGraw-Hill, New York, 1967.

5-13 AN EARLY LOOK AT SOLUBILITY PRODUCT CALCULATIONS

Solubility Product Constant

Of the compounds determined gravimetrically, only the solubility of inorganic compounds is characterized by an equilibrium constant. Here we will introduce this equilibrium constant, called the solubility product constant, and in Section 7-1 we will complete the story.

The solubility product constant is generally given the special symbol of K_{sp} to set it apart from other constants. By convention it is usually written for a reaction in which the insoluble compound is the reactant and the various ions or molecules resulting from dissolution are the products. Strictly speaking, such a reaction should be written as involving water and energy; for example, for the dissolution of silver chloride the reaction is strictly

$$AgCl(s) + xH_2O + energy \rightleftharpoons Ag(OH_2)_4^+ + Cl(H_2O)_{x-4}^- \tag{5-2}$$

Equation 5-2 is very informative. First of all, it indicates that the dissolving process involves water molecules which bond to the metal ion through the oxygens and to the anion through the hydrogens. These interactions are what make it possible for the silver and chloride ions to dissolve. Secondly, the equation also indicates that it takes energy for the bonds in the insoluble silver chloride to break. At room temperature not enough energy is available to break many of the very strong silver-chloride bonds, and as a result very little silver chloride dissolves. (Increasing the temperature does increase the solubility.) The point of equilibrium lies on the left, and this is indicated by the relative lengths of the two arrows in the equation.

Equation 5-2 can be simplified as follows for the purposes of defining K_{sp}:

$$AgCl(s) = Ag^+ + Cl^- \tag{5-3}$$

The K_{sp} expression can now be written and given a numerical value.

$$K_{sp} = [Ag^+][Cl^-] = 1.8 \times 10^{-10} M^2 \tag{5-4}$$

Note that since the reactant is an insoluble salt its concentration cannot be specified so it is defined to be unity. Hence it does not appear in the above expression. Since the numerical value of K_{sp} is equal to the product of the molarities of the silver and chloride ions, it must have the units of M^2. Use of these units in calculations and in writing the ionic terms in the solubility product expression will be very helpful to you.

The value given for K_{sp} is at 25°C (room temperature). Since the dissolving of silver chloride requires energy (5-2), raising the temperature increases the solubility and K_{sp}. Some typical values are:

10°C $4 \times 10^{-11} M^2$
25°C $1.8 \times 10^{-10} M^2$
50°C $1.3 \times 10^{-9} M^2$

SEC. 5-13 An Early Look at Solubility Product Calculations

Formulation of the K_{sp} Expression. The formulation of a K_{sp} expression (such as 5-4) for a *known* salt $A_n B_m$ is straightforward. In general the K_{sp} expression is written

$$K_{sp} = [A^{+m}]^n [B^{-n}]^m = r \times 10^{-s} M^{(n+m)} \tag{5-5}$$

The molarity of each ion is always raised to a power equal to its subscript. The units of molarity following the numerical value of K_{sp} will always be the sum of all of the exponents of all of the ions in the expression. Values for a number of K_{sp}s are listed in Appendix 1; units are given for all values.

Formulating the K_{sp} expression for an *unknown* salt may be done by one of two approaches. If the units are given you can usually make a good hypothesis (guess) by writing down an expression which is reasonable and which has the same total power for molarity as the M units of the K_{sp} value. For example, consider AgN_3 whose $K_{sp} = 2.9 \times 10^{-9} M^2$. Without the units given for the K_{sp} value you may have guessed that there were three separate -1 nitrogen anions in AgN_3. However, the M^2 units lead to the reasonable expression

$$K_{sp} = [Ag^+][N_3^-] = 2.9 \times 10^{-9} M^2$$

If the units were not given for the K_{sp}, you would have had to consult an inorganic reference for compounds of nitrogen to discover that AgN_3 was the silver azide salt in which N_3^- is the azide ion.

Other important health-related insoluble compounds are $Ca_3(PO_4)_2$ and $CaHPO_4 \cdot 2H_2O$, both of which precipitate on the surface of the teeth as dental calculus. The K_{sp} expression for calcium phosphate is

$$K_{sp} = [Ca^{+2}]^3 [PO_4^{-3}]^2 = 1 \times 10^{-26} M^5$$

The units of M^5 are consistent with the formulation containing the square of the molarity of calcium and the cube of the molarity of phosphate.

Solubility in "Pure" Water

Since most precipitates in quantitative work are washed with pure water or with a solution which contains no ions in common with the precipitate, it is important to be able to calculate the molar solubility, S, of the insoluble compound in the wash solution. Such a solution is not strictly pure, of course, because it always contains a tiny amount of the ions of the insoluble compound.

In a pure water solution the concentrations of all ions are unknown, so that *each concentration* must be represented by a multiple of the symbol S. This is readily done if a balanced equation is written for dissolving using multiples of S as coefficients in front of each species. For example, for the dissolving of $Ca_3(PO_4)_2$ the equation would be

$$S Ca_3(PO_4)_2(s) = 3S Ca^{+2} + 2S PO_4^{-3}$$

The balanced equation automatically gives the correct symbolic concentrations to substitute into the K_{sp} expression.

$$K_{sp} = [3S]^3[2S]^2 = 108S^5$$

This equation is rearranged to give the relation between S and K_{sp}.

$$S = \sqrt[5]{\frac{K_{sp}}{108}} \tag{5-6}$$

Although a numerical value for S must still be obtained from **5-6**, the derivation of **5-6** is the most important part of the calculation; the rest is plain mathematics.

It is important to be able to calculate any of the possible roots that may arise from relations such as **5-6**. *Roots such as fifth or seventh roots cannot be calculated on most inexpensive electronic calculators or on slide rules, but they can be calculated with logs.* (*See Appendix 3 for how to use logs and an inexpensive calculator to calculate any root.*) *In general, proceed as follows: For* $S = \sqrt[n]{K_{sp}/x}$,

$$\log S = \frac{\log(K_{sp}/x)}{n}$$

$$S = \text{antilog}\left[\frac{\log(K_{sp}/x)}{n}\right] \tag{5-7}$$

Effect of Inert Salts on Solubility in Pure Water. Inert salts such as $NaNO_3$, KNO_3, and $LiClO_4$ increase the ionic strength, μ, of a solution and the solubility of insoluble salts such as silver(I) chloride. (The ionic strength of inert salts of univalent cations and anions is the same as the total molarity of such salts; ionic strength for other inert salts may be calculated as shown in Sec. 1-1). The solubility increase is reflected by an increase in the value of K_{sp} for a given insoluble compound. For example, Appendix 1 lists values of K_{sp} for two different ionic strengths—$\mu = 0$ (pure water) and $\mu = 0.1$ (for example, a $0.1 M$ solution of an inert salt such as KNO_3).

Calculation for Wash Water Solutions. The per cent of an insoluble compound that dissolves in a pure water wash (Sec. 5-9) may be calculated by deriving and using an equation such as **5-6**. If roughly *equal volumes of wash water and the unknown solution of the ion to be precipitated* are used, the per cent is calculated as follows:

$$\% \text{ dissolved} = \frac{\text{Molar solubility of compound in wash water}}{\text{Molarity of ion in unknown solution}} (100) \tag{5-8}$$

No more than 0.1% should dissolve for a quantitative (99.9%) precipitation.

EXAMPLE 5-3. The chloride ion is precipitated as silver chloride from 100 ml of a $0.0050M$ solution. The insoluble silver(I) chloride is washed with about 90 ml of water containing $0.03M$ nitric acid. Will the precipitation be quantitative?

Solution: Since the volumes of the unknown solution and the wash solution are about the same, **5-8** may be used to calculate the per cent of the chloride ion that dissolves in the wash solution. The solubility of silver chloride is

$$S = \sqrt{K_{sp(\mu=0)}} = \sqrt{1.8 \times 10^{-10} M^2} = 1.3 \times 10^{-5} M \text{ (pure water)}$$

Strictly speaking, the wash solution is not pure water since the nitric acid gives it an ionic strength of 0.03: however, the increase in the molar solubility is too small to be significant for the calculation of the per cent dissolved.

The per cent dissolved is calculated as follows:

$$\% \text{ dissolved} = \frac{1.3 \times 10^{-5} M \text{ chloride dissolved}}{5 \times 10^{-3} M \text{ chloride unknown}} (100) = 0.2_6\%$$

Since 0.3% is greater than 0.1% the precipitation is not quantitative. The error could be reduced by washing with a smaller volume of water, concentrating the unknown solution, or lowering the solubility by lowering the temperature.

The Common Ion Effect

When an ion is precipitated for gravimetric analysis the counter ion added to form an insoluble compound is usually added in approximately a 10% to 20% excess. The excess ion lowers the molar solubility; this effect is called the common ion effect. It is important to be able to understand this effect and to make calculations where it is present.

The common ion effect acts to shift the point of equilibrium in any precipitation equilibrium to the left; for example, for silver ion being precipitated by chloride as common ion

$$AgCl(s) = Ag^+ + Cl^-$$

pt. of equilibrium ← excess Cl⁻
lies here

The excess chloride lowers the silver ion concentration, as well as the solubility of silver chloride, below that in a saturated solution in pure water by precipitating more of the silver.

Since the excess chloride is added as HCl or KCl the molar solubility of the silver chloride will be approximately equal to the molarity of the silver ion, rather than the molarity of the chloride ion. The K_{sp} expression can be solved for the molar solubility of silver chloride as follows:

$$S_{AgCl} = [Ag^+] = \frac{K_{sp}}{[Cl^-]} \cong \frac{K_{sp}}{M_{\text{excess Cl}^-}} \qquad (5\text{-}9)$$

Equation **5-9** contains an approximation in the right-hand term because the equilibrium concentration of the chloride is, strictly speaking, the sum of the excess chloride ion plus the chloride resulting from a small amount of insoluble silver chloride dissolving. The latter is usually small compared to the excess chloride which has been added to the solution as HCl or KCl.

A more general equation than **5-9** can be written for calculating the solubility of any insoluble compound, M_xA_y, in the presence of an excess of the anion A^{-x}:

$$S_{M_xA_y} = \frac{1}{x}[M^{+y}] = \frac{1}{x}\sqrt[x]{\frac{K_{sp}}{[A^{-x}]^y}} \cong \frac{1}{x}\sqrt[x]{\frac{K_{sp}}{[M_{excess\,A^{-x}}]^y}} \quad (5\text{-}10)$$

This equation can be used to calculate the approximate molar solubility for any compound after precipitation while it is still in equilibrium with the excess common ion. (It cannot be used once the compound has been washed free of the common ion.) The following examples will illustrate common ion calculations.

EXAMPLE 5-4. Silver chloride is precipitated from a solution to which an excess of 4 ml of 0.5M hydrochloric acid has been added. The final volume of the solution is 90–100 ml so that a rough value of 100 ml can be used for the volume. What is the molar solubility of silver chloride, and is it less than that in pure water?

Solution: First, the final molarity of the excess chloride ion must be calculated.

$$M_{excess\,Cl^-} \cong \frac{4\text{ ml} \times 0.5M}{100\text{ ml}} \cong \frac{2\text{ mmoles}}{100\text{ ml}} \cong 2 \times 10^{-2}M$$

Now the molar solubility of the silver chloride can be calculated using the K_{sp} expression, and the molarity of the excess chloride.

$$S_{AgCl} \cong [Ag^+] \cong \frac{K_{sp}}{M_{excess\,Cl^-}} \cong \frac{1.8 \times 10^{-10}M^2}{2 \times 10^{-2}M} \cong 9 \times 10^{-9}M$$

Since the molar solubility of silver chloride in pure water is $1.3 \times 10^{-5}M$ the common ion effect has lowered the concentration of silver chloride below that in pure water.

EXAMPLE 5-5. Calcium phosphate is precipitated from a solution to which $1 \times 10^{-3}M$ excess phosphate has been added. What is the molar solubility of calcium phosphate, and has it been lowered below its solubility in pure water?

Solution: Calculate the molar solubility using **5-10**, the K_{sp} expression for calcium phosphate, and the molarity of the excess phosphate.

$$S_{Ca_3(PO_4)_2} \cong \frac{1}{3}[Ca^{+2}] \cong \frac{1}{3}\sqrt[3]{\frac{K_{sp}}{[PO_4^{-3}]^2}} \cong \frac{1}{3}\sqrt[3]{\frac{1 \times 10^{-26}M^5}{(1 \times 10^{-3}M)^2}} \cong 7 \times 10^{-8}M$$

Since the solubility of calcium phosphate in pure water is of the order of $10^{-6}M$ the common ion effect has lowered the concentration of calcium phosphate below that in pure water.

SEC. 5-13   An Early Look at Solubility Product Calculations

Self-Test 5-2. Calculation of Solubility in Pure Water

Directions: Work one problem at a time, checking your answers with those below.

A. The precipitation of Ca^{+2} as $CaSO_4$ under certain conditions will be quantitatively complete if the molar solubility of $CaSO_4$ in pure water (used to wash the precipitate) is less than $4.8 \times 10^{-3}M$.
   a. Using $K_{sp} = 2.4 \times 10^{-5}M^2$ calculate the molar solubility of $CaSO_4$ in pure water.
   b. Is the precipitation quantitatively complete?
B. The precipitation of Pb^{+2} will be complete if the molar solubility of the lead(II) salt in pure water is less than $1 \times 10^{-3}M$. Given that $K_{sp} = 1.8 \times 10^{-14}M^2$ for $PbCrO_4$ and that $K_{sp} = 1.6 \times 10^{-5}M^3$ for $PbCl_2$,
   a. Calculate the molar solubilities of $PbCrO_4$ and $PbCl_2$ in pure water.
   b. Decide whether either compound is insoluble enough for complete precipitation of Pb^{+2}.
C. Milk of magnesia consists of suspended $Mg(OH)_2$ in pure water. Normally a solution of an alkali or alkaline earth metal hydroxide would be too basic to be safely ingested. Assuming such a suspension is a satured solution,
   a. Calculate the molar solubility of $Mg(OH)_2$ using $K_{sp} = 1.8 \times 10^{-11}M^3$.
   b. Calculate the $[OH^-]$ of the solution.
   c. Compare the $[OH^-]$ with the $[OH^-]$ of $0.10M$ NaOH as to basicity and safety.
D. Maalox antacid tablets consist of a maximum of 0.8 g of an insoluble base such as $Al(OH)_3$ per tablet. Assuming you chewed one tablet to form a saturated solution of $Al(OH)_3$ in your mouth,
   a. Calculate the molar solubility of $Al(OH)_3$ using $K_{sp} = 4.6 \times 10^{-33}M^4$.
   b. Calculate the $[OH^-]$ of the solution and comment on its safety.

Answers to Self-Test 5-2

A. a. $S = 4.9 \times 10^{-3}M$ b. No
B. a. $PbCrO_4$ sol. $= 1.3_4 \times 10^{-7}M$; $PbCl_2$ sol. $= 1.5_9 \times 10^{-2}M$
   b. Only $PbCrO_4$
C. a. $1.6_5 \times 10^{-4}M$ b. 3.2 (or 3.3) $\times 10^{-4}M$ c. Much less basic; safe
D. a. $3.6 \times 10^{-9}M$ b. Neglecting OH^- from water $[OH^-] = 1.0_8 \times 10^{-8}M$ (safe)

QUESTIONS AND PROBLEMS

(Answers to most even-numbered problems are found in Appendix 5.)

Concepts and Definitions
1. List
   a. Three major advantages of gravimetric methods.
   b. Three major disadvantages of gravimetric methods.

2. List the main steps in a gravimetric analysis.
3. What are a few of the preliminary adjustments that might have to be made to a dissolved sample before it can be precipitated?
4. Name the four different types of precipitates.
5. What is the purpose of digestion?
6. What are the advantages of
   a. A sintered glass crucible over a Gooch crucible?
   b. Filter paper over a sintered glass crucible?
   c. A sintered glass crucible over filter paper?
   d. A Gooch crucible over a sintered glass crucible?
7. Define a gravimetric factor.
8. Explain why the gravimetric factor for the calculation of the percentage chloride in impure aluminum chloride is calculated with the formula weight of silver(I) chloride when the gravimetric factor for the calculation of the percentage of aluminum chloride in impure aluminum chloride is calculated using a multiple of three times the formula weight of silver(I) chloride.
9. Why is it important to be able to calculate the solubility of an insoluble salt in the solution used to wash the precipitate of the salt?
10. Explain in detail how cholesterol is determined gravimetrically using digitonin.

Gravimetric Factor Calculations
11. Milk of magnesia can be analyzed by dissolving the magnesium hydroxide and then precipitating the magnesium(II) ion with 8-hydroxyquinoline to form $Mg[(C_9H_6NO)_2 \cdot 2H_2O]$. Calculate the gravimetric factor for reporting the per cent magnesium hydroxide from the weight of the precipitate.
12. An iron ore sample is analyzed by precipitation of $Fe(OH)_3$ (form wt = 314.03) and ignition of the $Fe(OH)_3$ to Fe_2O_3, which is weighed. Calculate the gravimetric factor for finding the percentage of
    a. Fe (at wt = 55.85)
    b. Fe_2O_3 (form wt = 159.7)
    c. Fe_3O_4 (form wt = 231.54)
    d. $2Fe_2O_3 \cdot 3H_2O$ (form wt = 337.38)
13. Carbon-hydrogen compounds are analyzed by measuring the weight of carbon dioxide (form wt = 44.01) and water (form wt = 18.02) released during combustion.
    a. Calculate the gravimetric factor for converting the weight of water to the weight of hydrogen.
    b. Calculate the gravimetric factor for converting the weight of carbon dioxide to the weight of carbon.
14. A certain fish containing mercury in the form of $(CH_3)_2Hg$ was analyzed for mercury and found to contain 3.5 ppm of mercury. Use the appropriate gravimetric factor to calculate the ppm of $(CH_3)_2Hg$ in the fish.

Calculation of Results
15. The FDA tolerance for mercury in food is 0.5 ppm (0.5 mg/kg). A 120 g fish sample is found to contain 0.14 mg of mercury.
    a. Calculate the ppm of mercury to see whether it exceeds FDA standards.
    b. Calculate the ppm of $(CH_3)Hg$ in the fish.
16. A 1.000 g impure cholesterol sample yields 0.5000 g of the cholesterol-digitonin precipitate (form wt = 1618.6). Calculate the percentage of cholesterol (form wt = 386.6) in the sample.

17. A 0.6000 g sulfate (form wt = 98.07) sample was dried and found to contain 5.00% water. The sample yields 1.167 g of barium sulfate (form wt = 233.40). Calculate
    a. The percentage of sulfur (at wt = 32.06) in the moist sample.
    b. The percentage of sulfur in the dry sample.
18. A 0.1500 g pure sample of an organic compound of the formula $C_6OCl_xH_{6-x}$ was decomposed to give 0.4040 g of silver(I) chloride. Calculate the numerical value of x and the correct formula.

Solubility Calculations

19. Using a log table to find the root (see Appendix 4), calculate the molar solubility of each compound below in pure water. You must show the use of the logs even though you use a calculator (see Appendix 3).
    a. $Ca_3(PO_4)_2$; $K_{sp} = 1 \times 10^{-26} M^5$
    b. $Th(IO_3)_4$; $K_{sp} = 2.5 \times 10^{-15} M^5$
    c. $CoHg(SCN)_4$; $K_{sp} = 1.5 \times 10^{-6} M^6$
    d. $Fe_4[Fe(CN)_6]_3$; $K_{sp} = 3.0 \times 10^{-41} M^7$
20. Derive a relationship between the molar solubility of each insoluble compound in pure water and the K_{sp}. Then using the K_{sp} value in Appendix 1, calculate a numerical value for S to the correct number of significant figures.
    a. Ag_2CrO_4
    b. Hg_2Cl_2
    c. $CaMg(CO_3)_2$
    d. $MgNH_4PO_4$
21. Assuming that each of the cations in the insoluble compounds in the previous problem is precipitated from about 100 ml of a $0.001 M$ solution, calculate the per cent of each dissolved in a 100 ml wash solution.
22. The chloride ion is precipitated from a solution to which an excess of 2.0 ml of $0.50 M$ silver nitrate has been added. The final volume of the solution is about 100 ml.
    a. Calculate the molar solubility of silver chloride and the molarity of the chloride in the final solution.
    b. If the chloride in the original sample was $0.0010 M$, is the precipitation of the chloride quantitative (99.9%)?
23. Magnesium(II) is precipitated from a solution to which hydroxide ion is added so that the final pH is 10.3. Magnesium hydroxide precipitates.
    a. Calculate the molar solubility of $Mg(OH)_2$ using $K_{sp} = 1.8 \times 10^{-11} M^3$.
    b. If the magnesium(II) ion in the original sample was $0.01 M$, is the precipitation quantitative (99.9%)?

Challenging Problems

24. An iron ore was incorrectly calculated to contain 10.0% of Fe_3O_4 instead of the % Fe_2O_3. Calculate the % Fe_2O_3 without knowing the sample weight or any other measurements.
25. Calculate the molar solubility of each metal hydroxide below in pure water assuming the water is neutral.
    a. $Mg(OH)_2$; $K_{sp} = 1.8 \times 10^{-11} M^3$
    b. $Fe(OH)_3$; $K_{sp} = 2.5 \times 10^{-39} M^4$
26. An organic compound containing only carbon and oxygen is found to contain 50.0% C and 50.0% O. Calculate:
    a. Its empirical formula.
    b. Its molecular formula, assuming a molecular weight of 289 ± 2.

27. Calculate the empirical formula of an inorganic compound containing 36.4% P, 37.6% O, 2.37% H, and 23.5% Ca. Also decide if the empirical formula is a reasonable molecular formula.
28. A mixture containing only sodium chloride (form wt = 58.44) and potassium chloride (form wt = 74.56) without any other substances present can be analyzed by a single gravimetric analysis. Calculate the percentage of sodium chloride in 191.44 mg of the mixture if it yields 429.96 mg of silver chloride (form wt = 143.32).
29. A mixture containing sodium bromide (form wt = 102.9), potassium bromide (form wt = 119.0), and inert matter can be analyzed by two gravimetric analyses. Calculate the percentage of each present in 400.0 mg of the mixture if it yields 563.34 mg of silver bromide (form wt = 187.78) in one analysis and 340.9 mg of sodium bromide plus potassium bromide (without inert matter) in the second analysis.

6. Fundamentals of Volumetric Methods of Analysis

> *"Let us have some fresh blood,"* Holmes
> said, ... *drawing off the resulting drop of*
> *blood in a* chemical pipette. *"Now I*
> *add this small quantity of blood to a litre*
> *of water."*
>
> ARTHUR CONAN DOYLE
> A Study in Scarlet, Chapter 1

A volumetric method of analysis is based on a titration in which the volume of a reagent (titrant) reacting with a measured amount of a sample is measured. The sample itself may be measured by weighing it on the balance or by taking a known volume with a volumetric pipet (as the great Sherlock Holmes did with his blood sample).

The fundamentals to be mastered include certain basic concepts, a number of calculations, and some laboratory techniques. All of these must be used together for a successful volumetric analysis, but of course we will discuss each fundamental separately in its turn.

6-1 BASIC CONCEPTS OF VOLUMETRIC ANALYSIS

To begin with, you must understand some basic concepts of volumetric analysis. These are the standard solution, the primary standard, the equivalence point, the end point, the standardization process, and the concept of quantitative reaction. To illustrate these concepts the following general titrant reaction will be used.

$$tT + sS \rightarrow tT_p + sS_p \tag{6-1}$$

In this reaction, T and T_p are the reactant and product forms of the titrant, and S and S_p are the reactant and product forms of the sample, respectively.

The Titrant: A Standard Solution

In a titration the sample, S, is measured into a flask, dissolved if necessary, and then titrated with a solution of the titrant, T (**6-1**). The volume of the titrant added to the flask is measured with a *buret* (Sec. 6-3). The solution of T usually reacts completely as it is added to S. The point at which the reaction is complete is called the equivalence point. It is essential that the solution of T be a *standard solution* for which there are several requirements.

We might ask just what characterizes a standard solution. First of all, *the concentration of such a solution must be known accurately, usually to four significant figures.* There are two ways of calculating the concentration. One is to add a known weight of a primary standard reagent (a reagent of known purity) to a volumetric flask (Sec. 6-3) and dilute to the known volume of the flask. The other way is to standardize a solution of approximate concentration by titrating it against a weighed amount of a primary standard reagent.

Some of the other requirements for a standard solution containing T are:

2. The solution should be stable for as long as the analysis and standardization require; that is, its concentration should not change significantly during this time.
3. The reaction of the titrant T with the sample S should be stoichiometric; that is, there should be a whole-number ratio between t and s in the reaction.
4. The reaction of the titrant T and the sample S should be rapid and quantitatively complete. Usually this means a 99.9% reaction.

Primary Standards

Preparing a standard solution depends very much on using the proper primary standard chemical reagent. For example, a standard solution of sodium hydroxide cannot be prepared by weighing sodium hydroxide into a volumetric flask because the purity of the sodium hydroxide varies with time. If you know the requirements for a primary standard, you can readily decide whether a chemical meets such requirements.

The main requirements for a primary standard chemical are:

1. The chemical must be of known composition and highly pure. It is desirable that it be a minimum of 99.9% pure or purer. It is difficult to prepare an absolutely pure reagent so a minimum purity of 99.9+% is specified. (Sometimes a range, such as 99.95–100.05%, is given on bottles.) In addition to the purity, the label on the bottle should state that the reagent is a primary standard material.
2. The chemical should be stable at room temperature and should dry to a constant weight and formula, in an oven if necessary. It should not absorb gases such as water, carbon dioxide, or sulfur dioxide from the air.
3. To minimize the relative error from weighing (Sec. 4-1), the amount weighed should be relatively large. Thus a large formula and/or equivalent weight is desirable. Although this is not the most important requirement, it could be the deciding factor between two primary standards of otherwise equal properties.
4. The chemical should react rapidly and stoichiometrically with the titrant. The latter implies that the reaction be quantitative (99.9% complete).

Now you should be able to understand why sodium hydroxide is not a primary standard material. It is not stable at room temperature since it reacts with both water vapor and carbon dioxide. It is also not available initially in 99.9% purity, and its formula weight is small.

Equivalence Point and End Point

The equivalence point is the *theoretical* point at which the amount of added titrant T is exactly equivalent to the amount of dissolved sample S. Assuming that the reaction is quantitative (99.9%), it is also the point at which the reaction is complete. Beyond this point an insignificant amount ($\leq 0.1\%$) of the sample will react.

The end point is our perception of the equivalence point. Usually some physical property is continuously monitored either visually or electrically to indicate when the first excess of the titrant has been added beyond the equivalence point. An indicator which changes color is used frequently to detect the end point. A small portion of indicator is used so that the amount of titrant that reacts with the indicator is negligible. Under ideal conditions the color change at the end point coincides with the equivalence point. Frequently there is a small error, but this can usually be compensated for by standardizing the titrant under the same conditions as the sample titration.

Standardization

Unless a titrant can be prepared by weighing out a primary standard chemical, it must be standardized before or during use. This is done by titrating a weighed amount of a primary standard chemical.

As an example, let us consider the standardization of sodium hydroxide titrant. *Solid* sodium hydroxide is not a primary standard chemical. A *standard solution* of sodium hydroxide is an acceptable titrant if kept closed to contact with the air. It is made by diluting a saturated solution of sodium hydroxide to an approximate concentration in the 0.01–0.5M range. In this range sodium hydroxide solutions are stable provided they are not unduly exposed to carbon dioxide.

A primary standard acid is then weighed out for the standardization. The preferred primary standard acid is potassium acid phthalate, $KH[C_8H_4O_4]$, or just KHP. (The P^{-2} symbolizes the phthalate anion, $[C_8H_4O_4]^{-2}$.) The primary standard acid is then titrated with the sodium hydroxide to an end point.

$$KHP + NaOH \rightarrow P^{-2} + H_2O + K^+ + Na^+$$

The true concentration of the sodium hydroxide may then be calculated by the method described in the next section. The solution is now a standard solution since its concentration is accurately known. Provided it is not contaminated by overexposure to carbon dioxide, it can be used for weeks without a change in the concentration of sodium hydroxide.

6-2 CALCULATIONS OF VOLUMETRIC ANALYSIS

Definitions

A few of the definitions of concentrations you should have mastered previously are reviewed below. The mole-molarity system of calculations will be discussed first, to be followed by the equivalent-normality system.

The Mole-Molarity System. The mole, mmole, and often the μmole (Sec. 1-2) are used to measure quantities of chemical species. The operational definition of a mole is the formula weight of a species expressed in grams; the formula weight in turn is expressed in g/mole. Thus a mole of calcium carbonate contains 100.1 g, and the formula weight of calcium carbonate is 100.1 g/mole. It is also convenient to talk about a mole of an ion, so that we can say a mole of chloride ion contains 35.45 g.

The operational definition of a mmole is the formula weight of a species expressed in milligrams (mg); the formula weight in turn is expressed in mg/mmole. Thus a mmole of calcium carbonate contains 100.1 mg, and its formula weight is also 100.1 mg/mmole.

The molarity, M, of a solution can also be expressed using moles and liters or mmoles and milliliters (ml).

$$M = \frac{\text{moles}}{\text{liter}} = \frac{\text{mmoles}}{\text{ml}}$$

If the molarity of a solution is known and its volume is known, then the total quantity of a solute in the solution may be calculated in either of two ways:

$$\text{moles} = (M)(\text{liters}) \quad \text{or} \quad \text{mmoles} = (M)(\text{ml})$$

The Equivalent-Normality System. Many chemical substances contain more than one ion per mole that reacts with a reagent. For example, sulfuric acid, H_2SO_4, may lose one or both of its ionizable hydrogens to a base, depending on the strength of the base. Thus one mole of sulfuric acid is *equivalent* to either one or two moles of a monoprotic acid. To correct the formula weight for this the concept of *equivalent weight* is used.

The equivalent weight is calculated using the formula weight and the variable n, which depends on the chemical reaction of the substance in question. For acid-base reactions n is the number of moles of H^+ (or OH^-) neutralized per mole of acid (or base) in the reaction. For oxidation-reduction reactions n is the number of moles of electrons lost or gained per mole of reactant. For precipitation reactions n is the number of moles of univalent anions needed to precipitate the cation. For complex-forming reactions, n is the number of moles of ligand needed to complex the metal ion. Regardless of the reaction, the calculation of the equivalent weight is the same.

$$\text{eq wt} = \frac{\text{form wt}}{n} \tag{6-2}$$

The calculation of equivalent weight will be illustrated during the discussion of the calculation of titration results.

Analogous to the mole and the mmole, the equivalent (eq) and the milliequivalent (meq) are used to measure quantities in the equivalent system. The operational definition of an equivalent is the equivalent weight of a species expressed in grams; the equivalent weight is expressed in g/eq. Thus an equivalent of sulfuric acid contains 98.08/2 or 49.04 g, and the equivalent weight of sulfuric acid is 49.04 g/eq. We can also talk about an equivalent of an ion; for example, we can say that an equivalent of hydrogen ion contains 1.008 g.

The operational definition of a milliequivalent is the equivalent weight of a species expressed in mg; the equivalent weight is expressed in mg/meq. Thus a milliequivalent of sulfuric acid contains 49.04 mg, and the equivalent weight of sulfuric acid is 49.04 mg/meq.

The molarity of a solution must be corrected for the same reason that the formula weight is corrected by converting it to an equivalent weight. When the molarity has been adjusted for the reaction of more than one ion per mole, it is called the normality, N. The adjustment is done mathematically using the variable n, which depends on the chemical reaction of the substance. For example, sulfuric acid usually loses 2 moles of H^+ per mole of sulfuric acid in acid-base reactions. Thus a $1M$ solution of sulfuric acid is actually $2M$ in H^+; its normality is therefore 2. To calculate normality the following relation is used:

$$N = (n)M \qquad (6\text{-}3)$$

The way n is calculated will depend on the reaction, just as for the equivalent weight.

The normality of a solution can also be calculated directly from equivalents and liters or milliequivalents and milliliters.

$$N = \frac{\text{eq}}{\text{liters}} = \frac{\text{meq}}{\text{ml}} \qquad (6\text{-}4)$$

If the normality of a solution is known and its volume is known, then the total quantity of solute in the solution may be calculated in either of two ways:

$$\text{eq} = (N)(\text{liters}) \quad \text{or} \quad \text{meq} = (N)(\text{ml})$$

Calculations with Molarity

To calculate the results of a titration using molarity you must know the volume and the molarity of the titrant used as accurately as is necessary for the analysis. (Usually three or four significant figures are needed.) You must also know or obtain the balanced equation for the reaction occurring during the titration. If the reacting ratio of the titrant and the sample are one-to-one, then the calculation is straightforward. If not, a reaction ratio must be calculated.

Calculating the Titrant Concentration. The first calculation necessary in most analyses is that of calculating the concentration of the titrant after standardization.

Since the standardization can be done either using standard solution or a weighed amount of a primary standard, both types of calculations should be mastered. Let us use **6-1** to represent any standardization reaction. If a standard solution (S) is used and a 1:1 reacting ratio exists, then the relation below can be used to calculate the molarity (M_T) of the titrant (T).

$$(ml_T)(M_T) = mmoles_T = mmoles_S = (ml_S)(M_S) \qquad (6\text{-}5)$$

If the reacting ratio is not 1:1, then the mmoles of S will not equal the mmoles of T. The product of ml_S and M_S must be multiplied by the reacting ratio (t/s) from the balanced equation (**6-1**) to give the correct number of mmoles of T. The correct relation for calculating the molarity of T is

$$(ml_T)(M_T) = mmoles_T = (t/s)\,mmoles_S = (t/s)(ml_S)(M_S) \qquad (6\text{-}6)$$

EXAMPLE 6-1. Calculate the molarity of HCl titrant which is standardized against a standard sodium hydroxide solution and against a standard barium hydroxide solution. The reactions are:

$$HCl + NaOH = H_2O + NaCl$$

$$2HCl + Ba(OH)_2 = 2H_2O + BaCl_2$$

Titration of 25.00 ml of $0.2000M$ NaOH requires 24.00 ml of HCl. Titration of 25.00 ml of $0.1000M$ Ba(OH)$_2$ requires 24.01 ml of HCl.

Solutions: To calculate the molarity from the titration of NaOH rearrange **6-5**:

$$M_T = M_{HCl} = \frac{(25.00 \text{ ml NaOH})(0.2000M \text{ NaOH})}{24.00 \text{ ml HCl}} = 0.2083_3 M$$

To calculate the molarity from the titration of Ba(OH)$_2$ rearrange **6-6** using $(t/s) = 2$:

$$M_T = M_{HCl} = \frac{(2)(25.00 \text{ ml Ba(OH)}_2)(0.1000M \text{ Ba(OH)}_2)}{24.01 \text{ ml HCl}} = 0.2082_4 M$$

Note that the calculated molarities agree closely even though the reactions are different.

If the standardization is done using a weighed amount of a primary standard, then the weight in mg and the formula weight in mg/mmole must be substituted for the volume and molarity of the standard. Since the weight in mg divided by the formula weight in mg/mmole will give mmoles of the standard, it follows from **6-6** that

$$(ml_T)(M_T) = mmoles_T = \frac{(t/s)\,mg_S}{(mg_S/mmole_S)} \qquad (6\text{-}7)$$

EXAMPLE 6-2. Calculate the molarity of HCl titrant which is standardized against primary standard sodium carbonate powder. The reaction is

$$2HCl + Na_2CO_3 = H_2O + CO_2 + 2NaCl$$

Titration of 212.0 mg of sodium carbonate requires 22.00 ml of hydrochloric acid.

Calculations of Volumetric Analysis

Solution: To calculate the molarity from the titration of sodium carbonate, whose formula weight is 106.0 mg/mmole, rearrange **6-7**:

$$M_{\text{T}} = M_{\text{HCl}} = \frac{(2)\ 212.0\ \text{mg Na}_2\text{CO}_3}{(106.0\ \text{mg/mmole})(22.00\ \text{ml HCl})} = 0.1818 M$$

Note that the reacting ratio is 2, which corrects for the fact that two mmoles, or moles, of hydrochloric acid titrant react with only one mmole, or mole, respectively, of sodium carbonate.

In oxidation-reduction standardizations the primary standard may be dissolved and react in a different form than it is weighed. To obtain the correct reacting ratio in such cases it is recommended that the formula of the primary standard as it is weighed be written in the equation. As an example consider the standardization of iodine titrant against primary standard arsenic(III) oxide, As_2O_3. After it is dissolved this oxide is converted to two moles of arsenious acid, H_3AsO_3, and it reacts with iodine as follows:

$$I_2 + H_3AsO_3 + H_2O = 2I^- + H_3AsO_4 + 2H^+$$

To obtain the correct reacting ratio it is recommended that the reaction between arsenic(III) oxide and iodine be written as though it is the reaction occurring in solution.

$$2I_2 + As_2O_3 + 5H_2O = 2H_3AsO_4 + 4I^- + 4H^+ \qquad (6\text{-}8)$$

EXAMPLE 6-3. Calculate the molarity of iodine titrant which is standardized against primary standard arsenic(III) oxide according to **6-8**. Titration of 395.6 mg of arsenic(III) oxide requires 22.00 ml of iodine titrant.

Solution: To calculate the molarity of iodine from the titration of arsenic(III) oxide, whose formula weight is 197.8, rearrange **6-7**:

$$M_{\text{T}} = M_{\text{I}_2} = \frac{(2)\ 395.6\ \text{mg As}_2\text{O}_3}{(197.8\ \text{mg/mmole})(22.00\ \text{ml I}_2)} = 0.1818 M$$

Calculating Percentage Composition. After the concentration of the titrant has been calculated, the composition of the sample in terms of percentage may then be calculated. Equation **6-1** can also be used to represent any reaction between a titrant and the substance sought, S. At the end point in the titration

$$(t/s)\ \text{mmoles}_S = \text{mmoles}_T = (\text{ml}_T)(M_T) \qquad (6\text{-}9)$$

The reacting ratio of t/s corrects for the difference in the number of mmoles.

The true weight of the substance sought is equal to $(\%S)(\text{mg}_{IS})/100$, the product of the percentage purity and mg_{IS}, the weight of the impure sample. Dividing the product of these by the formula weight in mg/mmole will then give the mmoles of S. This quotient can then be substituted into **6-9** to give

$$\frac{(t/s)(\%S)(\text{mg}_{IS})/100}{\text{mg}_S/\text{mmole}_S} = \text{mmoles}_T = (\text{ml}_T)(M_T) \qquad (6\text{-}10)$$

This is easily rearranged to calculate the percentage of the substance sought in the sample.

$$\%S = \frac{(ml_T)(M_T)(mg_S/mmole_S)\,100}{(t/s)\,mg_{IS}} \tag{6-11}$$

Note that because of the rearrangement necessary to obtain **6-11** the reacting ratio appears in the denominator. If it seems more logical to you, you may use the equivalent form

$$\%S = \frac{(ml_T)(M_T)(s/t)(mg_S/mmole_S)\,100}{mg_{IS}} \tag{6-12}$$

In either case, check your calculation setup by writing **6-9** and **6-10** and satisfying yourself that the units are correct. The following examples will help.

EXAMPLE 6-4. The citric acid ($H_3C_6H_5O_7$) content of the pharmaceutical caffeine citrate $H_3C_6H_5O_7(C_8H_{10}O_2N_4)$ can be determined by titration with standard NaOH:

$$3NaOH + H_3C_6H_5O_7(C_8H_{10}O_2N_4) \rightarrow C_6H_5O_7^{-3} + 3H_2O + 3Na^+ + C_8H_{10}O_2N_4$$

A 1.0000 g sample of N.F. grade caffeine citrate requires 39.10 ml of 0.2000M NaOH for neutralization. Does the per cent citric acid meet the N.F. requirement of 48–52%?

Solution: The formula weight of citric acid is 192.1 mg/mmole. To calculate the percentage purity use either **6-11** or **6-12**.

$$\%H_3C_6H_5O_7 = \frac{(39.10\ ml\ NaOH)(0.2000M\ NaOH)(192.1\ mg/mmole)\,100}{(3)\,1000.0\ mg} = 50.05\%$$

The citric acid content of 50.05% does meet the N.F. requirement.

EXAMPLE 6-5. Calculate the percentage of arsenic (As) in an impure sample which is titrated with a standard iodine solution as follows:

$$I_2 + H_3AsO_3 + H_2O = 2I^- + H_3AsO_4 + 2H^+$$

Titration of 1.0000 g of impure sample requires 31.00 ml of 0.1000M iodine (I_2).

Solution: The formula weight of arsenic is the atomic weight of 74.92 mg/mmole. To calculate the percentage purity note that the reacting ratio is 1:1 and need not be shown. Use either **6-11** or **6-12**

$$\%As = \frac{(31.00\ ml\ I_2)(0.1000M\ I_2)(74.92\ mg/mmole)\,100}{(1)\,1000\ mg} = 23.22_5\%$$

Calculations with Normality

To calculate the results of a titration using normality you must know the volume and the normality of the titrant as accurately as is necessary for the analysis. (Usually three or four significant figures are needed.) You must decide what value *n* has and use it in **6-2** to calculate the equivalent weight for the standardization of the titrant and the analysis. Finally, the weight of the impure sample is used to calculate percentage, or the weight of the pure sample is used to calculate the equivalent weight.

Calculating the Titrant Concentration. The first calculation necessary in most titrimetric analyses is the calculation of the normality of the titrant after standardization. Let us use **6-1** to represent any standardization reaction and assume that the primary standard (S) is in solution. At the end point of the titration, by definition

$$\text{meq}_S = \text{meq}_T \quad \text{or} \quad \text{eq}_S = \text{eq}_T \tag{6-13}$$

If the primary standard is added in the form of a standard solution, then to calculate normality **6-4** is rearranged to give meq = (*N*)(ml). This can be substituted into **6-13** to give

$$(N_T)(\text{ml}_T) = (N_S)(\text{ml}_S) \tag{6-14}$$

Equation **6-14** is easily rearranged to calculate N_T, the normality of the titrant from N_S, the normality of the standard solution.

If the primary standard is added in the form of a weighed amount, then you should recognize that dividing the weight by the equivalent weight in mg/meq gives

$$\text{meq} = \frac{\text{mg}}{\text{mg/meq}} \tag{6-15}$$

Equations **6-14** and **6-15** can then be substituted into **6-13** to give

$$(\text{ml}_T)(N_T) = \frac{\text{mg}_S}{\text{mg}_S/\text{meq}_S} \tag{6-16}$$

Equation **6-16** is easily rearranged to calculate N_T; however, the equivalent weight of the primary standard must first be calculated using **6-2**.

The calculation of the equivalent weight depends on the type of reaction and the value of n *in the reaction.* Consider the following example of an acid-base reaction:

$$\text{NaH}_3\text{P}_2\text{O}_7 + 2\text{NaOH} = \text{Na}_3\text{HP}_2\text{O}_7 + 2\text{H}_2\text{O} \tag{6-17}$$

The equivalent weight of monosodium pyrophosphate would be equal to the formula weight divided by two in the above reaction because the number of moles of H^+ lost in the reaction per mole of acid is two. If a reaction is not given, then you can usually assume that all of the hydrogen ions are neutralized by base. For example, the equivalent weight of sulfuric acid is generally 98.08/2 in acid-base reactions because both hydrogen ions are usually neutralized by common bases.

Note that for both monosodium pyrophosphate and sulfuric acid the equivalent weight is *smaller* than the formula weight, thus correcting for the fact that both lose more than one proton per mole to base. It can be seen from **6-15** that the number of meq of either acid would be twice the number of mmoles of either acid. (The number of mmoles of H^+ neutralized by base is of course the same whether molarity or normality is used.)

For other types of reactions n is calculated differently. For example, in oxidation-reduction reactions n is the number of electrons gained or lost per mole of reactant. This value can usually be obtained from a *half reaction* involving only the reactant in question. Consider the following half reaction involving oxidation of arsenious acid to arsenic acid:

$$H_3AsO_3 + H_2O = H_3AsO_4 + 2H^+ + 2e^- \qquad (6\text{-}18)$$

The value of n in the above reaction is two so the equivalent weight of arsenious acid is 125.94/2.

Arsenic(III) oxide, a primary standard, can be weighed, dissolved to form arsenious acid, and oxidized to arsenic acid. Its equivalent weight can be obtained by writing the oxidation of arsenic(III) oxide directly to arsenic acid, as though it had occurred.

$$As_2O_3(s) + 5H_2O = 2H_3AsO_4 + 4H^+ + 4e^-$$

The value of n in the above reaction is four so the equivalent weight of arsenic(III) oxide is 197.84/4.

EXAMPLE 6-6. Calculate the normality of iodine titrant which is standardized in both of the following ways:
a. Titration of 25.00 ml of 0.1454N arsenious acid requires 10.00 ml of iodine titrant.
b. Titration of 395.6 mg of primary standard arsenic(III) oxide requires 22.00 ml of the same iodine titrant.

Solution to a: To calculate the normality of the iodine titrant rearrange **6-14** and substitute the given values.

$$N_{I_2} = \frac{(25.00 \text{ ml } H_3AsO_3)(0.1454N \text{ } H_3AsO_3)}{10.00 \text{ ml } I_2} = 0.3635N$$

Solution to b: The equivalent weight of arsenic(III) oxide is 197.84/4 mg/meq. To calculate the normality of the iodine titrant rearrange **6-16** and substitute the given values.

$$N_{I_2} = \frac{395.6 \text{ mg } As_2O_3}{(197.84/4 \text{ mg/meq})(22.00 \text{ ml } I_2)} = 0.3636N$$

Note that both standardizations give results that agree closely.

Calculating Percentage Composition

After the concentration of the titrant has been calculated the percentage composition of a sample may be calculated. Since **6-13** applies to the titration of any

SEC. 6-2   Calculations of Volumetric Analysis   109

impure sample as well as to pure samples we can substitute the product $(ml_T)(N_T)$ for the meq of the titrant.

$$meq_S = (ml_T)(N_T) \tag{6-19}$$

The true weight of the substance sought, mg_S, is equal to the product of the percentage purity, %S, and the weight of the impure sample, mg_{IS}:

$$mg_S = \frac{(\%S)(mg_{IS})}{100} \tag{6-20}$$

The meq of S is equal to the quotient of the mg of S (right-hand side of **6-20**) and its equivalent weight. Substituting this quotient for the left-hand side of **6-19** gives

$$\frac{(\%S)(mg_{IS})/100}{mg_S/meq_S} = (ml_T)(N_T) \tag{6-21}$$

Equation **6-21** is easily rearranged for the calculation of the percentage of S.

$$\%S = \frac{(ml_T)(N_T)(mg_S/meq_S)\,100}{mg_{IS}} \tag{6-22}$$

Equation **6-21** may be rearranged to calculate any other unknown factor as well. For example, to calculate the equivalent weight of a pure unknown compound ($mg_{IS} = mg_S$) it rearranges to

$$mg_S/meq_S = \frac{mg_S}{(ml_T)(N_T)} \tag{6-23}$$

EXAMPLE 6-7. The purity of pharmaceutical grade citric acid, $H_3C_6H_5O_7$, may be found by titration with standard sodium hydroxide.

$$3NaOH + H_3C_6H_5O_7 \rightarrow C_6H_5O_7^{-3} + 3H_2O + 3Na^+$$

A 0.3000 g sample of U.S.P. grade citric acid requires 46.60 ml of 0.1000N for neutralization. Does it meet the U.S.P. requirement of 99.5% purity?

Solution: The formula weight of citric acid is 192.1; its equivalent weight is 192.1/3 since three protons per molecule are neutralized. To calculate the percentage purity, use **6-22**.

$$\%H_3C_6H_5O_7 = \frac{(46.60 \text{ ml NaOH})(0.1000N \text{ NaOH})(64.03 \text{ mg/meq})\,100}{300.0 \text{ mg}} = 99.46\%$$

The purity is just short of the requirement.

EXAMPLE 6-8. Calculate the percentage of arsenic (As) in an impure sample of arsenious acid which is titrated with iodine.

$$I_2 + H_3AsO_3 + H_2O = 2I^- + H_3AsO_4 + 2H^+$$

Titration of 1.0000 g of impure sample requires 31.00 ml of 0.0500N iodine.

Solution: The equivalent weight of arsenic is 74.92/2 mg/meq since arsenious acid loses two electrons per molecule (**6-18**). Use **6-22** to calculate the percentage purity.

$$\%As = \frac{(31.00 \text{ ml } I_2)(0.0500N \text{ } I_2)(37.46 \text{ mg/meq}) \, 100}{1000.0 \text{ mg}} = 5.80_6\%$$

Clinical Calculations

Most clinical chemists prefer to report results as milligram per cent (mg%). On a weight per volume basis this is mg/100 ml of liquid sample (Sec. 1-2). Electrolytes in blood are also reported in terms of meq/liter.

For calculating the results of titrations in mg% the concentration of the titrant can be expressed in molarity since most of the titrations involve a reaction ratio of 1:1. An equation similar to **6-11** can then be used to calculate the mg%. This equation can be derived by first writing an equation to calculate the mg of a substance per 1 ml of liquid sample.

$$\frac{mg_S}{\text{ml sample}} = \frac{(ml_T)(M_T)(mg_S/\text{mmole}_S)}{\text{ml sample}}$$

Multiplying the right-hand side of the above equation by a factor of 100 gives an equation to calculate mg%:

$$\frac{mg_S}{100 \text{ ml sample}} = mg_S\% = \frac{100(ml_T)(M_T)(mg_S/\text{mmole}_S)}{\text{ml sample}} \quad (6\text{-}24)$$

If the volume of the sample is given in μl, then the volume in μl must be divided by a factor of 1000 to convert it to ml.

EXAMPLE 6-9. Calcium(II) ion can be titrated with ethylenediaminetetraacetic acid (EDTA) according to the following reaction (Sec. 7-1):

$$Ca^{+2} + EDTA \rightarrow Ca\text{-}EDTA$$

Calculate the mg% of calcium(II) in a 100 μl sample of blood serum which is titrated with 2.00 ml of $1.00 \times 10^{-4}M$ EDTA. State whether it is in the normal serum range, below it, or above it (Fig. 1-1).

Solution: First, convert the volume of the sample to ml.

$$\frac{100 \text{ } \mu l}{1000 \text{ } \mu l/\text{ml}} = 0.100 \text{ ml}$$

Next we use **6-24** to calculate the mg% of calcium.

$$mg\% \text{ Ca} = \frac{100(2.00 \text{ ml})(1.00 \times 10^{-4}M \text{ EDTA})(40.1 \text{ mg Ca/mmole Ca})}{0.100 \text{ ml sample}} = 8.02 \text{ mg}\%$$

We see that the calculated value of 8.02 mg% is lower than the normal range of 8.5–10.5 mg% calcium in blood serum (Fig. 1-1).

For calculating the results of an analysis in meq/liter the concentration of the titrant is best expressed in molarity so that the mmoles of the substance can be calculated.

$$(ml_T)(M_T) = mmoles_S \tag{6-25}$$

For the purpose of calculating an electrolyte balance in blood the number of meq of a substance is calculated from the number of mmoles by

$$meq_S = (\text{charge on ion})(mmoles_S) \tag{6-26}$$

By using this concept a check can be kept on the total number of meq of -1 charges and the total number of meq of $+1$ charges. (Obviously, the number of meq of $+1$ charges must equal that of -1 charges, but the number of mmoles of differently charged cations will not equal the number of mmoles of differently charged anions.) Once the number of meq of a substance is calculated it is divided by the volume in liters to obtain meq/liter.

EXAMPLE 6-10. Calculate the concentration of calcium(II) ion in meq/liter from the results of the analysis in Example 6-9.

Solution: First, calculate the mmole of Ca^{+2} using **6-25**.

$$(2.00 \text{ ml EDTA})(1.00 \times 10^{-5} M \text{ EDTA}) = 2.00 \times 10^{-4} \text{ mmole } Ca^{+2}$$

Next, we use **6-26** to calculate the meq of Ca^{+2}.

$$meq\ Ca^{+2} = (2)(2.00 \times 10^{-4} \text{ mmole}) = 4.00 \times 10^{-4} \text{ meq}$$

Finally, we calculate meq/liter.

$$\frac{meq\ Ca^{+2}}{\text{liter}} = \frac{4.00 \times 10^{-4} \text{ meq}}{1.00 \times 10^{-4} \text{ liter}} = 4.00 \text{ meq/liter}$$

6-3 USING VOLUMETRIC GLASSWARE IN THE LABORATORY

Thus far you have mastered some basic concepts of volumetric analysis and learned how to handle several types of calculations. Now you should become familiar with volumetric glassware and the techniques necessary to use it for a successful analysis.

The three basic pieces of volumetric glassware are the volumetric flask, the volumetric pipet, and the buret. Each of these items is manufactured to contain (TC) or to deliver (TD) a certain volume. Volumetric flasks are made to contain the volume written on them. Burets and volumetric pipets are made to deliver the stated volume by draining. Clinical (serological) pipets must be blown out to deliver the stated volume.

Because glassware is mass-produced its volume cannot be assumed to be completely accurate. You should therefore understand something about the accuracy and precision of glassware to use it intelligently.

Accuracy and Precision

There is a difference between the accuracy and precision of glassware. The *accuracy* of a given piece of volumetric glassware may be expressed in terms of the maximum allowable error, or *tolerance*. The *precision* may be expressed in terms of the uncertainty of a buret reading or the uncertainty of filling a flask or a pipet to the mark. By definition there is no precision of a single reading or a single filling, but it is proper to speak of the uncertainty (see Ch. 2).

Table 6-1 gives some of the National Bureau of Standards tolerances for volumetric glassware. The less expensive equipment found in instructional analytical laboratories does not meet these tolerances and may actually have tolerances double those in the table.

TABLE 6-1. Tolerances for Volumetric Glassware

Capacity, ml	Maximum error allowable, ml		
	Volumetric flasks	Volumetric pipets	Burets
5	—	0.01	0.01
10	—	0.02	0.02
25	0.03	0.03	0.03
50	0.05	0.05	0.05
100	0.08	0.08	0.10
500	0.15	—	—
1000	0.30	—	—

The Kimball brand Kimax, Class A, and the Corning brand Pyrex glassware conform to these specifications (National Bureau of Standards).

These tolerances are *absolute* values of the maximum allowable error. For example, the tolerance of 0.05 ml for a 50 ml buret means that the absolute error in the volume delivered may be as large as 0.05 ml. If a volume of 40 ml were used from the buret, the *relative* value of the maximum allowable error in parts per hundred (pph) would be

$$\frac{0.05 \text{ ml}}{40.00 \text{ ml}} (100) = 0.1_{25} \text{ pph (50 ml buret only)}$$

Since it is the most easily characterized only the precision of a buret will be discussed. The uncertainty of a single 50 ml buret reading is somewhat subjective, but the *absolute* value of the uncertainty is thought to be ± 0.02 ml. Since a buret is always read twice (at the beginning and at the end of a titration) the *absolute* value of the total uncertainty may be as much as ± 0.04 ml. The absolute uncertainty is the same for any volume delivered from a 50 ml buret, but the *relative* uncertainty in parts per hundred or parts per thousand differs. The calculation of relative

SEC. 6-3   Using Volumetric Glassware in the Laboratory   113

uncertainty for 10.00 ml and 40.00 ml volumes delivered from a 50 ml buret is shown below:

$$\frac{0.04 \text{ ml}}{10.00 \text{ ml}}(100) = 0.40 \text{ pph}$$

$$\frac{0.04 \text{ ml}}{40.00 \text{ ml}}(100) = 0.10 \text{ pph}$$

For the 10 ml volume the relative uncertainty is greater than the desirable relative precision of 0.1 to 0.2 pph for a volumetric method. It is obvious that a titration with a 50 ml buret should involve 35–40 ml of titrant for good precision and accuracy. A 10 ml buret would be more proper for measuring volumes of 10 ml or less because the 0.02 ml subdivisions on the buret permit a lower relative uncertainty in the reading.

Volumetric Flasks

Almost any size flask may be obtained for laboratory work; all are made to contain (TC) the exact volume of liquid when the bottom of the meniscus just touches the etched line across the neck. Most volumetric flasks employ ground-glass stoppers, but some are equipped with screw caps lined with polyethylene. If necessary, volumetric flasks may be cleaned by scrubbing with a dilute detergent solution.

Volumetric flasks are used to prepare standard solutions by exact dilution after weighing out and dissolving a primary standard. Another important use of the volumetric flask is in diluting samples to an exact volume prior to taking an aliquot for analysis.

A sample that requires heat for dissolving should be dissolved in a beaker and the resulting solution transferred quantitatively to a volumetric flask after cooling. (A volumetric flask should never be heated since heating and cooling may change its volume.) The flask is filled to a point slightly below the etched line with distilled water (or other solvent) without contacting the ground-glass portion of the flask. A minute is allowed for draining from the upper portion of the neck and then the flask is carefully filled to the mark by means of a long-barrelled medicine dropper or a pipet. The flask finally is stoppered, inverted, and agitated to mix thoroughly.

Volumetric flasks are not used for the storage of solutions. Solutions, especially alkaline ones, should be transferred immediately to plastic bottles after being made up in a volumetric flask.

Pipets and Their Use

The two types of pipets are the measuring pipet and the volumetric, or transfer, pipet. Both are shown in Figure 6-1. The measuring pipet is calibrated, but it

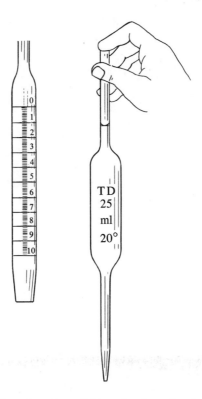

FIGURE 6-1. *Measuring pipet (left) and volumetric pipet (right) showing meniscus.*

does not deliver a given volume of liquid as accurately or as reproducibly as the buret or volumetric pipet. The volumetric pipet, not the measuring pipet, should be used when an aliquot must be taken from a standard solution (Table 6-1).

Cleaning the Pipet. If distilled water does not drain uniformly from the pipet but leaves water breaks or droplets of water adhering to the sides, the pipet must be cleaned. The following solutions are recommended in the order listed.

1. Fill with a hot dilute (2%) detergent solution and rotate the pipet to cover the inside thoroughly. Drain and rinse with distilled water.
2. Fill with a hot dilute (0.004M) alkaline (pH 12) solution of EDTA and soak for no more than 15 minutes. Drain, rinse with dilute acid, and rinse with distilled water.
3. Fill the pipet halfway *cautiously* with a hot (60°C) dichromate-sulfuric acid cleaning solution using a rubber suction bulb and rotate the pipet to wet the inside thoroughly. Return the cleaning solution to the storage bottle and rinse the pipet thoroughly with distilled water.

Use. In using the volumetric pipet, you should observe all of the following points.

1. *Use of rubber suction bulb to fill pipet.* Never fill a pipet using mouth suction; always use a rubber bulb to supply suction. This is especially important for concentrated acids, solutions of arsenic, and ammonia solutions.

2. *Rinsing*. Rinse the pipet with distilled water before using. Next rinse the pipet with the solution to be pipetted to avoid dilution by the water adhering to the inside of the pipet. Pour a small amount of the solution to be pipetted into a beaker and use this solely to rinse the pipet. Never insert an unrinsed pipet into the container of solution. (If an alkaline solution is to be pipetted, it is preferable to take aliquots also from a beaker of the solution.) Rinse the pipet, not by filling completely, but by drawing in about one fifth of the pipet's volume and twirling the pipet horizontally two or three times. Rinse above the mark by tipping the pipet slightly. Rinse at least twice in this manner with the solution to be pipetted. The pipet should then drain uniformly, or else it needs cleaning or further rinsing.
3. *Filling*. Fill the pipet about an inch above the etched line (place the fleshy part of a forefinger over the top of the pipet to stop the overflow). Then place the tip of the pipet against the inside of the vessel and rotate the pipet, allowing the solution to drain until the bottom of the meniscus just touches the etched line at eye level, as in Figure 6-1. There should be no air bubbles anywhere in the pipet.
4. *Carrying*. The pipet may be conveniently carried by tilting it slightly so that the solution flows back away from the tip slightly towards the other end.
5. *Draining*. Wipe the outside of the pipet tip free of any liquid with a tissue before draining. Place the tip against the inside of the vessel to which the solution is to be transferred and allow the pipet to discharge. Keep the tip against the inside for 20 seconds after the pipet has emptied for complete drainage. Remove the pipet from the side of the container with a rotating motion to completely remove any of the drop on the tip. The small quantity of liquid inside the tip is not to be blown out even though it appears to grow larger after a time. [The pipet has been calibrated to deliver (TD) a certain accurate volume.]

Storage. The volumetric pipet should not be allowed to remain unrinsed after use, especially after the transfer of alkaline solutions. It is good practice to fill the pipet with distilled water and to cap both ends with rubber bulbs from droppers. If this is not possible, the pipet should be thoroughly rinsed and stored in a rack or drawer where it will not easily be scratched or chipped.

Experiment B. Use of the Volumetric Pipet

The objectives of this experiment will be to acquaint you with the proper procedure in using a 25 ml volumetric pipet and to check the precision of your measurement of a 25 ml volume three times with the same pipet. If your pipetting technique is reproducible, the weight of 25 ml of distilled water that you measure with the pipet should agree within the limits of the tolerance for the 25 ml pipet.

PROCEDURE

1. Weigh three weighing bottles to the nearest 0.01 g. (It is not necessary to weigh them more accurately because you will be checking the weight of water delivered from a pipet to within ±0.01 g.

2. Rinse a 25 ml pipet with distilled water and allow it to drain to check for cleanliness. If drops cling to the inside of the pipet, clean as directed above.
3. Fill a 100 ml beaker with distilled water to use in pipetting.
4. Obtain a rubber suction bulb to fill the pipet. Review steps 1 to 5 for using the pipet. Then fill the pipet using the rubber bulb to about an inch above the etched line. Dry the outside of the pipet with a tissue. Then place the tip of the pipet against the wall of the beaker and drain it carefully until the meniscus just touches the etched line. Then place the tip of the pipet against the inside of one of the weighing bottles and allow the water to drain. Keep the tip against the side of the bottle for 20 seconds after the pipet has emptied for complete drainage. Repeat the process two more times using the other two weighing bottles.
5. Weigh each weighing bottle again to obtain the weight of water by difference to the nearest 0.01 g.
6. Report the weight of water delivered for all three times. Compare the difference in weights with the tolerance of the 25 ml pipet (Table 6-1) to see if your precision is acceptable or not.

Burets and Their Use

Burets are made and calibrated to deliver variable volumes of liquid. The essential parts are shown in Figure 6-2. Most modern burets are equipped with a glass stopcock lubricated with hydrocarbon greases or with a Teflon plastic stopcock

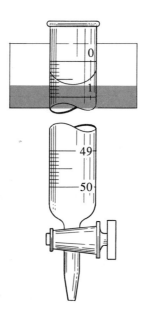

FIGURE 6-2. *Essential parts of a buret, showing the correct position of the meniscus illuminator (black portion just below meniscus).*

which requires no lubrication. Teflon stopcocks can be used for nonaqueous solvents and will not freeze even after long contact with basic solutions.

Care of the Buret. Observe the following points.

1. *Lubrication of buret stopcock.* Remove all the old grease from the stopcock and stopcock hole (use a fine wire). Apply a thin uniform layer of grease (do not use silicone lubricant); use even less grease near the holes in the stopcock. Insert the stopcock and rotate several times. If too much grease has been applied, some of it will be forced from between the stopcock and barrel or may eventually work itself into the buret tip. If too little grease has been applied, the lubricant layer will not appear uniform and transparent.
2. *Cleaning the buret.* If distilled water does not drain uniformly from the buret but leaves drops of water clinging to the sides, the buret must be cleaned. The following solutions are recommended in the order listed.

    (a) Use a hot, dilute (ca. 2%) detergent solution and scrub with a long-handled brush. Concentrated detergent may affect the glass and is difficult to rinse out completely. Rinse the detergent out with plenty of distilled water.

    (b) Soak for 10–15 minutes in a hot, dilute (0.004M, pH 12) solution of EDTA to remove metal ions. Rinse with dilute acid and then with distilled water.

    (c) Use a dichromate-sulfuric acid cleaning solution. Since a cold cleaning solution works slowly the buret may require overnight soaking. Cleaning solution may disperse more stopcock lubricant than it removes and is properly drawn into the buret inverted, by suction; hence, it is inconvenient to use in a student laboratory. Cleaning solution should not be thrown away but returned to the storage bottle for reuse.

Use. The lab instructor may check your buret technique during the standardization of sodium hydroxide against potassium acid phthalate or in the titration of acetic acid or potassium acid phthalate unknowns. In any titration the following points are important.

1. *Filling.* The buret must be first rinsed with titrant to remove water adhering to the inside of the buret. Do not rinse by filling the buret and draining, but by pouring about 10 ml of the titrant around the inside three times. Rotate the buret to wet the inside thoroughly; leave the stopcock open while rinsing and allow the buret to drain completely between rinses. Close the stopcock and fill the buret at least an inch above the zero mark.
2. *Cleanliness.* During the rinsing check whether the titrant drains uniformly from the buret. If it does, it is clean and may be used. If drops of titrant form on the inside after rinsing, the buret must be cleaned.
3. *Bubbles.* Check for bubbles in the tip of the buret by rapidly draining a milliliter or so of titrant after the buret is full. Check for bubbles in the entire buret when titrating with a dark solution, such as iodine or potassium permanganate.
4. *Reading the buret.* Allow the level of titrant to drain slowly to the zero mark. Using the meniscus illuminator shown in Figure 6-2, take an initial reading by estimating to 0.01 ml. The initial reading can be exactly 0.00 ml or larger. Record the initial reading. Bring the meniscus illuminator up so that the black half is just below the meniscus; be sure your eye is at the same level as the meniscus to avoid parallax error. (Perception of the meniscus from an angle causes the parallax error.)

5. *Titrating.* Fold a white index card or piece of paper and place under and behind the flask for a white background. Position the tip of the buret within the neck of the flask. Swirl the flask with the right hand and manipulate the stopcock with the left hand from behind the buret. This maintains a slight pressure on the stopcock and avoids leakage. (For more efficient stirring use a magnetic stirrer and stirring bar.) Add titrant rapidly at first. As the color of the indicator changes more slowly, signalling the approach of the end point, add the titrant by drops.
6. *The end point.* Just before the end point rinse down the sides of the flask with distilled water from a wash bottle. Split drops of titrant by allowing only a partial drop to form on the tip and washing it into the flask with distilled water. Since the buret can be read to 0.01 ml and since a drop is 0.05 ml, splitting drops is essential for accuracy. Allow a minute for drainage before the final reading.
7. *The "squirt" technique.* An alternative to dropwise titration and splitting drops just before the end point is the "squirt" technique. Hold the barrel of the stopcock steady with the left hand and quickly twist the stopcock 180° with the right hand. This will deliver a squirt of titrant of 0.01–0.05 ml.
8. *Vague end points.* If the color change at the end point is uncertain, a useful general method is to record the volume of titrant added for successive additions (0.01–0.05 ml) of titrant noting the color change with each addition. Usually, the point of maximum color change will be obvious after one or two further additions of titrant.

Storage. Burets filled with titrant should not be left standing long, especially if they are filled with sodium hydroxide. To avoid freezing the stopcock discard the titrant (do *not* return it to the original container) and rinse the buret several times with distilled water. It is good practice to fill the buret with distilled water after rinsing and cap it to keep dust out. If this is not possible, the buret should be stored upside down. Burets used for nonaqueous work may be rinsed with acetone and stored dry and upside down.

QUESTIONS AND PROBLEMS

(Answers to most even-numbered problems are found in Appendix 5.)

Definitions and Concepts
1. Define the following terms.
    a. Standard solution
    b. Primary standard
    c. Equivalence point and end point
    d. Molarity
    e. Normality
2. List at least four requirements for a primary standard material. Is it necessary that a primary standard be 100% pure?
3. Tell whether the following reagents are primary standard materials.
    a. Sodium hydroxide
    b. Hydrochloric acid

SEC. 6-3   Using Volumetric Glassware in the Laboratory   119

    c. Potassium acid phthalate
    d. Sulfuric acid
4. Explain the difference between a pipet that is stamped *TC* and one that is stamped *TD*.
5. What is the tolerance of a pipet or buret? Does it have anything to do with their accuracy? with their precision?
6. Describe the important steps that should be taken to locate an end point accurately with a buret, and then to read the buret properly at the end point.

Acid-Base Calculations
7. Exactly 23.16 ml of sodium hydroxide is used to titrate a 600.0 mg sample of primary standard potassium acid phthalate (KHP); its formula weight is 204.2. Calculate the molarity (or normality) of the sodium hydroxide.
8. Exactly 27.08 ml of hydrochloric acid is required to titrate 25.00 ml of $0.1079M$ hydroxide. Calculate the molarity (or normality) of the hydrochloric acid.
9. Exactly 24.60 ml of hydrochloric acid is needed to neutralize a 278.0 mg sample of primary standard tris(hydroxymethyl)aminomethane (form wt = 121.1) according to the following reaction:

$$HCl + (CH_2OH)_3CNH_2 \rightarrow (CH_2OH)_3CNH_3{}^+Cl^-$$

    Calculate the molarity (or normality) of the hydrochloric acid.
10. A 500.0 mg sample contains either impure H_3PO_4 or impure NaH_2PO_4. It is titrated with 21.00 ml of $0.1000M$ sodium hydroxide to the phenolphthalein end point to give entirely the HPO_4^{-2} ion.
    a. Assuming the sample is H_3PO_4 calculate the $\%H_3PO_4$ (form wt = 98.00).
    b. Assuming the sample is NaH_2PO_4 calculate the $\%NaH_2PO_4$ (form wt = 119.98).
11. Calculate the purity of a 500.0 mg sample of impure sodium carbonate which requires 22.00 ml of $0.1800M$ hydrochloric acid.

Precipitation Titration Calculations
12. A 0.3000 g sample of impure magnesium chloride (form wt = 95.23) is titrated with 45.00 ml of $0.1000M$ silver nitrate according to the following equation.

$$MgCl_2 + 2AgNO_3 \rightarrow 2AgCl(s) + Mg(NO_3)_2$$

    a. Calculate the percentage chloride (at wt = 35.45) in the sample.
    b. Calculate the percentage magnesium chloride in the sample.
13. A 100.0 mg sample of impure fluoride sample is titrated with 16.40 ml of $0.0120M$ thorium(IV) nitrate according to the following equation.

$$Th^{+4} + 4F^- \rightarrow ThF_4(s)$$

    Calculate the percentage of fluoride (at wt = 19.0) in the sample.
14. A 1.000 g sample of impure aluminum chloride (form wt = 133.34) requires 44.00 ml of $0.1000M$ silver nitrate for complete precipitation of all the chloride. Calculate the percentage of aluminum chloride in the sample.

Oxidation-Reduction Calculations
15. A 93.0 mg sample of primary standard arsenic(III) oxide is dissolved according to the following equation.

$$As_2O_3 \rightarrow 2H_3AsO_3$$

The resulting arsenious acid requires 18.40 ml of cerium(IV) for oxidation according to the following equation.

$$H_3AsO_3 + 2Ce^{IV} - 2H_3AsO_4 + 2Ce^{III}$$

Using the formula weight of 197.84 for As_2O_3 calculate either
a. The molarity of the Ce^{IV}, or
b. The normality of the Ce^{IV}.

16. An impure sample of 1.000 g of arsenious acid is titrated with 45.00 ml of $0.0800N$ ($0.0400M$) iodine. Calculate
    a. The percentage of arsenious acid (form wt = 125.9).
    b. The percentage of arsenic (at wt = 74.92).

Clinical Calculations

17. Calculate the mg% of calcium(II) ion in a 0.200 ml sample of blood serum which is titrated with 1.50 ml of $0.000100M$ EDTA. Is the concentration of calcium within, below, or above the normal range (Fig. 1-1)?
18. Calculate the mg% of calcium(II) ion in a 50 μl serum sample which requires 200 μl of $1.000 \times 10^{-4}M$ EDTA for titration. Is the concentration of calcium within or outside the normal range (Fig. 1-1)?
19. Calculate the meq/liter chloride in a 100 μl serum sample which requires 2.00 ml of $0.00500M$ silver nitrate. If the average serum contains 104 meq/liter, is this sample above, at, or below this level?

Challenging Problems

20. A 10.00 ml aliquot of a sulfuric acid solution requires 28.16 ml of $0.1000M$ sodium hydroxide for titration. What volume of $0.1000M$ barium chloride will be required to titrate a second 10.00 ml aliquot of sulfuric acid to produce barium sulfate?
21. A 345.0 mg sample of a pure unknown monoprotic acid is dissolved and titrated with 27.40 ml of $0.1000M$ sodium hydroxide. Calculate the formula weight of the monoprotic acid.
22. The sum of the meq/liter of chloride and bicarbonate ions in blood serum should be within 10% of the meq/liter of the sodium ion. Check this by calculating the meq/liter of each from the following data:

$$Cl^- = 383 \text{ mg\%}$$
$$HCO_3^- = 17.7 \text{ mg\%}$$
$$Na^+ = 329 \text{ mg\%}.$$

7. Introduction to Titration Equilibria and End Point Detection

> *"Detection is, or ought to be, an exact science...,"* said Holmes.
>
> ARTHUR CONAN DOYLE
> The Sign of the Four, Chapter 1

In the following chapters the analytical use of a number of different types of titration reactions will be discussed. Before you read about each type of analytical reaction you should understand thoroughly the equilibrium concepts involved as well as the methods of end point detection. All of these concepts and end point detection certainly ought to be as close as possible to the "exact science" Sherlock Holmes strove to achieve.

7-1 TYPES OF TITRATION EQUILIBRIA AND EQUILIBRIUM CONSTANTS

In this section we will describe the four main types of titration equilibria:

1. Precipitation equilibria
2. Acid-base equilibria
3. Complex ion equilibria
4. Oxidation-reduction equilibria

In our discussion we'll assume that you have encountered each type of equilibrium in previous chemistry courses and will provide a good review only at this point.

In each type of equilibrium the concentration of a species will be specified by the molarity, not activity. Recall from Section 1-2 that activity is related to molarity by the activity coefficient, f.

$$\text{activity} = f[M^{+n}]$$

where $[M^{+n}]$ is the molarity of an ion with a charge of $+n$. The activity coefficient corrects for the interionic attraction among the ions and is usually less than one. This means that the activity of an ion is usually less than the molarity, so the activity is a sort of molarity that has been corrected for interionic attraction. We will neglect this correction in our discussions to focus on the more important concept of equilibrium. The errors involved will vary and range from less than 1% in many cases to more than 10% in a few cases. Conditions in analytical methods are always adjusted to compensate for these errors.

Precipitation Equilibria

Precipitation equilibria are always described by writing the insoluble compound as the reactant and the various ions as the products of the dissolving of the insoluble compound. A simplified example would be the dissolving of silver chloride to form silver ion and chloride ion:

$$AgCl(s) = Ag^+ + Cl^- \tag{7-1}$$

The (s) term denotes that the silver chloride is an insoluble solid. The equilibrium constant for such a reaction is known as the solubility product constant, K_{sp}. The insoluble compound is not written as part of the equilibrium constant expression so for **7-1** K_{sp} is defined as follows:

$$K_{sp} = [Ag^+][Cl^-] = 1.8 \times 10^{-10} M^2 \tag{7-2}$$

Since the numerical value of K_{sp} is equal to the product of the molarities of the silver ion and the chloride ion it must have the units of M^2.

Since the general formulation of any K_{sp} expression has been given in Section 5-13 it will not be necessary to discuss this any further. It should be noted that titration reactions involving precipitation are just the opposite of reactions such as **7-1**. For example, the titration of sodium chloride with silver nitrate is described by the reaction

$$Cl^-(NaCl) + Ag^+(AgNO_3) = AgCl(s) \tag{7-3}$$

The equilibrium constant for the titration reaction in **7-3** is defined as

$$K_{rxn} = \frac{1}{[Ag^+][Cl^-]} = \frac{1}{K_{sp}} \tag{7-4}$$

Note that the numerical value of the equilibrium constant is equal to the reciprocal of the solubility product constant. This will be discussed further in the next section.

Acid-Base Equilibria

Since most acid-base analyses occur in aqueous solution the ionization of water is always of prime importance in considering acid-base equilibria. This ionization is

SEC. 7-1 Types of Titration Equilibria and Equilibrium Constants

generally written using water as the reactant and the hydroxide and hydrogen (or H_3O^+) ions as the products.

$$H_2O = H^+ + OH^-$$

The equilibrium constant for this reaction is denoted as K_w, the ion product for water. It is defined as follows:

$$K_w = [H^+][OH^-] = 1.00 \times 10^{-14} M^2 \text{ (25°C)}$$

Since the concentration of water is constant in dilute solutions its concentration is not written in the definition of K_w; instead it is included in K_w.

Ionization of Acids. Acids are classified as strong (completely ionized) or weak (less than 100% ionized). The common strong acids are hydrochloric, nitric, and perchloric acid. The first hydrogen of sulfuric acid is also completely ionized. When any of these acids are added to water the complete ionization may be written as

$$HCl \rightarrow H^+ + Cl^-$$

$$H_2SO_4 \rightarrow H^+ + HSO_4^-$$

Thus a hydrochloric acid solution, such as that found in the gastric fluid, contains no HCl molecules, only H^+ and Cl^-. It is therefore impractical to measure an equilibrium constant for a strong acid; instead it is usually said that such constants approach infinity.

In contrast to strong acids most weak acids ionize 1% or less under the usual conditions. Typical weak acids are hydrofluoric acid, acetic acid, carbonic acid, and phosphoric acid. The ionization of acetic acid may be written as

$$CH_3CO_2H = CH_3CO_2^- + H^+$$

Note that organic acids such as acetic acid are written so that the ionizable hydrogen is last in the formula rather than first. For this and other reasons such acids are also written in an abbreviated form. For example, acetic acid is abbreviated as HOAc. Carbonic acid, an important constituent in the blood's buffer system, is a diprotic acid; its ionization is written in two steps:

$$H_2CO_3 = H^+ + HCO_3^-$$

$$HCO_3^- = H^+ + CO_3^{-2}$$

The ionization constant for weak monoprotic acids is symbolized as K_a; ionization constants for diprotic acids may be symbolized as K_1 and K_2 or K_{a_1}

and K_{a_2}. The ionization constants for acetic acid and carbonic acid are defined as follows:

$$K_a = \frac{[CH_3CO_2^-][H^+]}{[CH_3CO_2H]} = \frac{[H^+][OAc^-]}{[HOAc]} = 1.8 \times 10^{-5}$$

$$K_1 = \frac{[H^+][HCO_3^-]}{[H_2CO_3]} = 4.3 \times 10^{-7}$$

$$K_2 = \frac{[H^+][CO_3^{-2}]}{[HCO_3^-]} = 4.8 \times 10^{-11}$$

Note that the numerical value of K_2 of carbonic acid is much less than that of K_1. This is generally the case for all diprotic, triprotic, and other polyprotic acids.

Ionization of Bases. Bases are also classified as strong (completely ionized) or weak (less than 100% ionized). The common strong bases are sodium hydroxide, potassium hydroxide, and barium hydroxide. When any of these bases are added to water the complete ionization may be written as

$$NaOH \rightarrow Na^+ + OH^-$$

$$Ba(OH)_2 \rightarrow Ba^{+2} + 2OH^-$$

Note that in contrast to sulfuric acid *both* of the hydroxides of barium hydroxide are completely ionized.

A solution of sodium hydroxide thus contains no NaOH molecules, only Na^+ and OH^-. It is therefore impractical to measure an equilibrium constant for a strong base; again, such constants are usually said to approach infinity just as for the strong acids.

In contrast to strong bases most weak bases ionize 1% or less under the usual conditions. The most common weak base is ammonia. It is correctly represented as NH_3, but is also written as NH_4OH. Actually very few molecules of NH_4OH exist in an ammonia solution, but most chemists will use either formula. The ionization of ammonia may be written with either formula to give the same products.

$$NH_3 + H_2O = NH_4^+ + OH^-$$

$$NH_4OH = NH_4^+ + OH^-$$

Other common weak bases are any of the various organic amines such as methylamine, CH_3NH_2; aniline, $C_6H_5NH_2$; and ethylenediamine, $NH_2C_2H_4NH_2$. Other common weak bases are anions of weak acids such as acetate ion (in sodium acetate), fluoride ion (in sodium fluoride), and carbonate ion (in sodium carbonate). Ionization for such bases may be written as follows:

$$CH_3NH_2 + H_2O = CH_3NH_3^+ + OH^-$$

$$CO_3^{-2} + H_2O = HCO_3^- + OH^-$$

(The carbonate ion can ionize further to carbonic acid but this reaction is not significant compared to the above reaction.)

The ionization constant for weak bases is symbolized as K_b; ionization constants for dibasic compounds may be symbolized as K_1 and K_2, or K_{b_1} and K_{b_2}. The ionization constant for ammonia may be defined as follows:

$$K_b = \frac{[NH_4^+][OH^-]}{[\text{ammonia}]} = 1.8 \times 10^{-5}$$

where [ammonia] may be written as $[NH_3]$ or as $[NH_4OH]$ depending on the formula used for ammonia.

The ionization constants for ethylenediamine are defined as follows:

$$K_1 = \frac{[NH_2C_2H_4NH_3^+][OH^-]}{[NH_2C_2H_4NH_2]} = 1.28 \times 10^{-4}$$

$$K_2 = \frac{[NH_3C_2H_4NH_3^{+2}][OH^-]}{[NH_2C_2H_4NH_3^+]} = 2.0 \times 10^{-7}$$

Note that the numerical value of K_2 for ethylenediamine is much less than that of K_1. This is generally the case for all such bases.

The K_b for anions of weak acids is found by dividing K_w by the K_a for conjugate acid. Thus for sodium acetate,

$$K_b = \frac{K_w}{K_a} = \frac{1.00 \times 10^{-14}}{1.8 \times 10^{-5}} = 5.6 \times 10^{-10} = \frac{[HOAc][OH^-]}{[OAc^-]}$$

The K_b for sodium carbonate is found by using the K_2 of carbonic acid; K_2 is used because the bicarbonate ion is the conjugate acid of the carbonate ion.

$$K_b = \frac{K_w}{K_2} = \frac{1.00 \times 10^{-14}}{4.8 \times 10^{-11}} = 2.1 \times 10^{-4} = \frac{[HCO_3^-][OH^-]}{[CO_3^{-2}]}$$

Acid-Base Reactions. Titration reactions involve primarily the reaction of an acid with a base rather than just the ionization of either one. For example, the titration of an acid HA with a base BOH is described by the reaction

$$HA + BOH = H_2O + A^- + B^+ \tag{7-5}$$

The equilibrium constant for the titration reaction in 7-5 is defined as

$$K_{rxn} = \frac{[B^+][A^-]}{[BOH][HA]} = \frac{K_b K_a}{K_w} \tag{7-6}$$

Thus whether an acid-base reaction occurs completely or not will depend on both the strength of the acid and the strength of the base. This equation will be derived and discussed further.

The derivation of **7-6** for an acid-base reaction such as **7-5** follows. K_a for HA and K_b for BOH are defined as

$$K_a = \frac{[H^+][A^-]}{[HA]}$$

$$K_b = \frac{[B^+][OH^-]}{[BOH]}$$

An identity for the left side of the middle term of **7-6** can be written by rearranging the K_b expression above.

$$\frac{[B^+]}{[BOH]} = \frac{K_b}{[OH^-]} \tag{7-7}$$

Similarly, an identity for the right side of the middle term of **7-6** can be written by rearranging the K_a expression:

$$\frac{[A^-]}{[HA]} = \frac{K_a}{[H^+]} \tag{7-8}$$

Both **7-7** and **7-8** can be substituted into **7-6** to give

$$K_{rxn} = \frac{K_b K_a}{[OH^-][H^+]} \tag{7-9}$$

Since K_w for water is equal to the denominator of the right-hand side of **7-9** this can be substituted for the product of the two terms to give

$$K_{rxn} = \frac{K_b K_a}{K_w} \tag{7-10}$$

This is the derivation of **7-6**; specifically, it shows how the equilibrium constant for any acid-base reaction may be derived from the ionization constant for the acid and that for the base.

Equilibria of Complexes

Complex ions or complex compounds are formed by the reaction of a metal ion, M, with a ligand, L. (Ligands may be either molecules or anions.) Such reactions occur in stepwise fashion with one ligand at a time reacting with the metal ion. The equilibrium reaction is usually described by writing the metal and the ligand as reactants and the complex ion or compound (ML_n) as the product.

$$M + L = ML$$

$$ML + L = ML_2$$

$$\vdots$$

$$ML_{(n-1)} + L = ML_n$$

Depending on the conditions the metal may complex with one, two, or more ligands. It can complex with some maximum number, n, of ligands that is characteristic of the particular ligand complexing with it. The value of n may vary from two, as in $Ag(NH_3)_2^+$, to four, as in $Cu(NH_3)_4^{+2}$, or to six, as in $Cu(H_2O)_6^{+2}$.

Chelates. Certain types of ligands can complex with more than one atom per ligand to more than one coordination position of the metal ion. This type of ligand is known as a multidentate ligand, in contrast to ligands such as ammonia which are *monodentate*, or *unidentate*, ligands. The complex ion formed is known as a chelate because so-called chelate rings are formed between the metal ion and the ligand.

One of the simplest of the multidentate ligands is ethylenediamine, $NH_2C_2H_4NH_2$. It can form a complex with both of the nitrogen atoms at either end. A common chelate formed by it is $Cu(NH_2C_2H_4NH_2)_2^{+2}$, or $Cu(en)_2^{+2}$ where *en* is an abbreviation for ethylenediamine. The structure of this chelate is

$$\begin{bmatrix} & NH_2 & NH_2 & \\ H_2C & \diagdown & \diagup & CH_2 \\ & & Cu & \\ H_2C & \diagup & \diagdown & CH_2 \\ & NH_2 & NH_2 & \end{bmatrix}^{+2}$$

Note that each of the ethlenediamine molecules forms a five-membered ring with copper(II) ion. The fact that ethylenediamine is able to complex with two nitrogen atoms rather than one, as in ammonia, makes its chelate with copper(II) more stable than the $Cu(NH_3)_2^{+2}$ complex ion.

One of the most useful multidentate ligands is known commonly as EDTA (ethylenediaminetetraacetic acid); this ligand is discussed in more detail in Chapter 10. At this point it is only important to know that it is a hexadentate ligand and can form 1:1 chelates with most metal ions. Such chelates are very stable so EDTA can be used for the titration of numerous metal ions. For example, it can be used to titrate calcium(II) ions in water.

$$Ca(H_2O)_6^{+2} + EDTA = Ca\text{-}EDTA + 6H_2O \tag{7-11}$$

Note that calcium(II) ion exists as a hydrated metal ion in water and that when EDTA reacts with it all of the water molecules are displaced by the EDTA in forming the Ca-EDTA chelate.

Stability Constants. The equilibrium constant for complexes and chelates is usually written for the formation of the complex or chelate rather than for the dissociation of either. The constant is termed the *stability constant* or also the formation constant. (If the reaction is written as a dissociation, the constant is termed the instability constant.) For each step in the reaction of a metal M with a ligand L there is a corresponding stepwise stability constant, $K_1, K_2, ..., K_n$, where n is the maximum number of Ls coordinating to M.

The constants for the $Cu(NH_3)_4^{+2}$ are a good example of the stepwise stability constants. They are defined as follows:

$$K_1 = \frac{[Cu(NH_3)^{+2}]}{[Cu^{+2}][NH_3]} = 1.3 \times 10^4 \qquad K_2 = \frac{[Cu(NH_3)_2^{+2}]}{[Cu(NH_3)^{+2}][NH_3]} = 3 \times 10^3$$

$$K_3 = \frac{[Cu(NH_3)_3^{+2}]}{[Cu(NH_3)_2^{+2}][NH_3]} = 8 \times 10^2 \qquad K_4 = \frac{[Cu(NH_3)_4^{+2}]}{[Cu(NH_3)_3^{+2}][NH_3]} = 1.3 \times 10^2$$

Note that the four stability constants describe the equilibria involved starting with the reaction of Cu^{+2} with the first ammonia ligand and ending with the fourth and final ammonia ligand reacting to form $Cu(NH_3)_4^{+2}$. This is obviously a very complicated equilibrium situation; handling it mathematically is beyond the scope of this book.

The 1:1 chelates are not difficult to handle because only one stability constant is involved. The very important reactions of EDTA with various metals can readily be understood because only one step and one constant are involved. Thus a "conditional" stability constant for the calcium-EDTA reaction (7-11) at a constant pH is defined as follows:

$$K_{\text{M-EDTA}} = \frac{[\text{Ca-EDTA}]}{[Ca^{+2}][\text{EDTA}]} = 5 \times 10^{10}$$

The conditional stability constant for any 1:1 metal-EDTA chelate is also the same as the equilibrium constant for the titration reaction of the metal and EDTA. There is no need to derive an equation to calculate the equilibrium constant as has already been done for precipitation reactions and acid-base reactions.

Oxidation-Reduction Equilibria

Oxidation-reduction equilibria involve an oxidizing agent which gains electrons and a reducing agent which loses electrons. General and specific examples of an oxidizing agent are

$$A_{ox} + ye^- = A_{red}$$
$$I_2 + 2e^- = 2I^- \qquad E° = +0.535 \text{ V}$$

The half reaction for the iodine-iodide couple is characterized by a standard reduction potential of $+0.535$ volts (see Sec. 11-3 for full details).

General and specific examples of a reducing agent are

$$B_{red} = B_{ox} + xe^-$$
$$C_4H_6O_4C(OH)C=C(OH) = C_4H_6O_4C(=O)-C=O + 2H^+ + 2e^-$$

The specific reducing agent above is ascorbic acid which is oxidized to dehydroascorbic acid. Because the above half reaction is written as an oxidation it does not have a standard reduction potential. However, if the direction of the half reaction is reversed, the resulting reduction has an $E°$ of $+0.39$ V (Sec. 11-3).

The standard potentials above are a measure of the equilibria for each half reaction. They are not the same as equilibrium constants; they can be used to obtain an equilibrium constant for a complete oxidation-reduction reaction. First, suppose we give the general equation assuming out general oxidizing agent and general reducing agent react as follows:

$$x A_{ox} + y B_{red} = x A_{red} + y B_{ox} \tag{7-12}$$

The general oxidizing agent gains y electrons and the general reducing agent loses x electrons in **7-12** just as in the respective half reactions.

As shown in Section 11-3, a general equation for calculating the equilibrium constant for any complete reaction can be derived to give

$$\log \frac{[A_{red}]^x [B_{ox}]^y}{[A_{ox}]^x [B_{red}]^y} = \log K = \frac{n(E_A° - E_B°)}{0.0590} \tag{7-13}$$

The n term in **7-13** is the lowest common denominator of x and y; that is, it is the total number of electrons involved in the complete reaction.

To see if you can use **7-13** properly, try to calculate the log of the equilibrium constant for the reaction of iodine and ascorbic acid. You should obtain a value of 4.9_{15}.

7-2 TITRATION CURVES AND EQUILIBRIA AT THE END POINT

In this section we will discuss the general shape of a so-called titration curve and then consider the various types of equilibria that can exist at the end point of the titration.

Shape of the Titration Curve

Let us consider the general titration reaction below to illustrate the titration curve.

$$T + S = T_p + S_p \tag{7-14}$$

In this reaction T and T_p are the reactant and product forms of the titrant, respectively, and S and S_p are the reactant and product forms of the sample, respectively. We will simplify the situation by assuming that the reaction is a 1:1 reaction.

A titration curve can be constructed by plotting the concentration of one of the reactants or one of the products against the volume of titrant added. Because the change in concentration is so large the negative log of the concentration is usually plotted against the volume of titrant. This is symbolized with a small p in front of the formula of the species whose negative log of concentration is being plotted. Three common examples are

$$\text{Precipitation of Cl}^-: \quad pCl = -\log[Cl^-]$$
$$\text{Neutralization of HCl:} \quad pH = -\log[H^+]$$
$$\text{Complexation of Ca}^{+2}: \quad pCa = -\log[Ca^{+2}]$$

By use of a pair of electrodes connected to a potentiometer the pCl, pH, pCa, and so on, can be measured for each volume of titrant added. These readings can then be plotted to obtain a titration curve.

General Titration Curve. We will first consider a titration in general using **7-14** as the titration reaction. The titration will be conducted by adding measured volumes of T from a buret to a $0.1M$ solution of S in a flask. Let us assume that pS can be measured using a pair of electrodes and a potentiometer. Figure 7-1

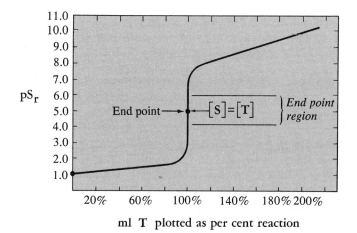

FIGURE 7-1. A general titration curve showing the change in concentration of sample S plotted as pS as the titrant is added. The end point occurs at pS = 5.0. Volume changes are neglected.

shows the titration curve that will be obtained in this hypothetical situation if volume changes are neglected.

Before any titrant is added the pS will be 1.0; that is,

$$pS = -\log 0.1M = -\log(1 \times 10^{-1}) = 1.0$$

This is typical in that many samples for titration are adjusted to be $0.1M$. As the titrant is added the concentration of S will drop below $0.1M$; at 90% reaction only 10% of S will be left and its concentration will be $0.01M$. At this point pS will be 2.0; that is,

$$pS = -\log(10\%)(0.1M) = -\log 0.01M = -\log(1 \times 10^{-2}) = 2.0$$

After this point the concentration of S begins to drop significantly and the value of pS increases significantly. For example, at 99.0% reaction only 1.0% of S will be left and its concentration will be $0.001 M$. Using the method above pS is calculated to be 3.0. Then the concentration of S and the value of S change very suddenly as the reaction passes through the 100% point. In the hypothetical case shown in Figure 7-1 the equivalence point (100.0% reaction) occurs at a pS of 5.0. At this point the small equilibrium concentration of S remaining in the solution equals the equilibrium concentration of T remaining in the solution. The region on either side of a pS of 5.0 (about 2 pS units in length) is called the end point region because of the large change in pS occurring per volume of titrant added. (About 0.1 ml of titrant or less is all that is required for this change to occur.)

The detection of the end point will be discussed in detail in the last section; it is sufficient for you to know at this time that this point (100.0% reaction) is in the middle of the end point *region*.

Addition of the titrant in this case is continued past the end point. In this region the concentration of S is so small that it does not change much. As an example, let's consider a hypothetical change from 101% reaction with a pS of 7.0 to 150% reaction. Suppose that the concentration of S also decreases 50% at that point. Then pS will be calculated as follows:

$$pS = -\log(50\%)(1 \times 10^{-7} M) = -\log(5 \times 10^{-8}) = 7.3$$

A change from pS of 7.0 to 7.3 is of course a small change for such a large change in the per cent reaction.

In general, a titration curve increases in slope slightly until near the equivalence or end point. Then its slope increases sharply. After the end point the slope decreases so there is only a slight increase in pS as each increment of titrant is added.

A pH Titration Curve. Now that you understand a titration in general you should also try to comprehend a specific titration. The best example is that of an acid-base titration curve.

A typical acid-base titration is conducted by titrating a sample of an acid in a flask using standard $0.1 M$ sodium hydroxide titrant. The titration can be conducted by following the change in pH with a pair of electrodes and a pH meter (Secs. 7-3 and 17-2). (We will neglect volume changes except at the end point.)

Suppose that 50 ml of $0.1 M$ hydrochloric acid is the sample to be titrated. This is a strong acid so it is completely ionized to hydrogen ions and chloride ions. Before any sodium hydroxide titrant is added the pH will be

$$pH = -\log 0.1 M\ H^+ = -\log(1 \times 10^{-1} M\ H^+) = 1.0$$

Note that in Figure 7-2 the pH for zero ml of titrant is 1.0. As the sodium hydroxide

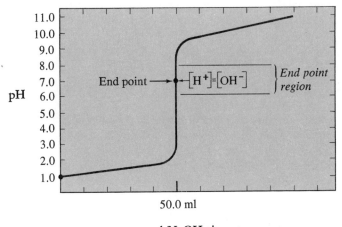

FIGURE 7-2. A titration curve showing the change in concentration of 50 ml of 0.1M HCl plotted as pH as the NaOH titrant is added. The end point occurs at pH = 7.0.

titrant is added the [H⁺] will drop below $0.1M$ and the pH will increase. At 90% reaction only 10% of the hydrogen ion will be left and the pH will be

$$\text{pH} = -\log(10\%)(0.1M) = -\log 0.01M = 2.0$$

After this point the pH begins to increase significantly. For example, at 99.0% reaction only 1.0% of the original hydrogen ion is left and its concentration will be $0.001M$. The pH at this point is 3.0. Then the concentration of the hydrogen ion and the pH change very drastically as the reaction passes through the end point region. As shown in Figure 7-2, the theoretical end point occurs at a pH of 7.0. The pH has this value because by definition the end point indicates the point at which the equilibrium concentration of the titrant equals that of the sample; that is, [OH⁻] = [H⁺]. The only pH at which this can be true for mixtures of a strong acid and strong base is 7.0. This is readily calculated.

$$K_w = [\text{H}^+][\text{OH}^-]$$

$$K_w = [\text{H}^+][\text{H}^+]$$

$$[\text{H}^+] = \sqrt{K_w} = 1.0 \times 10^{-7}$$

The region on either side of the end point pH of 7.0 (about three to four pH units in length) is called the end point region because of the large pH change occurring per volume of titrant added. (About 0.1 ml of titrant is all that is required for this change to occur.)

When the end point is detected using electrodes and a pH meter the titrant is usually added beyond the end point. In this region the concentration of hydrogen ion is so small that it does not change much. As an example, consider the change from 101% reaction with a pH of 11 to 150% reaction (75 ml of titrant). The pH is controlled at this point by the amount of sodium hydroxide added. The concentration of hydroxide present is now

$$\frac{(25 \text{ ml OH}^-)(0.1 M \text{ OH}^-)}{50 + 25 \text{ total ml}} = \frac{2.5 \text{ mmoles OH}^-}{75 \text{ ml}} = 0.033 M \text{ OH}^-$$

The pOH of this solution is calculated as

$$\text{pOH} = -\log 0.033 M = 1.5$$

The pH is therefore $14 - 1.5$, or 12.5. A change from pH 11 to pH 12.5 is of course a small change for such a large amount (25 ml) of titrant.

You have now seen how the course of a titration may be followed by plotting a titration curve. In particular you should see the sharp changes in concentration that occur at the end point *in favorable reactions*. Not all reactions will be favorable; that is, not all reactions are necessarily quantitative and give a favorable end point. Next we shall consider how to predict whether a reaction is quantitative or not. If it is, then we can expect to find a favorable end point.

Predicting Quantitative Reactions at a Titration End Point

Each of the four main types of titration reactions we have been considering is somewhat unique. Therefore we will have to consider each separately in terms of whether that type of reaction will be quantitative or not at the end point. However, the definition of quantitative is the same for each and this should be discussed first.

The accuracy and precision of the measurement device places a limit on the definition of a quantitative reaction. In Section 6-3 both the accuracy and precision of a 50 ml buret were of the order of 0.1 pph, or 0.1% (for a 40–50 ml volume). Therefore the upper limit for a quantitative reaction using a 50 ml buret will be $100.0\% - 0.1\%$, or 99.9% reaction. If the size of the buret changes or the measurement is performed using an instrument, then the upper limit of the definition may change.

Using this definition of a quantitative reaction, we will now examine each of the four different types of reactions. We will always use the equilibrium constant for the titration reaction which was discussed for each in Section 7-1.

Predicting Quantitative Precipitation Titrations. We saw in Section 7-1 for the precipitation of silver chloride that the equilibrium constant for the titration reaction was equal to the reciprocal of the solubility product constant. This is true for all precipitation reactions; that is, for the reaction

$$n\text{A} + m\text{B} = \text{A}_n \text{B}_m(s) \tag{7-15}$$

where n ions of cation A are precipitated by m ions of anion B. The solubility product is defined as

$$K_{sp} = [\text{A}]^n [\text{B}]^m$$

(See also the discussion in Sec. 5-13.)

The equilibrium constant for the titration reaction (7-15) is

$$K_{rxn} = \frac{1}{[A]^n[B]^m} = \frac{1}{K_{sp}} \qquad (7\text{-}16)$$

The *minimum theoretical value* of K_{rxn} for a quantitative titration reaction will be different for each type of insoluble compound. You will readily understand this once you recognize that precipitation of a typical $0.1M$ solution of cation A will be quantitative when [A] is reduced to or below a limit calculated as follows:

$$\text{limit of } [A] = (0.1\%)(0.1M) = 1 \times 10^{-4} M$$

The same is true for titration of $0.1M$ solutions of anion B.

We will illustrate the calculation of the minimum theoretical value of K_{rxn} for simple 1:1 salts such as the silver halides, AgX(s). For the titration of a $0.1M$ solution of any halide ion, X^-, with silver nitrate titrant quantitative precipitation will occur at the end point if $[X^-]$ is reduced to or below the limit of $1 \times 10^{-4} M$. The $[Ag^+]$ by definition must equal $[X^-]$ at the end point. The minimum theoretical value of K_{rxn} is then calculated by substituting these concentrations into the proper form of **7-16** for silver halide salts.

$$\text{min theo } K_{rxn} = \frac{1}{[1 \times 10^{-4} M\ Ag^+][1 \times 10^{-4} M\ X^-]} = 1 \times 10^8 \qquad (7\text{-}17)$$

By comparing the *actual value* of K_{rxn} for a specific 1:1 salt with the above value you can decide whether a precipitation titration will be quantitative. The guideline is summarized for you in the box below.

If the actual K_{rxn} for a 1:1 salt is $\geq 1 \times 10^8$ (the minimum theoretical value), the titration reaction will be quantitative at the $0.1M$ level.

For example, consider the titration of a $0.1M$ sodium iodate solution with $0.1M$ silver nitrate. Since the K_{sp} of silver iodate is 3.0×10^{-8}

$$K_{rxn} \text{ for AgIO}_3 = \frac{1}{3.0 \times 10^{-8}} = 3.3 \times 10^7$$

This is less than 1×10^8 so the titration will not be quantitative at the $0.1M$ level. However, the two values are so close that a change in concentration of the sample or a change in method could achieve a quantitative reaction.

The above guideline is valid for all types of 1:1 salts, such as barium sulfate. Consider the titration of sodium sulfate with $0.1M$ barium nitrate to produce insoluble barium sulfate. Since the K_{sp} of barium sulfate is 6.3×10^{-10}

$$K_{rxn} \text{ for BaSO}_4 = \frac{1}{6.3 \times 10^{-10}} = 1.6 \times 10^9$$

SEC. 7-2 Titration Curves and Equilibria at the End Point

This is more than 1×10^8 so the titration will be quantitative. Indeed, such titrations are done routinely for sulfate analysis.

The minimum theoretical value of K_{rxn} will be different for salts which are not 1:1. For example, for the titration of a $0.1 M$ solution of a -2 anion, say B^{-2}, with $0.1 M$ silver nitrate, the $[B^{-2}]$ is reduced to $1 \times 10^{-4} M$, but the $[Ag^+]$ is reduced to $2 \times 10^{-4} M$. The minimum theoretical value of K_{rxn} is then calculated by substituting these concentrations into the proper form of **7-16** for an Ag_2B (2:1) type salt.

$$\text{min theo } K_{rxn} = \frac{1}{[2 \times 10^{-4} M \, Ag^+]^2 [1 \times 10^{-4} M \, B^{-2}]} = 2.5 \times 10^{11} \qquad (7\text{-}17)$$

By comparing the actual value of K_{rxn} for a specific Ag_2B salt or other 2:1 salt you can decide whether such a precipitation titration will be quantitative. The guideline is summarized in the box below.

If the actual K_{rxn} for a 2:1 salt such as Ag_2B is $\geqslant 2.5 \times 10^{11}$ (the minimum theoretical value), the titration reaction will be quantitative at the $0.1 M$ level.

For example, consider the titration of $0.1 M$ sodium chromate, Na_2CrO_4, with $0.1 M$ silver nitrate titrant. Since the K_{sp} of silver chromate is 5×10^{-12}

$$K_{rxn} \text{ for } Ag_2CrO_4 = \frac{1}{5 \times 10^{-12}} = 2 \times 10^{11}$$

This is less than 2.5×10^{11} so that the titration will not be quantitative at the $0.1 M$ level. However, the two values are so close that a change in concentration of the sample or a change in the method could achieve quantitative titration.

The above guideline is of course valid for all types of 2:1 salts, such as copper pyrophosphate, $Cu_2P_2O_7$. Consider the titration of $0.1 M$ sodium pyrophosphate with $0.1 M$ copper nitrate to produce insoluble copper pyrophosphate. Since the K_{sp} of copper pyrophosphate is 8.4×10^{-16}

$$K_{rxn} \text{ for } Cu_2P_2O_7 = \frac{1}{8.4 \times 10^{-16}} = 1.2 \times 10^{15}$$

This is more than 2.5×10^{11} so the titration will be quantitative.

Other types of precipitation reactions that you should consider are included in the self-test at the end of this section.

Predicting Quantitative Acid-Base Reactions. We saw in Section 7-1 that the equilibrium constant for the general acid-base reaction in **7-5** was defined as

$$K_{rxn} = \frac{K_a K_b}{K_w}$$

In theory then, any titration with a strong base titrant or a strong acid titrant should be quantitative because the equilibrium constants for either are essentially infinity. This means that K_{rxn} should also be infinity. However, the anion A^- of a weak acid or the cation B^+ of a weak base does react with water in the end point region to a sometimes significant degree. The equilibrium constant K_b for the reaction of A^- with water at the end point is defined as

$$K_b = \frac{K_w}{K_a} = \frac{[HA][OH^-]}{[A^-]} \qquad (7\text{-}18)$$

To calculate the *maximum theoretical value* of K which corresponds to 0.1% or less reaction of A^- at the end point we must first assume that $[A^-]$ will have a typical value of $0.1M$. Then the limiting value of $[HA]$ will be 0.1% of that, or $1 \times 10^{-4}M$. Since HA and OH^- are produced in equal quantities by the reaction of A^- with water, $[OH^-]$ will also equal $1 \times 10^{-4}M$. The maximum theoretical value of K_b is then calculated.

$$\text{max theo } K_b = \frac{[1 \times 10^{-4} M \text{ HA}][1 \times 10^{-4} M \text{ OH}^-]}{[0.1M \text{ A}^-]} = 1 \times 10^{-7} \qquad (7\text{-}19)$$

By comparing the actual value of K for a specific acid you can decide whether a sodium hydroxide titration will be quantitative at the end point. The guideline is summarized in the box below.

If the actual K for the anion-water reaction at the end point is $\leq 1 \times 10^{-7}$ (maximum theoretical value), the sodium hydroxide titration of the weak acid will be quantitative at the $0.1M$ level.

For example, consider the sodium hydroxide titration of phenol whose K_a is 1.6×10^{-10} and whose formula is C_6H_5OH. The titration reaction is

$$NaOH + C_6H_5OH = C_6H_5O^- + Na^+ + H_2O$$

Note that the $C_6H_5O^-$ anion is produced at the end point; we will assume its concentration is a typical $0.1M$. Using **7-18** the actual value for the equilibrium constant for its reaction with water is

$$K_b = \frac{K_w}{K_a} = \frac{1.0 \times 10^{-14}}{1.6 \times 10^{-10}} = 6.3 \times 10^{-5}$$

This is larger than 1×10^{-7} so that the titration will not be quantitative at the $0.1M$ level. Near the end point the reaction of the $C_6H_5O^-$ anion with water will regenerate phenol in amounts so that 99.9% of the phenol can never be neutralized to the anion. The fundamental reason for this is that phenol is too weak an acid in water for quantitative neutralization. This is also the reason phenol is used medicinally as a disinfectant. It is just acidic enough to disinfect by killing microorganisms, but not too acid at low concentrations to be harmful to mammals.

Predicting Quantitative Complexation Reactions. We saw in Section 7-1 that the only important complexation titrations involved formation of chelates with EDTA (ethylenediaminetetraacetic acid). The conditional stability constant, $K_{\text{M-EDTA}}$, for such a chelate is defined so that it is the same as the equilibrium constant for a general metal (M)-EDTA titration reaction.

$$\text{M} + \text{EDTA} = \text{M-EDTA}$$

The minimum theoretical value of the equilibrium constant can be calculated after calculating concentrations of M and EDTA at the end point. Assuming a typical value of $0.01M$ for the M-EDTA chelate at the end point the limit of both concentrations for a 99.9% reaction is

$$\text{limit of [M]} = \text{limit of [EDTA]} = (0.1\%)(0.01M) = 1 \times 10^{-5} M$$

The minimum theoretical value of K_{rxn} is then calculated as follows:

$$\text{min theo } K_{\text{rxn}} = \frac{[0.01M \text{ M-EDTA}]}{[1 \times 10^{-5} M \text{ M}][1 \times 10^{-5} M \text{ EDTA}]} = 1 \times 10^8 \quad (7\text{-}20)$$

By comparing the actual value of a conditional stability constant for a specific metal-EDTA chelate with the minimum theoretical value you can decide whether an EDTA titration will be quantitative at the end point. Use the guideline in the box below.

> If the actual K for a metal-EDTA reaction is $\geqslant 1 \times 10^8$ (the theoretical minimum value), the EDTA titration of the metal ion will be quantitative at the $0.01M$ level.

For example, consider the titration of $0.01M$ Ba^{+2} with EDTA at pH 12. The conditional stability constant (actual value of K) for Ba-EDTA is 6.3×10^7. This is smaller than 1×10^8 so the EDTA titration of barium(II) ion will not be quantitative at the $0.01M$ level. However, the two values are so close that a change in the concentration of the sample or a change in the method could achieve a quantitative reaction.

The titration of most metal ions with EDTA is quantitative; for example, the conditional stability constant of calcium-EDTA at pH 12 is 5×10^{10}. This is much larger than the theoretical minimum value of 1×10^8.

Predicting Quantitative Oxidation-Reduction Reactions. We saw in Section 7-1 that an oxidation-reduction reaction in general involved a reaction between A_{ox}, an oxidizing agent, and B_{red}, a reducing agent, to give A_{red}, the product of A_{ox}, and B_{ox}, the product of B_{red}. The general reaction is

$$xA_{\text{ox}} + yB_{\text{red}} = xA_{\text{red}} + yB_{\text{ox}}$$

The general definition of the equilibrium constant for such a reaction is

$$K_{rxn} = \frac{[A_{red}]^x [B_{ox}]^y}{[A_{ox}]^x [B_{red}]^y} \tag{7-21}$$

The minimum theoretical value of K_{rxn} will vary depending on the values of x and y. Let's just consider the case where $x = y = 1$. Assume that typical values of A_{red} and B_{ox} at the end point are both $0.1 M$. For a 99.9% reaction the limit of both the concentration of A_{ox} and B_{red} is

$$\text{limit of } [A_{ox}] = \text{limit of } [B_{red}] = (0.1\%)(0.1 M) = 1 \times 10^{-4} M$$

The minimum theoretical value of K_{rxn} is then calculated using **7-21**.

$$\text{min theo } K_{rxn} = \frac{[0.1 M \; A_{red}][0.1 M \; B_{ox}]}{[1 \times 10^{-4} M \; A_{ox}][1 \times 10^{-4} M \; B_{red}]} = 1 \times 10^6 \tag{7-22}$$

By comparing the actual value of an oxidation-reduction equilibrium constant with the minimum theoretical value you can decide whether the reaction will be quantitative at the end point. Use the guideline in the box below.

If the actual K_{rxn} for an oxidation-reduction 1:1 reaction is $\geqslant 1 \times 10^6$ (the theoretical minimum value), the titration will be quantitative at the $0.1 M$ level.

For example, the oxidation of iron(II) to iron(III) by cerium(IV), which is reduced to cerium(III), has an equilibrium constant of about 10^{14}. Since this is much larger than 1×10^6 the oxidation of iron(II) with cerium(IV) titrant is quantitative. Most common oxidation-reduction reactions have large equilibrium constants and are therefore quantitative. However, you cannot assume that all such reactions are quantitative. The self-test at the end of this section will illustrate this.

Self-Test 7-1. Deciding Whether a Reaction is Quantitative

Directions: Calculate the K_{rxn} for each problem to answer it.

A. Decide whether a precipitation titration producing the respective 1:1 insoluble salt will be quantitative or not at the $0.1 M$ level. If not, suggest whether a change in concentration might achieve a quantitative reaction.
  a. $Ag^+ + Cl^- = AgCl(s)$; $K_{sp} = 1.8 \times 10^{-10}$
  b. $Pb^{+2} + SO_4^{-2} = PbSO_4$; $K_{sp} = 1.0 \times 10^{-7}$

SEC. 7-3 **End Point Detection Methods** 139

B. Decide whether a precipitation titration producing the respective 2:1 insoluble salt will be quantitative or not at the 0.1M level.
   a. $2Ag^+ + C_2O_4^{-2} = Ag_2C_2O_4(s)$; $K_{sp} = 4.0 \times 10^{-11}$
   b. $2Mn^{+2} + Fe(CN)_6^{-4} = Mn_2Fe(CN)_6(s)$; $K_{sp} = 7.9 \times 10^{-13}$
C. Decide whether a precipitation titration producing the respective 1:1 insoluble salt will be quantitative or not at the 0.01M level. First derive the minimum theoretical value for K_{rxn}.
   a. $Ag^+ + Cl^- = AgCl(s)$; $K_{sp} = 1.8 \times 10^{-10}$
   b. $Ba^{+2} + SO_4^{-2} = BaSO_4(s)$; $K_{sp} = 6.3 \times 10^{-10}$
D. Derive the minimum theoretical value for K_{rxn} at the 0.1M level for the reaction $3Ag^+ + B^{-3} = Ag_3B(s)$.
E. After calculating K for the anion-water reaction at the end point decide whether sodium hydroxide titration of the following acids will be quantitative or not at the 0.1M level.
   a. Acetic acid; $K_a = 1.8 \times 10^{-5}$
   b. Arsenious acid; $K_1 = 8.0 \times 10^{-10}$, $K_2 = 8.0 \times 10^{-13}$, $K_3 = 4 \times 10^{-14}$
F. Decide whether the EDTA titration of the following metal ions will be quantitative or not at the concentrations listed.
   a. 0.01M and 0.001M Mg^{+2}; $K = 5 \times 10^8$ (stability constant)
   b. 0.01M and 0.0001M Ca^{+2}; $K = 5 \times 10^{10}$ (stability constant)
G. Decide whether the oxidation-reduction reactions below are quantitative.
   a. $Ag^+ + Fe^{+2} = Ag°(s) + Fe^{+3}$; $K = 3.2$
   b. Ascorbic acid + I_2 = Dehydroascorbic acid + $2I^-$; $K = 9 \times 10^4$ (You must calculate the minimum theoretical K_{rxn} for this.)

Answers to Self-Test 7-1

A. a. Quant; $K_{rxn} = 5.6 \times 10^9$ b. Not quant; $K_{rxn} = 1 \times 10^7$
B. a. Not quant; $K_{rxn} = 2.5 \times 10^{10}$ b. Quant; $K_{rxn} = 1.3 \times 10^{12}$
C. Min theo $K_{rxn} = 1 \times 10^{10}$
   a. Not quant   b. Not quant; $K_{rxn} = 1.6 \times 10^9$
D. For 0.1M B^{-3}, min theo $K_{rxn} = 3.7 \times 10^{14}$
E. a. Quant   b. $H_3AsO_3 \rightarrow H_2AsO_3^-$ not quant; $K_b = 1.2_5 \times 10^{-5}$
F. a. 0.01M: quant; 0.001M: not quant ($K < 1 \times 10^9$)   b. 0.01M: quant; 0.001M: quant
G. a. Not quant   b. Min theo $K_{rxn} = 5 \times 10^5$; not quant

7-3 END POINT DETECTION METHODS

In this section we will discuss in general the various methods of detection of the titration end point. All the other general details of titrations have been discussed, particularly the end point equilibria and the concentration changes that occur in the end point region (Figs. 7-1 and 7-2).

In a *general way* we have pointed out that the end point can be located in the middle of the end point region, the region where there is a sharp change in concentration per volume of titrant added. However, there are three *specific* types of end point detection you should be aware of—visual methods, spectral methods, and electrical methods. Let's begin by discussing the visual methods.

Visual Methods for End Points

In this type of method the eye detects a color change that coincides closely with the true end point (equivalence point). The *indicator* used is a chemical reagent generally exhibiting one color before the true end point and another color after the true end point. The indicator exhibits the most pronounced *color change* in the end point region; when the eye perceives the color change the titration is stopped and the volume of titrant recorded.

A few visual end points are based on precipitation. In the Mohr method for halide ions (Ch. 10) a soluble chromate salt is added as indicator. When the silver nitrate titrant has reacted with all of the halide ion being titrated the following end point reaction occurs.

$$2Ag^+ + CrO_4^{-2} = Ag_2CrO_4(s)$$
(yellow)   (red ppt)

The true end point is the first permanent reddening of the yellow color of the chromate ion; the reddening is caused by the formation of a small amount of suspended silver chromate.

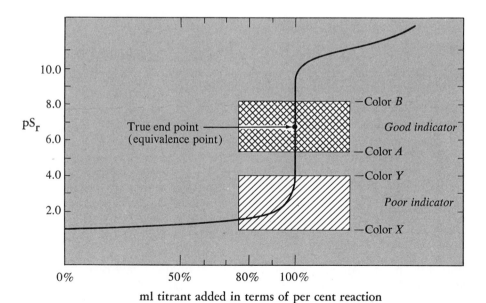

FIGURE 7-3. *Color changes of a poor indicator and a good indicator.*

Most visual end points are based on the color change of a soluble indicator. Indicators for acid-base titrations are either weakly ionized acids or bases. Indicators for complexation titrations are colored dyes which form colored complexes with the metal ion to be titrated with EDTA. Indicators for oxidation-reduction titrations are usually colored oxidizing or reducing agents.

An indicator should be chosen so that its most perceptible color change occurs in the end point region. This is shown in Figure 7-3 for the titration of a sample ion, S. The concentration of S is plotted as pS on the left axis against the per cent reaction. Note that the poor indicator begins to change color at about 80% reaction, well before the true end point. It has changed to its second color, color Y, even before the true end point is reached. In contrast, the good indicator begins to change from color A to color B, its second color, in the end point region.

Spectral Methods for End Points

In this type of method a spectral change is followed by an instrument rather than the eye. The same types of indicators may be used, but the color change is measured on a spectrophotometer (see Ch. 13). The measurements are plotted against the volume of titrant added at each point. A titration curve such as that in Figure 7-1 may be obtained. It is also possible to measure the fluorescence (see Ch. 15) of a fluorescent indicator and make the same kind of plot.

Electrical Methods for End Points

In the most common adaptation of this method a pair of electrodes and a potentiometer are used to measure the potential (voltage) of the solution as the titration is conducted. For acid-base titrations a special type of potentiometer called a pH meter is used. The so-called glass electrode is used to measure pH changes. Such a procedure is described in Chapter 17; the titration curve in Figure 7-2 is obtained by this means.

Other types of electrical methods for end point detection are based on measuring changes in conductance (conductometric titration) and on measuring the change in current passing through a cell in which the voltage applied across the electrodes is kept constant (amperometric titration).

QUESTIONS AND PROBLEMS

(Answers to most even-numbered problems are in Appendix 5.)

The Four Types of Titration Equilibria
1. Write a solubility product expression similar to 7-2 for the following insoluble salts. Also calculate K_{rxn} for each.
   a. AgBr

b. AgN_3
   c. Hg_2Cl_2
   d. $Ca_3(PO_4)_2$
2. Write an ionization constant expression for each of the following acids and calculate K_{rxn} (7-6) for the neutralization of each with sodium hydroxide titrant.
   a. Acetic acid
   b. HF
   c. C_6H_5OH (phenol)
   d. HNO_3
3. Calculate K_{rxn} for the reaction of acetic acid and ammonia. Does the fact that both are weakly ionized imply that they will not react to a significant degree?
4. The stability constants for $Cu(ethylenediamine)_2^{+2}$, or $Cu(en)_2^{+2}$, are $K_1 = 3.6 \times 10^{10}$ and $K_2 = 1 \times 10^9$.
   a. Write out stability constant expressions for each K.
   b. By comparing K_1 with the product $K_1 K_2$ for $Cu(NH_3)_2^{+2}$ decide whether the $Cu(en)^+$ chelate is more or less stable than the $Cu(NH_3)_2^{+2}$ complex ion. Why is K_1 compared with this product, and not K_1 for $Cu(NH_3)^{+2}$?
   c. Do the same as part b for $Cu(en)_2^{+2}$ and $Cu(NH_3)_4^{+2}$.
5. Write a conditional stability constant expression for magnesium-EDTA ($K = 5 \times 10^8$). Is this the same expression as the expression for K_{rxn} for the titration of magnesium with EDTA?

Titration Curves
6. Sketch a titration curve for the titration of $1 \times 10^{-2} M$ metal ion U^{+4}, assuming that in the end point region $[U^{+4}]$ changes from $1 \times 10^{-5} M$ to $1 \times 10^{-7} M$. Plot pU vs. per cent reaction. Indicate on the graph the exact value of pU at the equivalence point.
7. Sketch a titration curve for the titration of 50 ml of $0.1M$ sodium hydroxide using $0.1M$ hydrochloric acid as titrant. Plot pH vs. per cent reaction and locate the end point.
8. Sketch a titration curve for the titration of 50 ml of $0.001M$ hydrochloric acid using $0.001M$ sodium hydroxide as titrant. Plot pH vs. per cent reaction and locate the end point.

Quantitative Reactions
9. Decide whether the following precipitation titrations will be quantitative or not at the $0.1M$ level. If not, suggest whether a change in concentration might achieve a quantitative reaction.
   a. $Ag^+ + N_3^- = AgN_3(s)$; $K_{sp} = 2.9 \times 10^{-9}$
   b. $Ag^+ + VO_3^- = AgVO_3(s)$; $K_{sp} = 5 \times 10^{-7}$
   c. $2Ag^+ + SO_3^{-2} = Ag_2SO_3(s)$; $K_{sp} = 1.5 \times 10^{-13}$
   d. $3Ag^+ + PO_4^{-3} = Ag_3PO_4(s)$; $K_{sp} = 1.3 \times 10^{-20}$
10. Decide whether the precipitation titrations in the preceding problem will be quantitative at the $0.01M$ level. Be sure to derive the minimum theoretical value for K_{rxn} for each type.
11. Decide whether sodium hydroxide titration of the following acids will be quantitative at the $0.1M$ level after calculating K for the anion-water reaction at the end point.
    a. Hydrogen cyanide; $K_a = 5.0 \times 10^{-10}$
    b. Boric acid; $K_1 = 5.9 \times 10^{-10}$
    c. Hydrofluoric acid; $K_a = 6.7 \times 10^{-4}$

SEC. 7-3 End Point Detection Methods 143

12. Decide whether the EDTA titration of the following metal ions at the concentrations listed will be quantitative or not.
   a. $0.01 M$ and $0.0001 M$ Mg^{+2}; conditional stability constant $= 5 \times 10^8$
   b. $0.01 M$ and $0.0001 M$ Mn^{+2}; conditional stability constant $= 1 \times 10^{12}$

Challenging Problems

13. Derive an equation to calculate a numerical value for K_{rxn} for the following reaction.

$$HC^- + BOH \rightarrow C^{-2} + B^+ + H_2O$$

Assume that HC^- is the acid anion of diprotic acid H_2C and that K_1 and K_2 are the ionization constants for this acid. Also assume that K_b is the ionization constant for BOH.

14. Suppose that you wished to derive a simple rule of thumb to predict whether an acid will react with ammonia to more or less than 9%.

$$HA + NH_3 \rightarrow NH_4^+ + A^- + H_2O$$

Begin by calculating a theoretical value for K_{rxn} assuming that [HA] and [A$^-$] are roughly equal. Then use a value of 10^{-5} for K_b for ammonia in connection with **7-6**.

15. Calculate a minimum theoretical value for K_{rxn} for a precipitation reaction involving the same number of ions as the reaction below.

$$4Fe^{+3} + 3Fe(CN)_6^{-4} \rightarrow Fe_4[Fe(CN)_6]_3(s)$$

8. Elementary Acid-Base Equilibria

"Interesting though elementary,"
said Holmes.

ARTHUR CONAN DOYLE
The Hound of the Baskervilles, Chapter 1

In this chapter your goal should be to master the fundamentals of acid-base equilibria so that you have a foundation for the next chapter on acid-base titration methods. We will not present rigorous methods of calculation, but instead methods that are useful approximations and somewhat elementary (in the hope that you will not imitate Dr. Watson, who usually didn't understand anything "elementary"). The discussion below is based on an understanding of the Bronsted acid-base theory. If necessary, you should review the Bronsted definition of a conjugate acid and a conjugate base, the key being that they differ by only one proton.

8-1 THE IONIZATION OF WATER AND pH

In this section the emphasis will be placed on the definition of pH and pOH and on the importance of the ionization of water in defining pH. First, it is necessary to consider the ionization of water when it is the solvent.

Ionization of Water

Any proton-containing solvent like water undergoes an autoprotolysis, or self-ionization. This can be written in two ways:

$$2H_2O = H_3O^+ + OH^- \quad \text{or} \quad H_2O = H^+ + OH^-$$

In either case we understand that the hydrogen ion is solvated by one or more water molecules acting as Bronsted bases towards the proton.

In pure water we can speak accurately of the activity (Sec. 7-1) of the hydrogen ion, a_{H^+}, or the molarity of the hydrogen ion, $[H^+]$. The same is true for the hydroxide ion. Thus in pure water

$$a_{H^+} = [H^+] \quad \text{and} \quad a_{OH^-} = [OH^-]$$

However, when ionic compounds are dissolved in the water the equalities above are not valid. To simplify, we will use molarities rather than activities. Using this approximation the equilibrium constant, K_w, for the ionization of water is defined as

$$K_w = [H^+][OH^-] = [H_3O^+][OH^-] \tag{8-1}$$

Although a value of 1.0×10^{-14} is commonly used for K_w this value is valid only at 25°C. This is because the ionization is endothermic; that is,

$$2H_2O + \text{heat} = H_3O^+ + OH^- \tag{8-2}$$

Thus as the temperature is increased, the degree of ionization and the hydrogen ion concentration of water increase. Some values of K_w for other temperatures are given in Table 8-1.

TABLE 8-1. *Values of K_w at Other Temperatures*

Temperature, °C	K_w	pK_w
0	1.1×10^{-15}	14.94
10	2.9×10^{-15}	14.53
20	6.8×10^{-15}	14.17
25	1.0×10^{-14}	14.00
30	1.47×10^{-14}	13.83
37 (body temp.)	2.5×10^{-14}	13.60
40	2.9×10^{-14}	13.53
60	9.6×10^{-14}	13.02

Note that K_w gradually *increases* as the temperature increases. Even at body temperature, which is only 12°C above room temperature, K_w is significantly larger than it is at room temperature. Also note that the negative log of K_w, pK_w, gradually *decreases* as the temperature increases. Either constant may be used to calculate pH, as will be shown next.

The pH Scale

Strictly speaking, pH is defined in terms of the activity of the hydrogen ion.

$$\text{pH} = -\log a_{H^+}$$

146　　　　　　　　　　Elementary Acid-Base Equilibria　　　　　　　　CHAP. 8

The same is true for pOH. Since we are assuming that molarity is approximately the same as activity, we will use the following approximate definitions of pH and pOH.

$$\text{pH} = -\log[\text{H}^+]$$
$$\text{pOH} = -\log[\text{OH}^-]$$

It is also true that

$$\text{p}K_w = \text{pH} + \text{pOH} \tag{8-3}$$

At 25°C where the $\text{p}K_w = 14.00$,

$$\text{pH} = 14.00 - \text{pOH}$$
$$\text{pOH} = 14.00 - \text{pH}$$

The *usual* pH scale ranges from the pH of a $1M\ \text{H}^+$ solution to the pH of a $1M\ \text{OH}^-$ solution. The pH of $1M\ \text{H}^+ = 0.0$. The pOH of a $1M\ \text{OH}^-$ solution is 0.0; the pH is 14.0 at 25°C (**8-3**). Obviously it is possible to have a negative pH if the $[\text{H}^+]$ is larger than $1.0M$, just as it is possible to have a pH larger than 14 if the $[\text{H}^+]$ is smaller than $1.0 \times 10^{-14} M$.

A *neutral solution* at any temperature is one in which $[\text{H}^+] = [\text{OH}^-]$, or the pH = pOH. Substituting this equality into **8-1** we obtain for any temperature:

$$K_w = [\text{H}^+][\text{OH}^-] = [\text{H}^+][\text{H}^+] = [\text{H}^+]^2$$
$$[\text{H}^+] = \sqrt{K_w}, \text{ or} \tag{8-4}$$
$$\text{pH} = \frac{1}{2}\text{p}K_w \tag{8-5}$$

Thus a neutral solution at 25°C is one in which the $[\text{H}^+] = 1.0 \times 10^{-7}$, and the pH = 7.00.

It is interesting to compare the pH scale at 25°C with that at the body temperature of 37°C. At 37°C the pH scale would again encompass the range from a pH of $1M\ \text{H}^+$ to the pH of $1M\ \text{OH}^-$. At this temperature the pH of $1M\ \text{H}^+$ would still be 0.0. The pH of $1M\ \text{OH}^-$ would not be 14.0, however. It is true that the pOH of $1M\ \text{OH}^-$ would still be 0.0; however, using **8-3** and substituting the appropriate value of $\text{p}K_w$ from Table 8-1,

$$\text{pH}_{37°C} = \text{p}K_{w\,37°C} - \text{pOH} = 13.60 - 0 = 13.60$$

A neutral solution (NS) at 37°C does not have a pH of 7.00. Using **8-5** the pH would be

$$\text{pH} = \frac{1}{2}\text{p}K_w = \frac{1}{2}(13.60) = 6.80(\text{NS})$$

From these values we can compile the side-by-side comparison in Table 8-2.

TABLE 8-2. *pH Scales at 25° and 37°C (Body Temperature)*

Solution	pH at 25°C	pH at 37°C (body)
$1M\ H^+$	0.0	0.0
$0.02M$ HCl in stomach	1.7	1.7
Neutral: $[H^+] = [OH^-]$	7.00(NS)	6.80(NS)
A blood pH of 7.4	7.4	7.4
$1M\ OH^-$	14.00	13.60

Note in Table 8-2 that the pH of the stomach has the same meaning at both temperatures; the pH of $0.02M$ H^+ from stomach hydrochloric acid is the same for each. This is true by definition for any acid solution at body temperature. In the basic pH range there is a difference. A blood pH of 7.4 is basic by only 0.4 pH units at 25°C; however, at body temperature a blood pH of 7.4 is basic by 0.6 pH units. Thus at 37°C a blood pH of 7.4 is a more basic solution than at room temperature by 0.2 pH unit. The pOH of a pH 7.4 blood sample at 37°C would be

$$pOH_{37°C} = pK_{w\,37°C} - pH = 13.60 - 7.4 = 6.2$$

whereas at room temperature the pOH would be $14.00 - 7.4 = 6.6$.

8-2 CALCULATION OF pH OF STRONG AND WEAK ACIDS AND BASES

In Section 7-1 strong acids and bases were defined as those acids and bases which were completely ionized in water. Weak acids and bases were defined as those acids and bases which were quite a bit less than 100% ionized. With that in mind let's see how the pH of such solutions is calculated.

The pH of Strong Acids and Bases

Recall that strong acids and bases have no equilibrium constants because they are 100% ionized to protons and anions or cations and hydroxide ions. We will discuss strong acids first.

Calculations for Strong Acids. The common strong acids are nitric acid, hydrochloric acid, perchloric acid ($HClO_4$), and the first hydrogen of sulfuric acid. The second hydrogen of sulfuric acid has a K_2 value of 1×10^{-2} so it is not a strong acid.

To calculate the pH of any strong acid simply assume the molarity of the hydrogen ion is the same as the molarity of the strong acid and take its negative log. The correct number of significant figures should be used; recall from Section 2-1 that a pH value should have the same number of significant figures to the right of the decimal point as the total number of significant figures of the $[H^+]$. The following examples will illustrate these points.

EXAMPLE 8-1. Calculate the pH of a stomach fluid containing $0.020M$ hydrochloric acid.

Solution: Assume that the acid is 100% ionized and that the hydrogen ion concentration equals that of the hydrochloric acid. Write the $[H^+]$ in exponential form.

$$[H^+] = 0.020M = 2.0 \times 10^{-2}M$$

Calculate the negative log of the $[H^+]$ using two significant figures to the right of the decimal point in the pH.

$$pH = -\log(2.0 \times 10^{-2}) = 2 - 0.3010 = 1.69_9 \text{ or } 1.70$$

Note that the two is an exact number because it comes from an exact exponent; it has no effect on the number of significant figures in the pH because significant figure rules apply only to measured numbers, not to exact numbers. The log of 2.0 was given to four digits because it was read from the four-place log table in Appendix 4. The correct number of significant figures in the pH was obtained by rounding off after subtraction.

EXAMPLE 8-2. Calculate the pH of a $0.100M$ solution of H_2SO_4, sulfuric acid ($K_1 = $ infinity; $K_2 = 1.0 \times 10^{-2}$).

Solution: To obtain an answer valid to one significant figure assume the first hydrogen is 100% ionized and neglect ionization of the HSO_4^- ion to H^+ and SO_4^{-2}. Write the $[H^+]$ in exponential form.

$$[H^+] \cong 0.1M \cong 1 \times 10^{-1}M$$

Calculate the pH using one significant figure to the right of the decimal point.

$$pH = -\log(1 \times 10^{-1}M) = 1 - 0.0000 = 1.0$$

To obtain a more exact answer substitute the $0.1M$ H^+ from the first ionization into the second ionization constant expression using $x = [SO_4^{-2}]$:

$$K_2 = 1.0 \times 10^{-2} = \frac{[H^+][SO_4^{-2}]}{[HSO_4^-]} = \frac{[0.100+x][x]}{[0.100-x]}$$

Using the solution to the quadratic equation (inside front cover), $x = 8.4 \times 10^{-3}M$ and the exact $[H^+] = 0.108_4M$. The pH is 0.965.

Calculation for Strong Bases. The common strong bases are sodium hydroxide, potassium hydroxide, and barium hydroxide. Unlike sulfuric acid, both the hydroxide ions of barium hydroxide are 100% ionized.

To calculate the pH of any strong base simply assume the molarity of the hydroxide is the same as the molarity of the strong bases sodium hydroxide and potassium hydroxide. (The molarity of the hydroxide ion is twice the molarity of barium hydroxide.) Then calculate the pOH. Finally use **8-3** to calculate the pH. As for strong acids, the correct number of significant figures should be used. The following examples will illustrate the calculation of pH from a given concentration and the calculation of the [OH⁻] from a given pH.

EXAMPLE 8-3. Calculate the pH of a $0.10M$ $Ba(OH)_2$ solution.

Solution: Assume that $0.010M$ barium hydroxide is 100% ionized to $0.020M$ OH^- and $0.010M$ Ba^{+2}. Write the $[OH^-]$ in exponential form.

$$[OH^-] = 0.020M = 2.0 \times 10^{-2}M$$

Calculate the negative log of the $[OH^-]$ using two significant figures to the right of the decimal point in the pOH.

$$pOH = -\log(2.0 \times 10^{-2}M) = 2 - 0.3010 = 1.70 \text{ or } 1.69_9$$

Because the two is an exact number it has no effect on the number of significant figures in the pOH. The log of 2.0 was given to four digits because it was taken from the four-place log table in Appendix 4. The correct number of significant figures was obtained by rounding off after subtraction. Now **8-3** is used to find the pH.

$$pH = pK_w - pOH = 14.00 - 1.70 = 12.30$$

EXAMPLE 8-4. The pH of a certain blood sample at 37°C is 7.4. Calculate (a) the $[H^+]$ of this blood sample and (b) the $[OH^-]$ of this sample.

Solution to a: Substitute the pH value of 7.4 into the definition for pH and solve for the $[H^+]$.

$$pH = 7.4 = -\log[H^+]$$
$$-7.4 = \log[H^+]$$
$$+0.6 - 8 = \log[H^+]$$

Now take the antilog of both sides of the equation.

$$4 \times 10^{-8}M = [H^+]$$

Solution to b: Use **8-3** to calculate the pOH at 37° from the pH.

$$pOH_{37°C} = pK_{w\,37°C} - pH$$
$$pOH_{37°C} = 13.60 - 7.4 = 6.8$$
$$pOH = 6.8 = -\log[OH^-]$$
$$+0.2 - 7 = -\log[OH^-]$$
$$1._6 \times 10^{-7} = [OH^-]$$

Calculations for Weak Acids

Most weak acids ionize 1% or less in contrast to strong acids. The definition of ionization constants (K_a) for weak monoprotic and polyprotic acids was given in Section 7-1. The values for a number of weak acids are found in the appendix. Some references also give the pK_a values; therefore you should be able to calculate the K_a from the pK_a. This is done by simply solving for the K_a from the definition of the pK_a ($= -\log K_a$). For example, the pK_a of acetic acid is 4.74. The K_a is found as follows:

$$pK_a = 4.74 = -\log K_a$$
$$-4.74 = \log K_a$$
$$+0.26 - 5 = \log K_a$$
$$1.8 \times 10^{-5} = K_a$$

To calculate the pH of a solution of a pure weak acid, HA, simply write the definition of the ionization constant.

$$K_a = \frac{[H^+][A^-]}{[HA]} \tag{8-6}$$

and assume that $[H^+] = [A^-]$. To use the approximate method assume that [HA] is approximately equal to the initial molarity of HA. This means that the ionization of HA to H^+ and A^- is neglected. To calculate the approximate $[H^+]$ use the equation

$$[H^+] \cong \sqrt{M_{HA}(K_a)} \tag{8-7}$$

where M_{HA} is the initial molarity of HA in the solution before ionization. To check this approximation calculate the per cent error using the following equation.

$$\% \text{ error} = \frac{[H^+]}{M_{HA}}(100) \tag{8-8}$$

If the per cent error is less than 1%, the approximate $[H^+]$ is valid to two or less significant figures. The following examples will illustrate the use of **8-6** and **8-8**.

EXAMPLE 8-5. The pain-killing ingredient in aspirin tablets is acetylsalicylic acid (ASA), $C_8H_7O_2CO_2H$. This is a weak monoprotic acid with $K_a = 3.3 \times 10^{-4}$. Assume two tablets, each containing 315 mg of ASA, are dissolved in 100 ml of aqueous body fluid. (a) Calculate the pH of this solution. (b) If this fluid were the stomach gastric fluid, would the ASA make the stomach significantly more acid?

Solution to a: First calculate the molarity of ASA using a formula weight of 180.

$$M = \frac{\text{mmoles ASA}}{100 \text{ ml}} = \frac{630 \text{ mg ASA}/180 \text{ mg/mmole}}{100 \text{ ml}} = 0.0350M \text{ ASA}$$

Next, use the initial molarity of 0.0350M ASA to set up an *exact* equation for calculation of the $[H^+]$. Assume that $[H^+] = [C_8H_7O_2CO_2^-]$, the equilibrium concentration of the anion of ASA. Equation **8-6** becomes

$$K_a = \frac{[H^+][H^+]}{0.0350M - [H^+]}$$

where the terms in the denominator represent the initial amount of ASA minus the amount ionized, or the amount of un-ionized ASA at equilibrium.

Then we neglect the ionization of ASA to $[H^+]$ and obtain the *approximate* equation

$$K_a \cong \frac{[H^+]^2}{0.0350M}$$

Substituting the value of K_a and rearranging we obtain

$$H^+ \cong \sqrt{0.0350M(3.3 \times 10^{-4})} \cong 3.4 \times 10^{-3}M$$

(To calculate a square root using logarithms, see Appendix 3.) Using **8-8** the per cent error is calculated.

$$\% \text{ error} = \frac{3.4 \times 10^{-3}M}{0.035M \text{ ASA}} (100) = 9.7\%$$

Since the per cent error is more than 1% we cannot report the $[H^+]$ to two significant figures. However, the per cent error is less than 10% and we can estimate the $H[^+]$ to one significant figure.

$$[H^+] = 3._4 \times 10^{-3}M$$

$$pH = 2.4_7 \text{ or } 2.5$$

Solution to b: Since the $[H^+]$ of the stomach is about $2 \times 10^{-2}M$, the $3 \times 10^{-3}M$ hydrogen ion produced by ionization of ASA will not significantly affect the H^+ of the stomach. In fact, the acid in the stomach gastric fluid will repress the ionization of ASA to its ions.

In addition to organic acids like ASA, certain types of nitrogen salts can act as acids in the Bronsted sense. Examples are $NH_4^+Cl^-$, ammonium chloride; $NH_3OH^+Cl^-$, hydroxylammonium chloride; and $CH_3NH_3^+Cl^-$, methylammonium chloride. The K_as for these acids are calculated from the K_b of the corresponding bases: NH_3, ammonia; NH_2OH, hydroxylamine; and CH_3NH_2, methylamine, respectively.

$$K_a = \frac{K_w}{K_b} \tag{8-9}$$

The following example will illustrate the calculation of the $[H^+]$ of such acids.

EXAMPLE 8-6. The blood plasma can contain as high as 0.040M ammonium ion. Assuming there are no other acids or bases present, calculate the pH of such a solution, given that the pK_b of the conjugate base NH_3 is 4.74.

Solution: First calculate the K_b of NH_3 from the pK_b in a similar fashion to the method used for acetic acid at the beginning of the discussion of weak acids.

$$pK_b = 4.74 = -\log K_b$$
$$+0.26 - 5 = \log K_b$$
$$1.8 \times 10^{-5} = K_b$$

Now use **8-9** to calculate the K_a of NH_4^+.

$$K_a = \frac{1.00 \times 10^{-14}}{1.8 \times 10^{-5}} = 5.5_5 \times 10^{-10}$$

Write the ionization of NH_4^+.

$$NH_4^+ = H^+ + NH_3$$

The ionization constant for this acid is defined as follows:

$$K_a = 5.5_5 \times 10^{-10} = \frac{[H^+][NH_3]}{[NH_4^+]}$$

This is similar to the general definition for any K_a given in **8-6** except that NH_3 is uncharged rather than being an anion. Therefore **8-7** can be used to calculate the approximate $[H^+]$. Substituting the value of $0.040M$ for initial concentration, M_{HA}, and the value of $5.5_5 \times 10^{-10}$ for K_a in **8-7** gives

$$[H^+] \cong \sqrt{0.040M\ NH_4^+\ (5.5_5 \times 10^{-10})} \cong 4.7_1 \times 10^{-6} M$$

Using **8-8** the per cent error is calculated.

$$\%\ \text{error} = \frac{4.7_1 \times 10^{-6} M}{0.040M\ NH_4^+} (100) = 0.01\%$$

Since the per cent error is less than 1% we can report the $[H^+]$ to two significant figures.

$$[H^+] = 4.7_1 \times 10^{-6} M\ \text{or}\ 4.7 \times 10^{-6} M$$
$$pH = 5.32_7\ \text{or}\ 5.33$$

Calculations for Weak Bases

Most weak bases ionize 1% or less in contrast to strong bases. The definition of the ionization constant (K_b) for weak bases was given in Section 7-1 and typical values are given in the appendix. The pK_b value is also listed in some references; this can be converted to a K_b value in the same manner as was done in Example 8-6.

To calculate the pH or pOH of a pure weak base, B, first write its ionization reaction with water.

$$B + H_2O = BH^+ + OH^-$$

SEC. 8-2   Calculation of pH of Strong and Weak Acids and Bases

Then write the definition of the ionization constant.

$$K_b = \frac{[BH^+][OH^-]}{[B]} \tag{8-10}$$

To use an approximate method for calculating pH first assume that $[BH^+] = [OH^-]$. Then assume that [B] is approximately equal to the initial molarity of B. This means that we neglect the ionization of B to BH^+ and OH^-. To calculate the approximate $[OH^-]$ use the equation

$$OH^- \cong \sqrt{M_B(K_b)} \tag{8-11}$$

where M_B is the initial molarity of B in the solution before ionization. To check this approximation calculate the per cent error as follows:

$$\% \text{ error} = \frac{[OH^-]}{M_B}(100) \tag{8-12}$$

If the per cent error is less than 1%, the approximate $[OH^-]$ is valid to two or less significant figures. A per cent error of 1%–10% implies that the $[OH^-]$ may be estimated to only one significant figure. The following examples will illustrate the use of **8-11** and **8-12**.

EXAMPLE 8-7. Calculate the pH of a 0.10M solution of ammonia. Ammonia is a weak base whose $K_b = 1.8 \times 10^{-5}$ and whose ionization may be written as

$$NH_3 + H_2O = NH_4^+ + OH^- \tag{8-13}$$

or as

$$NH_4OH = NH_4^+ + OH^- \tag{8-14}$$

Solution: Using **8-13** to represent the ionization of ammonia we define K_b in a manner similar to **8-10**.

$$K_b = \frac{[NH_4^+][OH^-]}{[NH_3]} \tag{8-15}$$

Next assume that $[NH_4^+] = [OH^-]$ and substitute this equality into **8-15**:

$$K_b = \frac{[OH^-][OH^-]}{0.10M - [OH^-]} \tag{8-16}$$

where the terms in the denominator represent the initial amount of ammonia minus the amount ionized, or the amount of un-ionized ammonia at equilibrium.

Then we neglect the ionization of ammonia to OH^- and obtain from **8-16** the approximate equation

$$K_b \cong \frac{[OH^-]^2}{0.10M}$$

Substituting the value of K_b and rearranging, we obtain

$$[OH^-] \cong \sqrt{0.10M(1.8 \times 10^{-5})} \cong 1.3_4 \times 10^{-3} M$$

Using **8-12** the per cent error is calculated as follows:

$$\% \text{ error} = \frac{1.3 \times 10^{-3} M}{0.10 M} (100) = 1.3\%$$

Since the per cent error is essentially 1% we can report the [OH$^-$] to two significant figures.

$$[OH^-] = 1.3 \times 10^{-3} M$$
$$pOH = 2.87$$
$$pH = 14.00 - 2.87 = 11.13$$

In addition to organic nitrogen bases and ammonia, certain types of anions can act as bases in the Bronsted sense. These are the anions of weak acids such as the acetate ion, the fluoride ion, the anion of ASA, and the benzoate ion, $C_6H_5CO_2^-$ (the anion of benzoic acid). The K_bs for these bases are calculated from the K_a of the corresponding acids: acetic acid, hydrofluoric acid, ASA, and benzoic acid, respectively.

$$K_b = \frac{K_w}{K_a} \tag{8-17}$$

EXAMPLE 8-8. Calculate the pH of a 0.10M solution of sodium acetate, $CH_3CO_2^-Na^+$. The K_a of acetic acid is 1.8×10^{-5}.

Solution: First calculate K_b of sodium acetate using **8-17**.

$$K_b = \frac{1.00 \times 10^{-14}}{1.8 \times 10^{-5}} = 5.5_5 \times 10^{-10}$$

Write the ionization of sodium acetate omitting the sodium ion since it does not affect the pH.

$$CH_3CO_2^- + H_2O = OH^- + CH_3CO_2H$$

The ionization constant for this base is defined as

$$K_b = 5.5_5 \times 10^{-10} = \frac{[OH^-][CH_3CO_2H]}{[CH_3CO_2^-]}$$

This is similar to the general definition for any K_b given in **8-15** except that the acetate ion is charged. Therefore **8-11** can be used to calculate the approximate [OH$^-$]. Substituting the value of 0.10M for the initial concentration, M_B, and the value of $5.5_5 \times 10^{-10}$ for K_b in **8-15** gives

$$[OH^-] \cong \sqrt{0.10M \ CH_3CO_2^- (5.5_5 \times 10^{-10})} \cong 7.4_5 \times 10^{-6} M$$

Using **8-12** the per cent error is calculated as follows:

$$\% \text{ error} = \frac{7.5 \times 10^{-6} M}{0.10 M} (100) = 7.5 \times 10^{-3} \%$$

SEC. 8-3   Mixtures and Simple Buffers   155

Since the per cent error is less than 1% we can report the [OH$^-$] to two significant figures.

$$[OH^-] = 7.4_5 \times 10^{-6} M$$

$$pOH = 5.13$$

$$pH = 8.87$$

Self-Test 8-1. Strong and Weak Acids and Bases

Directions: Check your answers with those given at the end of the self-test.

A. Calculate the pH of the following strong acids or bases.
   a. Gastric fluid containing 0.025M HCl
   b. Gastric fluid containing 0.0250M HCl
   c. A 0.0010M Ba(OH)$_2$ solution
B. Calculate the pH of 0.100M sulfuric acid ($K_2 = 1.0 \times 10^{-2}$)
   a. To one significant figure.
   b. To three significant figures.
C. Calculate the K_a given
   a. A pK_a of 7.93 for hemoglobin (H-Hb).
   b. A pK_a of 6.68 for oxyhemoglobin (H-HbO$_2$).
D. Calculate the pH of each of the solutions below to see whether they will make blood more acid or more basic; also calculate the % error.
   a. A 0.0010M hemoglobin (H-Hb) solution (pK_a = 7.93)
   b. A 0.0010M oxyhemoglobin (H-HbO$_2$) solution (pK_a = 6.68)

Answers to Self-Test 8-1

A. a. 1.60   b. 1.602   c. 11.30
B. a. 0.1M b. 0.108M (see Ex. 8-2)
C. a. 1.1$_7 \times 10^{-8}$ b. 2.1 × 10^{-7}
D. a. 5.46$_5$ (% error in approximation = 0.34%)   b. 4.99 (% error in approximation = 1%)

8-3 MIXTURES AND SIMPLE BUFFERS

In the preceding section we have discussed the calculation of pH of solutions of *pure* acids and *pure* bases. No mixtures were treated. However, a mixture of a weak acid and its anion, or a mixture of a weak base and its cation, are very common occurrences. Such mixtures are called *conjugate acid-base pairs* according to the Bronsted acid-base theory. They are made by mixing an acid with a salt containing a common anion or by mixing a base with a salt containing a common cation. Frequently mixtures of such pairs are used as *buffers* to control the pH. We will first discuss the calculation of the pH of such mixtures and then their use as buffers.

Calculation of pH

First, you should recognize that a mixture of an acid and its conjugate anion is not as acidic as a solution of the pure acid alone. This may be understood by considering the ionization equilibrium of a general weak acid HA.

$$HA = H^+ + A^-$$

$$\underset{\text{equilibrium}}{\text{point of}} \xleftarrow{} \underset{\text{anion}}{\text{excess}}$$

In a solution of a pure weak acid at equilibrium the anion and the hydrogen ion are present in nearly equal concentrations. In a mixture of acid plus its conjugate anion the large excess of the anion displaces the point of the equilibrium farther to the left. This lowers the hydrogen ion concentration as compared to a solution of the pure weak acid alone. For actual comparison the $[H^+]$ of a $0.10M$ solution of acetic acid alone is $1.3 \times 10^{-3}M$ whereas the $[H^+]$ of an equimolar mixture of acetic acid and the acetate ion is $1.8 \times 10^{-5}M$.

Similarly, a mixture of a base and its conjugate cation is not as basic as a solution of the pure base alone. Consider the ionization equilibria of a general weak base B:

$$B + H_2O = B^+ + OH^-$$

$$\underset{\text{equilibrium}}{\text{point of}} \xleftarrow{} \underset{\text{cation}}{\text{excess}}$$

In a solution of a pure weak base at equilibrium the cation and the hydroxide ion are present in nearly equal concentrations. In a mixture of the base plus its conjugate cation the large excess of the cation displaces the point of equilibrium farther to the left. This lowers the hydroxide ion concentration as compared to a solution of the pure weak base alone. For an actual comparison the $[OH^-]$ of a $0.10M$ solution of ammonia alone is $1.3 \times 10^{-3}M$, whereas the $[OH^-]$ of an equimolar mixture of ammonia and ammonium chloride is $1.8 \times 10^{-5}M$.

The pH of Acid-Anion Mixtures. To calculate the pH of a mixture of a weak acid, HA, and a salt, NaA, containing its conjugate anion A^- we start by writing the definition of the ionization constant.

$$K_a = \frac{[H^+][A^-]}{[HA]} \qquad (8\text{-}18)$$

SEC. 8-3 Mixtures and Simple Buffers

To make an exact calculation we recognize that there are two sources of A^-: the salt NaA and the HA molecules that ionize to H^+ and A^-. Thus

$$[A^-] = M_{A^-} + [H^+] \tag{8-19}$$

where M_{A^-} is the initial molarity of A^- from the salt NaA. To calculate the pH from **8-18** we substitute the right-hand side of **8-19** for $[A^-]$.

$$K_a = \frac{[H^+](M_{A^-} + [H^+])}{M_{HA} - [H^+]} \tag{8-20}$$

where M_{HA} is again the initial molarity of HA before ionization. To make an approximate calculation we assume that the $[H^+]$ can be neglected in both the numerator and denominator of **8-20** and solve for $[H^+]$.

$$[H^+] \cong K_a \frac{M_{HA}}{M_{A^-}} \cong K_a \frac{\text{mmoles HA}}{\text{mmoles A}^-} \tag{8-21}$$

Note that because **8-21** involves a ratio of the molarity of HA to the molarity of A^-, the volume terms in a molarity definition such as mmoles/ml drop out of the ratio. Thus the pH can be calculated without knowing the volume of the mixture.

To check the approximations in **8-21** calculate the per cent error using the following equation,

$$\% \text{ error} = \frac{[H^+]}{M}(100) \tag{8-22}$$

where M is the molarity of either HA or the salt NaA. If the per cent error is less than 1%, the approximate $[H^+]$ is valid to two or less significant figures. If it is between 1% and 10%, the approximate $[H^+]$ may be estimated to one significant figure. The following example will illustrate all of the above concepts.

EXAMPLE 8-9. Calculate the pH of a solution prepared by mixing 20 ml of $0.30M$ acetic acid and 10 ml of $0.30M$ sodium acetate, $CH_3CO_2^-Na^+$. The K_a of acetic acid is 1.8×10^{-5}.

Solution: First calculate the molarity of the acetic acid and sodium acetate.

$$M_{CH_3CO_2H} = \frac{\text{mmoles}}{\text{final vol}} = \frac{(20 \text{ ml})(0.30M)}{20 + 10 \text{ ml}} = \frac{6.0 \text{ mmoles}}{30 \text{ ml}} = 0.20M$$

$$M_{CH_3CO_2^-Na^+} = \frac{\text{mmoles}}{\text{final vol}} = \frac{(10 \text{ ml})(0.30M)}{20 + 10 \text{ ml}} = \frac{3.0 \text{ mmoles}}{30 \text{ ml}} = 0.10M$$

Next substitute these numbers into approximate **8-21**.

$$[H^+] \cong 1.8 \times 10^{-5} \frac{6.0 \text{ mmoles } CH_3CO_2H/30 \text{ ml}}{3.0 \text{ mmoles } CH_3CO_2^-Na^+/30 \text{ ml}} \cong 3.6 \times 10^{-5} M$$

Note that 30 ml final volume appears in both the numerator and denominator and does not need to appear in the calculation. To check the approximation the per cent error is calculated using **8-22** twice.

$$\% \text{ error} = \frac{3.6 \times 10^{-5} M}{0.20 M \text{ CH}_3\text{CO}_2\text{H}} (100) = 1.8 \times 10^{-2} \%$$

$$\% \text{ error} = \frac{3.6 \times 10^{-5} M}{0.10 M \text{ CH}_3\text{CO}_2^-\text{Na}^+} (100) = 3.6 \times 10^{-2} \%$$

Since the per cent error in either case is less than 1% we can report the $[\text{H}^+]$ to two significant figures.

$$[\text{H}^+] = 3.6 \times 10^{-5} M$$

$$\text{pH} = 4.44$$

The pH of Base-Cation Mixtures. To calculate the pH of a mixture of a weak base, B, and a salt, BH^+Cl^-, containing its conjugate cation BH^+ we start by writing the definition of the ionization constant.

$$K_b = \frac{[\text{BH}^+][\text{OH}^-]}{[\text{B}]} \tag{8-23}$$

To make an exact calculation you should see that there are two sources of BH^+ — the salt BH^+Cl^- and the molecules of B that ionize to BH^+ and OH^-. Thus

$$[\text{BH}^+] = M_{\text{BH}^+} + [\text{OH}^-] \tag{8-24}$$

where M_{BH^+} is the initial molarity of BH^+ from the salt BHCl. To calculate the pH from **8-23** we substitute the right-hand side of **8-24** for $[\text{BH}^+]$.

$$K_a = \frac{(M_{\text{BH}^+} + [\text{OH}^-])[\text{OH}^-]}{M_B - [\text{OH}^-]} \tag{8-25}$$

where M_B is again the initial molarity of B before ionization. To make an approximate calculation we assume that the $[\text{OH}^-]$ can be neglected in both terms in the numerator and denominator of **8-25** and solve for $[\text{OH}^-]$.

$$[\text{OH}^-] \cong K_b \frac{M_B}{M_{\text{BH}^+}} \cong K_b \frac{\text{mmoles B}}{\text{mmoles BH}^+} \tag{8-26}$$

Note that the volume terms in a molarity definition such as mmoles/ml drop out of the ratio because the volumes are the same.

To check the approximations in **8-26** calculate the per cent error using the following equation,

$$\% \text{ error} = \frac{[\text{OH}^-]}{M} (100) \tag{8-27}$$

where M is the molarity of either B or the salt BHCl. If the per cent error is less than 1%, the approximate $[\text{OH}^-]$ is valid to two or less significant figures. The following example will illustrate the above.

Mixtures and Simple Buffers

EXAMPLE 8-10. Calculate the pH of a solution prepared by mixing 5.0 ml of 0.50M ammonia with 20 ml of 0.25M ammonium chloride. The K_b of ammonia is 1.8×10^{-5}.

Solution: First calculate the molarity of the ammonia and ammonium ion.

$$M_{NH_3} = \frac{\text{mmoles}}{\text{final vol}} = \frac{(5.0 \text{ ml})(0.50M)}{20+5 \text{ ml}} = \frac{2.5 \text{ mmoles}}{25 \text{ ml}} = 0.10M$$

$$M_{NH_4^+} = \frac{\text{mmoles}}{\text{final vol}} = \frac{(20 \text{ ml})(0.25M)}{20+5 \text{ ml}} = \frac{5.0 \text{ mmoles}}{25 \text{ ml}} = 0.20M$$

Next substitute these numbers into approximate **8-26**.

$$[OH^-] \cong 1.8 \times 10^{-5} \frac{2.5 \text{ mmoles NH}_3/25 \text{ ml}}{5.0 \text{ mmoles NH}_4^+/25 \text{ ml}} \cong 9.0 \times 10^{-6} M$$

Note that the 25 ml final volume appears in both the numerator and denominator and does not need to be used in the calculation. To check the approximation the per cent error is calculated using **8-27** twice.

$$\% \text{ error} = \frac{9.0 \times 10^{-6} M}{0.1M \text{ NH}_3}(100) = 9.0 \times 10^{-3} \%$$

$$\% \text{ error} = \frac{9.0 \times 10^{-6} M}{0.20M \text{ NH}_4^+}(100) = 4.5 \times 10^{-3} \%$$

Since the per cent error in either case is less than 1% we can report the $[OH^-]$ to two significant figures.

$$[OH^-] = 9.0 \times 10^{-6} M$$
$$pOH = 5.05$$
$$pH = 8.95$$

An interesting observation can be made after making another calculation for the ammonia-ammonium ion system of Example 8-10. If we calculate the pH of a mixture of 5.0 mmoles of ammonia and 5.0 mmoles of ammonium chloride, we obtain a pH of 9.26. Thus even though the amount of weak base has *doubled*, the pH has increased only about 0.3 pH units! The same small change in pH is also noted with similar changes in the amount of weak acid from the conditions of Example 8-9. The ability of mixtures of conjugate acids and conjugate bases to control the change in pH has led to their use as *buffers*; this will be discussed below.

Buffers

A buffer is a mixture of a conjugate acid and conjugate base that will resist a change in pH from:

1. Addition of a strong or weak acid,
2. Addition of a strong or weak base, or
3. Dilution with pure water.

The buffer action is the result of the conversion of all other acids or bases to two species in equilibrium with each other; for example,

$$HA = A^- + H^+ \qquad (8\text{-}28)$$
$$\text{(acid)} \quad \text{(base)}$$

$$B + H_2O = BH^+ + OH^- \qquad (8\text{-}29)$$
$$\text{(base)} \qquad \text{(acid)}$$

The $[H^+]$ or $[OH^-]$ is thus controlled by the relative amounts of acid and base present in the equilibrium.

Note that a buffer *resists* a pH change; it does *not* keep the pH constant to two significant figures to the right of the decimal point. Generally a buffer only maintains a pH constant to the digit(s) to the left of the decimal point; it cannot prevent small pH changes of 0.01–0.3 units. For example, buffers in the blood maintain the pH within 7.3 to 7.5, but do not hold the pH at exactly pH 7.40.

Simple buffers consist of a *relatively high* concentration of a weak acid, such as acetic acid, and its conjugate base, the acetate ion, or *high* concentrations of a weak base, such as ammonia, and its conjugate acid, the ammonium ion. The buffering action destroys added acid, producing more of the conjugate weak acid and leaving less of the conjugate base. Only two species remain after buffering action occurs, and a *single equilibrium*, such as that in **8-28** or **8-29**, *is in control of the pH*. The buffer must have a relatively high concentration of both the conjugate acid and conjugate base so that neither is used up by the added acid or base. If either is used up, the pH will be affected by more than one equilibrium.

The addition of a strong acid to a buffer mixture of the type in **8-28** results in the reaction,

$$H^+ \quad + \quad A^- \quad \rightarrow \quad HA$$
$$\text{strong acid} \quad \text{buffer base} \quad \text{buffer acid}$$

Addition of a strong base to the same buffer mixture involves this reaction,

$$OH^- \quad + \quad HA \quad \rightarrow \quad A^-$$
$$\text{strong base} \quad \text{buffer acid} \quad \text{buffer base}$$

Note that in each case the strong acid or base is neutralized completely, producing a weak acid or base which is part of the single equilibrium in **8-28**. The addition of strong acid or base to a buffer mixture of the type in **8-29** involves the same type of reactions except that B is the buffer base and BH^+ is the buffer acid.

Calculation of pH Changes During Buffering. In making calculations of the pH of a buffer mixture you should use the K_a or K_b equilibrium expression of the weak acid or weak base of the buffer. Begin by calculating the total mmoles of strong acid or strong base added. Then calculate the total mmoles of buffer acid and buffer base. Then subtract the mmoles of strong acid (or base) from the mmoles

SEC. 8-3 Mixtures and Simple Buffers 161

of the buffer base (or acid). The reaction will also produce an equivalent amount of the buffer acid (from a strong acid) or the buffer base (from a strong base), and this must be added to the amount of the buffer acid, or buffer base, present at the start. After these calculations are finished the mmoles of buffer acid and buffer base are substituted into the K_a or K_b equilibrium expression to calculate the $[H^+]$.

EXAMPLE 8-11. A $0.30M$ acetic acid-$0.30M$ sodium acetate buffer has a pH of 4.74. Calculate the change in pH when 10 ml of $0.050M$ sulfuric acid is added to 40 ml of the buffer. The K_a is 1.8×10^{-5}.

Solution: The $0.050M$ sulfuric acid furnishes $0.10M$ H^+. The number of mmoles of H^+ (total acid) added is

$$(10 \text{ ml})(0.10M \text{ H}^+) = 1.0 \text{ mmoles H}^+$$

The number of mmoles of acetic acid and sodium acetate present at the start in 40 ml of the buffer is

$$(40 \text{ ml})(0.30M) = 12 \text{ mmoles each of } CH_3CO_2H \text{ and } CH_3CO_2^-Na^+$$

Next write the reaction between the buffer base and the strong acid (H^+), and underneath write the amount of each at the start, the amount of change in each, and the amount of each present at equilibrium.

	$CH_3CO_2^-$	+	H^+	$\rightarrow$	CH_3CO_2H
Start:	12 mmoles		1.0 mmole		12 mmoles
Change:	-1.0 mmole		-1.0 mmole		$+1.0$ mmole
Equilibrium:	11 mmoles		0 mmole		13 mmoles

Next substitute these numbers into the approximate equilibrium constant expression (**8-21**).

$$[H^+] \cong 1.8 \times 10^{-5} \frac{13 \text{ mmoles } CH_3CO_2H/50 \text{ ml}}{11 \text{ mmoles } CH_3CO_2^-/50 \text{ ml}} \cong 2.1 \times 10^{-5} M$$

Note that the final volume of 50 ml appears in both the numerator and denominator and drops out of the expression. To check the approximation the per cent error is calculated using **8-22**. Since the largest error will involve sodium acetate only that error will be calculated.

$$\% \text{ error} = \frac{2.1 \times 10^{-5} M \text{ H}^+}{0.22M \text{ CH}_3\text{CO}_2^-\text{Na}^+}(100) = 9.5 \times 10^{-3}\%$$

Since the per cent error is less than 1% we can report the $[H^+]$ to two significant figures.

$$[H^+] = 2.1 \times 10^{-5} M$$

$$pH = 4.68$$

Choosing the Proper Buffer

To choose the proper buffer follow the general rule:

> For optimum buffer action use a buffer pair in which the pK_a of the *buffer acid* is as close as possible to the specified pH.

The reason this rule is valid is that a buffer has its maximum effect when the ratio of the amount of buffer acid to that of buffer base is 1:1. Consider the following logarithmic equation relating pH to the concentrations of buffer acid and buffer base.

$$pH = pK_a - \log \frac{M \text{ of HA}}{M \text{ of A}^-} \tag{8-30}$$

If the ratio of the molarity of HA to the molarity of A^- is 1:1, then the pK_a of the buffer acid will equal the pH of the buffer pair. You can see by inspection that addition of a given amount of a strong acid or strong base will cause the smallest change in pH for a 1:1 ratio compared to any other ratio, such as 2:1, 1:2, and so on. The following example will illustrate this point.

EXAMPLE 8-12. Addition of 1.0 mmole of strong acid (in 10 ml solution) to 40 ml of 0.30M acetic acid-0.30M sodium acetate buffer results in a change of pH from 4.74 to 4.68, or 0.06 pH unit (see Ex. 8-11). Calculate the change in pH when 1.0 mmole of strong acid in 10 ml is added to 40 ml of a 0.30M acetic acid-0.15M sodium acetate buffer which has an initial pH of 4.44. (See Ex. 8-9 for a similar calculation.)

Solution: First calculate the number of mmoles of acetic acid and sodium acetate.

$$(40 \text{ ml})(0.30M) = 12 \text{ mmoles of } CH_3CO_2H$$

$$(40 \text{ ml})(0.15M) = 6.0 \text{ mmoles of } CH_3CO_2^-Na^+$$

Next write the reaction between the buffer base and the strong acid (H^+). Underneath write the amount of each at the start, the amount of change in each, and the amount of each present at equilibrium.

	$CH_3CO_2^-$	+	H^+	$\rightarrow$	CH_3CO_2H
Start:	6.0 mmoles		1.0 mmole		12 mmoles
Change:	-1.0 mmole		-1.0 mmole		$+1.0$ mmole
Equilibrium:	5.0 mmoles		0 mmole		13 mmoles

Next substitute these numbers into the approximate equilibrium constant expression **(8-21)**.

$$[H^+] = 1.8 \times 10^{-5} \frac{13 \text{ mmoles } CH_3CO_2H/50 \text{ ml}}{5.0 \text{ mmoles } CH_3CO_2^-/50 \text{ ml}} = 4.6_8 \times 10^{-5} M$$

SEC. 8-3   Mixtures and Simple Buffers   163

Checking the approximation by using **8-22** gives a per cent error that is less than 1% so the [H$^+$] can be reported to two significant figures.

$$[H^+] = 4.7 \times 10^{-5} M$$

$$pH = 4.33$$

The pH change from the addition of 1.0 mmole of strong acid is 4.44−4.33, or 0.11 pH unit. This is a *larger change* than the 0.06 pH change when the same amount of strong acid was added to a 0.30M acetic acid-0.3M sodium acetate buffer. The two situations can be summarized as follows:

M acid-M base	Ratio, acid:base	Initial pH	pH of buffer +1 mmole H$^+$	pH change
0.30M-0.30M	1:1	4.74	4.68	−0.06
0.30M-0.15M	2:1	4.44	4.33	−0.11

Calculation of Buffer Acid-Base Ratio. After the buffer acid whose pK_a is closest to the specified pH has been chosen it is usually necessary to calculate the ratio of the buffer acid to the base to achieve the exact value of the specified pH. The pK_a will usually be within ±0.3 unit of the pH but it will seldom be identical to the pH to two digits to the right of the decimal point. The buffer ratio can be calculated by using a rearranged form of **8-30**.

$$\log \frac{M \text{ of HA}}{M \text{ of A}^-} = pK_a - pH \tag{8-31}$$

The following example will illustrate the use of this equation.

EXAMPLE 8-13. It is desired to prepare a buffer for a pH of 3.48. The following acids are available: acetic acid ($K_a = 1.8 \times 10^{-5}$), formic acid ($K_a = 1.70 \times 10^{-4}$), and nitrous acid ($K_a = 5.0 \times 10^{-4}$). Which should be used for optimum buffer action? What ratio of buffer acid to buffer base should be used?

Solution: First calculate the pK_a of each of the acids to see which is closest to the specified pH of 3.48. The pK_as are as follows:

$$\text{acetic acid, } pK_a = 4.74$$

$$\text{formic acid, } pK_a = 3.77$$

$$\text{nitrous acid, } pK_a = 3.30$$

Since nitrous acid is 0.18 unit away from the pH of 3.48 and formic acid is 0.29 unit away, nitrous acid should be chosen.
To calculate the ratio of buffer acid to buffer base use **8-31**.

$$\log \frac{M \text{ of HNO}_2}{M \text{ of Na}^+\text{NO}_2^-} = 3.30 - 3.48 = -0.18$$

Multiplying both sides of the equation by -1 and then inverting the log term to eliminate the negative sign we obtain

$$\log \frac{M \text{ of } Na^+NO_2^-}{M \text{ of } HNO_2} = 0.18$$

$$\frac{M \text{ of } Na^+NO_2^-}{M \text{ of } HNO_2} = \frac{1.5}{1}$$

Thus we need 1.5 times as much of the buffer base, the nitrite ion, as the buffer acid. For example, a $0.15M$ sodium nitrite-$0.10M$ nitrous acid solution would make a suitable buffer provided that this is sufficient to neutralize the strong acid or base to be added.

Although the buffer ratio should be as close as possible to 1:1 there are special situations such as in the blood buffers where a ratio of 20:1 still provides effective buffering. This will be discussed in the next section.

8-4 POLYPROTIC ACIDS AND PHYSIOLOGICAL BUFFERS

In the preceding two sections we have discussed the equilibria of monoprotic acids and simple buffers made from monoprotic acids and their salts. Polyprotic acids exhibit more complicated equilibria and we have waited until the end to discuss some of these. The discussion will not be exhaustive; only a few key differences will be discussed as an *introduction* to the equilibria involved.

Two of the most important polyprotic acids in chemistry and in the body are carbonic acid, H_2CO_3, and phosphoric acid, H_3PO_4. Each of these acids ionizes in steps and each step has an ionization constant with a subscript that refers to the number of the step involved.

Since carbonic acid is really hydrated carbon dioxide, $CO_2(aq)$, its first ionization step can be written either of the following two ways.

$$H_2CO_3 = H^+ + HCO_3^-$$

or

$$CO_2(aq) + H_2O = H^+ + HCO_3^-$$

Note that the products are the same in either case. The equilibrium expression for K_1, the first ionization constant, can therefore be defined in two ways.

$$K_1 = 4.3 \times 10^{-7} = \frac{[H^+][HCO_3^-]}{[H_2CO_3]} = \frac{[H^+][HCO_3^-]}{[CO_2(aq)]}$$

The second step can only be defined in one way since this is the ionization of the bicarbonate ion to hydrogen ion and carbonate ion.

$$K_2 = 4.8 \times 10^{-11} = \frac{[H^+][CO_3^{-2}]}{[HCO_3^-]}$$

Polyprotic Acids and Physiological Buffers

As for most polyprotic acids the value of K_1 is much larger than that of K_2. Phosphoric acid ionizes in three steps which are defined as follows:

$$K_1 = 7.5 \times 10^{-3} = \frac{[H^+][H_2PO_4^-]}{[H_3PO_4]}$$

$$K_2 = 6.2 \times 10^{-8} = \frac{[H^+][HPO_4^{-2}]}{[H_2PO_4^-]}$$

$$K_3 = 4.8 \times 10^{-13} = \frac{[H^+][PO_4^{-3}]}{[HPO_4^{-2}]}$$

Phosphoric acid is more typical of polyprotic acids than carbonic acid is in that the first ionization constant is fairly large, that is, of the order of 10^{-3} to 10^{-2}. The second ionization constant of phosphoric acid is about 10^{-4} smaller than the first, just as is the case for carbonic acid.

Calculation of the pH of a Solution of a Pure Polyprotic Acid

To calculate the pH of a solution containing only a polyprotic acid you should first recognize that K_1 for almost all of the common polyprotic acids is significantly larger than K_2 and K_3. This means that K_2 and K_3 can be neglected when considering the ionization of H_2CO_3 or H_3PO_4. In other words, less than a 1% error is incurred by neglecting the ionization of HCO_3^- in a solution of H_2CO_3 (or by neglecting the ionization of $H_2PO_4^-$ in a solution of H_3PO_4). For a solution of H_2CO_3 alone the following equality can be written.

$$[H^+] = [HCO_3^-] \tag{8-32}$$

Similarly for a solution of phosphoric acid alone this equality is valid:

$$[H^+] = [H_2PO_4^-] \tag{8-33}$$

To calculate the pH of such solutions the procedure for weak monoprotic acid outlined in Section 8-2 is used. The ionization of the polyprotic acid to H^+ and the corresponding anion is neglected, and an equation for the approximate $[H^+]$ analogous to **8-7** is written.

$$[H^+] \cong \sqrt{M_{H_nA}(K_1)} \tag{8-34}$$

where M_{H_nA} is the initial molarity of polyprotic acid H_nA in the solution before ionization. To check this approximation calculate the per cent error using the following equation.

$$\% \text{ error} = \frac{[H^+]}{M_{H_nA}}(100) \tag{8-35}$$

If the per cent error is less than 1%, the approximate $[H^+]$ is valid to two or less significant figures. If the per cent error is less than 10%, the $[H^+]$ may be estimated to one significant figure. The example below will illustrate the use of the above equations.

EXAMPLE 8-14. Rapid breathing changes the carbon dioxide concentration of a hypothetical unbuffered bloodstream in the lungs to 0.010M CO_2 (aq). Assuming no other acidic species are present calculate the pH of this bloodstream.

Solution: First, you should recognize that the predominate ionization is that of carbon dioxide to hydrogen ion and bicarbonate ion.

$$CO_2(aq) + H_2O = H^+ + HCO_3^- \qquad K_1 = 4.3 \times 10^{-7}$$

This is essentially the first ionization step of carbonic acid whose K_1 value is given above. Since K_1 is much larger than K_2 for this acid the equality in **8-32** is valid, and we can use approximate **8-34** to calculate the approximate $[H^+]$.

$$[H^+] \cong \sqrt{0.010M\ CO_2(4.3 \times 10^{-7})} \cong 6.5_6 \times 10^{-5} M$$

Checking the approximation by using **8-35** gives

$$\%\ \text{error} = \frac{6.6 \times 10^{-5} M\ H^+}{0.010M\ CO_2}(100) = 0.66\%$$

Since this is less than 1% error the $[H^+]$ can be expressed to two significant figures.

$$[H^+] = 6.6 \times 10^{-5} M$$

$$pH = 4.18$$

(The pH of normal, buffered blood is between 7.3 and 7.5.)

The above example should not be taken as typical of all polyprotic acids. Many polyprotic acids, such as phosphoric acid, have large K_1 values and the per cent error for most concentrations is more than 1%. In such cases the solution to the quadratic equation must be used to calculate the exact value of the $[H^+]$.

$$[H^+] = \frac{-(K_1) \pm \sqrt{(K_1)^2 - 4(K_1)(M\ \text{of acid})}}{2} \qquad (8\text{-}36)$$

A few polyprotic acids, such as H_2SO_4 and $H_4P_2O_7$, are strong acids with respect to the ionization of the first proton. The pH of such acids is calculated exactly by using the K_2 ionization (see Ex. 8-2).

Calculation of the pH of Acid Salts of Polyprotic Acids

Acid salts of polyprotic acids are *amphiprotic*; that is, they can ionize to give up H^+ or they can act as bases. Examples of such salts are $NaHCO_3$, sodium bicarbonate; KH_2PO_4, potassium dihydrogen phosphate; Na_2HPO_4, sodium hydrogen phosphate; $KHC_8H_4O_4$, potassium acid phthalate; and KHC_2O_4, potassium hydrogen oxalate.

Consider the reactions of a general acid salt, NaHA, the acid salt of the general diprotic acid, H_2A. The anion of this salt can ionize to give up a proton to water.

$$HA^- = H^+ + A^{-2} \qquad (8\text{-}37)$$

SEC. 8-4   Polyprotic Acids and Physiological Buffers

It can also act as a base and remove small amounts of H^+ from the solution.

$$HA^- + H^+ = H_2A \tag{8-38}$$

At the same time water is also ionizing.

$$H_2O = H^+ + OH^- \tag{8-39}$$

Using a lengthy derivation an *exact equation* relating the $[H^+]$ to $[HA^-]$, the molarity of the salt containing the amphiprotic anion, can be obtained.

$$[H^+] = \sqrt{\frac{K_2[HA^-]+K_w}{1+[HA^-]/K_1}} \tag{8-40}$$

Equation **8-40** *can be derived by first writing an equation summing up the contributions to the* $[H^+]$ *from 8-37, 8-39, and 8-38.*

$$[H^+] = H^+ + H^+ - H^+$$

$$(8\text{-}37) \quad (8\text{-}39) \quad (8\text{-}38)$$

Then we substitute equalities from each equation for the H^+ *in each.*

$$[H^+] = [HA^-] + [OH^-] - [H_2A]$$

$$(8\text{-}37) \quad (8\text{-}39) \quad (8\text{-}38)$$

Next we substitute identities for each term from the three corresponding equilibrium expressions.

$$[H^+] = \frac{K_2[HA^-]}{[H^+]} + \frac{K_w}{[H^+]} - \frac{[H^+][HA^-]}{K_1}$$

Rearrangement of the above equation leads to **8-40**.

Exact **8-40** can be simplified by two approximations. In the first we assume that K_w is less than 10% of the $K_2[HA^-]$ term in the numerator and neglect K_w. This is always true for acid salts where K_2 is about 10^{-8}. For salts like $NaHCO_3$ where K_2 is 10^{-11} or smaller it is true for solutions more concentrated than $0.01 M$. Neglecting K_w in **8-40** gives

$$[H^+] \cong \sqrt{\frac{K_2[HA^-]}{1+[HA^-]/K_1}} \tag{8-41}$$

In the second approximation we assume that in the denominator the value of one is less than 10% of the $[HA^-]/K_1$ term. This gives

$$[H^+] \approx \sqrt{\frac{K_2[HA^-]}{[HA^-]/K_1}} \approx \sqrt{K_1K_2} \tag{8-42}$$

The double approximation in **8-42** is valid only if $[HA^-]$, the concentration of the acid salt, is at least five times the value of K_1. This should be checked before using **8-42** by making the following calculation.

$$\frac{[HA^-]}{K_1} \geq \frac{5}{1} \tag{8-43}$$

If **8-43** is true, then the $[HA^-]/K_1$ term in **8-41** is large enough to neglect the value of one with an error of 9.5% or less.

Note that the double approximation in **8-42** implies that the $[H^+]$ and the pH of acid salts *is independent of concentration*. Compared to the other types of acids and bases this appears to be unreasonable. Intuitively, we insist that increasing the concentration of an acid salt should change the $[H^+]$. A look at **8-37** and **8-38** reminds us that acid salts have a double action in solution so their tendency to act as bases counteracts their tendency to act as acids. In fact these salts act as a *type of buffer*. Recall that a buffer resists changes in pH; with that in mind **8-42** is not so unreasonable. The following examples illustrate the above ideas.

EXAMPLE 8-15. Calculate the pH of $0.020M$ and $0.20M$ solutions of $NaHCO_3$, sodium bicarbonate. For H_2CO_3, $K_1 = 4.3 \times 10^{-7}$ and $K_2 = 4.8 \times 10^{-11}$.

Solution: First check to see if the approximation to convert **8-41** to **8-42** is valid by using **8-43**.

$$\text{For } 0.020M \text{ NaHCO}_3: \quad \frac{0.020M \text{ HCO}_3^-}{4.3 \times 10^{-7}} = \frac{4.4 \times 10^4}{1}$$

$$\text{For } 0.20M \text{ NaHCO}_3: \quad \frac{0.20M \text{ HCO}_3^-}{4.3 \times 10^{-7}} = \frac{4.4 \times 10^5}{1}$$

In each case the $[HA^-]/K_1$ term from **8-43** is far greater than 10:1, so the double approximation **8-42** can be used to calculate $[H^+]$.

$$[H^+] \approx \sqrt{(4.3 \times 10^{-7})(4.8 \times 10^{-11})} \approx 4.5_4 \times 10^{-8} M$$

$$\text{pH} = 7.34$$

Since the $[H^+]$ is independent of concentration the pH of $0.020M$ and of $0.20M$ sodium bicarbonate is the same, 7.34.

EXAMPLE 8-16. Calculate the pH of $0.015M$ KH_2PO_4, potassium dihydrogen phosphate. For H_3PO_4, $K_1 = 7.5 \times 10^{-3}$ and $K_2 = 6.2 \times 10^{-8}$.

Solution: First check to see if the approximation needed to convert **8-41** to **8-42** is valid by using **8-43**.

$$\frac{0.015M \text{ H}_2\text{PO}_4^-}{7.5 \times 10^{-3}} = \frac{2}{1}$$

Since the $[HA^-]/K_1$ term is much less than 5:1, **8-42** is not valid, and **8-41** will have to be used.

$$[H^+] \cong \sqrt{\frac{6.2 \times 10^{-8}[0.015M \text{ H}_2\text{PO}_4^-]}{1 + [0.015M \text{ H}_2\text{PO}_4^-]/7.5 \times 10^{-3}}}$$

$$[H^+] \cong \sqrt{\frac{9.3 \times 10^{-10}}{1 + 2.0}} \cong 1.7_6 \times 10^{-5} M$$

$$\text{pH} = 4.75$$

SEC. 8-4   **Polyprotic Acids and Physiological Buffers**   169

Calculation of the [H$^+$] using the double approximation in **8-42** would give $2.1_6 \times 10^{-5} M$, a value which is 23% high.

EXAMPLE 8-17. It is desired to buffer a solution near pH 9.6. Would any of the following acid salts be satisfactory: KH_2PO_4, $NaHCO_3$, or Na_2HPO_4?

Solution: From Examples 8-15 and 8-16 we see that the pH of $NaHCO_3$ is 7.34 and the pH of KH_2PO_4 is 4.75. Since neither of these are satisfactory we will use **8-42** to calculate the approximate pH of Na_2HPO_4. Since this is a -2 anion K_2 and K_3 must be used instead of K_1 and K_2, which are only used for -1 anions.

$$[H^+] \approx \sqrt{(6.2 \times 10^{-8})(4.8 \times 10^{-13})} \approx 1.7_2 \times 10^{-10} M$$

$$pH = 9.76$$

Since the calculated pH is close to pH 9.6 the Na_2HPO_4 can be used as a buffer. Any proposed concentration should be checked using **8-43** to see if **8-42** is valid for that concentration.

Mixtures and Physiological Buffers

Thus far we have discussed solutions in which only one species was present—a polyprotic acid alone or an acid salt of a polyprotic acid. Now we will discuss mixtures of a polyprotic acid and an acid salt or an acid salt and its salt. Examples of such mixtures are H_2CO_3 and $NaHCO_3$ or $NaHCO_3$ and Na_2CO_3. Such mixtures also function as buffers.

Calculation of pH of Mixtures. In a mixture of a general diprotic acid H_2A and acid salt NaHA, the H_2A functions as the acid and the HA$^-$ anion functions as the base. The presence of an excess of the HA$^-$ anion serves to repress the ionization of the diprotic acid.

$$H_2A = H^+ + HA^-$$

point of equilibrium $\longleftarrow$ excess anion

This lowers the hydrogen ion concentration of the solution as compared to a solution of the pure diprotic acid alone. At the same time the solution is more acidic than a solution of the acid salt NaHA alone. The following example will illustrate this.

EXAMPLE 8-18 Compare the pH of a 0.10M solution of H_2CO_3 with that of a mixture of 0.10M H_2CO_3 and 0.10M $NaHCO_3$, and with that of a solution of 0.10M $NaHCO_3$ alone.

Solution: First we calculate the [H$^+$] of 0.10M H_2CO_3 alone using **8-34**.

$$[H^+] = \sqrt{0.10M \ H_2CO_3(4.3 \times 10^{-7})} \cong 2.0_7 \times 10^{-4} M$$

Since **8-35** shows that the per cent error is only 0.2% the $[H^+]$ may be expressed to two significant figures.

$$[H^+] = 2.1 \times 10^{-4} M$$

$$pH = 3.68$$

Next we use approximate **8-21** to calculate the $[H^+]$ of the $0.10M$ H_2CO_3-$0.10M$ $NaHCO_3$ mixture.

$$[H^+] \cong 4.3 \times 10^{-7} \frac{0.10M \; H_2CO_3}{0.10M \; HCO_3^-} \cong 4.3 \times 10^{-7} M$$

Use of **8-22** shows that the error is negligible in this calculation so the $[H^+]$ may be reported to two significant figures.

$$[H^+] = 4.3 \times 10^{-7} M$$

$$pH = 6.37$$

From Example 8-15 we see that a solution of $0.10M$ $NaHCO_3$ has a $[H^+]$ of $4.5 \times 10^{-8} M$ and a pH of 7.34. We can thus summarize the acidities of the three solutions in chart form.

Solution	$[H^+]$	pH
$0.10M \; H_2CO_3$	$2.1 \times 10^{-4} M$	3.68
$0.10M \; H_2CO_3 + 0.10M \; NaHCO_3$	$4.3 \times 10^{-7} M$	6.37
$0.10M \; NaHCO_3$	$4.5 \times 10^{-8} M$	7.34

We can readily see that a solution of a diprotic acid alone is more acidic than a solution of a diprotic acid plus the conjugate acid salt, which in turn is more acidic than a solution of the conjugate acid salt alone.

Mixtures of a general acid salt NaHA and a general salt Na_2A can be calculated in a manner similar to that used in Example 8-18 for mixtures of H_2A and NaHA, except that K_2 must be used.

EXAMPLE 8-19. Calculate the pH of a mixture of $0.10M$ $NaHCO_3$ and $0.10M$ Na_2CO_3.

Solution: Use approximation **8-21** substituting K_2 of H_2CO_3 for the ionization constant.

$$[H^+] \cong 4.8 \times 10^{-11} \frac{0.10M \; HCO_3^-}{0.10M \; CO_3^{-2}} \cong 4.8 \times 10^{-11} M$$

Use of **8-22** gives a per cent error of 4.8×10^{-8} % so the $[H^+]$ can be expressed to two significant figures.

$$[H^+] = 4.8 \times 10^{-11} M$$

$$pH = 10.32$$

Physiological Buffers. The living tissues of the body are very sensitive to changes in the surrounding fluids; in particular, they are very sensitive to pH changes. Because of this buffers exist in body fluids to protect the body against excess acid buildup, or *acidosis*, and against excess base buildup, or *alkalosis*. Buffers are very important both in the blood and in the muscles. A normal adult is said to have sufficient buffer in his blood to neutralize about 0.150 moles (150 mmoles) of hydrogen ion [1]. The buffering bases in the muscles have the capacity to neutralize five times as much acid.

There are three main buffer systems in the blood. They are the $CO_2(aq)/HCO_3^-$ buffer pair, the $H_2PO_4^-/HPO_4^{-2}$ buffer pair, and certain protein pairs including hemoglobins. The hydrated carbon dioxide/bicarbonate ion buffer pair is most important in buffering alveolar (lung) blood; it reacts with acids or bases as follows:

$$HCO_3^- + H^+ \underset{\text{excess base}}{\overset{\text{excess acid}}{\rightleftharpoons}} CO_2(aq) + H_2O$$

buffer base   buffer acid

(Recall that we said at the beginning of this section that H_2CO_3 was really hydrated carbon dioxide, $CO_2(aq)$.)

The $H_2PO_4^-/HPO_4^{-2}$ buffer pair has a pK_a of 7.2 which is quite close to the normal blood pH of 7.4. This pair reacts with acids or bases as follows:

$$HPO_4^{-2} + H^+ \underset{\text{excess base}}{\overset{\text{excess acid}}{\rightleftharpoons}} H_2PO_4^-$$

buffer base   buffer acid

The buffer pairs of the protein function in a similar way.

In Section 7-3 it was stated that the maximum buffer action is obtained when the pK_a of the buffer acid is as close as possible to the specified pH. It is interesting that this is not true for the $CO_2(aq)/HCO_3^-$ buffer pair and yet it functions as an effective buffer. The reason is that the concentration of the bicarbonate ion is much higher in the blood than is that of the hydrated carbon dioxide. The example below will give you a typical situation.

EXAMPLE 8-20. Calculate the buffer ratio for the carbon dioxide/bicarbonate ion buffer system in alveolar (lung) blood whose pH is 7.40.

Solution: Following the method of Example 8-13 we first calculate the pK_a from the K_1 of the carbonic acid system. Several authors [1, 2] use a value of 8.0×10^{-7} for K_1 under body conditions. The pK_a would then be 6.10.

To calculate the ratio of bicarbonate to hydrated carbon dioxide we use a modified form of **8-31**.

$$\log \frac{[HCO_3^-]}{[CO_2(aq)]} = pH - pK_a = 7.40 - 6.10 = 1.30$$

$$\frac{[HCO_3^-]}{[CO_2(aq)]} = \frac{20}{1}$$

This ratio will vary somewhat depending on the actual blood pH.

[1] R. A. Day and A. L Underwood, *Quantitative Analysis*, 3d ed., Prentice-Hall, Englewood Cliffs, N.J., 1974, p. 100.
[2] H. W. Davenport, *The ABC of Acid-Base Chemistry*, University of Chicago Press, Chicago, 1969, pp. 44–49.

The reactions that occur in the alveolar blood as oxygen combines with hemoglobin (HHb) and forms the more acidic oxyhemoglobin (HHb-O_2) are a good illustration of how the 20:1 buffer ratio in Example 8-20 is maintained. First inhaled oxygen combines with hemoglobin in the lung blood to form the more acidic oxyhemoglobin.

$$HHb + O_2 = HHb\text{-}O_2$$

The oxyhemoglobin ionizes, raising the [H^+].

$$HHb\text{-}O_2 = H^+ + Hb\text{-}O_2^-$$

To remove the excess H^+, bicarbonate ion migrates to the region and reacts with the excess H^+.

$$HCO_3^- + H^+ \rightarrow H_2CO_3$$

(A shift in chloride ion away from this region maintains a balance in the positively and negatively charged ions.) To prevent a buildup in carbonic acid, which is really hydrated carbon dioxide, it is rapidly decomposed by the enzyme carbonic anhydrase.

$$H_2CO_3(CO_2\text{-}H_2O) = CO_2(g) + H_2O$$

The gaseous carbon dioxide, $CO_2(g)$, diffuses into the lungs and is exhaled. This keeps the buffer ratio constant, near 20:1 or whatever ratio exists at the pH of the blood.

QUESTIONS AND PROBLEMS

(Answers to most even-numbered problems are in Appendix 5.)

Ionization of Water
1. Explain why water undergoes *self-ionization*.
2. Give a mathematical equation which describes a neutral solution at *any* temperature.
3. Explain why the ionization of water is endothermic.
4. Calculate the pH of a neutral solution at
   a. 37°C.   b. 0°C.   c. 60°C.
5. Construct a pH scale for the temperatures of 0° and 60°C. Locate the pH of $1M\ H^+$, $1M\ OH^-$, neutral solution, and the blood pH of 7.4.

Strong and Weak Acids and Bases
6. Calculate the pH of the following.
   a. Gastric fluid containing $0.020M$ HCl
   b. Gastric fluid (after being mixed with food in the stomach) containing $1.5 \times 10^{-4}M$ HCl
   c. Rationalize the difference between the two pH values above
7. Calculate the pH of the following.
   a. $0.002M$ sodium hydroxide
   b. $0.00200M$ sodium hydroxide
   c. $0.0020M$ barium hydroxide

SEC. 8-4   Polyprotic Acids and Physiological Buffers

8. Calculate the pH of $0.200M$ sulfuric acid ($K_2 = 1.0 \times 10^{-2}$)
   a. To one significant figure.
   b. To three significant figures.
9. Calculate the pH of a $0.10M$ solution of phenol, C_6H_5OH, a weak monoprotic acid with a K_a of 1.4×10^{-10}. Also calculate the per cent error in your approximation. (To find a square root using logs, see Appendix 3.)
10. Calculate the $[OH^-]$ of a $0.100M$ solution of sodium benzoate, $C_6H_5CO_2^-Na^+$. The K_a of benzoic acid is 6.3×10^{-5}.

Mixtures and Simple Buffers
11. Calculate the pH of a solution prepared by mixing 10 ml of $0.30M$ acetic acid and 20 ml of $0.30M$ sodium acetate.
12. Calculate the pH of a solution prepared by mixing 25 ml of $0.30M$ ammonia with 5.0 ml of $0.15M$ ammonium chloride.
13. A $0.20M$ acetic acid-$0.20M$ sodium acetate buffer has a pH of 4.74. Calculate the change in pH when 10 ml of $0.10M$ hydrochloric acid is added to 40 ml of the buffer. The K_a of acetic acid is 1.8×10^{-5}.
14. Choose the proper pair and calculate the ratio of the acid to the base necessary to achieve the following pHs.
    a. 4.60   b. 9.26   c. 9.10
15. According to recent research [R. C. Plumb, *J. Chem. Ed.* **49**, 179 (1972)] pain is produced when cells are ruptured, releasing their acidic buffers into the nerve areas so that the pH of the nerve endings falls below 6.2. What would the $[H^+]$ of a cell solution have to be to cause pain?

Polyprotic Acids
16. Calculate the pH of a solution containing only $0.0050M$ carbon dioxide.
17. Calculate the pH of the following solutions of phosphoric acid.
    a. $5.0M\ H_3PO_4$
    b. $0.10M\ H_3PO_4$
18. Calculate the pH of the following acid salt solutions.
    a. $0.30M\ NaHCO_3$
    b. $0.010M\ KH_2PO_4$
19. Select a suitable buffer for a solution near pH 9.70.

Challenging Problems
20. Calculate the pH of a $1.0 \times 10^{-5}M$ solution of the monoprotic acid phenol whose $K_a = 1.4 \times 10^{-10}$. Be sure your answer is reasonable.
21. Show that the double approximation of **8-42** is valid with exactly 9.5% error if the ratio in **8-43** is exactly 5:1.
22. Suppose that 10 mmoles of hydrogen ion are to be absorbed by the hydrated carbon dioxide/bicarbonate buffer pair with only a 0.5% change in the blood pH of 7.40. What number of mmoles of each would have to be present at the start to limit the change to 0.5%?

9. Acid-Base Titrations

"Well, I gave my mind a thorough rest by plunging into a chemical analysis,"
Holmes said.

ARTHUR CONAN DOYLE
The Sign of the Four, Chapter 10

Acid-base titrations are so convenient and so accurate that scientists in many fields outside of chemistry will not hesitate to employ them, even though they may not "plunge" into them with Holmes' idea of resting the mind! Hundreds of inorganic and organic acids and bases are readily titrated in aqueous solution, and hundreds more are easily titrated in nonaqueous solvents. Both direct and indirect acid-base titration methods are used widely in chemistry, biology, pharmacy, and so on. We will discuss mainly titrations in aqueous solution, but we will devote some space to nonaqueous titrations at the end of the chapter.

The basic concepts of volumetric analysis were described in Section 6-1. A primary standard is needed to standardize a titrant, and the end point in the titration of a given sample is located by means of an indicator or pH meter. Reference will also be made to some of the acid-base equilibria discussed in the previous chapter. We will begin by discussing titrants, then indicators, and finally applications.

9-1 CHOOSING AND STANDARDIZING A TITRANT

In general strong acids or strong bases are used as titrants. Since this choice follows from the concept of quantitative acid-base reactions, we'll discuss this concept before talking about the specific titrants.

Quantitative Acid-Base Reactions in Water

In Section 7-2 the equilibrium constant for a general acid-base reaction was defined as

$$K_{rxn} = \frac{K_a K_b}{K_w} \qquad (9\text{-}1)$$

In theory any titration with a strong acid or base titrant should have a value of infinity for K_{rxn} because K_a for a strong acid or K_b for a strong base approaches infinity (**9-1**). This is much larger than the minimum theoretical value of 10^6 for a quantitative acid-base reaction.

In contrast consider the use of a weak acid with a K_a of 1×10^{-5} as a titrant. It *appears* that such a weak acid will be adequate for the titration of any weak base with a $K_b \geqslant 1 \times 10^{-3}$ since the K_{rxn} will be

$$K_{rxn} = \frac{(1 \times 10^{-5})(\geqslant 1 \times 10^{-3})}{1 \times 10^{-14}} \geqslant 1 \times 10^6$$

However, if such an acid is used to titrate any of the many weak bases with a K_b smaller than 10^{-3}, the K_{rxn} will be less than the theoretical minimum needed for quantitative reaction. For example, for $K_b = 10^{-4}$

$$K_{rxn} = \frac{(1 \times 10^{-5})(1 \times 10^{-4})}{1 \times 10^{-14}} = 1 \times 10^5$$

Thus, strong acids such as hydrochloric acid, nitric acid, or perchloric acid are the only generally useful titrants for all bases in water; similarly, strong bases like sodium hydroxide are the only useful titrants for all acids in water.

Limits to Titrations in Water. It appears from **9-1** that all acid-base titrations involving a strong acid or a strong base are quantitative in water. However it was pointed out in Section 7-2 that the anion of a weak acid or the cation of a weak base does react with water at the end point to a sometimes significant degree. The equilibrium constant, K, for this *back reaction* was defined in **7-18**, and the maximum theoretical value, below which reactions are quantitative in water, was given as 1×10^{-7} (**7-19**) for typical $0.1 M$ samples.

In practice this means that weak acids or bases can be titrated quantitatively in water at the $0.1 M$ level only if K_a or K_b is $\geqslant 1 \times 10^{-7}$. If the K_a or K_b is slightly less than 1×10^{-7}, titration of a more concentrated sample than $0.1 M$ can be tried. If K_a or K_b is significantly less than 1×10^{-7}, then a nonaqueous titration must be used. The reason is that the reaction of water with the anion of the weak acid, A^-, or the cation of the weak base is significant:

$$A^- + H_2O = HA + OH^-$$

The only way to prevent the above reaction is to avoid using water as a solvent and to use a nonaqueous solvent. This will be discussed at the end of the chapter.

Choosing and Standardizing Strong Base Titrants

Sodium hydroxide is the most common strong base titrant. As discussed in Section 6-1 sodium hydroxide is not a primary standard, but a solution of sodium hydroxide, once standardized and protected from CO_2, is a stable titrant. It is usually made from a saturated solution which has been allowed to stand to permit precipitation of sodium carbonate impurity. Once such a solution is diluted it is stable provided it is not exposed to carbon dioxide for too long a time. If it is, the following reaction becomes significant:

$$2OH^- + CO_2 \rightarrow CO_3^{-2} + H_2O$$

This reaction will of course lower the concentration of the sodium hydroxide.

Primary Standard Acids for Sodium Hydroxide. To standardize sodium hydroxide any of a number of primary standard weak acids may be used. They must all meet the requirements for a primary standard (Sec. 6-1). Some primary standard acids are listed below.

Sulfamic acid, $K_a = 1 \times 10^{-1}$
Benzoic acid, $K_a = 6.3 \times 10^{-5}$
Potassium acid phthalate, $K_a = 3.9 \times 10^{-6}$

Sulfamic acid, HSO_3NH_2, is felt by some authors [1] to be the best primary standard for the standardization of strong bases. It is close to being a strong acid, is anhydrous and nonhygroscopic, and is stable up to 130°C. Unfortunately, its formula weight of 97.09 is not as large as is desirable for a primary standard.

Benzoic acid, $C_6H_5CO_2H$, is not nearly as strongly ionized as sulfamic acid but is commonly available. It is stable, anhydrous, and nonhygroscopic. Its formula weight of 122.12 is also not as large as some other primary standards but it is satisfactory. Its low solubility in water (0.3 g/100 ml) and high solubility in organic solvents makes it more suitable as a primary standard for nonaqueous titrations.

Potassium acid phthalate, $KHC_8H_4O_4$ or KHP, is weakly ionized but is generally preferred for standardization of sodium hydroxide. Not only is it stable at 110°C drying, anhydrous, and nonhygroscopic, but it has a large formula weight of 204.22. This of course reduces the effect of weighing errors because it requires that a larger amount be weighed to obtain the usual 4 mmoles of primary standard for standardization of $0.1 M$ sodium hydroxide.

Potassium hydroxide may also be employed as a titrant; it is standardized in the same manner that sodium hydroxide is, using the same primary standards.

Choosing and Standardizing Strong Acid Titrants

In theory hydrochloric acid (HCl), nitric acid (HNO_3), and perchloric acid ($HClO_4$) are all equally useful as strong acid titrants. Nitric acid has the advantage that it forms almost no insoluble salts to obscure end points and the disadvantages that it

[1] H. A. Diehl and G. F. Smith, *Quantitative Analysis*, Wiley, New York, 1952.

oxidizes many species and is frequently photodecomposed to a brown product. For these reasons hydrochloric acid is usually chosen; however, it forms insoluble salts in the presence of silver(I), lead(II), and mercury(I) ions so it would not be used in the presence of these cations. Perchloric acid has none of the above disadvantages and can be used as a $0.1M$ titrant without problems. None of these strong acids are obtained commercially as primary standards, although constant-boiling hydrochloric acid, being of known concentration, can be prepared; all must be standardized using primary standards.

Primary Standard Bases for Acid Titrants. Acid titrants may be standardized using standard sodium hydroxide; this reaction has a sharper end point than any of the titrations involving the primary standards discussed below. The only disadvantage is that any error made in standardizing the sodium hydroxide is incorporated into the standardization of the acid titrant.

The primary standards for strong acid titrants must meet all the requirements for a primary standard (Sec. 6-1). The two common primary standard bases are listed below.

Sodium carbonate, $K_{b_2} = 2.0 \times 10^{-8}$
THAM, Tris(hydroxymethyl)aminomethane, $K_b = 1.26 \times 10^{-6}$

Sodium carbonate, Na_2CO_3, is the classic primary standard; it is available in pure form except for a trace of sodium bicarbonate which is converted to sodium carbonate by drying at 300°C [1]. When it is titrated to carbonic acid its formula weight of 106.00 must be divided by two to obtain an equivalent weight (Sec. 6-2). The end point is not sharp unless the carbonic acid is boiled off as carbon dioxide, a tedious process.

THAM, tris(hydroxymethyl)aminomethane, $(CH_2OH)_3CNH_2$, is now preferred to sodium carbonate, but it has a low formula weight (121.1) and its basicity is not that much larger than sodium carbonate. Thus the end point is fairly sharp but not excellent.

Gravimetric standardization of hydrochloric acid by precipitation as silver chloride (Ch. 5) is of course the most accurate but slowest standardization.

9-2 ACID-BASE INDICATORS

Acid-base indicators function by changing colors just after the equivalence point of a titration; this color change is called the *end point*. The end point is most often detected visually (Sec. 7-3). Most acid-base indicators are organic dye molecules which are either acids or bases. Let's discuss the acid-base nature of indicators before showing how they are used.

Acid-Base Nature of Indicators

Indicators may be monoprotic (HIn) or diprotic (H_2In) acids. The acid form of an indicator is usually colored; when it loses a proton the resulting anion (In^-), or base form of the indicator, exhibits a different color. The two colored forms exist in equilibrium with one another just as any weak monoprotic acid (Sec. 8-2).

$$HIn = H^+ + In^- \quad (9\text{-}2)$$
$$(\text{color } A) \qquad\qquad (\text{color } B)$$

A monoprotic acid indicator has an ionization constant, K_{in}, which is analogous to the K_a of a weak monoprotic acid. The ionization constant expression is

$$K_{in} = \frac{[H^+][In^-]}{[HIn]} \quad (9\text{-}3)$$

A diprotic acid indicator, H_2In, ionizes in two steps giving the HIn^- and In^{-2} forms, each of which may exhibit a different color than the H_2In form.

$$H_2In = H^+ + HIn^- = H^+ + In^{-2} \quad (9\text{-}4)$$
$$(\text{color } A) \qquad (\text{color } B) \qquad (\text{color } C)$$

Such an indicator would have two ionization constants just as a diprotic acid (Sec. 8-4).

A simple example of a monoprotic acid indicator is *p*-nitrophenol, which ionizes to the *p*-nitrophenoxide anion as follows:

$$O_2N\text{-}C_6H_4\text{-}OH = H^+ + O_2N\text{-}C_6H_4\text{-}O^-$$
$$(O_2NC_6H_5OH) \qquad\qquad (O_2NC_6H_5O^-)$$

In dilute solution the acid form of this indicator is colorless; the base form, however, is yellow. The ionization constant for *p*-nitrophenol has a value of 1×10^{-7}.

$$K_{in} = 1 \times 10^{-7} = \frac{[H^+][O_2NC_6H_5O^-]}{[O_2NC_6H_5OH]} \quad (9\text{-}5)$$

This indicator exhibits a *pH transition range* of 6.2 to 7.5. Below pH 6.2 it exists in the colorless acid form; slightly above pH 6.2 a *faint yellow* color develops because of the presence of a significant fraction of the yellow phenoxide anion form. In the middle of this transition range (pH 7.0) a *definite yellow* color is observed. It can be seen by rearranging 9-5, that in the middle of the transition range for *p*-nitrophenol the ratio of the base form to the acid form will be 1:1.

$$\frac{[O_2NC_6H_5O^-]}{[O_2NC_6H_5OH]} = \frac{K_{in}}{[H^+]} = \frac{1 \times 10^{-7}}{1 \times 10^{-7} M\, H^+} = \frac{1}{1}$$

The *definite yellow* color observed at this point is the result of a mixture of equal amounts of the *intense yellow* phenoxide anion form and the *colorless* acid form of *p*-nitrophenol. Finally, above a pH of 7.5 an *intense* yellow color is observed as the indicator is completely converted to the phenoxide anion. These changes can be summarized in the *bar graph* form below.

p-Nitrophenol
pH 6.2–7.5

Colorless	Definite yellow	Intense yellow
All HIn	1:1 HIn/In⁻	All In⁻

pH: 6.2   7.0   7.5

(Middle)

An example of a diprotic acid indicator is the thymol blue molecule; it has two pH transition ranges (Table 9-1). Using the terminology of **9-4** the colors can be ascribed to the following forms.

Red: H_2In form
Yellow: HIn^- form
Blue: In^{-2} form

The color changes in the two pH transition ranges can be described using the two bar graphs below (recall that red *and* yellow appear orange).

Thymol blue
pH 1.2–2.8

Red	Orange	Yellow
All H_2In	1:1 H_2In/HIn^-	All HIn^-

pH: 1.2   2.0   2.8

(Middle)

Thymol blue
pH 8.0–9.6

Yellow	Green	Blue
All HIn^-	1:1 HIn^-/In^{-2}	All In^{-2}

pH: 8.0   8.8   9.6

(Middle)

From the observations for thymol blue and for *p*-nitrophenol, we can make the following generalization for all indicators.

> In the middle of the pH transition range an indicator exhibits a color usually consisting of equal amounts of the color of the acid form and the color of the base form.

TABLE 9-1. Some Acid-Base Indicators

Common Name	Transition Range (pH)	Color Change Acid	Base
Crystal violet	0.1–1.5	yellow	blue
Thymol blue	1.2–2.8	red	yellow
Methyl yellow	2.9–4.0	red	yellow
Methyl orange	3.1–4.4	red	yellow
Bromcresol green	3.8–5.4	yellow	blue
Methyl red	4.2–6.3	red	yellow
Chlorophenol red	4.8–6.4	yellow	red
Bromothymol blue	6.0–7.6	yellow	blue
Phenol red	6.4–8.0	yellow	red
Neutral red	6.8–8.0	red	yellow-orange
Cresol purple	7.4–9.0	yellow	purple
Thymol blue	8.0–9.6	yellow	blue
Phenolphthalein	8.0–9.7	colorless	red
Thymolphthalein	9.3–10.5	colorless	blue
Alizarin yellow	10.1–12.0	colorless	violet

Selecting and Using Indicators

In Section 7-3 the selection of indicators in general was discussed. Essentially, the idea is to choose an indicator that undergoes a distinct color change at the equivalence point (true end point) of an acid-base titration. A color change, rather than the appearance of a particular color or shade of a color, is used because it is easier for the eye to perceive.

After reviewing the above discussion of the acidic nature and color changes of indicators we can readily conclude that the *greatest* color change must occur in the *middle half* of the pH transition range, from 25%–75% color change. This is shown in the bar graph below.

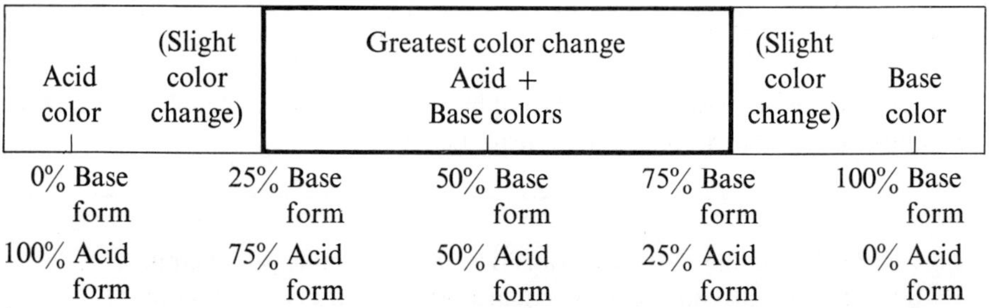

Acid color	(Slight color change)	Greatest color change Acid + Base colors	(Slight color change)	Base color
0% Base form	25% Base form	50% Base form	75% Base form	100% Base form
100% Acid form	75% Acid form	50% Acid form	25% Acid form	0% Acid form

It is in the middle half of the transition range that we observe the inversion in the ratio of the base form to acid form, going from 25% base/75% acid to 75%

base/25% acid. In terms of **9-3** the inversion starts at the following ratio of indicator ionization constant to $[H^+]$,

$$\frac{K_{in}}{[H^+]} = \frac{[In^-]}{[HIn]} = \frac{25\% \text{ Base}}{75\% \text{ Acid}} = \frac{1}{3}$$

and ends at the following ratio.

$$\frac{K_{in}}{[H^+]} = \frac{[In^-]}{[HIn]} = \frac{75\% \text{ Base}}{25\% \text{ Acid}} = \frac{3}{1}$$

This means the $[H^+]$ has decreased to one-ninth, or almost one-tenth, of its value at the start of the middle half of the transition range.

To choose an indicator two things are necessary: we must know the pH transition range of the indicator (Table 9-1), and we must know the pH at the equivalence point, or at the steepest part of the titration curve. We can then use the following generalization.

> Choose a pH indicator whose middle half of the pH transition range encompasses the pH at the equivalence point, or the pH at the steepest part of the titration curve.

Finding the pH at the Equivalence Point. The pH at the equivalence point can either be estimated by calculation or by performing a pH titration and plotting a titration curve. The former method is faster, but the latter method gives a better picture of the pH change. The latter method may show that several indicators may be used.

The estimation of the pH at the equivalence point by calculation depends on the type of reaction involved. For the titration of strong acid with strong base, or vice versa, we saw in Section 7-2 that at the end point

$$[H^+] = [OH^-]$$

$$[H^+] = \sqrt{K_w}$$

$$\text{pH} = \frac{pK_w}{2} = 7.00 \ (25°C) \tag{9-6}$$

EXAMPLE 9-1. Choose an indicator for the titration of 50 ml of 0.10M HCl with 0.10M NaOH.

Solution: Regardless of the concentrations the pH at the end point of a titration of a strong acid with a strong base will always be 7.00 (**9-6**). There are three indicators in Table 9-1 whose transition ranges include pH 7.0; they are, with the *middle half of their transition ranges*

	Middle half of transition range
Bromothymol blue	6.4–7.2
Phenol red	6.8–7.6
Neutral red	7.1–7.7

You can see that both bromothymol blue and phenol red are suitable indicators in that the middle half of their transition ranges includes the pH of 7.0 at the end point. Theoretically, neutral red would be rejected as an indicator because the middle half of its range does not include pH 7.0. A glance at Figure 7-2 will indicate that the experimental pH change at the end point is broad enough to allow neutral red to be used. (Other indicators whose middle half of the transition range does not include pH 7.0 can also be used; see Self-Test 9-1.)

For the titration of a weak acid with a strong base the pH at the end point is controlled entirely by the concentration of the anion of the weak acid. For example, consider the titration of $0.20M$ acetic acid with $0.20M$ sodium hydroxide titrant. The titration reaction is

$$CH_3CO_2H + NaOH \rightarrow CH_3CO_2^- + H_2O + Na^+$$

(acetic acid)　　　　　(acetate ion)

At the end point the acetate ion produced acts as a Bronsted base and reacts with water to make the solution slightly basic.

$$CH_3CO_2^- + H_2O = OH^- + CH_3CO_2H \tag{9-7}$$

All that is needed is to calculate the pH of the solution using the methods in Section 8-2 and to consult Table 9-1.

EXAMPLE 9-2. Choose an indicator for the titration of $0.2M$ acetic acid with $0.2M$ sodium hydroxide titrant.

Solution: At the end point of the titration the concentration of the acetate ion will be one half that of the acetic acid because of dilution. It is now necessary to calculate the [H$^+$] of a $0.1M$ solution of acetate which ionizes according to **9-7**. The K_b of the acetate ion is equal to K_w/K_a of acetic acid, and the calculation is done following the approach in Example 8-8 where the pH was calculated to be 8.87. There are three indicators in Table 9-1 whose transition ranges include pH 8.87; they are, with the middle half of their transition ranges

	Middle half of transition range
Cresol purple	7.8–8.6
Thymol blue	8.4–9.2
Phenolphthalein	8.4–9.2

You can see that cresol purple is unsatisfactory because the pH of 8.87 at the equivalence point falls outside the middle half of its transition range. Both thymol blue and phenolphthalein would be satisfactory because pH 8.87 falls inside the middle half of their transition ranges (see Fig. 9-1).

If a pH titration curve is available, the steepest part of the curve should be used to locate the equivalence point pH, and the middle half of the indicator transition range should include this pH.

EXAMPLE 9-3. Choose an indicator for the titration of 0.2M acetic acid, the pH titration curve of which is shown in Figure 9-1.

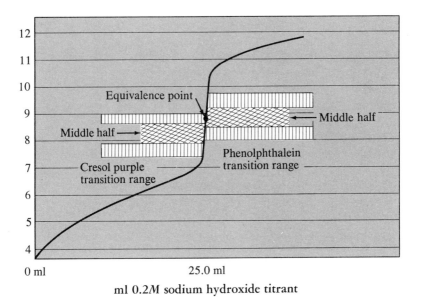

FIGURE 9-1. *Titration curve for 25 ml of 0.2M acetic acid with 0.2M sodium hydroxide.*

Solution: The middle of the steepest part of the titration curve in Figure 9-1 is at approximately pH 8.9 (see dot). The same three indicators considered in Example 9-2 can be considered here: cresol purple (pH 7.4–9.0), thymol blue (pH 8.0–9.6), and phenolphthalein (pH 8.0–9.7). Although the transition range of each does enclose the steepest part of the curve, it can be seen from Figure 9-1 that the middle half of the range for cresol purple does not encompass the steepest part of the titration curve (see dot). The middle half of the range for phenolphthalein and for thymol blue (not shown) does, however. Either of the latter two would be a much better choice for an indicator than cresol purple.

Self-Test 9-1. Choosing Indicators

Directions: Check your answers with those given at the end of the self-test.
A. The indicator methyl red has a pH transition range of 4.2–6.3.
   a. Calculate half of the difference between these two pH values, carrying a nonsignificant figure along if necessary.
   b. Since the middle half of this transition range must have a quarter of the range on either side calculate a quarter of the range.
   c. Subtract a quarter of the range from 6.3 and add a quarter of the range to 4.2 to obtain the middle half of the range.
B. Repeat the process in problem A for the indicator bromcresol green which has a pH transition range of 3.8–5.4.
C. Decide whether either one, both, or neither of the above indicators are suitable for equivalence points with the following calculated pH values.
   a. pH = 5.5   b. pH = 4.8   c. pH = 4.0   d. pH = 5.0
   e. pH = 4.3
D. The pH at the equivalence point in the titration of 0.1M HCl with 0.1M NaOH changes sharply from 4.0 to 10.5. Suggest several indicators from Table 9-1 that will be suitable.

Answers to Self-Test 9-1

A. a. 1.0_5 b. 0.5_3 c. 4.7 to 5.8
B. 0.8; 0.4; 4.2 to 5.0
C. a. Methyl red   b. Both   c. Neither   d. Both   e. Bromcresol green
D. Methyl red, bromothymol blue, and phenolphthalein

9-3 AQUEOUS TITRATION METHODS

In this section we will discuss the application of acid-base titration methods to the analysis of various types of samples in water only. Nonaqueous titration methods will be discussed in the next section.

The types of samples that may be analyzed by acid-base methods are simple one-component samples, simple two-component mixtures, and complex mixtures. Both direct and back titrations may be employed in the analysis. A back titration is one in which an excess of a standard solution is added to react with all of the acid or base; the remaining standard solution is then back-titrated with a second titrant.

Analysis of Simple One-Component Samples

A one-component sample may be either a solid or liquid sample containing just one acid or base, either strong or weak. Because the analysis of strong acids or strong bases is relatively simple compared to the analysis of weak acids or bases, the former will be discussed first.

Titration of Strong Acids or Bases. A one-component strong acid sample may be simply a solution of hydrochloric acid, nitric acid, sulfuric acid, or perchloric acid in water. A one-component strong base sample may be just a solution of sodium hydroxide, potassium hydroxide, or barium hydroxide in water. Sodium hydroxide is the usual titrant for strong acids, and a typical titration reaction is

$$\text{NaOH} + \text{HCl} \rightarrow \text{Na}^+ + \text{Cl}^- + \text{H}_2\text{O} \qquad (9\text{-}8)$$

(titrant) (hydrochloric acid)

Note that in **9-8** there is no Bronsted acid or base produced; neither the sodium ion nor the chloride ion reacts with water to produce hydrogen ion or hydroxide ion. The only source of either at the titration end point is from the ionization of water; therefore

$$[\text{H}^+] = [\text{OH}^-] \qquad (9\text{-}9)$$

$$[\text{H}^+] = \sqrt{K_w} \qquad (9\text{-}10)$$

At room temperature the $[\text{H}^+]$ will always be $1.0 \times 10^{-7} M$ at the equivalence point for all strong acids. However, the steepest part of the titration curve will vary in length depending on the initial concentration (see Figure 9-2, p. 186). Because of this several indicators may be used with equal accuracy to locate the end point. Note in Figure 9-2 that although the transition range of bromothymol blue falls in the middle of the steepest part of the titration curve of $0.1 M$ hydrochloric acid, both methyl red and phenolphthalein indicators can also be used. Neither of the latter indicators is satisfactory for the titration of $0.001 M$ strong acid, however. Therefore if the concentration of the strong acid is not known, it is safest to choose an indicator where the middle half of the transition range will include pH 7.00.

The titration of most strong bases is just the reverse of the titration of strong acids. As long as the base is soluble it is titrated directly with a strong acid titrant such as hydrochloric acid. A typical titration reaction for a strong base such as barium hydroxide is

$$\text{Ba(OH)}_2 + 2\text{HCl} \rightarrow \text{H}_2\text{O} + \text{Ba}^{+2} + 2\text{Cl}^- \qquad (9\text{-}11)$$

(barium hydroxide) (titrant)

Note that there is no Bronsted acid or base produced in **9-11**, just as there was none in **9-8**. The barium(II) ion, like the sodium ion, does not react with water.

Cations in Groups IA and IIA of the periodic table are known not to react with water in contrast to other cations such as Al^{+3}, which reacts slightly to give small amounts of $Al(OH)^{+2}$ and H^+.

At the end point in the titration of any strong base the only source of hydrogen ion or hydroxide ion is the ionization of water. Equations **9-9** and **9-10** give the $[\text{H}^+]$ and the pH, which is 7.00 at the equivalence point. The same choice of indicators is available for the titration of strong bases at $0.1 M$ concentration or at different concentrations, such as those shown in Figure 9-2 for acids.

Certain insoluble, medically important "strong" bases such as magnesium

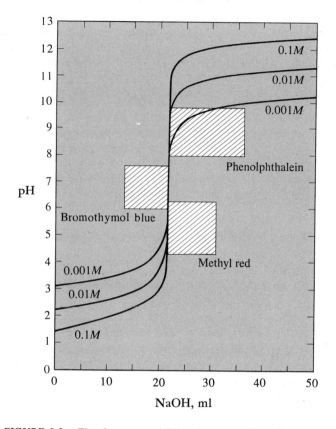

FIGURE 9-2. *Titration curves of different concentrations of strong acid with the corresponding concentration of sodium hydroxide, showing the feasibility of using three different indicators.*

hydroxide and aluminum hydroxide cannot be analyzed by direct titration because they form white suspensions and obscure the indicator color change.

Magnesium hydroxide is used in a liquid suspension as Milk of Magnesia and is frequently mixed with aluminum hydroxide in solid antacids such as Maalox.

Such insoluble bases are analyzed by a *back titration* procedure instead of a direct titration, using standard solutions of hydrochloric acid and sodium hydroxide. For example, Milk of Magnesia can be analyzed for magnesium hydroxide content by the following steps. A measured excess of standard hydrochloric acid is added to the Milk of Magnesia.

$$2HCl(\text{excess}) = Mg(OH)_2(s) \rightarrow 2H_2O + Mg^{+2} + 2Cl^-$$

Then the unreacted acid is back titrated with standard sodium hydroxide.

$$HCl + NaOH \rightarrow H_2O + Na^+ + Cl^-$$
$$\text{(titrant)}$$

To calculate the percentage of magnesium hydroxide the amount of $Mg(OH)_2$ in mmoles or meq is calculated. If the molarity system of calculation (Sec. 6-2) is used, calculate mmoles.

$$\text{mmoles } Mg(OH)_2 = 1/2[(M \text{ HCl})(\text{ml HCl}) - (M \text{ NaOH})(\text{ml NaOH})]$$

If the normality calculation system (Sec. 6-2) is used, calculate meq.

$$\text{meq } Mg(OH)_2 = (N \text{ HCl})(\text{ml HCl}) - (N \text{ NaOH})(\text{ml NaOH})$$

Weak Monoprotic Acids and Bases. A one-component weak acid sample may be any one of solutions or solids, such as those official pharmaceuticals listed in Table 9-2. These include both inorganic and organic pharmaceuticals in tablet, solid,

TABLE 9-2. *Acidic Pharmaceuticals Determined by Titration*

Pharmaceutical	Formula (Organic acidic hydrogen written last)
Acetic acid, U.S.P.	CH_3CO_2H
Benzoic acid, U.S.P.	$C_6H_5CO_2H$
Boric acid, N.F.	H_3BO_3
Citric acid, U.S.P.	$C_6H_5O_7H_3$
Citrated caffeine, N.F.	Caffeine plus $C_6H_5O_7H_3$
Glutamic acid hydrochloride, N.F.	$C_5H_9NO_4 \cdot HCl$
Niacin, N.F.	$C_6H_4NO_2H$
Phenobarbital, U.S.P.	$C_{12}H_{11}N_2O_3H$
Potassium bitartrate, N.F.	$KHC_4H_4O_6$
Saccharin, U.S.P.	$C_7H_4N_2OSH$
Salicylic acid, U.S.P.	$C_7H_4O_3H_2$
Tartaric acid, N.F.	$C_6H_4O_6H_2$

suspension, or liquid form. One-component weak base samples are similar: typical weak bases are ammonia, methylamine (CH_3NH_2), and salts of weak acids.

The usual titrant for weak acids is sodium hydroxide; a typical titration reaction is the neutralization of acetic acid.

$$NaOH + CH_3CO_2H \rightarrow H_2O + CH_3CO_2^- + Na^+ \qquad (9\text{-}12)$$

Note that in **9-12**, there is a Bronsted base produced—the acetate ion. At the equivalence point it reacts with water as follows:

$$CH_3CO_2^- + H_2O = OH^- + CH_3CO_2H$$

This makes the solution slightly basic. For example, if $0.2M$ acetic acid were titrated with $0.2M$ sodium hydroxide, the pH of the resulting $0.1M$ acetate ion at the end point would be 8.87 (see Ex. 8-8 and the titration curve in Fig. 9-1). Other weak acids can be treated similarly.

EXAMPLE 9-4. Phenobarbital ($C_{12}H_{11}N_2O_3H$) is an important pharmaceutical which is a weak acid ($K_a = 5 \times 10^{-8}$) that can barely be titrated in water. Calculate the pH at the end point in the titration of a $0.4M$ solution with $0.40M$ sodium hydroxide.

Solution: First, calculate the K_b of the $C_{12}H_{11}N_2O_3^-$ anion at the end point.

$$K_b = \frac{K_w}{K_a} = \frac{1.00 \times 10^{-14}}{5 \times 10^{-8}} = 2 \times 10^{-7}$$

Then calculate the $[OH^-]$ at the end point following the method of Example 8-8.

$$[OH^-] \cong \sqrt{M\, C_{12}H_{11}N_2O_3^-(K_b)} \cong \sqrt{0.2M(2 \times 10^{-7})} \cong 2 \times 10^{-4} M$$

The per cent error in the approximation is calculated as follows:

$$\% \text{ error} = \frac{2 \times 10^{-4} M\ OH^-}{0.2 M}(100) = 0.1\%$$

Since the per cent error is less than 1% the $[OH^-]$ *may* be reported to two significant figures; however, only one significant figure is justified because the base concentration was given to only $0.2M$.

$$[OH^-] = 2 \times 10^{-4} M$$

$$pH = 10.3$$

Since the pH at the end point is quite basic none of the indicators in Table 9-1 would be perfectly suitable. Neither thymolphthalein nor alizarin yellow indicators would be satisfactory since the middle half of each of their ranges is outside the calculated pH.

In Section 9-1 it was stated that titrations of $0.1M$ weak acids are quantitative only if the K_a is 1×10^{-7} or greater. It is also true that the titration curve does not exhibit a reasonably accurate break at the equivalence point so it cannot be located graphically. Figure 9-3 illustrates this point for $0.1M$ solutions. Note that the titration curves for weak acids through one with a K_a equal to 1×10^{-7} exhibit steep portions such that the equivalence point can be located accurately.

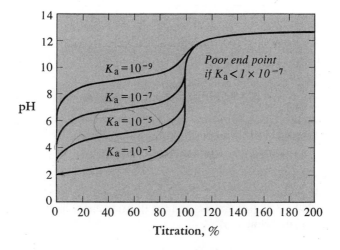

FIGURE 9-3. *Theoretical titration curves of $0.1M$ solutions of various weak acids each having the K_a values shown with $0.1M$ sodium hydroxide.*

An acid with a smaller K_a, such as 1×10^{-9}, exhibits no steep portion at all but merely a "wiggle" near the equivalence point. Of course these titration curves apply only to $0.1M$ concentrations; at lower concentrations, such as $0.01M$, steep portions of the titration curve are observed only for weak acids with a K_a as low as 1×10^{-6}. The titration reaction is not quantitative if the K_a is less than 1×10^{-6}.

An example of a typical method for a single weak acid component is the analysis of vinegar for acetic acid. Commercial vinegar is distilled from the raw materials used in its manufacture; of the acids, only acetic acid (boiling point 118°) distills over. Other possibly acidic materials are left behind because they boil at higher temperatures. Thus the vinegar contains only one acid component. Because the vinegar is relatively concentrated (5% or $1M$) it is either diluted before analysis or a very small sample is taken. The calculation of the concentration of such samples is usually done in terms of weight per cent acetic acid, which is generally expressed in grams of acetic acid per milliliter of solution. See the example below.

EXAMPLE 9-5. Calculate the weight per cent of acetic acid in 5.00 ml of vinegar which requires 40.00 ml of $0.1000M$ sodium hydroxide. Use 60.0 mg/mmole as the formula weight of acetic acid, CH_3CO_2H.

Solution: First calculate the number of mg of CH_3CO_2H.

mg CH_3CO_2H = (40.00 ml NaOH)(0.1000M NaOH)(60.0 mg/mmole CH_3CO_2H)

mg CH_3CO_2H = 240 mg

Convert the mg to g.

$$\frac{240 \text{ mg}}{1000} = 0.240 \text{ g } CH_3CO_2H$$

Then calculate the weight per cent in terms of g CH_3CO_2H/ml.

$$\text{Wt \% } CH_3CO_2H = \frac{0.240 \text{ g } CH_3CO_2H}{5.00 \text{ ml}} (100) = 4.80\%$$

Strictly speaking, 5.00 ml of water should be converted to grams of water, but since 5.00 ml of water is also 5.00 g of water it is not necessary.

Another example of a typical analysis for a single weak acid component is the titration of acid salts of diprotic acids. Typical salts are potassium acid oxalate, KHC_2O_4; potassium acid tartrate, $KHC_4H_4O_6$; and potassium acid phthalate, $KHC_8H_4O_4$. The reaction with the latter is

$$NaOH + KHC_8H_4O_4 \rightarrow H_2O + Na^+ + K^+ + C_8H_4O_4^{-2}$$

(titrant)

Such salts behave as monoprotic acids since they lose but one hydrogen per formula weight.

The titration of most weak bases is similar to the titration of weak acids. They are titrated directly with a strong acid titrant such as hydrochloric acid.

A typical titration reaction for ammonia, which can be written as NH_3 or NH_4OH, is:

$$NH_3 + \underset{\text{(titrant)}}{HCl} \rightarrow NH_4^+ + Cl^- \qquad (9\text{-}14)$$

(Writing ammonia as NH_4OH, the reaction is

$$NH_4OH + HCl \rightarrow NH_4^+ + H_2O + Cl^-.)$$

Note that in **9-14** a Bronsted acid is produced; this is the ammonium ion. At the equivalence point this is the only acidic species present; it reacts with water as follows:

$$NH_4^+ + H_2O = H_3O^+ + NH_3 \qquad (9\text{-}15)$$

This means the solution is slightly acidic, the pH being controlled by the concentration of the ammonium ion present. See the following example.

EXAMPLE 9-6. Calculate the pH at the end point in the titration of $0.20M$ ammonia with $0.20M$ hydrochloric acid. Choose an indicator.

Solution: First calculate the K_a of the ammonium ion in **9-15** using the general equation **(8-9)**.

$$K_a = \frac{K_w}{K_b} = \frac{1.00 \times 10^{-14}}{1.8 \times 10^{-5}} = 5.5_5 \times 10^{-10}$$

Then using the methods of Example 8-6 calculate the approximate $[H^+]$. Recall that the approximate relation of **8-7** is valid for any weak acid such as the ammonium ion.

$$[H^+] \cong \sqrt{M \text{ of } NH_4^+(K_a)} \cong \sqrt{0.10M(5.5_5 \times 10^{-10})} \cong 7.4_5 \times 10^{-6}M$$

Using **8-8** calculate the per cent error.

$$\% \text{ error} = \frac{7.4 \times 10^{-6}M}{0.10M}(100) = 7.4 \times 10^{-3}\%$$

Since the per cent error is less than 1% we can report the $[H^+]$ to two significant figures.

$$[H^+] = 7.4_5 \times 10^{-6}M$$

$$pH = 5.13$$

Because the middle half of the transition range of methyl red indicator is 4.7 to 5.8 this indicator would be satisfactory. Bromcresol green would not be satisfactory since the middle half of its range is 4.2 to 5.0.

The titration curve for $0.20M$ ammonia using $0.20M$ hydrochloric acid

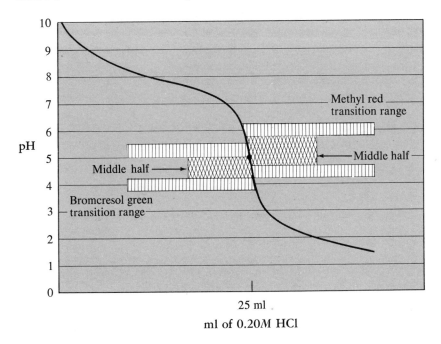

FIGURE 9-4. Titration curve for 25 ml of 0.20M ammonia with 0.20M hydrochloric acid.

titrant is shown in Figure 9-4. The middle of the steepest part of the titration curve is at approximately pH 5.1 (see dot). There are three indicators whose transition ranges include pH 5.1.

Indicator	Middle half of transition range
Bromcresol green	4.2–5.0
Methyl red	4.7–5.8
Chlorophenol red	5.2–6.0

Although the transition range of each indicator does enclose the steepest part of the curve, it can be seen in Figure 9-4 that the *middle half* of the range of bromcresol green as well as that of chlorophenol red (not shown) does not include the steepest part of the curve. That of methyl red indicator does, making this a better choice than either of the other two indicators.

In Section 9-1 we stated that the titration of $0.1M$ solutions of weak bases having a K_b of 1×10^{-7} or larger is quantitative. It is likewise true that such titration curves do exhibit reasonably accurate breaks at the equivalence point. Figure 9-5 illustrates this situation for $0.1M$ solutions. Note that the titration

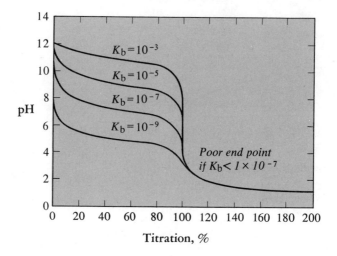

FIGURE 9-5. *Theoretical titration curves of 0.1M solutions of various weak bases having the K_b values shown with 0.1M hydrochloric acid.*

curves for weak bases from K_b of 1×10^{-3} through a K_b of 1×10^{-7} exhibit steep portions such that the equivalence point can be located accurately. A base with a smaller K_b, such as 1×10^{-9}, exhibits no steep portion at all, just a "wiggle." At lower concentrations than $0.1M$, such as $0.01M$, a steep portion is observed at the equivalence point only if the K_b is 1×10^{-6} or larger. The titration reaction is not quantitative if the K_b is less than 1×10^{-6} at $0.01M$ concentrations.

An example of a typical method for a single weak base component is the determination of ammonia in any number of cleaning products, such as household ammonia solutions. Many such solutions contain only water, ammonia, coloring matter, and perhaps a detergent. Equation **9-14** would then be the only acid-base reaction that could occur for such a sample. The calculation of the concentration of such samples is usually done in terms of weight per cent ammonia, which is usually expressed in terms of grams of ammonia per milliliter of solution. See below.

EXAMPLE 9-7. Calculate the weight per cent of ammonia in 5.00 ml of ammonia solution which requires 30.00 ml of $0.1000M$ hydrochloric acid. Use 17.0 mg/mmole as the formula weight for NH_3.

Solution: First calculate the number of mg of NH_3.

$$\text{mg } NH_3 = (30.00 \text{ ml HCl})(0.1000M \text{ HCl})(17.0 \text{ mg/mmole } NH_3)$$

$$\text{mg } NH_3 = 51.0 \text{ mg}$$

Next convert mg NH_3 to grams.

$$\frac{51.0 \text{ mg } NH_3}{1000} = 0.0510 \text{ g } NH_3$$

Then calculate the weight per cent NH_3 in terms of g NH_3/ml:

$$\text{wt \% } NH_3 = \frac{0.0510 \text{ g } NH_3}{5.00 \text{ ml}} (100) = 1.02\% \text{ } NH_3$$

A second example of an analysis for a single weak base component is the *direct* titration of sodium bicarbonate or potassium bicarbonate in tablet (pharmaceutical) form. The titration reaction is

$$HCl + NaHCO_3 \rightarrow H_2CO_3 + Na^+ + Cl^-$$

A number of other weakly basic pharmaceuticals cannot be titrated directly but can be determined by back titration. For example, the weak nitrogen base morpholine, C_4H_9ON, may be determined as follows:

$$C_4H_9ON + HCl(xs) \rightarrow C_4H_9ONH^+ + Cl^-$$

The unreacted acid is then back titrated with standard sodium hydroxide.

$$HCl + NaOH \rightarrow H_2O + Na^+ + Cl^-$$

An example of the calculations involved is below.

EXAMPLE 9-8. Calculate the percentage of morpholine, formula weight 87.1, in a 1.0000 g sample of morpholine pharmaceutical which was neutralized with 50.00 ml of 0.2000M hydrochloric acid and back titrated with 20.00 ml of 0.1000M sodium hydroxide.

Solution: First calculate the number of mmoles of morpholine by subtracting the number of mmoles of NaOH from the number of mmoles of excess HCl added.

mmoles C_4H_9ON = (50.00 ml HCl)(0.2000M HCl) − (20.00 ml NaOH)(0.1000M NaOH)

mmoles C_4H_9ON = 10.00 − 2.00 = 8.00 mmoles

$$\% \text{ } C_4H_9ON = \frac{(8.00 \text{ mmoles})(87.1 \text{ mg/mmoles})(100)}{1000 \text{ mg}} = 69.6_8\%$$

Analysis of Polyprotic Acids and Bases and Two-Component Samples

A two-component sample may be either a solid or liquid sample containing two acids or two bases, both strong, both weak, or one strong and one weak. Because there is a difference in the magnitude of K_1, K_2, and so on, of polyprotic acids, the titration of a polyprotic acid is much like the titration of a two-component mixture. We will therefore discuss the titration of polyprotic acids first and establish principles that can be used for the titration of two-component mixtures.

Titration of Polyprotic Acid One-Component Samples. The stepwise ionization of polyprotic acids has already been discussed in Section 8-4. Because there is a difference in the magnitude of K_1, K_2, and so on, the first hydrogen may be neutralized

100% before the second hydrogen is neturalized, or both hydrogens may be neutralized essentially together. The general rules are given in the box.

> If K_1/K_2 is $\geq 10^4$, the titration curve of a polyprotic acid will exhibit an end point for the first hydrogen. (If K_2 is $\geq 1 \times 10^{-7}$, an end point for the second hydrogen will also be observed.) If K_1/K_2 is $< 10^4$, the titration curve will not exhibit a large enough "break," or steep portion, for the first end point to be located accurately.

To illustrate the general rules let us consider the following acids.

maleic acid $K_1 = 1.2 \times 10^{-2}$ $K_2 = 6.0 \times 10^{-7}$
carbonic acid $K_1 = 4.3 \times 10^{-7}$ $K_2 = 4.8 \times 10^{-11}$
phosphoric acid $K_1 = 7.5 \times 10^{-3}$ $K_2 = 6.2 \times 10^{-8}$ $K_3 = 4.8 \times 10^{-13}$

The ratio of K_1/K_2 for maleic acid is

$$\frac{K_1}{K_2} = \frac{1.2 \times 10^{-2}}{6.0 \times 10^{-7}} = 2.0 \times 10^4$$

Since K_1/K_2 is greater than 10^4 the titration curve for maleic acid will exhibit an end point for the neutralization of the first hydrogen.

$$\text{NaOH} + \text{C}_2\text{H}_2(\text{CO}_2\text{H})_2 \rightarrow \text{C}_2\text{H}_2(\text{CO}_2\text{H})\text{CO}_2^- + \text{Na}^+ + \text{H}_2\text{O}$$

(titrant)   (maleic acid)

The value of K_2 for maleic acid is greater than 1×10^{-7} so the titration curve will also exhibit an end point for the neutralization of the second hydrogen.

$$\text{NaOH} + \text{C}_2\text{H}_2(\text{CO}_2\text{H})\text{CO}_2^- \rightarrow \text{C}_2\text{H}_2(\text{CO}_2)^{-2} + \text{Na}^+ + \text{H}_2\text{O}$$

(titrant)   (acid maleate anion)   (maleate anion)

The ratio of K_1/K_2 for carbonic acid is just slightly less than 10^4 so the titration curve for carbonic acid should exhibit an end point for the neutralization of the first hydrogen.

$$\text{NaOH} + \text{H}_2\text{CO}_3 \rightarrow \text{HCO}_3^- + \text{Na}^+ + \text{H}_2\text{O}$$

The value of K_2 for carbonic acid is less than 1×10^{-7} so the titration cannot be carried accurately beyond the neutralization to the bicarbonate ion above.

The titration of phosphoric acid is like the titration of two diprotic acids since K_1 must be compared to K_2 for the first end point, and K_2 must be compared to K_3 for the second end point. The ratio of K_1 to K_2 for phosphoric acid is as follows:

$$\frac{K_1}{K_2} = \frac{7.5 \times 10^{-3}}{6.2 \times 10^{-8}} = 1.2 \times 10^5$$

Since this is larger than 10^4 the titration curve will exhibit an end point for the neutralization of the first hydrogen.

$$NaOH + H_3PO_4 \rightarrow H_2PO_4^- + Na^+ + H_2O$$

This is shown in Figure 9-6. Since the only species present is the dihydrogen phosphate anion its pH can be calculated by using **8-42** (see Ex. 8-16).

$$[H^+] \approx \sqrt{K_1 K_2} \approx 2.1_6 \times 10^{-5} M$$

$$pH = 4.75$$

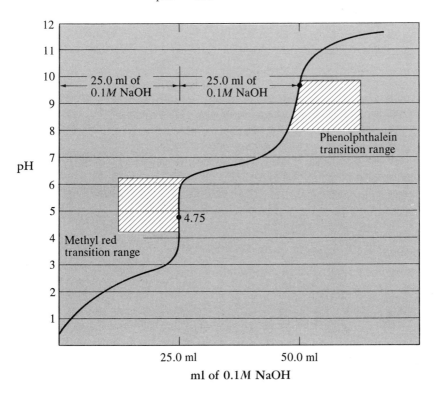

FIGURE 9-6. *Titration of 0.1M phosphoric acid with 0.1M sodium hydroxide.*

Because phosphoric acid is a triprotic acid we must also decide whether the titration curve will exhibit a second end point by calculating the ratio of K_2/K_3.

$$\frac{K_2}{K_3} = \frac{6.2 \times 10^{-8}}{4.8 \times 10^{-13}} = 1.3 \times 10^5$$

This verifies that the titration curve shown in Figure 9-6 should exhibit an end point for the neutralization of the second hydrogen.

$$NaOH + H_2PO_4^- \rightarrow HPO_4^{-2} + Na^+ + H_2O$$

Since the only species present at the second end point is the HPO_4^{-2} ion the pH can be calculated in the same manner as at the first end point (see Ex. 8-17).

$$[H^+] \approx \sqrt{K_2 K_3} \approx 1.7 \times 10^{-10}$$

$$pH = 9.76$$

The K_3 of phosphoric acid is less than 1×10^{-7} so the titration of the HPO_4^{-2} ion cannot be carried out accurately.

To calculate the amount of phosphoric acid present in a sample it is desirable to use the volume of titrant at the first end point because the titration curve is steeper there than at the second end point (see Fig. 9-6). However, if a second acid is present, whichever end point requires the smallest volume of titrant will give the only accurate results. Because a second acid may be titrated with H_3PO_4 or with $H_2PO_4^-$ two situations, summarized below, are possible.

Size of K_a of second acid	Relation of buret readings at end points	To calculate % H_3PO_4, use
$K_a \geqslant K_1$	$ml_{1st} > (ml_{2nd} - ml_{1st})$	$(ml_{2nd} - ml_{1st})$
$K_1 > K_a \geqslant K_2$	$(ml_{2nd} - ml_{1st}) > ml_{1st}$	ml_{1st}

Note that the buret reading at the second end point (ml_{2nd}) is actually the total volume of titrant added for the neutralization of H_3PO_4 and $H_2PO_4^-$. To calculate the volume required for neutralization of $H_2PO_4^-$ alone the ml of titrant used at the first end point must be subtracted from that used at the second end point. The calculation is best understood by considering the following example.

EXAMPLE 9-9. A 1.0000 g sample of phosphoric acid mixed with an unknown monoprotic acid is titrated with 0.1000M sodium hydroxide. The buret reading at the first end point is 30.00 ml, and the buret reading at the second end point is 55.00 ml. Calculate the percentage of phosphoric acid present. The formula weight of phosphoric acid is 98.0.

Solution: First decide what end point represents only the phosphoric acid by comparing the volumes.

$$ml_{1st} \text{ for neut. of } H_3PO_4 = 30.00 \text{ ml}$$

$$(ml_{2nd} - ml_{1st}) \text{ for neut. of } H_2PO_4^- = 55.00 - 30.00 = 25.00 \text{ ml}$$

Since the volume for the neutralization of $H_2PO_4^-$ is smaller this volume represents only the titration of phosphoric acid. The monoprotic acid is obviously being titrated along with H_3PO_4 to the first end point (contrast this with Fig. 9-6). Now calculate the percentage of phosphoric acid.

$$\% \; H_3PO_4 = \frac{(25.00 \text{ ml NaOH})(0.1000M)(98.0 \text{ mg/mmole})(100)}{1000.0 \text{ mg}} = 24.5\%$$

In conclusion, we can also say that the number of mmoles of the weak monoprotic acid is (5.00 ml)(0.1000M), or 0.500 mmoles. Unless the molecular weight is known, nothing further can be done.

Titration of Boric Acid. Although boric acid is a triprotic acid, all three hydrogens are very weakly ionized. The value of 5.9×10^{-10} for K_1 is less than 1×10^{-7} so even the first hydrogen cannot be titrated quantitatively (p.175).

The fact that boric acid is a very weakly ionized acid makes it ideal for use in eye wash solutions. For example, a 0.1M solution is acidic enough at $[H^+] = 8 \times 10^{-6} M$ for medicinal purposes, but not acidic enough to harm the cornea of the eye.

Fortunately, boric acid forms stable complexes with organic glycols; such complexes are acidic enough to be titrated quantitatively. For example, glycerine reacts with boric acid as follows:

$$H_3BO_3 + C_3H_5OH \rightarrow (HO)_2B\begin{matrix}OCH-CH_2OH\\|\\OCH_2\end{matrix} + H_2O \quad (9\text{-}16)$$

(boric acid) (glycerine)

The K_1 value of boric acid is increased by the reaction to a value larger than 1×10^{-7} [2]; the actual value depends on the amount of complexing glycol added. After the complex is formed it can be titrated as a monoprotic acid to a phenolphthalein end point [3].

$$(HO)_2B{=}O_2C_3H_5OH + NaOH \rightarrow O(HO)B{=}O_2C_3H_5OH^- + Na^+ + H_2O$$

(complex) (titrant)

$$(9\text{-}17)$$

The remaining second hydrogen is still too weak to be titrated so only one hydrogen can be titrated quantitatively from the complex.

Titration of Polybasic Base One-Component Samples. The most important polyfunctional base is sodium carbonate. Its two ionization constants are not listed in the appendix but can be calculated as K_b for any anion that is calculated by using **8-17**. Thus K_1 for the carbonate ion is calculated by inserting the ionization constant for the bicarbonate ion, its conjugate acid, into **8-17**.

$$K_1 = \frac{K_w}{K_{HCO_3^-}} = \frac{1 \times 10^{-14}}{4.8 \times 10^{-11}} = 2.1 \times 10^{-4}$$

The value of K_2 for the carbonate ion is similarly calculated from K_w and the ionization constant for carbonic acid.

$$K_2 = \frac{K_w}{K_{H_2CO_3}} = \frac{1 \times 10^{-14}}{4.3 \times 10^{-7}} = 2.3 \times 10^{-8}$$

The same general rules for the stepwise neutralization of polyprotic acids apply to the stepwise neutralization of polyfunctional bases. Thus if K_1/K_2 for

[2] G. Marinenko and C. E Champion, *Anal. Chem.* **41**, 1208 (1969).
[3] G. L. Jenkins, A. M. Knevel, and F. E. DiGangi, *Quantitative Pharmaceutical Chemistry*, McGraw-Hill, New York, 1967, pp. 103–4.

such bases is at least 10^4, the titration curve will exhibit an end point for the reaction of one proton of titrant per formula weight of the base. For sodium carbonate the ratio of K_1/K_2 is

$$\frac{K_1}{K_2} = \frac{2.1 \times 10^{-4}}{2.3 \times 10^{-8}} = 9.1 \times 10^3$$

Since this is almost 10^4 the titration curve will exhibit an end point for the neutralization of the carbonate ion to bicarbonate.

$$CO_3^{-2} + HCl \rightarrow HCO_3^- + H_2O$$

This is shown in Figure 9-7. Since the only species present at the end point is the

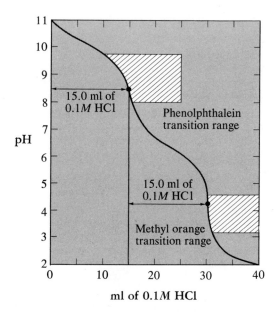

FIGURE 9-7. *Titration of 1.5 mmoles of sodium carbonate with 0.1M hydrochloric acid.*

bicarbonate ion the pH can be calculated by using **8-42** (see Ex. 8-15) and *the ionization constants for carbonic acid.*

$$[H^+] \approx \sqrt{K_1 K_2} \approx \sqrt{(4.3 \times 10^{-7})(4.8 \times 10^{-11})} \approx 4.5 \times 10^{-8} M$$

$$pH = 7.34$$

The K_2 of the carbonate ion is slightly less than 1×10^{-7} so the titration of $0.1M$ solutions should be just short of quantitative. Indeed the color change at the methyl orange end point (Fig. 9-7) is notorious for being difficult to perceive. Usually a pH 4 buffer is prepared and the color of methyl orange in the titration flask is matched with the color of methyl orange in the pH 4 buffer.

To calculate the amount of sodium carbonate present in a sample it is still relatively desirable to use the volume of titrant at the second end point rather than at the first end point, which is not as steep. If a second base is present, whichever end point requires the smallest volume of titrant will give the only accurate results. Because the second base may be a strong base like sodium hydroxide or a weak base like sodium bicarbonate two situations are possible. They are summarized below.

Strength of second base	Relation of buret readings at end points	To calculate % Na_2CO_3 use
Strong (NaOH, etc.)	$ml_{1st} > (ml_{2nd} - ml_{1st})$	$(ml_{2nd} - ml_{1st})$
Weak ($NaHCO_3$)	$(ml_{2nd} - ml_{1st}) > ml_{1st}$	ml_{1st}

Note that the buret reading at the second end point (ml_{2nd}) is actually the total volume of titrant added for neutralization to H_2CO_3. To calculate the volume required for neutralization of HCO_3^- to H_2CO_3, the ml of titrant used at the first end point must be subtracted from the buret reading at the second end point. See the example below.

EXAMPLE 9-10. A 1.0000 g sample of sodium carbonate mixed with an unknown base (monofunctional) is titrated with 0.1000M hydrochloric acid. The buret reading at the first end point is 25.00 ml; the reading at the second end point is 40.00 ml. Calculate the percentage of sodium carbonate present. Its formula weight is 106.0.

Solution: First decide which end point represents only the sodium carbonate by comparing the end point volumes.

$$ml_{1st} \text{ for neut. of } CO_3^{-2} = 25.00 \text{ ml}$$

$$(ml_{2nd} - ml_{1st}) \text{ for neut. of } HCO_3^- = 40.00 - 25.00 = 15.00 \text{ ml}$$

Because the volume for the neutralization of the CO_3^{-2} is smaller this volume must represent the titration of only sodium carbonate. The unknown base is obviously being titrated with the CO_3^{-2} to the first end point (contrast this to Fig. 9-7). Now calculate the percentage.

$$\% \, Na_2CO_3 = \frac{(15.00 \text{ ml HCl})(0.1000M \text{ HCl})(106.0 \text{ mg/mmole})(100)}{1000.0 \text{ mg}} = 15.90\%$$

The number of mmoles of the unknown monofunctional base is (10.00 ml)(0.1000M) or 1.000 mmole. Unless the molecular weight is known, nothing else can be calculated

Analysis for Two Components in a Sample. Two-component mixtures may consist of two monoprotic acids, two monofunctional bases, or more complex samples such as a mixture of a monoprotic acid and a polyprotic acid. The same general rules apply to these mixtures as to the titration of a single polyprotic acid or base. The K_a of the stronger acid is divided by the K_a of the weaker acid.

> If $K_{a(str)}/K_{a(wk)}$ is $\geq 10^4$, the titration curve will exhibit an end point for the stronger acid. If $K_{a(wk)}$ is $> 1 \times 10^{-7}$, an end point for the weaker acid will also be observed.

The same is true for a mixture of two bases.

If a mixture consists of a *strong* acid such as hydrochloric acid and a weak acid, the above rule cannot readily be applied. It appears from the titration curve of a mixture of hydrochloric and acetic acids that an end point for a *strong* acid will be observed if the K_a of the weak acid is 2×10^{-5} or less [4].

To calculate the amount of each acid or base present in a two-component mixture it is necessary to decide which acid or base is titrated first and then use the volume of titrant corresponding to it in the calculation. See the example below.

EXAMPLE 9-11. A 1.0000 g sample of trichloroacetic acid ($K_a = 2.2 \times 10^{-1}$) and acetic acid ($K_a = 1.8 \times 10^{-5}$) is titrated with 0.1000M sodium hydroxide. The buret reading at the first end point is 30.00 ml; the reading at the second end point is 55.00 ml. Calculate the percentage of each acid present. The formula weight of trichloroacetic acid is 163.4 and that of acetic acid is 60.0.

Solution: First decide which acid is titrated at the first end point. This should be trichloroacetic acid since its K_a is over 10^4 times as large as that of acetic acid. Now assign volumes at the end point to each acid.

$$\text{ml}_{1st} \text{ for neut. of } Cl_3CCO_2H = 30.00 \text{ ml}$$

$$(\text{ml}_{2nd} - \text{ml}_{1st}) \text{ for neut. of } CH_3CO_2H = 55.00 - 30.00 = 25.00 \text{ ml}$$

Finally calculate the percentage of each acid.

$$\% \ Cl_3CCO_2H = \frac{(30.00 \text{ ml})(0.1000M \text{ NaOH})(163.4 \text{ mg/mmole})(100)}{1000.0 \text{ mg}} = 49.02\%$$

$$\% \ CH_3CO_2H = \frac{(25.00 \text{ ml})(0.1000M \text{ NaOH})(60.0 \text{ mg/mmole})(100)}{1000.0 \text{ mg}} = 15.0_0\%$$

The calculations for mixtures containing a polyprotic acid or base and a monoprotic acid or base are more complicated. After deciding whether the monoprotic acid is titrated at the first or the second end point, the volume of titrant it consumes is found by subtracting the volume consumed by the polyprotic acid or base at the other end point. See the example below.

EXAMPLE 9-12. A 1.0000 g sample of sodium carbonate and sodium hydroxide is titrated with 0.1000M hydrochloric acid. As in Example 9-10 the buret reading at the first end point is 25.00 ml; the reading at the second end point is 40.00 ml. Calculate the percentages of sodium carbonate and sodium hydroxide (form wt = 40.00) present.

[4] J. S. Fritz and G. H. Schenk, *Quantitative Analytical Chemistry*, Allyn & Bacon, Boston, 1974, pp. 178–79.

Solution: First decide whether the sodium hydroxide is titrated at the first or second end point. Because it is a strong base it will be titrated at the first end point as the CO_3^{-2} ion is neutralized to the HCO_3^- ion. Now assign volumes at the end points to each acid.

$(ml_{2nd} - ml_{1st})$ for neut. of Na_2CO_3 = 40.00 − 25.00 = 15.00 ml (Ex. 9-10)

ml_{1st} − ml for neut. of Na_2CO_3 = ml for neut. of NaOH = 25.00 − 15.00 = 10.00

Finally calculate the percentage of each base.

From Example 9-10, % Na_2CO_3 = 15.00%

$$\% \text{ NaOH} = \frac{(10.00 \text{ ml})(0.1000M \text{ HCl})(40.00 \text{ mg/mmole})(100)}{1000.0 \text{ mg}} = 4.00\%$$

Analysis of Complex Mixtures: the Kjeldahl Method

Where certain samples are very complex and contain many components special methods may be needed before a titration can be employed. The Kjeldahl method for organic nitrogen is a good example of such a method.

The Kjeldahl Method for Nitrogen. The Kjeldahl method is used for the determination of the percentage nitrogen in many foods and fertilizers. Although much of the nitrogen in these substances is present as the basic amino group, $-\overset{|}{\underset{|}{C}}-NH_2$, some of the nitrogen is present in a form which is not basic. Thus, titration with acid will not give an accurate determination of the per cent nitrogen. The Kjeldahl method consists of three essential steps.

1. *Digestion.* The organic nitrogen is heated with concentrated sulfuric acid which converts the nitrogen to ammonium bisulfate.

$$H-\overset{|}{\underset{|}{C}}-NH_2 + 3H_2SO_4 \rightarrow CO_2(g) + NH_4HSO_4 + 2H_2SO_3[\rightarrow SO_2(g)]$$

The organic nitrogen is now in a readily titratable form.

2. *Distillation.* The digestion mixture is cooled and neutralized with sodium hydroxide.

$$NH_4HSO_4 + 2NaOH \rightarrow NH_3 + 2Na^+ + SO_4^{-2} + 2H_2O$$

The resulting ammonia is distilled into an excess of standard hydrochloric acid.

$$NH_3(g) + HCl(excess) \rightarrow NH_4^+ + Cl^- \quad \textbf{(9-18)}$$

3. *Titration.* The unreacted hydrochloric acid is back titrated with standard sodium hydroxide. The amount of ammonia is

$$\text{mmoles NH}_3 = (ml)(M \text{ HCl}) - (ml)(M \text{ NaOH})$$

In a variation of the method the ammonia is distilled into a standard solution of boric acid so instead of **9-18** the following reaction occurs.

$$NH_3(g) + H_3BO_3 \rightarrow NH_4^+ + H_2BO_3^- \tag{9-19}$$

The solution is then titrated with a standard solution of hydrochloric acid. Calculation of the equilibrium constant of the above reaction using **7-10** gives a value of only 1.1 so some ammonia does not react with the boric acid. It is titrated along with the $H_2BO_3^-$ ion by the acid, however.

Analysis of Other Mixtures. There are many other methods for analysis of complex mixtures that are found in advanced texts [1, 3, 4]. For example, nitrogen compounds containing the nitrate ion may be reduced by Devarda alloy to ammonia and then distilled. Other types of distillations are also possible.

Self-Test 9-2. Calculation of Acid-Base Titration Results

Directions: Compare answers with those given at the end.

A. A 0.4000 g sample of sodium bicarbonate tablets is analyzed by titration with 40.00 ml of 0.1000*M* hydrochloric acid. Calculate the percentage of $NaHCO_3$, formula weight = 84.01.
B. A 5.00 ml sample of Milk of Magnesia is analyzed by adding 50.00 ml of 0.2000*M* hydrochloric acid to it. The unreacted hydrochloric acid is back titrated with 10.00 ml of 0.1000*M* sodium hydroxide. Calculate the weight per cent of $Mg(OH)_2$, formula weight = 58.3.
C. A 1.0000 g sample of phosphoric acid mixed with an unknown monoprotic acid is titrated with 0.1000*M* sodium hydroxide. The buret reading at the first end point is 25.00 ml; the reading at the second end point is 51.12 ml. Decide whether the unknown acid is titrated at the first or second end point and calculate the percentage H_3PO_4, formula weight = 98.00.
D. A 1.0000 g sample of sodium carbonate mixed with an unknown monofunctional weak base is titrated with 0.1000*M* hydrochloric acid. The buret reading at the first end point is 15.00 ml; the reading at the second end point is 31.22 ml. Decide whether the unknown base is titrated at the first or second end point and calculate the percentage Na_2CO_3, formula weight = 106.0.
E. A 1.0000 g sample of food containing nitrogen is treated by the Kjeldahl method and the resulting ammonia is distilled into 50.00 ml of 0.1000*M* hydrochloric acid. The unreacted acid is back titrated with 15.00 ml of 0.1000*M* sodium hydroxide. Calculate the percentage N, formula weight = 14.00.

Answers to Self-Test 9-2

A. 84.01%
B. $5.24_7\%$ (0.0524_7 g/ml)
C. Unknown acid titrated at second end point; 24.50%
D. Unknown base titrated at second end point; 15.90%
E. 4.900% N

9-4 NONAQUEOUS TITRATION METHODS: PHARMACEUTICALS

In the previous section the applications of acid-base titrations in aqueous solution were discussed. In this section we shall discuss acid-base titrations in nonaqueous solvents. Such solvents consist mainly of pure organic solvents such as glacial acetic acid, dimethylformamide, acetone, pyridine, and acetonitrile.

One important advantage of nonaqueous titrations is that they are not limited to acids or bases with a K_a or K_b greater than 1×10^{-7} (Sec. 9-1). Usually acids or bases with a K_a or K_b greater than 10^{-11} can be titrated. The reason is that the nonaqueous solvent does not react with the anion or cation produced by the neutralization to the same extent that water does. Thus the end point is much sharper; that is, it has a steeper portion in the titration curve.

Another important advantage is that nonaqueous solvents can dissolve many organic compounds that are insoluble in water. Many important drugs or pharmaceuticals can thus be titrated.

Nonaqueous Solvents

Because the solvent is so important we will discuss the requirements for a nonaqueous solvent and the types of solvents before we discuss the applications.

Requirements. Fritz [5] has listed a number of essential requirements for nonaqueous solvents, some of which follow.

Good dissolving ability: A solvent should dissolve not only the reactants but also the products, if possible. Many salts produced in the titrations are insoluble in many solvents and are a potential source of error through coprecipitation of the substance being titrated.

Reasonably high dielectric constant: A reasonably high dielectric constant favors the attainment of stable potentiometer readings and good potentiometric titration curves. When a nonaqueous solvent is used instead of water the use of a

[5] J. S. Fritz, *Acid-Base Titrations in Nonaqueous Solvents*, Allyn & Bacon, Boston, 1973.

pH meter-potentiometer to locate the end point is often subject to problems, such as unstable readings, not encountered in water. Many of the solvents listed in Table 9-3 have dielectric constants of 12 or higher so these solvents have the potential

TABLE 9-3. Some Nonaqueous Solvents [5]

Solvent and Formula	Dielectric Constant, 25°C
Used for acids and bases	
Acetone, CH_3COCH_3	20.7
Acetonitrile, CH_3CN	36
Methanol, CH_3OH	33
Methyl isobutyl ketone, $CH_3COC_4H_9$	13
2-Propanol, $CH_3CHOHCH_3$	18.3
Sulfolane, $C_2H_4SOC_2H_4$	44
Used for acids	
tert-Butyl alcohol, C_4H_9OH	10.9
Dimethylformamide, $HCON(CH_3)_2$	27
Dimethylsulfoxide, CH_3SOCH_3	46.7
Ethylenediamine, $H_2NC_2H_4NH_2$	12.5
Pyridine, C_5H_5N	12.3
Used for bases	
Acetic acid, CH_3CO_2H	6.1
Acetic anhydride, $(CH_3CO)_2O$	20.7
1,4-Dioxane, $C_2H_4O_2$	2.2
Nitromethane, CH_3NO_2	36

to facilitate a potentiometric titration. A high dielectric constant is also another indication of good ability to dissolve the reactants and products of a titration.

Proper acidity or basicity: To facilitate the titration of a weak acid a solvent should be as weak an acid as possible; that is, it should be basic or *neutral*. (Neutral solvents include those listed for acids and bases in Table 9-3.) For the accurate titration of a weak base a solvent should be as weak a base as possible; that is, it should be acidic or *neutral*.

High purity: For the titration of bases a solvent should be as free as possible of basic impurities; for the titration of acids it should be as free as possible of acidic impurities. A solvent blank should be run by titrating the same amount of solvent used for a titration without any sample added. If necessary, this blank can then be subtracted from the volume of titrant used for the sample.

Stability: The solvent should not decompose during the titration or on standing in the pure state. Obviously, it should not react with the titrant or with any of the various samples to be titrated.

Hydrogen-bonding ability: In certain potentiometric titrations the neutralized product hydrogen bonds to the reactant causing the shape of the potentiometric titration curve to be abnormal. This makes it difficult to locate the end point, particularly when two end points are to be located for a mixture of two or more acids or bases. By using a solvent having an available hydrogen (such as methanol) or an oxygen to hydrogen bond to a hydrogen of the reactant (such as acetone), the solvent prevents the hydrogen bonding of the neutralized reactant to the product.

Reasonable volatility and viscosity: If a solvent is too volatile, its odor may be offensive or its fumes may be dangerous. Pyridine is a good example. It has a boiling point over 115°C, yet its odor is offensive and its fumes are not safe. It is best to use this solvent in a well-ventilated room or in a hood. If a solvent is too viscous, it will of course be difficult to stir efficiently during the titration.

Three different solvent types. The three different types of solvents for nonaqueous titrations are the solvents for acids, the solvents for bases, and the so-called neutral solvents which can be used for the titration of acids or bases (Table 9-3).

The neutral solvents are neutral only in the sense that both acids and bases can be titrated in them with good results. These solvents include mainly oxygenated solvents such as acetone and 2-propanol as well as acetonitrile. Weak organic acids such as phenols and weak organic bases such as amines have been titrated in each solvent [5]. For the titration of mixtures by potentiometric detection of the end point Fritz [5] recommends sulfolane which has the largest potential range reported so far. This range permits differentiation of a large number of acids or bases. Methyl isobutyl ketone also has a large range.

Neutral solvents are generally referred to as *differentiating* solvents because they permit mixtures of acids or bases to be differentiated according to their acid or base strength. Potentiometric titration curves of acids or bases may exhibit two or more end points if there is a factor of about 10^3 difference in their ionization constants.

The solvents for acids (Table 9-3) are generally basic to some degree with the exception of *tert*-butyl alcohol. The advantage of using a solvent with some basicity is that the basic titrant will react to a greater extent with the acid sample than with the solvent. Thus there is less competition of the solvent and the sample for the basic titrant.

With the exception of *tert*-butyl alcohol, the solvents for acids are *leveling* solvents. They react with acid samples and "level" them to nearly the same acidity. Titration curves of mixtures of such acids exhibit only one end point, not two or more. The advantage of using a leveling solvent is that it enhances the acidity of very weak acidic compounds more than the neutral, or differentiating, solvents so these compounds are more readily titrated.

The solvents for bases (Table 9-3) are all acidic to some degree. The advantage of using a solvent having some acidity is that the acid titrant will react to a greater extent with the base sample than with the solvent. With the possible exception of dioxane, the solvents for bases are leveling solvents with the same advantages and disadvantages given for the leveling solvents for acids.

Titrants and End Point Detection

We will first discuss the conditions for the titrations of acids and then proceed to do the same for bases.

Titration of Acids. There are two important types of titrants for weak acids—sodium, potassium, or lithium methoxide, and the tetraalkylammonium hydroxides. The methoxides are not readily available, but a methoxide such as sodium methoxide is made by carefully mixing a weighed amount of sodium metal with methyl alcohol.

$$2Na(s) + 2CH_3OH \rightarrow 2NaOCH_3 + H_2(g)$$

Such a solution is not a standard solution since the sodium is not pure; the sodium methoxide is standardized against primary standard benzoic acid. The preparation of tetraalkylammonium hydroxides such as tetra-*n*-butylammonium hydroxide is more complicated, but these compounds are available [5]. This type of titrant must also be standardized against primary standard benzoic acid.

End point detection is not much different for nonaqueous solvents than it is for water. Many of the same types of indicators are used, except that more weakly acidic indicators must be used for the titration of the weaker acids. We can summarize the choice of three indicators as follows:

Thymol blue: Acids whose $K_a \geqslant 10^{-9}$
Azo violet: Acids whose $K_a \geqslant 10^{-11}$
2-Nitroaniline: Acids whose $K_a < 10^{-11}$

These are general suggestions and differences may be noted in specific cases. Thymol blue (Table 9-1) is used for the standardization of sodium methoxide against benzoic acid and for the titration of most acids with a K_a larger than 10^{-9} in dimethylformamide solvent. The 2-nitroaniline indicator is used for very weak acids which must be titrated in ethylenediamine solvent.

End points may also be located potentiometrically using a pH meter with a potential readout. Special types of reference electrodes are used to overcome instability problems associated with the saturated aqueous calomel electrode. The best approach *for acids* is to use a salt bridge containing the nonaqueous solvent (*tert*-butyl alcohol, etc.) containing a soluble organic salt such as tetrabutylammonium bromide and achieve electrical contact between the solution and the saturated calomel electrode with the salt bridge [6]. The potentiometric titration of bases can be performed using a standard saturated calomel reference electrode without a salt bridge [5]. The glass electrode is usually used as the indicating electrode for titration of acids and bases. Fritz [5] recommends storing the glass electrode in water between periods of use to avoid possible dehydration of the outer layer.

[6] L. W. Marple and J. S. Fritz, *Anal. Chem.* **34**, 796 (1962).

Titration of Bases. The most important titrant is perchloric acid dissolved in glacial acetic acid. Since concentrated perchloric acid contains 28% water, acetic anhydride is added to the solution to react with the water and produce more acetic acid. The perchloric acid is standardized against primary standard potassium acid phthalate (KHP) which acts as a base in this case.

$$HClO_4 + KHC_8H_4O_4 \rightarrow H_2C_8H_4O_4 + KClO_4$$
$$\text{(KHP)} \qquad \text{(phthalic acid)} \quad \text{(potassium perchlorate)}$$

The most common indicators used are Methyl violet and crystal violet. Both change colors through several shades—from violet to blue to green to yellow, so the correct color change has to be established for each base. Again a pH meter and special electrodes may be used to locate the end point.

Application to Acids and Pharmaceuticals

Many types of organic compounds may be titrated as acids in nonaqueous solvents. These include carboxylic acids, phenols, imides (—CO—NH—CO—), aliphatic nitro compounds, and sulfa drugs [5]. The reaction of imides is

$$-CO-NH-CO- + NaOCH_3 \rightarrow (-CO-N-CO-)^{-1} + CH_3OH + Na^+$$

The above type of reaction also occurs with many sulfa drugs, such as sulfanilamide and sulfathizole, whose structures are as follows:

Sulfanilamide   Sulfathiazole

Most sulfa drugs such as sulfathiazole are fairly acidic and can be titrated in dimethylformamide solvent using thymol blue indicator. Sulfanilamide is very weakly acidic and does not react under these conditions. It can be titrated quantitatively using azo violet indicator and butylamine solvent. Mixtures of acidic sulfa drugs such as sulfathiazole and sulfanilamide can be analyzed by two titrations.

1. $NaOCH_3$ + azo violet: Total of sulfanilamide + acidic sulfa drug (sulfathiazole, etc.)

2. $NaOCH_3$ + thymol blue: Acidic sulfa drug only

The amount of sulfanilamide present is found by subtracting the mmoles of acidic sulfa drug from titration 2 from the mmoles of the total found in titration 1.

Application to Bases

In general, there are two types of weak organic bases that can be determined— organic amines (aliphatic and many aromatic) and salts of weak acids. The reactions are as follows:

$$RNH_2 + HClO_4 \rightarrow RNH_3{}^+ClO_4{}^-$$

$$CH_3CO_2{}^-Na^+ + HClO_4 \rightarrow CH_3CO_2H + NaClO_4$$

(sodium acetate)

Mixtures of bases which differ by a factor of 10^3 in their K_b values can also be titrated to obtain the amount of each [5].

QUESTIONS AND PROBLEMS

(Answers to most even-numbered problems are in Appendix 5.)

Concepts and Definitions
1. List at least two advantages of a direct titration over a back titration assuming that the end point can be clearly seen by each method.
2. What is the advantage of a back titration over a direct titration for a suspension such as Milk of Magnesia?
3. Define a primary standard for acid-base titrations. What are some requirements for an acid-base primary standard?
4. Explain why sodium hydroxide and hydrochloric acid cannot be considered primary standard reagents under normal conditions. Is hydrochloric acid ever used as a primary standard?
5. Define what is meant by the pH transition range of an acid-base indicator.
6. Of what use is the middle half of the pH transition range of an acid-base indicator?
7. Explain the three main steps of the Kjeldahl method for determining nitrogen.
8. When boric acid is substituted for hydrochloric acid in the Kjeldahl method, what are the two reactions that are used for the analysis?
9. What are the advantages of a nonaqueous titration over an aqueous titration?

Feasibility of Aqueous Titrations
10. Calculate K_{rxn} for the reaction of
    a. A strong acid with a weak base.
    b. A strong base with a weak acid.

SEC. 9-3   Nonaqueous Titration Methods: Pharmaceuticals   209

11. Do the values in the previous question imply that all weak acids or weak bases can be titrated in water? Why or why not?
12. Calculate K_{rxn} for the reaction of boric acid and ammonia (9-19) to two significant figures. Interpret this value's relevance to the Kjeldahl method.
13. Calculate K_{rxn} for the reaction of acetic acid with ammonia. Can this reaction be used for a quantitative acid-base titration?

Calculation of Titration Results
14. A 0.5000 g sample of NaH_2PO_4 is neutralized completely to Na_2HPO_4 with 16.10 ml of 0.0984M sodium hydroxide. Calculate the percentage of NaH_2PO_4 (form wt = 119.98) in the sample.
15. A 0.1000 g sample of impure sodium carbonate (form wt = 106.0) is neutralized to carbonic acid with 21.00 ml of 0.0450M hydrochloric acid. Calculate the per cent purity of the sodium carbonate.
16. A 1.0000 g impure sample containing sodium carbonate and sodium bicarbonate is dissolved in water and titrated with 0.1000M hydrochloric acid. The buret reading at the phenolphthalein end point is 15.5 ml, and at the methyl red end point it is 40.1 ml. Calculate the percentages of sodium carbonate and sodium bicarbonate in the sample.
17. A mixture of bases is titrated with hydrochloric acid first to a phenolphthalein end point, then the titration is continued to a methyl red end point. The buret reading at the first end point is 28.0, and the buret reading at the second end point is 40.0 ml. From this information,
    a. Decide whether the mixture is sodium hydroxide-sodium carbonate or sodium carbonate-sodium bicarbonate.
    b. Compute the number of milliliters of hydrochloric acid needed to titrate the sodium carbonate present.
18. A 1.000 g sample of food is analyzed for nitrogen by the Kjeldahl method. After digestion of the sample, the ammonia is distilled and collected in a receiver containing exactly 50.00 ml of 0.1000M hydrochloric acid. The *unreacted* hydrochloric acid requires 24.60 ml of 0.1200M sodium hydroxide for back titration. Calculate the percentage of nitrogen (N) in the sample.
19. Calculate the pH at the end point in the titration of 500 mg of potassium acid phthalate (form wt = 204.2) dissolved in 50 ml of water and titrated with 0.1000M sodium hydroxide. Choose an indicator suitable for detecting the end point. The ionization constants for phthalic acid are $K_1 = 1.2 \times 10^{-3}$ and $K_2 = 3.9 \times 10^{-6}$ ($\mu = 0$).
20. Calculate the pH at the end point in the titration of 0.2M potassium acid maleate (KHM) with 0.2M sodium hydroxide. Choose a suitable indicator for detecting the end point. The ionization constants for maleic acid (H_2M) are $K_1 = 1.2 \times 10^{-2}$ and 9×10^{-7}.

Challenging Problems
21. Titration of 664 mg of a pure unknown organic carboxylic acid gives a potentiometric break at 40.00 ml and at 80.00 ml of 0.1000M sodium hydroxide titrant. The pH at which 20.00 ml of titrant has been added is 2.78; the pH at which 60.00 ml of titrant has been added is 5.10. The ionic strength is essentially constant at 0.1 throughout.
    a. Is the acid monoprotic, diprotic, triprotic, or what?
    b. Calculate the value(s) of the ionization constant(s) of the acid.
    c. Calculate the formula weight of this acid.

d. If the acid is $C_xH_y(CO_2H)_n$, write a possible molecular formula. (*Hint:* assume in turn that it is *saturated*, monounsaturated, and so on, using general formulas such as C_xH_{2x+1} for a *saturated* carbon-hydrogen grouping.)
   e. Write a logical structure for the acid that is consistent with all facts.
22. Titration of 172.0 mg of a pure unknown organic acid, $R-(CO_2H)_n$ requires 20.00 ml of $0.1000M$ sodium hydroxide titrant. If the molecular weight of the acid is 172 ± 0.1,
   a. Calculate the value of n. (Is the acid monoprotic or diprotic?)
   b. Calculate a numerical value for R.
   c. Give at least two reasonable but different structures for R.
   d. Given that the acid does not react by addition with bromine or with ozone, suggest a reasonable structure for R.
   e. Given that the titration curve shows only one break and that the pH at 5.00 ml of titrant is 5.60, suggest a reasonable structure for the acid.
23. After consulting appropriate references, state whether each compound in each mixture can be determined by a nonaqueous differentiating titration, using either a potentiometric or a visual indicator end point. Also state which compound will be titrated first and give numerical reasons.
   a. Sulfathiazole and sulfanilamide
   b. Sulfathiazole and sulfapyridine
   c. Phenol and acetic acid
   d. Pyridine and *n*-butylamine
   e. Phenol and 2,4-dinitrophenol
   f. *m*-Nitrophenol, 2,4-dinitrophenol, and trinitrophenol (picric acid)
   g. 2,4,6-Trinitroaniline and 2,4-dinitroaniline

10. Precipitation and Complexation Titrations

> *"I've found it!"* Holmes shouted, running towards us with a test tube in his hand. *"I have found a reagent which is precipitated by hemoglobin* and by nothing else.... *Now we have the Sherlock Holmes test."*
>
> ARTHUR CONAN DOYLE
> A Study in Scarlet, Chapter 1

The reactions to be discussed in this chapter are aptly symbolized by Sherlock Holmes' experiment on hemoglobin. One class of reactions is based on a metal ion being *precipitated*, and the other on a metal ion being *complexed*, like iron(II) is complexed in hemoglobin. (Holmes of course may not have been aware of the fact that hemoglobin is a complex ion.)

Whether a metal ion is precipitated or kept soluble by complexation depends on the type of ligand involved (Sec. 7-1). *Unidentate* anionic ligands, being ligands that occupy one coordination position around the central metal ion, may precipitate certain $+1$ cations. For example, chloride ion can be used to precipitate silver(I) ion as silver(I) chloride. A large excess of chloride ion can also form $AgCl_2^-$ and other complex ions. *Bidentate* anionic ligands, or ligands that occupy two coordination positions around the metal ion, may precipitate certain $+2$ cations. For example, sulfate ion can be used to precipitate barium(II) ion as barium(II) sulfate. A large excess of sulfate ion can form complex ions with certain other metal ions. Few other types of anionic ligands are used for precipitation titrations, but *hexadentate* ligands such as ethylenediaminetetraacetic acid (EDTA) are used for complexation titrations. This of course involves chelate (Sec. 7-1) formation.

We will discuss precipitation titrations first and then proceed to examine complexation reactions, especially those involving EDTA.

10-1 PRECIPITATION TITRATIONS

Most of the precipitation reactions used in gravimetric analysis are unsuitable for the titrimetric determination of either the anion or the cation. The two types of reactions that are rapid and stoichiometric enough for analysis are the titration of halide ions with silver(I) ion, and the titration of sulfate ion with barium(II) ion. Even in the cases of the silver halides and barium sulfate the reverse titrations are not as widely used. Nevertheless, the titration of the halides with silver(I) nitrate is very important because of the widespread need for chloride analysis of body fluids.

General Characteristics of Silver Halide Titrations

The reactions of silver(I) ion with the various halide ions (Cl^-, Br^-, and I^-) and with SCN^- are usually quantitative at the $0.1M$ level. Recall from Section 7-2 that the equilibrium constant, K_{rxn}, for such titrations is calculated (**7-16**) as follows:

$$K_{rxn} = \frac{1}{K_{sp}} = \frac{1}{[Ag^+][X^-]} \tag{10-1}$$

For the precipitation of silver(I) chloride the value of K_{rxn} is

$$K_{rxn} = \frac{1}{1.8 \times 10^{-10}} = 5.6 \times 10^9$$

To decide whether this value corresponds to a quantitative reaction or not the minimum theoretical value of K_{rxn} must be calculated. For titrations at the $0.1M$ level where the halide ion is reduced to 0.1% of $0.1M$ at the end point, this was calculated to be (**7-17**)

$$\text{min theo } K_{rxn} = 1 \times 10^8 \quad (0.1M \text{ level})$$

Because the *actual value* of K_{rxn} for silver(I) chloride is greater than the above value we can say that the *direct titration* of $0.1M$ chloride ion with silver(I) ion is more than quantitative ($>99.9\%$). (*Back titrations* are more complicated and cannot be predicted by making just one comparison.)

Precipitation of Chloride in Body Fluids. The determination of chloride ion in many body fluids is very important. Chloride ion is measured routinely in the blood serum as one indication of the electrolyte balance in the blood (Sec. 22-1). The mean serum level of chloride is about $0.1M$ so that precipitation of chloride as silver chloride is quantitative.

Before chloride can be determined in most body fluids, any protein present must be removed from the fluid by precipitation and centrifugation. This avoids the side reaction of silver nitrate with the protein. If the protein is removed by adding nitric acid, the direct silver nitrate titration methods discussed in this section cannot be used since they require neutral or weakly acid pH. The Volhard (back titration) method can be used since it is carried out in nitric acid.

Chloride ion can also be determined by a coulometric method (Sec. 18-3) in which silver(I) ion is generated from an electrode and the end point is signaled by a sudden increase in current rather than by an indicator. There are also a few methods for chloride that are not based on precipitation of silver chloride. One of these is the titration of chloride with mercury(II) nitrate described in Section 10-2. Another is the specific ion electrode measurement of "sweat chloride" used as a means of detecting cystic fibrosis in infants (Sec. 17-3).

Since chloride in some body fluids as well as in other samples may be less concentrated than $0.1M$, we should ask whether the titration of $0.01M$ chloride will be quantitative. Using the approach in **7-17** the minimum theoretical K_{rxn} will involve reducing the chloride to 0.1% of $0.01M$, or $1 \times 10^{-5} M$:

$$\text{min theo } K_{rxn} = \frac{1}{[1 \times 10^{-5} M \text{ Ag}^+][1 \times 10^{-5} M \text{ Cl}^-]} = 1 \times 10^{10}$$

This of course applies to all halide ions, not just the chloride ion. The actual value of K_{rxn} for silver(I) chloride is only 5.6×10^9 so the titration of $0.01M$ levels of chloride is just short of quantitative. (It can be shown that titration of $0.015M$ and higher levels of chloride is quantitative, however.) Titration of 0.001–$0.1M$ bromide, iodide, and thiocyanate can also be shown to be quantitative.

Titration curves. It is helpful to consider the shape of silver halide titration curves even though most titration methods do not depend on plotting such a curve. Just as an acid-base titration curve is a plot of pH *vs.* milliliters of titrant, so a silver halide titration curve is a plot of pCl or pAg *vs.* milliliters of silver nitrate titrant. Recall that the definition of pCl is (Sec. 7-2)

$$\text{pCl} = -\log[\text{Cl}^-]$$

and pBr, pAg, and so on, are defined similarly.

At the start of a titration of, say, 50 ml of $0.10M$ chloride the $\text{pCl} = 1.00$. As silver nitrate titrant is added the $[\text{Cl}^-]$ decreases and pCl increases (Fig. 10-1, p. 124). At 90% reaction only 10% of the $[\text{Cl}^-]$ will be left, and its concentration will be $0.010M$; neglecting dilution, the pCl is

$$\text{pCl} = -\log(10\%)(0.10M) = -\log 1.0 \times 10^{-2} = 2.00$$

At 99.9% reaction, or 0.1% before the equivalence point, the pCl is

$$\text{pCl} = -\log(0.1\%)(0.10M) = -\log[1.0 \times 10^{-4} M] = 4.00$$

Again, we are neglecting dilution. At the equivalence point $[\text{Ag}^+] = [\text{Cl}^-]$ and

$$K_{sp} = [\text{Ag}^+][\text{Cl}^-] = [\text{Cl}^-][\text{Cl}^-]$$

$$[\text{Cl}^-] = \sqrt{K_{sp}} = \sqrt{1.8 \times 10^{-10}} = 1.3_4 \times 10^{-5} M \qquad (10\text{-}2)$$

$$\text{pCl} = 4.87$$

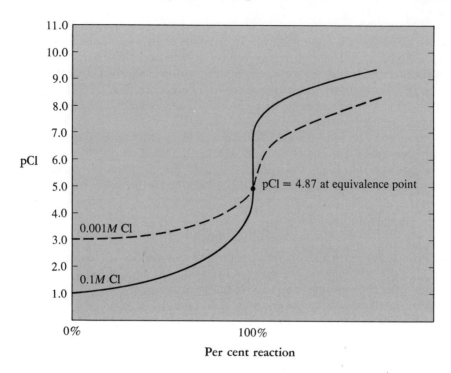

FIGURE 10-1. *Titration curves of 0.1M chloride with 0.1M silver nitrate and 0.001M chloride with 0.001M silver nitrate. Note that the pCl at the equivalence point is the same for both, but that the titration curve of the latter exhibits no steep portion in the region of the equivalence point.*

At 100.1% reaction, or 0.1% after the equivalence point, an excess of 0.00010 M silver(I) has been added, neglecting dilution. At this point, pCl is

$$[Cl^-] = \frac{K_{sp}}{[Ag^+]} = \frac{1.8 \times 10^{-10}}{1.0 \times 10^{-4} M} = 1.8 \times 10^{-6} M$$

$$pCl = 5.74$$

As you can see in Figure 10-1 the change in pCl in the vicinity of the equivalence point is from 4.00 to 5.74, or 1.74 (neglecting dilution). At the same time the pAg is decreasing by the same amount. Either change is large enough so that it can be detected by electrical means (Sec. 17-3) or by the use of indicators. The latter will be discussed below.

Direct Titration of Halide Ions

There are two direct titration methods for halide ions—the Mohr method and the adsorption indicator (Fajans) method. Since the chemistry of the indicator places different limitations on each method, they will be discussed separately. The Mohr method has the simpler chemistry, and we will begin with it.

The Mohr Method for Halide Ions. The Mohr method was published in 1856 [1] and is still widely used for the titration of chloride ion. The titration reaction is

$$AgNO_3 + Cl^- \rightarrow AgCl(s) + NO_3^- \tag{10-3}$$

(titrant)

Yellow sodium chromate is added at the beginning of the titration and the approach of the end point is signaled by temporary flashes of orange-red silver(I) chromate. The end point is signaled by the first permanent appearance of insoluble orange-red silver chromate.

$$2AgNO_3 + CrO_4^{-2} \rightarrow Ag_2CrO_4(s) + 2NO_3^- \tag{10-4}$$

(titrant)   (orange-red)

The silver(I) chromate is more soluble than silver(I) chloride so it does not precipitate until the equivalence point is reached. Unfortunately, the end point is somewhat difficult to see over the white silver(I) chloride precipitate and the yellow sodium chromate; a slight excess of titrant is usually necessary to obtain enough silver(I) chromate to see the color change. If the silver nitrate titrant is standardized against primary standard potassium chloride using the same conditions, this error may cancel out. However, it is best to run a blank and subtract this from each titration; the correction is usually in the 0.05–0.15 ml range.

The concentration of chromate is critical. If not enough is present, the end point is premature; if too much is present, the end point is late. The concentration can be estimated using the solubility product expression of silver(I) chromate and the concentration of silver(I) at the equivalence point. At this point, $[Ag^+] = [Cl^-]$. From **10-2** the $[Cl^-] = 1.3_4 \times 10^{-5} M$. Substituting this into the silver(I) chromate expression we have

$$[CrO_4^{-2}] = \frac{K_{sp}}{[Ag^+]^2} = \frac{1.1 \times 10^{-12}}{[1.3_4 \times 10^{-5} M]^2} = 6.1 \times 10^{-3} M$$

Theoretically the correct chromate ion concentration for **10-4** to indicate the end point is $0.006 M$; experimentally a range of $0.005–0.01 M$ has been found to be satisfactory. However, the pH is also critical; the pH should be in the range 6.5–10.3 [2]. Below pH 6.5 the acidity of the solution dissolves too much silver(I) chromate. The reactions are:

$$Ag_2CrO_4(s) = 2Ag^+ + CrO_4^{-2} + H^+ = HCrO_4^-$$

$$2HCrO_4^- = Cr_2O_7^{-2} + H_2O$$

(dichromate)

[1] F. Mohr, *Annalen der Chemie und Pharmacie* **97**, 335 (1856).
[2] R. Belcher, A. M. G. MacDonald, and E. Parry, *Anal. Chim. Acta* **16**, 524 (1957).

As the pH decreases the ratio of CrO_4^{-2} to $HCrO_4^-$ decreases. At pH 7 $[CrO_4^{-2}]/[HCrO_4^-] = 3/1$, but at pH 6 this ratio drops to $0.3/1$. Above pH 10.3 silver(I) begins to coprecipitate as silver(I) hydroxide. Thus, the pH cannot be acid because of the indicator, or too basic because of a side reaction of the solvent.

Applications: Theoretically, the Mohr method should be satisfactory for the determination of any anion which forms a more insoluble salt than silver chromate. In practice, it has been found that highly colored silver(I) salts prevent the indicator color change from being seen. It has also been found that some anions are adsorbed strongly on the surface of their precipitated silver salts so that not all of the anion is precipitated at the apparent end point.

Of the anions, chloride, bromide, and cyanide can be determined by the Mohr method. This includes hydrochloric acid and hydrobromic acid, which of course must first be neutralized, as well as alkali metal chlorides, bromides, and cyanides. (Metal ions other than alkali metals will tend to precipitate the chromate ion.) The thiocyanate ion and the iodide ion cannot be determined because they are too strongly adsorbed on AgSCN and AgI. In addition, silver iodide is yellow and would obscure the end point. Of the remaining white insoluble silver(I) salts, the only common ones are silver(I) iodate and silver(I) oxalate. Both salts have K_{sp}s which are larger than that of silver(I) chromate, indicating that the end point would not be accurate.

The Adsorption Indicator (Fajans) Method. In 1924 Fajans and his coworkers [3] observed that certain organic dyes changed colors when adsorbed on the surface of insoluble silver halide salts. The fluorescein family of dyes was found to work especially well for the titration of the halide ions. The indicator action involved the fluorescein anion Fl^- ionizing from the acid form, HFl, of the fluorescein molecule.

The titration reaction for the chloride ion is the same as given in **10-3** for the Mohr titration method. The *overall* reaction at the end point is

$$Ag^+ + AgCl(s) + Fl^- \rightarrow AgCl:Ag^+/Fl^-(s) \quad (10\text{-}5)$$

(excess titrant)   (yellow)   (red)

Note that the fluorescein anion is yellow before the end point and while it is in solution. At the end point it adsorbs on the surface of the silver chloride precipitate and changes color to a pink or red. Apparently the adsorption of the anion changes the type of light absorbed by the anion.

Because the color change occurs on the surface of the precipitate there must be enough surface for a reasonable amount of the indicator to be adsorbed and cause a noticeable change. Silver halide precipitates are known to coagulate, which reduces the amount of available surface. Therefore, dextrin or polyethylene glycol are usually added to keep the precipitate in the colloidal form and prevent coagulation. Coagulation is retarded by performing the titration quickly and avoiding excessive stirring. Dilute samples of halide ions also limit the amount of surface

[3] K. Fajans and H Wolff, *Z. Anorg. Allg. Chem.* **137**, 221 (1924).

available because the amount of precipitate is small. The most suitable concentrations are in the 0.005–0.025M range.

Because the fluorescein dyes are weakly ionized acids, too high an acidity reduces the concentration of the Fl^- anion to a point where the pink-red color is too faint to be seen. When fluorescein itself is used the pH must be above 7; dichlorofluorescein is less affected by acidity and can be used above pH 4.

Some of the fluorescein dyes are not suitable for the titration of chloride because their color changes occur before the equivalence point. At pH 7 dichlorofluorescein cannot be used for chloride for this reason. It apparently displaces the chloride ion from the surface of silver chloride somewhat before the equivalence point.

$$AgCl:Cl^-/Na^+(s) + Fl^- \rightarrow AgCl:Fl^-/Na^+(s) + Cl^-$$

By lowering the pH to 4 the concentration of the anion is reduced to the point that chloride can be titrated accurately.

One other limitation is that such titrations cannot be carried out in sunlight or strong room light because of photochemical decomposition which blackens the precipitate and interferes with the color change.

Applications: Theoretically, the adsorption indicator method should be satisfactory for the determination of any anion which is not displaced from the surface of its insoluble silver(I) salt by the indicator anion. In practice, this is true. Thus chloride, bromide, iodide, and thiocyanate anions can all be titrated readily with silver nitrate. The silver(I) ion can also be titrated with a standard solution of potassium chloride using a positively charged dye such as methyl violet. This is an advantage over the Mohr method which is impractical for the determination of silver(I) because the silver(I) chromate does not dissolve completely at the equivalence point. No methods appear to have been reported for the titration of iodate or oxalate [4]. Neither the adsorption indicator method nor the Mohr method appear to be satisfactory for the titration of chloride ion in blood serum or urine because of the presence of interfering protein and other organic material [5].

Back Titration of Halide Ions

There is only one widely used back titration method for halide ions; that is the Volhard method, published in 1874 [6]. This method is widely used because it permits halide ions to be titrated in acid solution. In the method a measured excess of standard silver(I) nitrate is added to an acid solution of the halide.

$$AgNO_3(excess) + Cl^- \rightarrow AgCl(s) + NO_3^- \tag{10-6}$$

[4] H. F. Walton, *Principles and Methods of Chemical Analysis*, Prentice-Hall, Englewood Cliffs, N.J., 1964, pp. 373–86.
[5] R. L. Searcy, *Diagnostic Biochemistry*, McGraw-Hill, New York, 1969, p. 156.
[6] J. Volhard, *J. Prakt. Chem.* 9, 217 (1874).

The unreacted silver nitrate is then back titrated with standard thiocyanate.

$$\text{KSCN} + \text{unreacted AgNO}_3 \rightarrow \text{AgSCN(s)} + \text{K}^+ + \text{NO}_3^- \qquad (10\text{-}7)$$

(titrant)

At the start of the back titration $0.02M$ iron(III) is added to the solution. At the equivalence point the excess thiocyanate complexes with the iron(III) to form the red thiocyanatoiron(III) ion.

$$\text{KSCN} + \text{Fe}^{\text{III}} \rightarrow \text{Fe(SCN)}^{+2} + \text{K}^+$$

(titrant) (yellow)   (red)

Since the K_{sp} for silver(I) thiocyanate is 1.1×10^{-12} the silver chloride, and certain other insoluble silver(I) salts, formed in **10-6** will be more soluble than the silver thiocyanate. The thiocyanate ion added in **10-7** can react with the insoluble silver chloride as follows:

$$\text{AgCl(s)} + \text{KSCN} \rightleftharpoons \text{AgSCN(s)} + \text{Cl}^- + \text{K}^+ \qquad (10\text{-}8)$$

If this occurs, too much thiocyanate titrant will be added in the back titration and the results for chloride will be low. This error may be avoided by filtering the silver chloride before back titrating, but this is slow and subject to possible errors from adsorption of some of the excess silver(I) ion. Another way of avoiding the error in **10-8** is to add a little liquid nitrobenzene before the back titration to coat the silver chloride and prevent the equilibrium in **10-8** from being established. It has been claimed that the nitrobenzene only slows down the reaction and does not prevent it [4], so the titration should be conducted as rapidly as possible.

Applications: The Volhard method is satisfactory for the determination of chloride, bromide, iodide, and thiocyanate. It can also be used for the determination of the anions of weak acids, such as oxalate and carbonate, by precipitating them in neutral solution, filtering away from the unreacted silver nitrate, and titrating the compound with potassium thiocyanate after dissolving in acid. Because the digestion of protein requires nitric acid, chloride ion in blood serum and urine can be analyzed after digestion in this acid using the Volhard method [5].

The number of mmoles (meq) of halide ion present must be calculated by subtraction.

$$\text{mmoles (or meq) halide} = (M\ \text{AgNO}_3)(\text{ml}) - (M\ \text{KSCN})(\text{ml})$$

The following example illustrates a typical Volhard calculation.

EXAMPLE 10-1. Calculate the percentage of sodium chloride in a 1.0000 g sample which requires 50.00 ml of $0.1000M$ silver nitrate for precipitation and 10.00 ml of $0.2000M$ potassium thiocyanate for back titration. Use 58.44 mg/mmole as the formula weight of sodium chloride.

SEC. 10-1   Precipitation Titrations

Solution: First calculate the number of mmoles (meq) of sodium chloride.

mmole NaCl = (0.1000M AgNO$_3$)(50.00 ml) − (0.2000M KSCN)(10.00 ml)

mmole NaCl = 5.000 − 2.000 = 3.000 mmoles

Then calculate the percentage.

$$\% \text{ NaCl} = \frac{(3.000 \text{ mmoles NaCl})(58.44 \text{ mg/mmole})(100)}{1000.0 \text{ mg}} = 17.53\%$$

Adsorption Indicator Method for Sulfate

There is also an adsorption indicator method for the titration of the sulfate ion that is quick and reasonably accurate when interfering cations are removed by ion-exchange [7]. The titration is carried out at pH 3.5 in a solvent of approximately 50% water and 50% methyl alcohol.

Alizarin red S, a dye, is used as the indicator. It is yellow in solution but forms a pink complex when it is adsorbed on the surface of the barium sulfate in the presence of a slight excess of barium(II) ion. Using AR$^-$ as the symbol for the indicator anion, the reaction at the end point is

$$\text{Ba}^{+2} + \text{BaSO}_4(s) + \text{AR}^- \rightarrow \text{BaSO}_4:\text{Ba}^{+2}/2\text{AR}^-(s)$$

(excess titrant)   (yellow)   (pink)

Cations such as aluminum(III), iron(III), and potassium(I) interfere seriously, but they can be removed by an ion exchange column (Ch. 21) before the titration. Chloride, bromide, and perchlorate anions cause only a small error, but nitrate causes a large positive error.

Self-Test 10-1. Calculation of Precipitation Titration Results

Directions: Compare your answers with those given at the end of the test.

A. Calculate the percentage bromide in a 1.0000 g sample which is analyzed by the Volhard method. 50.00 ml of 0.1000M silver nitrate is added; after precipitation the unreacted silver nitrate is back titrated with 15.00 ml of 0.0800M potassium thiocyanate. Use 79.9 mg/mmole for the formula weight of bromide.

B. Using either the molarity or normality system of calculation in Section 6-2 (see **6-2** for equivalent weight), calculate the percentage purity of each compound analyzed by direct titration.

[7] J. S. Fritz and M. Q. Freeland, *Anal. Chem.* **26**, 1593 (1954).

a. A 1.0000 g sample of impure magnesium chloride (form wt = 95.22) requires 35.00 ml of 0.1000M silver nitrate in the Mohr method.

b. A 1.0000 g sample of impure aluminum chloride (form wt = 133.34 mg/mmole) requires 44.00 ml of 0.1000M silver nitrate in the adsorption indicator method.

C. A 1.0000 g sample of chloride is analyzed by the Mohr method; it requires 31.00 ml of 0.1100M silver nitrate for titration. Another 1.0000 g sample is analyzed by the adsorption indicator method using dichlorofluorescein at pH 7; it requires 31.50 ml of 0.1050M silver nitrate.

a. Calculate the percentage chloride (form wt 35.453 mg/mmole) in the sample by both methods.

b. If there is a difference in the percentage chloride, suggest which of the two methods is the more accurate, and why.

D. A pure compound is either barium chloride (form wt 208.25) or barium chloride dihydrate ($BaCl_2 \cdot 2H_2O$, form wt 244.27). Titration of 732.8 mg of the pure compound requires 37.50 ml of 0.1600M silver nitrate. Which compound is it?

Answers to Self-Test 10-1

A. 30.36%
B. a. 16.66%   b. 19.56%
C. a. 12.09% (Mohr); 11.73% (adsorption)   b. The pH should be 4 for titration of chloride using dichlorofluorescein.
D. $BaCl_2 \cdot 2H_2O$

10-2 COMPLEXATION TITRATIONS

The ligands discussed in the previous section were mainly *unidentate* ligands which *precipitated* the metal ion as an insoluble salt or as a special type of insoluble 1:1 coordination (complex) compound. These precipitation reactions are exceptions to the usual situation in which a *soluble* complex ion, or coordination compound, is produced by the reaction of a metal ion with one or more ligands. The advantage of precipitation is that it terminates the reaction of the metal ion with the ligand at a *definite number* of ligands. Furthermore the reactions are quantitative because the precipitation favors this. The formation of complex ions, in contrast, does not necessarily possess either of these advantages. This is especially true in the case of unidentate ligands; see, for example, the reaction of copper(II) with ammonia (Sec. 7-1). The equilibrium constants for each step are not large and decrease with each successive step. It is therefore not surprising that there are very few complexation titrations based on reactions of unidentate ligands with metal ions.

Two exceptions to the above general trend involve the reactions of mercury(II) ion with chloride and bromide ions. The values of 2×10^5 and 3×10^7 for K_1 and K_2, respectively, for mercury complexing with chloride are so much larger than K_3 (1.4×10^1) and the other Ks that mercury(II) nitrate can be used to titrate chloride in body fluids and other samples.

$$Hg(NO_3)_2 + 2Cl^- \rightarrow HgCl_2 + 2NO_3^-$$

After mercury(II) chloride has formed quantitatively, the first excess of mercury(II) nitrate forms a purple color with the diphenylcarbazone indicator used. The same approach is used for the determination of bromide. Because the stepwise stability constants K_1 and K_2 are so much larger than K_3 for mercury(II) complexing with bromide ion, the bromide ion can be determined by adding an excess of standard mercury(II) perchlorate to it and using a modified Volhard method.

$$Hg(ClO_4)_2 + 2Br^- \rightarrow HgBr_2 + 2ClO_4^-$$
(excess)

Note that mercury(II) bromide is *soluble*; the reaction terminates at the $HgBr_2$ stage rather than at the $HgBr^+$ or $HgBr_3^-$ stages because of the great stability of this coordination compound. The unreacted mercury(II) ion is then back titrated using a standard solution of thiocyanate.

$$Hg(ClO_4)_2 + 2KSCN \rightarrow Hg(SCN)_2 + 2KClO_4$$
(titrant)

Because of the mentioned disadvantages of most of the unidentate ligands, *hexadentate* ligands have been employed to react with metal ions for analysis. These ligands form rings with the metal ions in which as many as six of the coordination positions around the metal ion are occupied by six different atoms of the *same* ligand. If we use A_1 through A_6 for the six atoms, the positions they occupy can be pictured as shown. This is the well-known *octahedral* structure in which

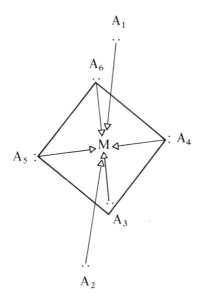

four of the atoms (A_3–A_6) are positioned in a plane around the central metal M, atom A_1 is above the plane, and atom A_2 is below the plane. We shall now discuss the most important of such hexadentate ligands, ethylenediaminetetraacetic acid (EDTA).

The Chemistry of EDTA

Ethylenediaminetetraacetic acid (EDTA) is a tetraprotic acid. Its structure is

$$\left[\begin{array}{c} H{-}OOCCH_2 \\ \\ H{-}OOCCH_2 \end{array} \right. \begin{array}{c} \ddots \\ N{-}CH_2CH_2{-}N \\ \end{array} \left. \begin{array}{c} CH_2COO{-}H \\ \\ CH_2COO{-}H \end{array} \right]$$

EDTA ($=H_4Y$)

Each of the protons outside the brackets is lost when the adjoining oxygen bonds to a metal ion through the pair of electrons freed by the loss of the proton. EDTA is thus a tetraprotic acid which can be represented as H_4Y, where Y^{-4} represents the anion inside the brackets. As many as four of the above oxygen atoms can occupy as many as four of the octahedral positions around a metal ion. The other two positions are occupied by the two nitrogen atoms which also bond to the metal via their electron pairs. Because the resulting complex ion consists of ring structures it is called a *chelate*. The general formula for any +2 metal-EDTA chelate is MY^{-2}. A simplified representation of the structure of such a chelate may be drawn by showing only the four oxygens and two nitrogens of EDTA as follows:

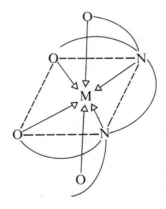

Note that the metal ion is completely surrounded by the EDTA; it is no wonder that such chelates are so stable and can be used for quantitative analysis.

The usual form of EDTA used in the laboratory is the disodium salt

Na_2H_2Y. When this form is used as the titrant the titration reaction is

$$Na_2H_2Y + M^{+2} \rightarrow MY^{-2} + 2H^+ + 2Na^+ \tag{10-9}$$

Note that the reaction produces hydrogen ion; near the equivalence point it is possible that this would prevent the reaction from being quantitative by breaking apart the MY^{-2} chelate to form HY^{-3}, H_2Y^{-2}, and so on. Therefore a buffer is used for the titration of many metals; for the titration of calcium(II) and/or magnesium(II) the solution is generally buffered at a pH of 10.

A large number of metal ions may be determined by the kind of direct titration described by **10-9**. The reason for this is that they form very stable 1:1 chelates with EDTA. Their stability constants, K_{MY}, (Sec. 7-1) are listed in Table 10-1.

TABLE 10-1. Stability Constants of Metal-EDTA Chelates

Metal Ion	Log K_{MY}	Metal Ion	Log K_{MY}
Fe^{+3}	25.1	Co^{+2}	16.3
Th^{+4}	23.2	Al^{+3}	16.1
Cr^{+3}	23	Ce^{+3}	16.0
Bi^{+3}	22.8	La^{+3}	15.4
VO^{+2}	18.8	Mn^{+2}	14.0
Cu^{+2}	18.8	Ca^{+2}	10.7
Ni^{+2}	18.6	Mg^{+2}	8.7
Pb^{+2}	18.0	Sr^{+2}	8.6
Cd^{+2}	16.5	Ba^{+2}	7.8
Zn^{+2}	16.5		

Essentially all of these constants are large enough for any metal to be titrated quantitatively with EDTA.

The calcium(II)-EDTA chelate also has an important medical use in treating patients with lead poisoning. When it is administered to such a person the lead(II) displaces the calcium(II) from the calcium(II)-EDTA because the lead(II)-EDTA chelate is more stable.

$$Body\text{-}Pb(II) + CaY^{-2} \rightarrow PbY^{-2} + Body\text{-}Ca(II)$$

(The stability constant for PbY^{-2} is $10^{18.0}$ compared to that for CaY^{-2} which is only $10^{10.7}$.) The lead(II) can no longer adhere to the body surfaces and is readily washed out of the body as PbY^{-2}. The EDTA is added as the calcium chelate to prevent it from reacting with calcium(II) already present in the body.

The indicators used for EDTA titrations are mainly weakly acidic organic dyes. The first of these was Eriochrome Black T (EBT) reported in 1948 [8].

[8] G. Schwarzenback and W. Biedermann, *Helv. Chim. Acta* **31**, 678 (1948).

An indicator like EBT functions by forming a colored metal complex, M-EBT$^-$, at the start of the titration. As long as some metal remains unchelated by EDTA the solution remains the color of the M-EBT$^-$ complex. At the equivalence point the EDTA removes the metal ion from the M-EBT$^-$ complex by chelating it, and the solution changes colors.

$$Na_2H_2Y + \text{M-EBT}^- \rightarrow HEBT^{-2} + MY^{-2} + 2Na^+ + H^+ \quad \text{(10-10)}$$

(titrant)   (color before end point)   (color at end point)

Formerly, the disodium salt of EDTA was not available in pure enough form for use as a primary standard. There is now available a 99.0% minimum purity A.C.S. grade which can be used as a primary standard for analyses that can tolerate one per cent error. Pure zinc metal can also be used as a primary standard after it has been dissolved in acid to form the zinc(II) ion. Calcium carbonate is also available in pure enough form to be used as a primary standard for calcium and/or magnesium analysis after it is dissolved in acid.

$$CaCO_3(s) + 2H^+ \rightarrow Ca^{+2} + CO_2(g) + H_2O$$

Calculation of Titration Results

The concentration of EDTA titrant may be expressed in molarity or in terms of the *titer* of EDTA for a certain ion or compound. The titer of EDTA for any species is defined as the weight of that species reacting with or equivalent to a unit volume of EDTA. Usually this is defined specifically as

$$\text{EDTA titer} = \frac{\text{mg of species}}{1 \text{ ml EDTA}} = \frac{\text{g of species}}{1 \text{ liter of EDTA}}$$

The titer may be calculated from the molarity of the EDTA, but is most often calculated from the standardization data. For example, if EDTA is standardized against calcium carbonate, its titer may be calculated in terms of mg of calcium carbonate reacting (as Ca^{+2}) with one ml of EDTA.

EXAMPLE 10-2. Exactly 1.0010 g of pure calcium carbonate (formula weight = 100.1 mg/mmole) is dissolved in 100.0 ml of water. A 10.0 ml aliquot is titrated with 9.00 ml of EDTA. Calculate the molarity of the EDTA and its titer for calcium carbonate.

Solution: First calculate the concentration of the calcium carbonate.

$$\text{mg CaCO}_3/\text{ml} = 1001.0 \text{ mg}/100.0 \text{ ml} = 10.01 \text{ mg CaCO}_3/\text{ml}$$

$$M \text{ of CaCO}_3 = \frac{10.01 \text{ mg CaCO}_3/\text{ml}}{100.1} = 0.0100 M$$

SEC. 10-2   Complexation Titrations

Then calculate the molarity of the EDTA.

$$M \text{ of EDTA} = \frac{(10.0 \text{ ml CaCO}_3)(0.0100M \text{ CaCO}_3)}{9.00 \text{ ml EDTA}} = 0.0111M$$

Finally, calculate its titer by either method below.

$$\text{EDTA titer} = \frac{(10.00 \text{ ml CaCO}_3)(10.01 \text{ mg CaCO}_3/\text{ml})}{9.00 \text{ ml EDTA}} = \frac{11.1_2 \text{ mg CaCO}_3}{\text{ml EDTA}}$$

$$\text{EDTA titer} = \frac{(10.00 \text{ ml CaCO}_3)(0.0100M \text{ CaCO}_3)(100.1 \text{ mg/mmole})}{9.00 \text{ ml EDTA}}$$

$$\text{EDTA titer} = \frac{11.1_2 \text{ mg CaCO}_3}{\text{ml EDTA}}$$

The most common EDTA *direct titration* is the water hardness determination (see below). This consists of titrating the *sum* of calcium(II) and magnesium(II) ions in one titration. The results are calculated as parts per million (ppm) of calcium carbonate since this is the major compound causing water hardness. One definition of ppm of any species is the number of mg of that species dissolved in one liter. Thus for calcium carbonate

$$\text{ppm CaCO}_3 = \frac{\text{mg CaCO}_3}{\text{liter water}}$$

EXAMPLE 10-3. A 50.00 ml water sample from the Los Angeles river requires 12.00 ml of 0.0100M EDTA. Calculate the hardness of this sample as ppm calcium carbonate (form wt = 100.1 mg/mmole).

Solution: First calculate the mg of calcium carbonate from the titration data, recognizing that mmoles of CaCO$_3$ equal mmoles of EDTA.

mg CaCO$_3$ = (12.00 ml EDTA)(0.0100M EDTA)(100.1 mg CaCO$_3$/mmole)

mg CaCO$_3$ = 12.01 mg

Then calculate ppm CaCO$_3$ after converting 50.00 ml to 0.0500 liter water.

$$\text{ppm CaCO}_3 = \frac{12.01 \text{ mg CaCO}_3}{0.0500 \text{ liter water}} = \frac{240.2 \text{ mg CaCO}_3}{\text{liter water}}$$

This is a very "hard" water since most city water supplies are below 100 ppm.

The most common EDTA *back titration* methods involve the determination of iron(III), aluminum(III), and titanium(IV). Few indicators will function in the presence of these metal ions so an excess of EDTA is added to chelate these metals. Then a standard solution of zinc(II) is used to back titrate the unreacted EDTA, using an indicator such as EBT. Since the above metal ions are already chelated by the EDTA they do not interfere with the action of the indicator. The calculation of back titrations is similar to the calculations for the Volhard method illustrated in Example 10-1. The number of mmoles of the zinc(II) titrant must be subtracted from the number of mmoles of standard EDTA added at the beginning to obtain the number of mmoles of iron(III), and so on. Self-Test 10-2 includes such a calculation.

Self-Test 10-2. Calculation of EDTA Titration Results

Directions: Compare your answers with those given at the end of the test.

A. A volume of 20.00 ml of EDTA is required to titrate 25.00 ml of standard 0.0100M calcium carbonate (form wt = 100.1 mg/mmole).
   a. Calculate the molarity of the EDTA.
   b. Calculate the titer of the EDTA for calcium carbonate.
   c. Calculate the titer of the EDTA for calcium(II) ion.
B. Titration of a 50.00 ml Detroit hard water sample requires 5.02 ml of 0.0100M EDTA.
   a. Calculate the hardness of the water as ppm calcium carbonate (form wt = 100.1 mg/mmole).
   b. Calculate the total molarity of calcium(II) and magnesium(II) ions in the hard water.
C. Titration of a 25.00 ml Des Moines (Iowa) hard water sample requires 7.50 ml of 0.0100M EDTA.
   a. Calculate the hardness of the water as ppm calcium carbonate.
   b. Calculate the hardness of the water as ppm calcium ion (form wt = 40.1 mg/mmole).
   c. Compare the hardness of the Detroit and Des Moines water supplies.
D. Back titration of a 50.00 ml Albuquerque hard water sample for iron(III) requires 20.80 ml of 0.0110M zinc(II) after the addition of 25.00 ml of 0.0100M EDTA. Calculate the ppm iron (form wt = 55.85 mg/mmole).

Answers to Self-Test 10-2

A. a. 0.0125M b. 1.25 mg/ml c. 0.500 mg/ml
B. a. 100.$_5$ ppm b. 0.00100M
C. a. 300.$_3$ ppm b. 120.$_3$ ppm c. Des Moines water is harder.
D. 23.6$_8$ ppm

The Effect of Acid on Quantitative Reaction

We have so far discussed the chemistry of EDTA titrations only enough to permit us to calculate EDTA titration results. Before we can discuss EDTA methods in depth we must emphasize that the pH of the solution can permit or prevent a quantitative metal-EDTA reaction. Recall from **10-9** that H^+ is a product of the EDTA titration so too low a pH will reverse the reaction near the end point. We recommend you read this section for whatever level of understanding your instructor recommends: a general grasp of the effect of pH, or a specific mastery of the calculations presented. Then go on to the following discussion of EDTA methods in depth.

We begin by recalling that EDTA is a weakly ionized tetraprotic acid (H_4Y). To calculate the effect of pH on metal-EDTA reactions we first define the four ionization constants of EDTA as follows:

$$H_4Y = H^+ + H_3Y^- \qquad K_1 = 10^{-2.00}$$
$$H_3Y^- = H^+ + H_2Y^{-2} \qquad K_2 = 10^{-2.67}$$
$$H_2Y^{-2} = H^+ + HY^{-3} \qquad K_3 = 10^{-6.16}$$
$$HY^{-3} = H^+ + Y^{-4} \qquad K_4 = 10^{-10.26}$$

Note that we have expressed the ionization constants in full exponential form instead of semiexponential form. This is done to simplify our calculations later on.

The important species of EDTA is the Y^{-4} anion; all of the stability constants of EDTA (Table 10-1) are defined in terms of this anion. Thus for the reaction of a $+2$ metal ion with EDTA the stability constant is defined as

$$K_{MY} = \frac{[MY^{-2}]}{[M^{+2}][Y^{-4}]} \tag{10-11}$$

If all of the EDTA is not in the form of the Y^{-4} anion, then K_{MY} will not give adequate information on the equilibrium. We can decide whether this is true by calculating the fraction of EDTA in the Y^{-4} form. This fraction is called α_4 and it is defined as follows:

$$\alpha_4 = \frac{[Y^{-4}]}{[EDTA]} \tag{10-12}$$

where [EDTA] is the total concentration of all forms of EDTA, including the Y^{-4} form. The mathematical relation needed to calculate α_4 can be derived using the definitions of the four ionization constants and is

$$\alpha_4 = \frac{K_1 K_2 K_3 K_4}{[H^+]^4 + [H^+]^3 K_1 + [H^+]^2 K_1 K_2 + [H^+] K_1 K_2 K_3 + K_1 K_2 K_3 K_4} \tag{10-13}$$

The variation of α_4 with pH is estimated in Figure 10-2, p. 228. You will notice that at pH 12 and above it has a value close to unity. Unfortunately, most EDTA titrations are conducted at pH 10 and below, so it is important to be able to calculate a value for α_4 at any given pH. This calculation usually can be simplified by using only the significant terms (generally two) in the denominator of **10-13**, as the following example will show.

EXAMPLE *10-4.* Calculate α_4 for EDTA at pH 10.0 and indicate what fraction of EDTA is present as Y^{-4} at this pH.

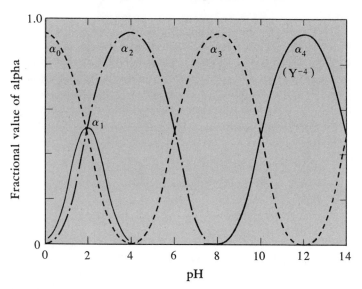

FIGURE 10-2. *A plot of the fraction of EDTA present in various forms at different pH values. The left-hand portion of each curve approaches, but does not become, zero. The subscript on each α value refers to the charge on the EDTA species; thus α_4 refers to the fraction of EDTA present in the Y^{-4} form.*

Solution: First calculate the products of $K_1 K_2 K_3 K_4$, and so on for the denominator of **10-13**. They are

$$K_1 K_2 K_3 K_4 = 10^{-21.09}$$

$$K_1 K_2 K_3 = 10^{-10.83}$$

$$K_1 K_2 = 10^{-4.67}$$

Then substitute these values into **10-13** with the $[H^+]$.

$$\alpha_4 = \frac{10^{-21.09}}{[10^{-10}]^4 + [10^{-10}]^3 \, 10^{-2.00} + [10^{-10}]^2 \, 10^{-4.67} + [10^{-10}] \, 10^{-10.83} + 10^{-21.09}}$$

Next decide which of the terms in the denominator is significant. Looking at Figure 10-2 we see that Y^{-4} and HY^{-3} are the important species present at pH 10, and since the last two terms in the denominator reflect these two species we evaluate these first. We find that they are indeed the two largest terms and we proceed to add them up.

$$[10^{-10}][10^{-10.83}] = 10^{-20.83} = 1.48 \times 10^{-21}$$

$$10^{-21.09} = 8.1 \times 10^{-22} = 0.81 \times 10^{-21}$$

$$\text{sum} = 2.29 \times 10^{-21} = 10^{-20.64}$$

Finally, we calculate α_4.

$$\alpha_4 = \frac{10^{-21.09}}{10^{-20.64}} = 10^{-0.45} = 3.5 \times 10^{-1}$$

Note that this is the value given in Table 10-2.

TABLE 10-2. Values of α_4 for EDTA at Different pHs

pH	α_4	
	Semiexponential	Exponential Value
2.0	3.7×10^{-14}	$10^{-13.44}$
3.0	2.5×10^{-11}	$10^{-10.60}$
4.0	3.3×10^{-9}	$10^{-8.48}$
5.0	3.5×10^{-7}	$10^{-6.45}$
6.0	2.2×10^{-5}	$10^{-4.66}$
7.0	4.8×10^{-4}	$10^{-3.33}$
8.0	5.1×10^{-3}	$10^{-2.29}$
9.0	5.1×10^{-2}	$10^{-1.29}$
10.0	0.35	$10^{-0.45}$
11.0	0.85	$10^{-0.07}$
12.0	0.98	$10^{-0.01}$
13.0	1.0	$10^{-0.00}$

Equation **10-13** can be used to calculate values for α_4 for all the pHs at which EDTA titrations may be conducted. These values are listed in Table 10-2.

Now that values for α_4 are known we can use them to calculate the effect of pH on the stability constant. First, we rearrange **10-12**.

$$[Y^{-4}] = \alpha_4 [\text{EDTA}] \tag{10-14}$$

Then we substitute the right side of **10-14** for $[Y^{-4}]$ in **10-11**.

$$K_{MY} = \frac{[MY^{-2}]}{[M^{+2}][\text{EDTA}]\alpha_4} \tag{10-15}$$

A careful look at this equation will reveal that it is close to being an equilibrium constant expression for the reaction of all forms of EDTA with a metal ion.

$$\text{EDTA} + M^{+2} \rightarrow MY^{-2} \quad (+xH^+)$$

Rearranging **10-15** gives the desired expression for the above reaction.

$$\alpha_4 K_{MY} = K_{\text{M-EDTA}} = \frac{[MY^{-2}]}{[M^{+2}][\text{EDTA}]} \tag{10-16}$$

The product of α_4 and K_{MY} yields a new stability constant, $K_{\text{M-EDTA}}$, which is a *conditional stability constant*, or conditional formation constant. It is constant only if the pH is constant during the titration. Since buffers are used in most EDTA titrations this is a very useful constant. It is more useful than K_{MY} because it can be used for accurate equilibrium calculations and to decide if a reaction is quantitative at a given pH.

The criteria used in Section 7-2 to decide whether a reaction is quantitative is as follows:

> If the *actual* conditional stability constant is $\geq$ the theoretical minimum value of $K_{M\text{-EDTA}}$, then direct EDTA titration of the metal ion will be quantitative.

The minimum theoretical value of $K_{M\text{-EDTA}}$ varies with concentration. A few values follow.

Concn	Min theo $K_{M\text{-}EDTA}$
0.01 M	1×10^8
0.001 M	1×10^9
0.0001 M	1×10^{10}

Since most EDTA titrations are conducted at the 0.01 M level the actual value of the conditional stability constant should be compared with the theoretical minimum of 1×10^8. The most acidic pH at which an EDTA titration can be conducted can also be calculated using the above values and a rearranged form of **10-16**.

$$\min \alpha_4 = \frac{\min \text{theo } K_{M\text{-EDTA}}}{K_{MY}} \qquad (10\text{-}17)$$

Suppose that we wish to know the most acidic pH at which a particular metal ion can be titrated at the 0.01 M level. We divide the value of K_{MY} (Table 10-1) into 1×10^8 to obtain a minimum value for α_4. Then we select the pH value closest to it in Table 10-2. This will be the most acidic pH at which the titration can be conducted and still be quantitative. Of course, the titration will still be quantitative at less acidic pHs, but other metal ions may interfere at these pHs. The more acidic the pH, the fewer the metal ions that will interfere by reacting with EDTA.

EXAMPLE *10-5.* It is proposed to titrate 0.01 M levels of magnesium(II) at a pH of 9. Will the titration be quantitative? If not, what is the most acidic pH at which the titration will be quantitative?

Solution: We see from Table 10-2 that α_4 is 0.051 or $10^{-1.29}$. We then use **10-16** to calculate the conditional stability constant at pH 9 and insert the value of $10^{8.7}$ for K_{MgY} as follows:

$$K_{M\text{-EDTA}} = (10^{-1.29})(10^{8.7}) = 10^{7.4}$$

Since this is less than the minimum theoretical value of 10^8, the titration will not be quantitative.

The most acidic pH at which the titration will be quantitative is found by solving for the minimum value of α_4 using **10-17**.

$$\min \alpha_4 = \frac{1 \times 10^8}{10^{8.7}} = 10^{-0.7}$$

We see in Table 10-2 that the above value is closest to the value of α_4 of $10^{-0.45}$ at pH 10. Therefore the most acidic pH at which titration will be quantitative is pH 10.

To titrate one metal ion, say M_1, in the presence of a second metal ion, M_2, it is necessary that the ratio of their stability constants (Table 10-1) be

$$\frac{K_{M_1Y}}{K_{M_2Y}} \geq \frac{10^6}{1}$$

It is also necessary to reduce the pH so that after M_1 has been titrated quantitatively, M_2 does not react appreciably with EDTA before the indicator color change for M_1. The safest thing to do is to adjust the pH so that the conditional stability constant for M_2-EDTA is 10^2 or less.

The Effect of Other Ligands. If there are other ligands, such as ammonia, present in a solution, they will reduce the ability of EDTA to chelate the metal ion. Their effect on the conditional stability constant can be calculated by means of the parameter β.

$$\beta = \frac{[M^{+2}]}{\text{total } M \text{ of } M^{II} \text{ not chelated by EDTA}}$$

A numerical value for β for any ML_4 species is calculated as follows:

$$\frac{1}{\beta} = 1 + K_1[L] + K_1 K_2 [L]^2 + K_1 K_2 K_3 [L]^3 + K_1 K_2 K_3 K_4 [L]^4 \quad \textbf{(10-18)}$$

where the Ks are the stepwise stability constants and $[L]$ is the molarity of the free ligand. The value of β is then multiplied times the conditional stability constant to obtain a new conditional stability constant which is valid only for that value of $[L]$.

EDTA Methods

Of all of the EDTA methods the titration of calcium(II) and magnesium(II) ions is the most common. We shall first discuss the titration of these ions separately, and then together as in the water hardness titration.

Titration of Magnesium(II). Magnesium is usually titrated at a pH of 10; at higher pHs, such as pH 12, it precipitates as magnesium hydroxide. Eriochrome Black T may be used as the indicator, but because it is somewhat unstable, Calmagite is preferred. Both indicators exhibit the same colors; for example, both form a pink-red magnesium complex so we will discuss the color change at the end point in terms of Eriochrome Black T.

Since Eriochrome Black T is a triprotic acid we will symbolize it as H_3EBT. Its first hydrogen is strongly ionized, but $pK_2 = 6.9$ and $pK_3 = 11.5$. Using these constants the predominant forms over the entire pH range can be deduced.

pH	Predominant form
1–6.9	Purple-red H_2EBT^-
6.9–11.5	Blue $HEBT^{-2}$
11.5–14	Orange EBT^{-3}

The color change at the titration end point for magnesium(II) at the usual pH of 10 is as follows:

$$MgEBT^- + Na_2H_2Y \rightarrow MgY^{-2} + HEBT^{-2} + H^+ \quad (10\text{-}19)$$

(pink-red)   (titrant)   (blue)

It is obvious that a distinct color change for this reaction can occur theoretically only in the pH range of 6.9–11.5. (In practice, the color change is found to be indefinite until the pH is above 8.) Below pH 6.9 the color change would be pink-red to purple-red; above pH 11.5 the change would be pink-red to orange. Both of these changes would be difficult to see.

Because magnesium(II) forms only a moderately stable EDTA chelate the most acidic pH at which it can be titrated is close to pH 10 (see Ex. 10-5). Therefore the titration of magnesium(II) is limited to a pH range of 10–11 by the indicator pH range, its EDTA chelate stability, and by the fact that it precipitates at pH 12. An ammonia buffer is usually used to maintain the pH around 10. Fortunately, magnesium(II) does not form complex ions with ammonia (see above) so its conditional stability constant is not affected by the ammonia buffer.

Titration of Calcium(II). The EDTA chelate of calcium(II) is more stable than that of magnesium(II); K_{CaY} is $10^{10.7}$ (Table 10-1). The most acidic pH at which it can be titrated can be calculated to be pH 8.0, using the method of Example 10-5. However, it is usually titrated at a pH of 10–13 because of the pH requirements of the available indicators.

Because the CaEBT$^-$ complex is unstable, calcium(II) cannot be titrated *alone* using Eriochrome Black T or Calmagite. It can be titrated if a small amount of magnesium(II) is present to form the stable MgEBT$^-$ complex. Therefore, a small amount of magnesium(II) is added to the titrant or a small amount of magnesium-EDTA is added to the solution. In either case the pink-red MgEBT$^-$ complex immediately forms. After the calcium(II) has been titrated, the reaction in **10-19** occurs to give a color change at the equivalence point.

Calcium(II) can also be titrated at pH 13 using Calcein or Calcein Blue indicators. The calcium chelate of these indicators exhibits a *fluorescent* yellow-green color; the indicators alone are a *nonfluorescent* light brown. The color change at the end point is therefore the disappearance of the fluorescent yellow-green color and the appearance of the light brown color. The titration can also be carried out in the dark using an ultraviolet lamp to cause the indicator to fluoresce yellow-green (see Ch. 15 for a discussion of fluorescence). At the end point the fluorescence ceases and the solution is dark.

Determination of Water Hardness. EDTA can be used to titrate the *total* of calcium(II) and magnesium(II) in hard water. The results are usually calculated as parts per million (ppm) of calcium carbonate (Ex. 10-3) rather than the total molarity of both ions. Such an analysis is very important since the amount of calcium and magnesium carbonate residue left behind by water which has evaporated affects the operation of equipment ranging from large industrial boilers to home hot-water heaters to steam irons.

The usual water hardness determination is conducted by titration at a pH of 10; an ammonia buffer is used to keep the pH at this value. Either Eriochrome Black T or Calmagite may be used as the indicator. After the magnesium(II) and calcium(II) are chelated, the MgEBT$^-$ reacts with EDTA (10-19) to provide the color change from pink-red to blue.

Small amounts of iron(III), aluminum(III), and copper(II) in the hard water are *potential* interferences. These ions react irreversibly with Eriochrome Black T and Calmagite forming stable pink-red complexes. If this is not avoided, the water hardness end point would be difficult or impossible to perceive since the color would not change to blue. The interference of small amounts of iron(III) and aluminum(III) can be avoided by adding the ammonia buffer *before* the indicator is added. This complexes these cations as hydroxo (OH$^-$) complexes so that they cannot react with the indicator. The addition of the cyanide ion, or hydroxylamine, prevents the interference of copper(II) by reduction to copper(I). These reagents also prevent the interference of small or large amounts of iron(III).

Titration of +3 and +4 Cations. Many +3 and +4 cations cannot be titrated in the basic solution needed for Eriochrome Black T and Calmagite because they precipitate as insoluble hydroxides. Such titrations can be accomplished using xylenol orange indicator which forms red complexes and is yellow in its uncomplexed form.

Thorium(IV) and bismuth(III), which form two of the most stable EDTA chelates (Table 10-1), can be titrated in acid solution at any pH in the range of 1.5–3.0. The color change at the end point for the titration of bismuth(III) is (XO = xylenol orange)

$$\text{Bi}^{\text{III}}(\text{XO}) + \text{Na}_2\text{H}_2\text{Y} \rightarrow \text{BiY}^- + \text{XO} + 2\text{H}^+ + 2\text{Na}^+ \qquad (10\text{-}20)$$

(red) (titrant) (yellow)

The color change is the same for the direct titration of thorium(IV). Theoretically, any metal ion with a stability constant of $10^{15.44}$ or less will have a conditional stability constant of 10^2 or less at pH 2 (see Table 10-2 for the value of α_4) and should not interfere in the titration of these two metal ions at that pH. Thus thorium(IV) and/or bismuth(III) can be titrated in the presence of lanthanum(III), calcium(II), and so on (Table 10-1).

Zirconium(IV) and iron(III) cannot be determined by direct titration using xylenol orange, but they can be determined by back titration. An excess of standard EDTA is added to either of these ions at a pH of 1–2. The unreacted EDTA is then back titrated with a standard solution of bismuth(III).

$$\text{H}_2\text{Y}^{-2} + \text{Bi}^{\text{III}} + \text{FeY} \rightarrow \text{BiY}^- + 2\text{H}^+ + \text{FeY}$$

(unreacted EDTA) (titrant)

Even though the iron(III)-EDTA [or zirconium(IV)-EDTA] is present at the end point, it does not react with the bismuth(III) because the iron(III)-EDTA chelate is more stable than the bismuth(III)-EDTA chelate (Table 10-1).

Other +2 cations such as lead(II) and cadmium(II) can also be titrated using xylenol orange at higher pHs, such as pH 5.

QUESTIONS AND PROBLEMS

(Answers to most even-numbered problems are found in Appendix 5.)

Concepts and Definitions
1. Write the titration reaction and the indicator reaction for the titration of potassium bromide with standard silver nitrate using each of the following.
   a. The Mohr method
   b. The adsorption indicator method
   c. The Volhard method
2. Suggest whether each of the three methods in Problem 1 can be used for titration of silver(I) ion with standard potassium chloride as titrant. If any method cannot be used, give a reason.
3. What advantage does the Volhard method have over the two direct titration methods for the determination of an anion whose insoluble silver(I) salt has a K_{sp} value appreciably less than 1×10^{-8}?
4. Decide whether each of the following anions can be determined by the Volhard method, with or without shaking with nitrobenzene before the final titration.
   a. IO_3^- iodate
   b. $C_2O_4^{-2}$ oxalate
   c. Br^- bromide
   d. N_3^- azide
5. Define each term and give an example.
   a. Unidentate ligand
   b. Hexadentate ligand
   c. Bidentate ligand
6. Explain why EDTA titrations are usually conducted in the presence of a buffer.
7. Explain why EDTA is a superior titrant to a unidentate ligand for the titration of metal ions.

Precipitation Titrations
8. For silver(I) bromide the $K_{sp} = 4.9 \times 10^{-13}$. For the titration of 0.10M bromide with 0.10M silver nitrate calculate the following, neglecting dilution.
   a. pAg at 99.9% reaction
   b. pAg at the equivalence point
   c. pAg at 100.1% reaction
9. Calculate the percentage iodide in a 1.0000 g sample to which is added 50.00 ml of 0.1000M silver nitrate. The unreacted silver nitrate is back titrated with 16.00 ml of 0.0800M potassium thiocyanate.
10. A 0.5000 g sample of chloride requires 15.50 ml of 0.1100M silver nitrate by the Mohr method. A 1.0000 g portion of the same sample requires 31.50 ml of 0.1050M silver nitrate using dichlorofluorescein at pH 7. Calculate the percentage chloride by both methods and rationalize any differences.

Complexation Titrations

11. Calculate values of α_4 at the following pHs.
    a. pH = 7.5   b. pH = 8.5   c. pH = 9.5   d. pH = 10.5
12. It is proposed to titrate $0.01M$ calcium(II) ion at each of the following pHs. Will each titration be quantitative?
    a. pH = 8.0   b. pH = 7.5   c. pH = 8.5
13. It is proposed to titrate $1 \times 10^{-4}M$ calcium(II) ion in urine samples. Is there a pH or a pH range at which the titration will be quantitative? What is the pH or pH range?
14. Titration of a 50.00 ml hard water sample requires 4.20 ml of $0.0100M$ EDTA. Calculate the ppm calcium(II) carbonate in the water.
15. Given the log of each stability constant for $Cu(NH_3)_4^{+2}$ below, calculate the new conditional stability constant for the titration of copper(II) ion with EDTA at pH 9 in $0.1M$ ammonia. Decide whether titration will be quantitative.
    $\log K_1 = 4.1$ $\log K_2 = 3.5$ $\log K_3 = 2.9$ $\log K_4 = 2.1$

Challenging Problems

16. What is the most acidic pH at which Cu^{+2} can be titrated with EDTA at
    a. The $0.01M$ level?
    b. The $1 \times 10^{-5}M$ level?
    c. The $5 \times 10^{-7}M$ level?
17. By comparing the solubilities of silver(I) salts with the solubility in pure water of silver(I) chromate and silver(I) thiocyanate, state for each anion whether it can be determined by the Mohr method and whether nitrobenzene must be or need not be used in the Volhard determination of the anion.
    a. Br^- b. IO_3^- c. CO_3^{-2}
    d. $Co(CN)_6^{-3}$ ($K_{sp} = 8.0 \times 10^{-21} M^4$ for the silver salt)
18. A mixture containing only sodium chloride (form wt = 58.44) and potassium chloride (form wt = 74.56) without any other substances present can be analyzed by a single direct titration with silver nitrate titrant. Calculate the percentage of sodium chloride in 191.44 mg of the mixture if 30.00 ml of $0.1000M$ silver nitrate are required for complete precipitation of the chloride ion.
19. Calculate the mg% of calcium(II) ion in a 0.200 ml blood sample which is titrated with 2.50 ml of $0.000100M$ EDTA. Is the concentration of calcium(II) within, below, or above the normal range (Fig. 1-1)?

11. Oxidation-Reduction Methods

"I flatter myself that I can distinguish at a glance the ash (oxides) *of any known brand either of cigar or tobacco,"*
said Holmes.

ARTHUR CONAN DOYLE
A Study in Scarlet, Chapter 4

In this chapter oxidation-reduction reactions and their standard potentials will be discussed. Our goal is not a complete mastery, as with Sherlock Holmes and his identification of tobacco ashes, but a basic understanding of oxidation-reduction methods and their applications to real analytical problems. These methods should complement the use of reactions such as acid-base, complexation, and precipitation discussed in previous chapters.

11-1 OXIDATION-REDUCTION CONCEPTS AND REACTIONS

You already may have been introduced to the fundamental concepts of oxidation-reduction reactions in general chemistry; if so, you may wish to proceed to Self-Test 11-1 to see whether or not you are able to apply the ion-electron method to the balancing of organic half reactions. In any case, a brief review will be given here. If you find that the material is still very difficult, you should probably go back to a general chemistry text and start from the beginning. No matter what your background, you need not review the balancing of oxidation-reduction equations since that will be discussed under the description of the *ion-electron method.*

Concepts

An oxidation-reduction (redox) reaction is one in which electrons are transferred from one species to another. It is easily determined that such a reaction has occurred if only inorganic ions are involved since the charges on the ions change, but if organic molecules are involved, then it is more difficult to recognize that a redox reaction has taken place. In either case, a redox reaction must involve a species that gains electrons (oxidizing agent, or oxidant) and a species that loses electrons (reducing agent, or reductant).

General and specific examples of an oxidizing agent (oxidant) are

$$A_{ox}^{+x} + ne^- = A_{red}^{+(x-n)}$$

$$2Cu^{+2} + 2OH^- + 2e^- = Cu_2O(s) + H_2O \qquad (11\text{-}1)$$

Note that both preceding reactions are reduction *half reactions* or *half equations*; an oxidation half reaction or half equation must be added to each to obtain a complete reaction.

In the *general* reduction half reaction A_{ox}^{+x} is the oxidizing agent, or oxidant; it *gains* a definite number n of electrons to give a product in which at least one atom is more negative, or less positive, than in the oxidizing agent. For example, in **11-1** two Cu^{+2} ions gain one electron each to give two copper(I) ions combined with the oxide ion as copper(I) oxide. The copper(I) is more negative, or less positive, than copper(II) ion.

General and specific examples of a reducing agent are

$$B_{red}^{+y} = B_{ox}^{+(y+n)} + ne^-$$

$$\underset{\text{CH}_3\overset{\displaystyle O}{\overset{\|}{\text{C}}}\text{H}}{} + 2OH^- = \underset{\text{CH}_3\overset{\displaystyle O}{\overset{\|}{\text{C}}}\text{OH}}{} + H_2O + 2e^- \qquad (11\text{-}2)$$

Again note that both reactions above are oxidation *half reactions* or *half equations*; they are not complete equations or reactions.

In the *general* oxidation half reaction B_{red}^{+y} is the reducing agent, or reductant; it *loses* a definite number n of electrons to give a product in which at least one atom is more positive, or less negative, than in the reducing agent. For example, in **11-2** acetaldehyde (CH_3CHO) loses two electrons to give acetic acid. Close inspection will reveal that the oxidation number of the carbonyl carbon has increased by a $+2$.

General and specific examples of a complete redox reaction are

$$A_{ox}^{+x} + B_{red}^{+y} = A_{red}^{+(x-n)} + B_{ox}^{+(y+n)}$$

$$2Cu^{+2} + \underset{\text{CH}_3\overset{\displaystyle O}{\overset{\|}{\text{C}}}\text{H}}{} + 4OH^- = Cu_2O(s) + \underset{\text{CH}_3\overset{\displaystyle O}{\overset{\|}{\text{C}}}\text{OH}}{} + 2H_2O \qquad (11\text{-}3)$$

Each complete reaction is obtained by *adding* together two half reactions which contain, or are multiplied to contain, the *same* number of electrons; this is necessary because no free electrons can appear in the complete reaction. If the number of electrons does not happen to be the same in each half reaction, then each must be cross multiplied by the number of electrons in the other half reaction, before adding together.

In the *general* reaction, B_{red}^{+y} gains the same number of electrons, n, that are lost by A_{ox}^{+x}. It is easy to diagnose that a redox reaction has occurred without knowing the half reactions because A_{ox}^{+x} has become more negative in oxidation number or charge and B_{red}^{+y} has become more positive.

The *specific* reaction (**11-3**) is the classic Fehling's or Benedict's test for aldehydes wherein red, insoluble copper(I) oxide is formed. In this reaction the two electrons lost by acetaldehyde are gained by the two copper(II) ions. If you look only at acetaldehyde and acetic acid in **11-3**, you may not recall enough organic chemistry to recognize that a redox reaction has occurred. However, by checking the other reactants and products, you should easily see that copper(II) ion has been *reduced* to copper(I) ion; therefore acetaldehyde must have been *oxidized*. (In general, aldehydes are *oxidized* to acids and *reduced* to alcohols.)

You also may have been wondering how the two-electron change was obtained in **11-2**. To find the electron change for *organic* half reactions such as **11-2**, it is better to use the so-called ion-electron method rather than the oxidation number method that is widely used. This method is usually applied only to inorganic half reactions in general chemistry texts, so a thorough discussion of its application to organic and biochemical half reactions follows.

The Ion-Electron Method

Simply stated, the ion-electron method of balancing half reactions consists of counting the charges on both sides of a balanced half reaction and then adding the correct number of electrons needed to balance the charges. This method is a completely general method of finding the electron change in any inorganic, organic, or biochemical half reaction. Although its name might be thought to imply that it can be used only for ionic reactions, this is not true. It is equally useful for molecular reactions.

To apply the ion-electron method to a half reaction requires that the half reaction first be balanced with respect to all atoms. Then the charges can be counted and electrons added. The entire method may involve all or most of the following steps.

1. Start by writing the half reaction with only the ions and/or molecules actually gaining or losing electrons.
2. Balance all the atoms, including O and H. Underlining identical large groups helps.
   a. If the solution is acid or neutral, on the side deficient in O atoms write H_2O with the proper coefficient needed to balance the O atoms. Only *after* you have balanced the O atoms should you try to balance the H atoms. On the side that is now deficient in H atoms, write H^+ with the proper coefficient needed to balance the H atoms.
   b. If the solution is basic, strictly speaking you should write OH^- on the side deficient in O atoms and H_2O on the side deficient in H atoms. However, this is somewhat more complicated than the method used in acid solution so some instructors may prefer to have you use step 2*a* for all situations. (Consult your instructor on this point.)

3. Sum the charges algebraically on each side of the half reaction.
4. Write a *charge-balance equation* to find the electron change. Write down the sum of the charges on each side and add the proper number of electrons to the more positive side so that the charges balance. For example, if the sum of the charges on the left is $+4$ and that on the right is $+2$, the charge-balance equation is $+4 + 2e^- = +2$.

To illustrate the use of the ion-electron method let us consider the oxidation of an aldehyde to an acid (**11-2**) and show that in general this is a two-electron change, no matter what aldehyde is oxidized. To show the power of the ion-electron method we will use the oxidation of vitamin A aldehyde, or retinal, to vitamin A acid as an example.

$$C_{16}H_{23}-\underset{\underset{CH_3}{|}}{C}=\underset{\underset{H}{|}}{C}-\overset{\overset{O}{\|}}{C}H \rightarrow C_{16}H_{23}-\underset{\underset{CH_3}{|}}{C}=\underset{\underset{H}{|}}{C}-\overset{\overset{O}{\|}}{C}OH \quad (11\text{-}4)$$

The molecules in **11-4** can be written more simply because they are virtually identical.

$$\underline{C_{19}H_{27}}-CHO \rightarrow \underline{C_{19}H_{27}}-CO_2H \quad (11\text{-}5)$$

Note that the $C_{19}H_{27}$ group in **11-5** is underlined on both sides, indicating that these groups are identical and need not be counted to balance the atoms on both sides.

To balance this half reaction steps 1–4 above should be followed. Step 1 has already been done in **11-5**. For step 2 we assume the solution is acid or neutral and first add water to the left side to balance the O atoms.

$$\underline{C_{19}H_{27}}-CHO + H_2O \rightarrow \underline{C_{19}H_{27}}-CO_2H$$

Then two H^+ are added to the right side to balance the H atoms.

$$\underline{C_{19}H_{27}}-CHO + H_2O \rightarrow \underline{C_{19}H_{27}}-CO_2H + 2H^+ \quad (11\text{-}6)$$

All atoms in **11-6** are now balanced, the underlined groups being identical. Since there are no charges on the left side and a total of $+2$ on the right (step 3), the *charge-balance equation* (step 4) is

$$0 = +2 + 2e^-$$

The complete half reaction is

$$\underline{C_{19}H_{27}}-CHO + H_2O \rightarrow \underline{C_{19}H_{27}}-CO_2H + 2H^+ + 2e^- \quad (11\text{-}7)$$

It can be seen from **11-7** that no matter what the underlined group is, any aldehyde with just one —CHO group will lose two electrons in being oxidized to an acid with just one —CO_2H group. This includes acetaldehyde (**11-2**). The reverse process of an acid such as vitamin A acid being reduced to an aldehyde such as retinal, or vitamin A aldehyde, is also a two electron process:

$$C_{19}H_{27}-CO_2H + 2H^+ + 2e^- \rightarrow C_{19}H_{27}-CHO + H_2O \qquad (11\text{-}8)$$

You may be interested to know that the body cannot reduce vitamin A acid to retinal (**11-8**). Experiments conducted with rats deficient in retinal needed for vision [1] proved this. The rats were fed only vitamin A acid and no other source of retinal. Not only did the rats remain deficient in retinal, but they eventually became night blind.

To become proficient at using the ion-electron method to balance organic and biochemical half reactions, you should work the following self-test. If you have not used the ion-electron method previously, you should balance other equations, such as those given at the end of the chapter, after you finish with the self-test.

Self-Test 11-1. The Ion Electron Method

Directions: Balance one half reaction at a time, giving the *charge-balance equation* for each in addition to following the special directions. Check each answer below.

A. Balance the oxidation of ethyl alcohol, CH_3CH_2OH, to acetic acid, CH_3CO_2H, using the following steps.
  a. The molecules actually involved in the oxidation are (identical groups underlined)
  $$\underline{CH_3}CH_2OH \rightarrow \underline{CH_3}CO_2H$$
  b. Balance all atoms, assuming an acid solution. Start with O atoms, not H atoms.
  $$\underline{CH_3}CH_2OH + \underline{} \rightarrow \underline{CH_3}CO_2H + \underline{}$$
  c. Sum charges on each side and write the charge-balance equation.
  $$\underline{} = \underline{} + \underline{}e^-$$

B. Balance the oxidation of $CH_3CH_2CHOHCH_2OH$ to $CH_3CH_2CO_2H + HCO_2H$ in acid.
  a. Underline the identical groups in the first and second molecules (omit third).
  b. Balance all atoms, sum charges, and write charge-balance equation.

C. Balance the reduction of gluconic acid, $CH_2OH(CHOH)_4CO_2H$, to glucose (blood sugar), $CH_2OH(CHOH)_4CHO$, after underlining identical groups.
$$CH_2OH(CHOH)_4CO_2H + \underline{} + \underline{}e^-$$
$$\rightarrow CH_2OH(CHOH)_4CHO + \underline{}$$

[1] D. C. Neckers, *J. Chem. Educ.* **50**. 166 (1973).

SEC. 11-2   Stoichiometric Redox Calculations

D. Balance the oxidation of glucose in acid to gluco-saccharic acid, $HO_2C(CHOH)_4CO_2H$.

$CH_2OH(CHOH)_4CHO +$ _____

$\rightarrow HO_2C(CHOH)_4CO_2H +$ _____ $+$ _____e^-

E. Balance the reduction of oxygen, O_2, in the body to water in acid solution.

$O_2 +$ _____ $+$ _____$e^- \rightarrow$ _____H_2O

Answers to Self-Test 11-1

A. a. Correct as written.   b. $CH_3CH_2OH + H_2O \rightarrow CH_3CO_2H + 4H^+$
    c. $0 = +4 + 4e^-$ ($4e^-$ change)
B. a. $CH_3CH_2\underline{CHOHCH_2OH} \rightarrow CH_3CH_2CO_2H$
    b. $CH_3CH_2\underline{CHOHCH_2OH} + 2H_2O \rightarrow CH_3CH_2CO_2H + HCO_2H + 6H^+ + 6e^-$
C. Charge-balance equation: $+2 + 2e^- = 0$
D. Charge-balance equation: $0 = +6 + 6e^-$
E. $O_2 + 4H^+ + 4e^- \rightarrow 2H_2O$

11-2 STOICHIOMETRIC REDOX CALCULATIONS

In Section 6-2 we discussed the calculations of volumetric analysis, including redox calculations. Both the mole-molarity system and the equivalent-normality system were discussed there. In this short section we intend to review the equivalent-normality system only. If you intend to use the mole-molarity system, you should review Section 6-2, particularly Examples 6-3 and 6-5. We feel that the equivalent-normality system is easier to use, once understood, than the other system and will therefore use this section for additional discussion.

The reason for mastering the ion-electron method in the previous section was to use it to calculate the equivalent weight and the normality N. A change in either will change the number of equivalents (eq) or milliequivalents (meq) of any oxidizing or reducing agent—thus the correct electron change is important.

The Equivalent Weight

The equivalent weight is expressed in units of g/eq or mg/meq and is defined for calculation purposes as

$$\text{Equivalent weight} = \frac{\text{Formula weight}}{e^- \text{ change per above weight}} \quad (11\text{-}9)$$

Note that the electron change must be the *total change* per formula weight of substance weighed out or to be calculated, not just the electron change in the simplest half reaction. The equivalent weight depends on the reaction involved. For example, ethyl alcohol can have at least two different equivalent weights: 46.07/4e for oxidation to acetic acid, and 46.07/2e (see below).

EXAMPLE 11-1. A simple half reaction is the oxidation of ethyl alcohol to acetaldehyde

$$CH_3CH_2OH \rightarrow CH_3CHO$$

Using the ion-electron method, the electron change is

$$CH_3CH_2OH \rightarrow CH_3CHO + 2H^+ + 2e^- \qquad (11\text{-}10)$$

Given that the formula weight of ethyl alcohol is 46.07, the equivalent weight (eq wt) is

$$\text{eq wt} = \frac{46.07}{2e^- \text{ change}} = 23.04 \text{ mg/meq}$$

*This oxidation (**11-10**) is the first step in the metabolism of alcohol in the body and is catalyzed by a zinc-containing enzyme. Failure of the body to oxidize acetaldehyde rapidly to carbon dioxide results in a state called acetaldehyde syndrome. Symptoms of this state include nausea, vomiting, sweating, and lowering of blood pressure [2].*

A more complicated example is analysis of the purity of Fe_2O_3, iron(III) oxide. In this analysis the oxide is dissolved and reduced to iron(II) before titrating it with an oxidizing agent to iron(III). The balanced titration half reaction is

$$Fe^{+2} \rightarrow Fe^{+3} + e^-$$

However, the electron change per Fe_2O_3 formula weight is two because each of the iron(II) ions produced from Fe_2O_3 loses an electron, for a total of two.

$$Fe_2O_3 + 6H^+ + 2e^- = 3H_2O + 2Fe^{+2} - (\text{oxidation}) \rightarrow 2Fe^{+3} + 2e^-$$

The equivalent weight is calculated:

$$\text{eq wt} = \frac{159.7}{2e^- \text{ total change}} = 79.85 \text{ mg/meq}$$

Milliequivalents and Normality

The number of milliequivalents (meq) or equivalents is found by dividing the appropriate weight by the equivalent weight expressed as mg/meq or g/eq.

$$\text{meq} = \frac{\text{mg}}{\text{eq wt in mg/meq}} \qquad \text{eq} = \frac{\text{g}}{\text{eq wt in g/eq}} \qquad (11\text{-}11)$$

The normality of a titrant in any titration method is always

$$N = \frac{\text{meq}}{\text{ml}} = \frac{\text{eq}}{\text{liter}} \qquad (11\text{-}12)$$

[2] D. A. Labianca, *Chemistry* **47**, 21 (Oct., 1974).

SEC. 11-2   Stoichiometric Redox Calculations   243

Iodine is a common titrant for both inorganic and organic analysis. As an oxidizing agent, its half reaction is

$$I_2 \rightarrow 2I^- + 2e^-$$

Suppose that iodine is weighed out and dissolved to make a titrant according to the following example.

EXAMPLE 11-2. The formula weight of iodine, I_2, is 253.8. Calculate first its equivalent weight, using the fact that it undergoes a total electron change of two.

$$\text{eq wt} = \frac{253.8}{2e^- \text{ total change}} = 126.9 \text{ mq/meq}$$

Now calculate the normality of 100 ml of solution to which is added exactly 1269 mg of pure iodine, I_2. The number of meq in 1269 mg is found using **11-11**.

$$\text{meq} = \frac{1269 \text{ mg}}{126.9 \text{ mg/meq}} = 10.00 \text{ meq}$$

The normality is then found by using **11-12**.

$$N = \frac{10.00 \text{ meq}}{100 \text{ ml}} = 0.100N$$

Per Cent Purity and Equivalent Weight Calculations

In general, the per cent purity of an impure sample S is found by titrating it with a standard solution of a titrant T. The weight, mg_S, of pure S is found from the ml of the titrant used to reach the titration end point; the normality, N_T, of the titrant; and the equivalent weight, eq wt_S, of the sample as follows:

$$mg_S = (ml_T)(N_T)(\text{eq wt}_S) \qquad (11\text{-}13)$$

The per cent purity, %S, of the impure sample is then calculated by dividing the weight of pure S by the weight, mg_{impure}, of the impure sample.

$$\%S = \frac{mg_S}{mg_{impure}}(100) \qquad (11\text{-}14)$$

EXAMPLE 11-3. Calculate the per cent purity of impure ascorbic acid, 1000 mg of which is titrated with 40.00 ml of 0.1000N iodine according to the following reaction.

$$I_2 + C_4H_6O_4(OH)C{=}C(OH) \rightarrow 2I^- + C_4H_6O_4C({=}O){-}C{=}O + 2H^+ \qquad (11\text{-}15)$$

The equivalent weight is 176.12/2, or 88.06 mg/meq of ascorbic acid. The weight of pure ascorbic acid is found using **11-13**.

$$mg_{asc.\,a.} = (40.00 \text{ ml } I_2)(0.1000N \text{ } I_2)(88.06 \text{ mg/meq asc. a.}) = 352.2 \text{ mg}$$

The per cent purity of ascorbic acid is found using **11-14**.

$$\% \text{ ascorbic acid} = \frac{352.2 \text{ mg asc. a.}}{1000 \text{ mg impure sample}}(100) = 35.22\% \text{ purity}$$

It is interesting to know that ascorbic acid, or vitamin C, is also oxidized to the same product, dehydroascorbic acid, in the body as in **11-15**. This half reaction will be considered in more detail in the following self-test.

The equivalent weight of a pure solid or liquid may also be found from a titration, knowing the ml of titrant, ml_T, used to reach the titration end point; the normality, N_T, of the titrant; and the weight, mg_S, of the *pure* sample of solid or liquid. (It is very important that the sample be as pure as possible or the equivalent weight will be in error.) Then **11-13** can be rearranged to give the appropriate equation.

$$\text{Eq wt} = \frac{mg_S}{(ml_T)(N_T)} \qquad (11\text{-}16)$$

Since the units on equivalent weight are mg/meq, you can see that the right-hand side of the preceding equation gives the correct units [recall that $(ml)(N) = \text{meq}$].

Since all of the above calculations are unique, you should work the following self-test to be sure you become somewhat familiar with all of them. Additional problems are given at the end of the chapter for areas where you need more practice.

Self-Test 11-2. Equivalent Weights and Other Calculations

Directions: In addition to following any special directions, find the electron change for each half reaction by the ion-electron method and then calculate the equivalent weight. Check each answer below (only the charge-balance equation, not the entire half reaction, will be given in the answer).

A. Ascorbic acid, vitamin C, is oxidized to dehydroascorbic acid by iodine.
  a. Use the ion-electron method to find the electron change for its half reaction.

  $C_4H_6O_4(OH)C=COH \rightarrow C_4H_6O_4C(=O)-C=O + \underline{\hspace{1cm}} + \underline{\hspace{1cm}} e^-$

  b. The formula weight of ascorbic acid is 176.12. What is its equivalent weight?

B. Glycerine, $CH_2OHCHOHCH_2OH$, has several different equivalent weights, depending on what oxidizing agent reacts with it. Using its formula weight of 92.1, calculate the equivalent weight in each half reaction below.
  a. Periodic acid, HIO_4, oxidizes it according to this half reaction (identical groups underlined).

  $CH_2OHCHOHCH_2OH + \underline{\hspace{1cm}} \rightarrow 2H_2CO + HCO_2H + \underline{\hspace{1cm}} + \underline{\hspace{1cm}} e^-$

  b. Cerium(IV) oxidizes it according to this half reaction.

  $CH_2OHCHOHCH_2OH + \underline{\hspace{1cm}} \rightarrow 3HCO_2H + \underline{\hspace{1cm}} + \underline{\hspace{1cm}} e^-$

C. Arsenious oxide, As_2O_3, dissolves to form two H_3AsO_3 molecules per As_2O_3.

SEC. 11-3   Standard Potentials and their Uses

a. Find the electron change per H_3AsO_3 molecule if it is oxidized to H_3AsO_4.
b. Using a formula weight of 197.8, calculate the equivalent weight of As_2O_3.
c. Calculate the normality of 100 ml of solution to which is added 395.6 mg of As_2O_3, which dissolves to form H_3AsO_3.

D. Calculate the per cent purity of 1000 mg of impure Fe_3O_4 (form wt = 231.55) which is dissolved to form three Fe^{+2} ions per Fe_3O_4. The resulting Fe^{+2} ions are all titrated with 20.00 ml of $0.1000N$ $K_2Cr_2O_7$, which oxidizes them to Fe^{+3}.

E. Calculate the equivalent weight of pure As_2O_x, 229.8 mg of which requires 40.00 ml of $0.1000N$ titrant.

Answers to Self-Test 11-2

A. a. $0 = +2 + 2e^-$ (2 e^- change)   b. 88.06
B. a. $0 = +4 + 4e^-$: eq wt = 23.0   b. $0 = +8 + 8e^-$; eq wt = 11.5
C. a. $2e^-$ b. 49.45   c. $0.0800N$
D. % = (154.36 mg/1000 mg)100 = 15.44%
E. eq wt = 229.8 mg/4.000 meq = 55.96 (x = 5)

11-3 STANDARD POTENTIALS AND THEIR USES

Each of the half reactions we have discussed previously has a definite tendency or potential to gain or lose electrons. Consider the reduction of dehydroascorbic acid to ascorbic acid.

$$C_4H_6O_4C(=O)-C=O + 2H^+ + 2e^- = C_4H_6O_4(OH)C=COH \quad (11\text{-}17)$$

As the reaction is written, dehydroascorbic acid has a certain tendency to gain electrons to form ascorbic acid; ascorbic acid also has a certain tendency to lose electrons to form dehydroascorbic acid. At equilibrium these two tendencies will be balanced at some point. (This is also true in the body; each of our bodies has a certain amount of dehydroascorbic acid and vitamin C in equilibrium with each other. Vitamin C tends to be oxidized to dehydroascorbic acid so that its supply must be replenished by dietary means daily.)

    If all of the known half reactions are written the same way, then their tendencies to gain or lose electrons can be compared by measuring them in the form of electrical potential or voltage. It has been recommended that all half reactions like **11-17** be written as reductions for this purpose by the International Union of Pure and Applied Chemistry (IUPAC). However, to complete an electrical cell to measure a potential, the potential must be measured relative to that of a second half reaction that will take up the electrons released by the half reaction being measured. If the same (standard) half reaction is used for all such measurements under a set of standard conditions, then a set of *standard potentials*, or $E°$ values, can be accumulated.

Measurement and Compilation of $E°$ Values

The standard half reaction chosen for the relative measurement of all $E°$ values is the reduction of hydrogen ions to hydrogen gas.

$$2H^+ + 2e^- = H_2(g) \qquad E° = 0.0 \text{ (defined)}$$

The $E°$ for this reaction is defined to be 0.0 volts at standard conditions of one atmosphere pressure and unit activity of hydrogen ion. (As usual, we will use molarity to approximate the activity of an ion such as the hydrogen ion.)

A special electrical cell called a *potentiometer* (Fig. 11-1) is used for measurement of the $E°$ values in relation to $E°$ of the reduction of hydrogen ion.

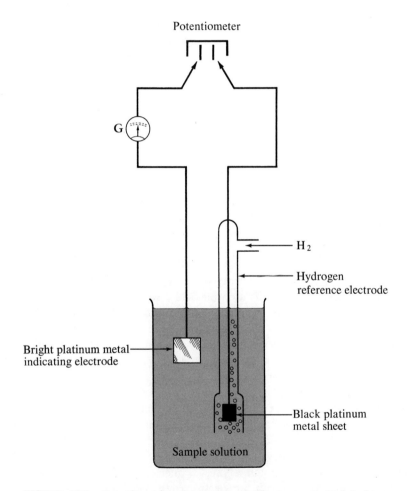

FIGURE 11-1. Potentiometric measurement of $E°$ versus standard hydrogen reference electrode (black platinum sheet covered with hydrogen gas in a 1M hydrogen ion solution).

Standard Potentials and their Uses

The complete cell consists of two electrodes. One is a standard hydrogen *reference* electrode in equilibrium with a solution of one molar hydrogen ion. The other electrode is typically a bright platinum *indicating* electrode in equilibrium with the species involved in the half reaction to be measured. The potentiometer measures the voltage difference between the two half reactions in such a way that no appreciable reaction occurs. Because of this, the voltage measured is referred to as a potential, or form of potential energy (Sec. 12-1). In general, the potential measured on the potentiometer, E_{pot}, is given by

$$E_{pot} = E_{ind\ el} - E_{ref} \qquad (11\text{-}18)$$

Since E_{ref} is defined to be 0.0 volts for a hydrogen reference electrode, the voltage reading (potential) of the potentiometer is equal to the $E°$ of the half reaction in equilibrium with the indicating electrode.

$$E_{pot} = E_{ind\ el} = E° \quad \text{(at } 1M \text{ concn)} \qquad (11\text{-}19)$$

Typical values measured using **11-19** are listed in Table 11-1, p. 248. These $E°$ values refer to a number of common oxidizing agents; the oxidizing agent is the reactant in each half reaction listed there. The $E°$ values apply to these half reactions only under standard conditions—one molar concentration of all species including H^+ and anions, one atmosphere pressure, and 25°C. Special exceptions are noted in the footnotes.

Interpretation of $E°$ Values

Note that all of the $E°$ values for the oxidizing agents in Table 11-1 are *positive* relative to those of the reduction of hydrogen ions. This implies that all have a tendency to gain electrons to a greater degree than H^+. In general, the oxidizing agent with the more positive $E°$ has the greater tendency to react and gain electrons. Thus the most powerful oxidizing agent in Table 11-1 would be cerium(IV) ion in $HClO_4$; its $E°$ of $+1.70$ volts is the most positive of those in the table.

As the $E°$ values become less positive and approach zero volts, the associated oxidizing agents have a lesser tendency to gain electrons and be reduced. Thus dehydroascorbic acid has a relative small positive $E°$ value and has a relative small tendency to gain electrons (**11-17**). In fact, compared to other oxidizing agents in the body, it is a much weaker oxidizing agent. Thus ascorbic acid tends to give up electrons in the body. A useful general rule is

> The oxidizing agent with the more positive $E°$ will react spontaneously with the reduced form of the oxidizing agent having the less positive $E°$.

In terms of Table 11-1 the oxidizing agent (species on left) will react with the reduced form (species on the right) of a half reaction *below* it in the table. This applies

TABLE 11-1. *Common Oxidizing Agents and Their Standard Potentials*

Oxidizing agent + ne^- → product	Standard potential, $E°$, in volts ($1M$ H^+)
$Ce(IV) + e^- = Ce^{+3}$ (in $1M$ $HClO_4$)	1.70[a]
$KMnO_4 + 4H^+ + 3e^- = MnO_2 + 2H_2O + K^+$	1.7 (neutral solution)
$H_5IO_6 + H^+ + 2e^- = IO_3^- + 3H_2O$ (periodic acid)	1.60
$KMnO_4 + 8H^+ + 5e^- = Mn^{+2} + 4H_2O + K^+$	1.51
$Ce(IV) + e^- = Ce^{+3}$ (in $1N$ H_2SO_4)	1.44[a]
$Cl_2 + 2e^- = 2Cl^-$	1.36
$Cr_2O_7^{-2} + 14H^+ + 6e^- = 2Cr^{+3} + 7H_2O$	1.33
$O_2 + 4H^+ + 4e^- = 2H_2O$	1.229 ($= +0.816$ at pH 7)
$IO_3^- + 6H^+ + 5e^- = \frac{1}{2}I_2 + 3H_2O$	1.195
$VO_2^+ + 2H^+ + e^- = VO^{+2} + H_2O$	1.00
$Cu^{+2} + I^- + e^- = CuI(s)$	0.86
$Ag^+ + e^- = Ag(s)$	0.800
$Fe^{+3} + e^- = Fe^{+2}$	0.771
$O_2 + 2H^+ + 2e^- = H_2O_2$	0.682
$H_3AsO_4 + 2H^+ + 2e^- = H_3AsO_3 + H_2O$	0.559 ($= 0.0$ at pH 7)
$I_2 + 2e^- = 2I^-$	0.535
Dehydroascorbic + $2H^+ + 2e^- =$ Ascorbic acid	0.39 ($= -0.023$ at pH 7)
$Fe(CN)_6^{-3} + e^- = Fe(CN)_6^{-4}$	0.36
Cytochrome-$a(Fe^{III}) + e^- =$ Cytochrome-$a(Fe^{II})$	0.290
$Hg_2Cl_2 + 2e^- = 2Hg + 2Cl^-$	0.268
$2H^+ + 2e^- = H_2(g)$	0.00

[a] These are so-called formal potentials in which the measurement is made in the acid in parentheses; a change in acid changes the potential.

strictly to standard conditions only, although later we will give a generalization for practical analytical conditions that is based on a difference in $E°$ values.

As an illustration of the above generalization, consider the reaction of iodine with ascorbic acid (**11-15**). To decide whether or not the two will react we first identify the half reaction that is the same in the table as in the complete reaction. This is the iodine half reaction, and it has the larger $E°$. Therefore the I_2 should react with the reduced form of dehydroascorbic acid, which is ascorbic acid, and the reaction in **11-15** does occur spontaneously.

To find the numerical difference in the $E°$ values we combine the half reactions and $E°$ values in the following manner. First we write down the half reaction that is the same in the table followed by its $E°$ value. Below it we write down the remaining half reaction after reversing the way it appears in the table and its $E°$ value with the opposite sign. Then we add the two half reactions and the two $E°$ values as follows:

$$I_2 + 2e^- = 2I^- \qquad +0.535 \text{ V}$$
$$\text{Ascorbic acid} = \text{Dehydroascorbic acid} + 2H^+ + 2e^- \qquad -0.39 \text{ V}$$
$$\overline{\text{Ascorbic acid} + I_2 = \text{Dehydroascorbic acid} + 2H^+ + 2I^- \qquad E°_{rxn} = +0.14_5 \text{ V}}$$

In general, the potential for the complete reaction, E°_{rxn}, will be $E^\circ_{ox} - E^\circ_{red}$, as is shown for this example. It should be noted that if the number of electrons in each half reaction was not equal, each would have to be multiplied by the number of electrons in the other half reaction before the addition is performed. (This multiplication does not affect E°_{rxn}.)

To illustrate the opposite situation, consider the oxidation of iodide to iodine by dehydroascorbic acid. To find whether this reaction would occur, the same general procedure as above would be used.

$$\text{Dehydroascorbic acid} + 2\text{H}^+ + 2e^- = \text{Ascorbic acid} \qquad +0.39 \text{ V}$$

$$2\text{I}^- = \text{I}_2 + 2e^- \qquad -0.535 \text{ V}$$

$$\overline{\text{Dehydroascorbic acid} + 2\text{H}^+ + 2\text{I}^- = \text{Ascorbic acid} + \text{I}_2 \quad E^\circ_{rxn} = -0.14_5 \text{ V}}$$

The E°_{rxn} for this reaction is negative, indicating that the reaction would not occur spontaneously under standard conditions.

Since many reactions occur under conditions that are not standard, it is helpful to have a generalization to predict whether such reactions will occur. One approach would be to calculate the equilibrium constant and use this as a guide. It can be shown, using the general equilibrium constant equation, that a certain minimum theoretical difference in E° values is necessary for a quantitative (99.9%) reaction. If both half reactions involve one electron change, then a reaction will be quantitative if E°_{rxn} is +0.354 V or greater. This assumes that equivalent amounts of the two reactants are mixed, not an excess of either. If both half reactions involve two electron changes, then the reaction will be quantitative if E°_{rxn} is +0.177 V or greater.

To illustrate the above generalizations, let us consider the reaction of iodine and ascorbic acid (**11-15**). The *actual* E°_{rxn} is +0.14$_5$ V. Since each half reaction involves two electrons, the *minimum theoretical* E°_{rxn} is +0.177 V. Since the actual E°_{rxn} is less than the minimum theoretical E°_{rxn}, the reaction of equivalent amounts of iodine and ascorbic acid is not quantitative (99.9%). This strictly applies only to 1M acid solutions. Since the reaction is usually conducted at lower acidities and since H$^+$ is a product, it will be quantitative at a higher pH. For example, consider the effect of conducting the reaction at pH 7. Using the potential of -0.023 V for ascorbic acid at that pH (Table 11-1), the actual E°_{rxn} is calculated to be +0.558 V. This is much greater than the minimum theoretical E°_{rxn} of +0.177 V so that the reaction is obviously quantitative at a pH of 7.

Calculation of the Equilibrium Constant

This discussion may or may not be appropriate for the level of the course you are taking. The instructor may direct that you omit this material in the latter case. This can be done without making the following sections difficult to understand.

Each oxidation-reduction reaction should in theory have a unique equilibrium constant, K. The general form of an oxidation-reduction reaction governed by K is

$$x A_{ox} + y B_{red} = x A_{red} + y B_{ox}$$

in which x molecules of an oxidizing agent, A_{ox}, react with y molecules of a reducing agent, B_{red}, to give the corresponding products. We will assume that each A_{ox} gains y electrons and that each B_{red} loses x electrons; we will also let the total number of electrons involved be n.

For each half reaction, the relationship between E, the potential of the half reaction, and the concentrations of the various species is given by the Nernst equation. The *general* form of this equation at 25°C is

$$E = E° + \frac{0.059}{\text{no. of } e^-} \log \frac{[\text{Ox}]}{[\text{Red}]} \qquad (11\text{-}20)$$

For the A_{ox}/A_{red} half reaction, **11-20** has the form

$$E_A = E_A° + \frac{0.059}{y} \log \frac{[A_{ox}]^x}{[A_{red}]^x} \qquad (11\text{-}21)$$

For the B_{red}/B_{ox} half reaction, **11-20** has the form

$$E_B = E_B° + \frac{0.059}{x} \log \frac{[B_{ox}]^y}{[B_{red}]^y} \qquad (11\text{-}22)$$

After A_{ox} and B_{red} have reacted and reached equilibrium, the potential of A to react further will equal the potential of B to react further; that is, $E_A = E_B$. We can then equate the right-hand side of **11-21** to that of **11-22**.

$$E_A° + \frac{0.059}{y} \log \frac{[A_{ox}]^x}{[A_{red}]^x} = E_B° + \frac{0.059}{x} \log \frac{[B_{ox}]^y}{[B_{red}]^y} \qquad (11\text{-}23)$$

Recognizing that the total number of electrons involved is the lowest common denominator of x and y, we combine terms in **11-23** to obtain

$$\log \frac{[A_{red}]^x [B_{ox}]^y}{[A_{ox}]^x [B_{red}]^y} = \log K = \frac{n(E_A° - E_B°)}{0.059} = \frac{nE_{rxn}°}{0.059} \qquad (11\text{-}24)$$

In the log term on the left-hand side of **11-24**, it can be recognized that this is the product of the concentration of the products divided by that of the reactants for our general oxidation-reduction reaction. In the two right-hand terms are given two different but equivalent terms for calculating the log of the equilibrium constant for any redox reaction.

It is important to use the correct value of n when using **11-24** for calculating $\log K$ values. If x and y are not equal, then $n = (x)(y)$. For example, if $x = 1$ and $y = 2$, then $n = 2$. If x and y involve 2 and 3 electrons, respectively, then $n = 6$. If $x = y = 2, 3, \ldots$, then in deriving **11-24** the terms simplify such that

$n = x = y = 2, 3, \ldots$ Thus if both half reactions involve two electrons, $n = 2$, not 4. An example to illustrate this is given below.

EXAMPLE 11-4. Calculate the equilibrium constant for the reaction of iodine and ascorbic acid to produce iodide and dehydroascorbic acid (**11-15**).

Solution: Identify iodine as the oxidizing agent and ascorbic acid as the reducing agent. This enables us to calculate the E°_{rxn}.

$$E^\circ_{rxn} = (E^\circ_A - E^\circ_B) = (E^\circ_{ox\,agt} - E^\circ_{red\,agt}) = +0.535 - (+0.39) = +0.14_5 \text{ V}$$

Using **11-24** and recognizing from Table 11-1 that $n = x = y = 2$, $\log K$ is found.

$$\log \frac{[I^-]^2[\text{Dehydro. a.}][H^+]^2}{[I_2][\text{Ascorbic a.}]} = \log K = \frac{2(+0.14_5)}{0.059} = 4.9_1 \quad \text{(two sig. figs.)}$$

Converting $\log K$ to a numerical value for K gives

$$\text{antilog of } 4.9_1 = 9._6 \times 10^4 \quad \text{(one sig. fig.)}$$

A review of significant figures in Section 2-1 will indicate why only one significant figure is justified in the constant when two significant figures were given for $\log K$.

It can be shown that the minimum theoretical value of K for a quantitative (99.9%) reaction should be 1×10^6, or greater. From this point of view, the reaction of iodine and ascorbic acid is not quantitative if the hydrogen ion concentration is one molar. Since the hydrogen ion is a product, lowering its concentration should be enough to make the reaction quantitative.

For further practice with calculation of the equilibrium constant, you should work the problems in that group at the end of the chapter.

11-4 REDOX METHODS OF ANALYSIS

This section is a general introduction to oxidation-reduction methods; it is not a complete survey of such methods since such a survey would be beyond the scope of this book. The basic requirements for redox methods and some examples of titrimetric and spectrophotometric methods are discussed to introduce you to this type of method.

General Principles

Several general principles are important to observe if a successful redox method is to be selected, or devised. One of these is to choose the proper titrant or reagent for spectrophotometric analysis. If the species to be determined is present in a lower oxidation state and can be oxidized quantitatively, then an oxidizing agent should be chosen as titrant or reagent. Table 11-1 lists many good oxidizing agents and the half reactions which they undergo. All the oxidizing agents are found on the *left* in each half reaction. Not all the oxidizing agents are written as compounds; just what compound to choose is also important and will be discussed later under the topic of primary standards.

If the species to be determined is present in its highest, or a higher, oxidation state and it is known that it can be reduced quantitatively, then a reducing agent should be chosen as titrant or reagent. Table 11-2 lists many good reducing agents and the half reactions which they undergo. Because the IUPAC recommends that all half reactions be written as reductions, all the reducing agents are found on the *right* in each half reaction. When each reacts it will give the product listed on the left. Not all the reducing agents are written as compounds; just what compound to use is also important and will be discussed under the topic of primary standards.

TABLE 11-2. *Common Reducing Agents and Their Standard Potentials*

Product $+ ne^- \leftarrow$ reducing agent	Standard potential, $E°$, in volts
$Fe^{+3} + e^- = Fe^{+2}$	$+0.771$
$H_3AsO_4 + 2H^+ + 2e^- = H_3AsO_3$	$+0.559$ ($=0.0$ at pH 7)
$I_2 + 2e^- = 2I^-$	$+0.535$
Dehydroascorbic $+ 2H^+ + 2e^- =$ Ascorbic acid	$+0.39$ ($=-.023$ at pH 7)
$Fe(CN)_6^{-3} + e^- = Fe(CN)_6^{-4}$	$+0.36$
$AgCl(s) + e^- = Ag(s) + Cl^-$	$+0.222$
$Sn^{+4} + 2e^- = Sn^{+2}$	$+0.15$
$S_4O_6^{-2} + 2e^- = 2S_2O_3^{-2}$ (thiosulfate)	$+0.15$
$TiO^{+2} + 2H^+ + e^- = Ti^{+3} + H_2O$	$+0.1$
$2H^+ + 2e^- = H_2(g)$	0.0
$Pb^{+2} + 2e^- = Pb(s)$	-0.126
$Cr^{+3} + e^- = Cr^{+2}$	-0.41
$2CO_2(g) + 2H^+ + 2e^- = H_2C_2O_4$	-0.49
$Zn^{+2} + 2e^- = Zn(s)$	-0.763
$Mg^{+2} + 2e^- = Mg(s)$	-2.37

Regardless of whether an oxidizing agent or a reducing agent is chosen as a titrant, it must be known or established that the reaction will be *quantitative and rapid*. If the $E°$ of the substance to be determined is known, then the $E°_{rxn}$ can be calculated as shown in the previous section. If the *actual* $E°_{rxn}$ is at least $+0.354$ V for a reaction where $n = 1$, then the reaction will be quantitative. (As also discussed in the previous section for a reaction where $n = 2$, the actual $E°_{rxn}$ need be only $+0.177$ V or greater for quantitative (99.9%) reaction.)

This means that the oxidizing agents with the larger positive $E°$ values will give the larger $E°_{rxn}$ values and that reactions involving these agents will tend to be 99.9%, or more, complete. However, it must also be known or established that the reaction is rapid as well. If a reducing agent is needed for a reaction, then one with as *negative* an $E°$ as possible should be selected. This does not necessarily imply that the $E°$ should always be negative; for example, iron(II) is a good reducing agent for the determination of many oxidizing agents like cerium(IV). The $E°$ of $+0.77$ V for the iron half reaction is sufficiently *negative* with respect to an $E°$ of 1.70, or 1.44, for cerium(IV) that the reaction between Ce(IV) and Fe(II) will be quantitative (99.9%).

It is difficult to predict whether a given reaction will be rapid or not. Several observations can be made on the basis of observed rates of reactions.

1. Reactions between cations and anions tend to be fast.
2. Reactions involving the exchange of unequal numbers of electrons tend to be slow (for example, $2Fe^{+3} + Sn^{+2}$ reacting to give $2Fe^{+2}$ and Sn^{+4} is slow).
3. Reactions involving the breaking of bonds in both reactants tend to be slow (for example, MnO_4^- oxidizing $C_2O_4^{-2}$ to CO_2 and forming Mn^{+2} as the other product is slow).

It should be stressed that the most important check on whether a reaction will be quantitative and rapid is *experimental*. No amount of theory or generalizations about reaction rates will replace actually performing the experiment in the laboratory. If you are following directions for a method that has already been published, you usually can be sure that it is quantitative and rapid. If you are modifying the method or devising a new method, you should check it by using 100% pure compounds, or primary standards.

Primary Standards. Although many of the general requirements discussed in Chapter 6 apply to primary standards for redox methods, redox primary standards must be chosen more carefully than those for other types of reactions. In many cases, a redox primary standard will not react quantitatively or will not react rapidly with the solution to be standardized. For example, suppose that it is desired to standardize a solution of iodine using one of the primary standard reducing agents listed in Table 11-3, p. 254. Sodium oxalate (no. 7) could not be used because it does not react rapidly enough with iodine. Ferrous ammonium sulfate (no. 10) could not be used either because the reaction would not be quantitative (the $E°_{rxn}$ would be negative).

To avoid the difficulties of choosing a proper primary standard, you are advised to consult Table 11-3, which lists many of the uses of each primary standard. Iodine is a special case and deserves further comment. Although it is a primary standard material, iodine is difficult to weigh and dissolve accurately, so that many methods call for standardizing it against arsenic(III) oxide after weighing it out approximately.

End Point Detection. As in most other methods, the end point may be detected by using an indicator, finding the end point by potentiometric means, or using a photometric titration. Because some titrants are intensely colored and form essentially colorless products, or vice versa, they can be used as *self indicators*. For example, potassium permanganate has an intense purple color and forms the nearly colorless manganese(II) ion as product. The first slight excess of potassium

TABLE 11-3. Redox Primary Standards and Their Uses

Primary standard	Uses in standardization
Oxidizing agents	
1. Iodine, I_2	1a. Used as a titrant
	1b. Standardization of reducing agents such as sodium thiosulfate
2. KIO_3	2a. Used as a titrant; titrated into acid and KI to produce I_2
	2b. Standardization of reducing agents ($Na_2S_2O_3$)
3. $(NH_4)_2Ce(NO_3)_6$	3. Used as a titrant
4. $K_2Cr_2O_7$	4. Used as a titrant
5. $KBrO_3$	5. Used as a titrant; titrated into acid and KBr to produce Br_2
Reducing agents	
6. As_2O_3	6. Standardization of I_2, Ce(IV), $KMnO_4$, and H_5IO_6 (periodic acid)
7. $Na_2C_2O_4$	7. Standardization of $KMnO_4$ and Ce(IV)
8. $K_4Fe(CN)_6$	8. Standardization of $KMnO_4$ and Ce(IV)
9. Fe wire	9. Standardization of $KMnO_4$, Ce(IV), and $K_2Cr_2O_7$
10. $Fe(NH_4)_2(SO_4)_2 \cdot 6H_2O$	10. Standardization of $KMnO_4$, Ce(IV), and $K_2Cr_2O_7$
11. Cu wire	11. Standardization of $Na_2S_2O_3$ after dissolution to Cu^{+2} and reduction to $CuI(s) + I_2$

permanganate titrant will therefore give the solution a light purple, or pink, color. Iodine is similar in that the first excess of iodine will cause aqueous solutions to appear light yellow, or nonaqueous solutions a light purple. The iodide ion product is colorless and will therefore not interfere with visual perception of the end point. In contrast to these is potassium dichromate titrant. It is an intense orange color

TABLE 11-4. Redox Indicators

$E°$, V	Indicator (reduced form)	Color	Color of oxidized form
—	Starch[a]	Colorless	Purple
+1.25	Tris(nitro-1,10-phenanthroline) iron(II) sulfate, or nitroferroin	Red	Weak blue
+1.14[b]	Tris(1,10-phenanthroline) iron(II) sulfate, or ferroin	Red	Weak blue
+1.06	p-Nitrodiphenylamine	Colorless	Violet
+0.97	Tris(2,2'-bipyridine) iron(II) sulfate	Red	Weak blue
+0.85[c]	Diphenylaminesulfonic acid	Colorless	Violet or purple
+0.76[c]	Diphenylamine	Colorless	Violet or purple
+0.54	1-Naphthol-2-sulfonic acid-indophenol	Colorless	Red
+0.53[b]	Methylene blue	Colorless	Blue

[a] Used only for iodine titrations.
[b] In $1M$ sulfuric or hydrochloric acids, the standard potential is 1.06 V.
[c] These potentials are for $1M$ sulfuric acid.

SEC. 11-4 Redox Methods of Analysis 255

but forms the light purple chromium(III) ion, or in HCl the light green $CrCl^{+2}$ ion, as a product. The concentration of either of these ions at the end point is sufficient in most cases to prevent perception of the first excess of potassium dichromate.

Table 11-4 lists a number of useful *redox indicators*. There is no difficulty in choosing an indicator for most standard titrants since redox indicators have been studied in detail. It can be shown that the $E°$ value listed for each indicator in the table should fall within $\pm 0.059/n$ volts of the potential at the end point. Since most indicators have already been experimentally checked for a particular method, it is usually only necessary to employ this rule when using a new or revised method or a new indicator.

The starch indicator action for iodine titrations is different than the action of the other indicators listed in Table 11-4 and needs explanation. Starch forms a *complex* with iodine and iodide ion as follows:

$$I_2 + \text{starch} + I^- = (\text{starch})I_3^-$$

The color of the starch-triiodide complex is blue to blue-purple. The first excess of iodine titrant added past the end point reacts with the starch and iodide ion to give the blue color. This complex is ionic and does not form readily in organic solvents so starch is used only in aqueous titrations.

Potentiometric detection of the end point is possible with a potentiometer, an indicating electrode, and a reference electrode in a manner similar to the potential measurements shown in Figure 11-1. Changes in concentration of the ion being titrated are indicated by the indicating electrode; the corresponding changes in potential are followed on the potentiometer. If a reducing agent is being titrated with an oxidizing agent, the potential will increase as shown in Fig. 11-2. At the

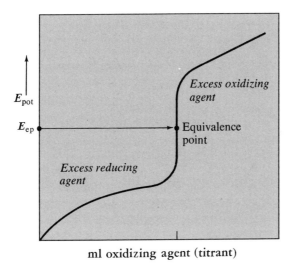

FIGURE 11-2. Potentiometric titration curve using an oxidizing agent (more positive $E°$) as titrant and reducing agent as the species titrated.

end point, a sharp increase in potential will occur, followed by a leveling off of the increase in potential. (If an oxidizing agent is titrated with a reducing agent, the opposite type of curve will be obtained with a sharp decrease in potential.) It can be shown that the potential at the end point, E_{ep}, of a redox titration can be calculated as follows:

$$E_{ep} = \frac{yE^{\circ}_{ox\ agt} + xE^{\circ}_{red\ agt}}{y+x} \tag{11-25}$$

in which y is the number of electrons gained in the half reaction of the oxidizing agent and x is the number of electrons gained in the half reaction of the reducing agent. Equation **11-25** is valid for all conditions except where the reaction involves hydrogen or hydroxide ions and the $[H^+]$ or $[OH^-]$ is not $1M$.

Titrations with Oxidizing Agents

Because so many elements are easily oxidized from a lower oxidation state to a higher one, and because so many organic compounds are readily oxidized to more oxidized forms, many methods have been based on direct or indirect titrations with oxidizing agents. Oxidizing agents involving some oxidation state of iodine are generally the most useful because they are milder, easier to handle, and have fewer side reactions. Some of these are potassium iodate, iodine itself, and periodic acid (HIO_4, or its hydrate H_5IO_6). Stronger oxidizing agents are cerium (IV) in sulfuric acid, potassium permanganate, and potassium dichromate. These reagents will oxidize more elements and compounds but undergo more side reactions and are more difficult to control because of this.

Iodine Titrations. Direct titration with iodine is a useful method for the determination of a fairly large number of inorganic and organic substances. The titration is simple, although in many cases the pH of the solution must be adjusted or controlled by buffering. If water is the solvent, starch is added at the beginning of the titration. Near the end point the purple color of the starch-triiodide complex appears momentarily, and as the end point is approached the mixing time required for its disappearance increases until a stable purple color appears.

Iodine oxidizes a number of inorganic substances via reactions in which the oxidation number of one element is increased by two, that is, via two-electron oxidations (Table 11-5). A good example is the reaction of iodine with arsenious acid, which is arsenic in the +3 oxidation state.

$$H_3AsO_3 + I_2 + H_2O \xrightarrow{NaHCO_3\ buffer} HAsO_4^{-2} + 2I^- + 4H^+$$

Because the reaction produces hydrogen ions, sodium bicarbonate buffer must be used to keep the pH slightly basic. If the pH was to fall below about 4, the reaction would not be quantitative. This reaction is also important because it can be used for the standardization of iodine. Primary standard arsenic(III) oxide (Table 11-3) is dissolved in basic solution and the solution is acidified to produce arsenious acid.

Iodine also oxidizes a number of inorganic and organic substances to dimers. The best example of this type of reaction is that with sodium thiosulfate, producing the tetrathionate ion.

$$2S_2O_3^{-2} + I_2 \rightarrow S_4O_6^{-2} + 2I^-$$
$$(Na_2S_2O_3)$$

The tetrathionate ion consists of two thiosulfate species bonded together as O_3S_2—$S_2O_3^{-2}$ and is essentially a dimer. This reaction can be used for the standardization or analysis of either reaction; since sodium thiosulfate is not a primary standard, it must be first standardized before using it to standardize iodine. Other pharmaceuticals listed in Table 11-5 are oxidized in similar fashion, as is the amino acid cysteine.

TABLE 11-5. Inorganic and Pharmaceutical Substances Determined by Iodine Titration

Inorganic or pharmaceutical[a]	Oxidation state change or reaction
Two-electron increase in oxidation number	
Sn(II)	Sn(II)–Sn(IV)
H_2SO_3 or SO_3^{-2}	SO_3^{-2}–SO_4^{-2}
As(III)	As(III)–As(V)
Oxophenarsine, $C_6H_6AsNO_2$	As(III)–As(V)
Carbarsone, $C_7H_9AsN_2O_4$	As(III)–As(V)
Sb(III)	Sb(III)–Sb(V)
Antimonic potassium tartrate, $KSbOC_4H_4O_6 \cdot \frac{1}{2}H_2O$	Sb(III)–Sb(V)
Oxidation to dimers	
Thiosulfate, $S_2O_3^{-2}$	$2S_2O_3^{-2}$–$S_4O_6^{-2}$
Mercaptans, RSH	$2RSH$–$RSSR + 2H^+$
Dimercaprol, $C_3H_8S_3$	$2RSH$–$RSSR + 2H^+$
Cysteine, $HSC_2H_3(NH_2)CO_2H$	Cysteine–Cystine
Other oxidations	
Ascorbic acid	Ascorbic acid–dehydroascorbic acid
$Fe(CN)_6^{-4}$	$Fe(CN)_6^{-4}$–$Fe(CN)_6^{-3}$

[a] Taken in part from G. L. Jenkins, A. Knevel, and F. E. DiGangi, *Quantitative Pharmaceutical Chemistry*, 6th ed., McGraw-Hill, New York, 1967, pp. 179–88.

A rather unusual oxidation is the determination of ascorbic acid by iodine titration (11-15). The iodine oxidizes the "enediol" functional group in ascorbic acid to a alpha-diketone group in dehydroascorbic acid as follows:

$$\begin{array}{c} | \\ C-OH \\ \| \\ C-OH \\ | \end{array} \rightarrow \begin{array}{c} | \\ C=O \\ | \\ C=O \\ | \end{array} + 2H^+$$

Ascorbic acid (vitamin C) can be determined by direct titration with iodine [3],

[3] J. W. Stevens, *Ind. Eng. Chem., Anal. Ed.* **10**, 269 (1938).

or a measured excess of standard potassium iodate along with potassium iodide and acid can be added to the ascorbic acid. The iodate and iodide react in acid to produce iodine.

$$IO_3^- + 5I^- + 6H^+ \rightarrow 3I_2 + 3H_2O$$

The iodine reacts with the ascorbic acid, and the unreacted iodine is determined by back titration with standard sodium thiosulfate [4]. Both of these approaches can be used for the determination of vitamin C in tablets (Experiment 10), in drinks such as Hi-C, and in dehydrated juice solids such as Tang.

Periodic Acid Oxidations. Oxidation with periodic acid is a very selective method for organic and biochemical molecules having a 1,2-diol grouping. The periodic acid cleaves the carbon-carbon bond and takes each remaining fragment to the next higher oxidation state. For example, it oxidizes ethylene glycol to two molecules of formaldehyde as follows:

$$\begin{array}{c} H_2C-OH \\ | \\ H_2C-OH \end{array} + HIO_4 \rightarrow \begin{array}{c} H_2C=O \\ \\ H_2C=O \end{array} + HIO_3 + H_2O$$

Glycerine, an important ingredient in many pharmaceutical preparations, is oxidized by periodic acid to two molecules of formaldehyde plus formic acid as follows:

$$\begin{array}{c} H_2C-OH \\ | \\ HC-OH \\ | \\ H_2C-OH \end{array} + 2HIO_4 \rightarrow \begin{array}{c} H_2C=O \\ \\ + HCOOH \\ \\ + H_2C=O \end{array} + 2HIO_3 + H_2O$$

Sugars such as glucose and fructose are also oxidized by periodic acid.

A measured excess of periodic acid is usually added to the compound to be determined, and the unreacted periodic acid is reduced to iodine, which is then titrated with standard sodium thiosulfate.

$$HIO_4 + 7I^- + 7H^+ \rightarrow 4H_2O + 4I_2 \text{ (titrated with } Na_2S_2O_3\text{)}$$

Ferricyanide Oxidations and Colorimetric Analysis. The ferricyanide ion, $Fe(CN)_6^{-3}$, is a mild oxidizing agent ($E° = +0.36$ V). It can therefore only be used for the quantitative oxidation of very easily oxidized substances; for example, aldehydes (RCHO) are oxidized to acids (RCO_2H). Because it has an intense yellow color it can be used for the *colorimetric determination* (Ch. 13) of such aldehydes. A measured excess of the ferricyanide is added, and the unreacted ferricyanide is determined colorimetrically. The difference between the amount of ferricyanide added before reaction and the amount unreacted is a measure of the amount of the aldehyde present.

[4] D. N. Bailey, *J. Chem. Educ.* **51**, 488 (1974).

A good example of this is the colorimetric determination of glucose in body fluids. A measured excess of ferricyanide oxidizes the glucose to gluconic acid.

$$2Fe(CN)_6^{-3} + C_5H_6(OH)_5CHO + H_2O$$
$$\xrightarrow{90°C} 2Fe(CN)_6^{-4} + C_5H_6(OH)_5CO_2H + 2H^+$$

The reaction is slow at room temperature and heat is needed to increase the rate of the reaction. This reaction is used for the automated colorimetric determination of glucose in many clinical and hospital laboratories. The details are discussed in Section 14-4.

Titrations with Strong Oxidizing Agents. Strong oxidizing agents are those which have an $E°$ greater than $+1$ V and which frequently can oxidize a substance with two or more oxidation products to the highest oxidation state or product. The most common strong oxidizing agents are potassium permanganate, cerium(IV), and potassium dichromate. All three can sometimes be used for the same determination, but each has special advantages and disadvantages.

Potassium permanganate, $KMnO_4$, is a convenient titrant in that it serves as its own self-indicator. The end point is indicated by the pink to light purple color imparted to the solution by the first excess of the permanganate ion. Since potassium permanganate is not available in primary standard purity, it must be standardized against one of the primary standards listed in Table 11-3.

Potassium permanganate can be used for the analysis of hydrogen peroxide solutions, such as the 3% solution sold for medicinal purposes. The hydrogen peroxide is oxidized to oxygen in the titration.

$$5H_2O_2 + 2KMnO_4 + 6H^+ \rightarrow 5O_2(g) + 2Mn^{+2} + 2K^+ + 8H_2O$$

Note that all of the products in the reaction are essentially colorless so the first excess of permanganate indicates the end point.

Potassium permanganate is also useful for the titration of iron(II), either in the form of ferrous sulfate tablets or after it has been produced following dissolution of iron ores or iron metal. The reaction is

$$5Fe^{+2} + KMnO_4 + 8H^+ \rightarrow 5Fe^{+3} + Mn^{+2} + K^+ + 4H_2O$$

The iron(III) ion is slighly colored in the sulfuric acid solvent used for the reaction, but the end point is readily seen.

Other uses for potassium permanganate are listed in Table 11-6, p. 260.

Cerium(IV), or the *ceric ion*, is a more useful titrant than potassium permanganate for the following reasons. Cerium(IV) is available in primary standard form as the ammonium hexanitratocerate(IV) salt (Table 11-3). It can be used in perchloric acid ($E° = +1.70$ V) or in sulfuric acid ($E° = +1.44$ V). Although cerium(IV) is yellow in sulfuric acid, its color is not suitable for indication of the end point. Instead, ferroin or one of the other indicators in Table 11-4 is used.

TABLE 11-6. *Inorganic and Pharmaceutical Substances Determined by Potassium Permanganate and Cerium (IV) Titration*

Inorganic or pharmaceutical[a]	Oxidation state change or reaction
$KMnO_4$ titrant	
Fe(II)	Fe(II)–Fe(III)
$Fe(CN)_6^{-4}$	$Fe(CN)_6^{-4}$–$Fe(CN)_6^{-3}$
As(III)	As(III)–As(V)
$Na_2C_2O_4$ or $H_2C_2O_4$	$C_2O_4^{-2}$–$2CO_2(g)$
H_2O_2 or Na_2O_2	O_2^{-2}–$O_2(g)$
Cerium(IV) titrant	
Fe(II)	Fe(II)–Fe(III)
$Fe(CN)_6^{-4}$	$Fe(CN)_6^{-4}$–$Fe(CN)_6^{-3}$
Ferrous fumarate tablets	Fe(II)–Fe(III)
Ferrous sulfate tablets	Fe(II)–Fe(III)
Ferrous gluconate tablets	Fe(II)–Fe(III)
As(III)	As(III)–As(V)
$Na_2C_2O_4$ or $H_2C_2O_4$	$C_2O_4^{-2}$–$2CO_2(g)$
Hydroquinone	Quinone
Tocopherol (vitamin E)	Quinone form of vitamin E

[a] Taken in part from G. L. Jenkins, A. Knevel, and F. E. DiGangi, *Quantitative Pharmaceutical Chemistry*, 6th ed., McGraw-Hill, New York, 1967, pp. 173–77.

Cerium(IV) can be used for the determination of a large number of substances, a few of which are listed in Table 11-6. The reactions for the determination of iron(II) in various forms is

$$Fe^{+2} + Ce(IV) \xrightarrow{H_2SO_4 \text{ solvent}} Fe^{+3} + Ce^{+3}$$

$$\text{Ferroin} + Ce(IV) \rightarrow \text{Ferriin} + Ce^{+3}$$

$$\text{(red)} \qquad \qquad \text{(weak blue)}$$

Although the oxidized form of the ferroin indicator, ferriin, is a weak blue color, the end point is easily perceived from the disappearance of the intense red color of ferroin itself.

Cerium(IV) can also oxidize many organic compounds; for example, it oxidizes hydroquinone to quinone.

$$\text{hydroquinone} + 2Ce(IV) \longrightarrow \text{quinone} + 2Ce^{+3}$$

Because vitamin E, or tocopherol, has a hydroquinone-type structure, cerium(IV) can oxidize vitamin E to a quinone form.

SEC. 11-4 Redox Methods of Analysis

Potassium dichromate, $K_2Cr_2O_7$, is a slightly weaker oxidizing agent ($E° = 1.33$ V) than either potassium permanganate or cerium(IV) and is therefore less useful. It is available in primary standard form (Table 11-3). Although it is orange in color, its color is not suitable for indication of the end point. It is commonly used with diphenylamine or diphenylaminesulfonic acid (Table 11-4) as indicators.

Potassium dichromate is especially useful for the determination of iron(II) following dissolution of metallic iron or iron ores in hydrochloric acid.

$$Fe°(s) + 2HCl = H_2(g) + Fe^{+2}$$

$$Fe_2O_3(s) + 6HCl = 3H_2O + 2Fe(III)$$

Stronger oxidizing titrants oxidize the chloride ion to chlorine, but potassium dichromate is too weak an oxidizing agent ($E° = +1.33$ V) to oxidize chloride to chlorine ($E° = +1.36$ V); that is, chlorine is a stronger oxidizing agent. If an iron ore is to be analyzed for per cent iron, the dissolution of the ore yields iron(III), which must then be reduced using tin(II) chloride.

$$2Fe(III) + SnCl_2 \xrightarrow{HCl\ solvent} 2Fe^{+2} + Sn(IV) + 2Cl^-$$

The titration of iron(II) is then the same whether metallic iron or iron ores are being analyzed.

$$6Fe^{+2} + K_2Cr_2O_7 + 14H^+ \rightarrow 6Fe^{+3} + 2Cr(III) + 7H_2O + 2K^+$$

The solution is somewhat green from the chloro complexes of chromium(III), but the purple to violet color of the diphenylamine indicator is readily seen.

Titrations Involving Reducing Agents

Not as many methods for titrations involving reducing agents are available as for oxidizing agents because the oxygen in the air oxidizes many reducing agents, making it difficult to prepare standard solutions. The most useful procedure to be discussed below is reduction with the iodide ion; it is used for the determination of many oxidizing agents.

Iodide Reductions. The iodide ion itself is not used for direct titration of many substances; instead an excess of the iodide ion is added to the substance, which must be capable of quantitatively oxidizing the iodide to iodine. The iodine is then titrated with a standard solution of sodium thiosulfate. The amount of sodium thiosulfate titrant is equivalent to the amount of the oxidizing substance originally present. The method for the determination of an oxidizing substance, A_{ox}, which is reduced to the form A_{red}, is summarized by the following equations.

$$A_{ox} + 2I^- (\text{excess}) = A_{red} + I_2$$
$$I_2 + 2Na_2S_2O_3 = 2I^- + 2Na^+ + Na_2S_4O_6$$
(titrant)

Starch is used as the indicator for the titration. It is usually added quite close to the end point because the starch-triiodide complex breaks apart somewhat slowly.

A number of substances that can be determined by the above method are listed in Table 11-7. An important example for both inorganic and pharmaceutical

TABLE 11-7. *Inorganic and Pharmaceutical Substances Determined by Reduction with Iodide Ion*

Inorganic or pharmaceutical[a]	Oxidation state change or reaction
Halogen-containing substances	
Cl_2	Cl_2–$2Cl^-$
OCl^-, hypochlorite (bleach)	OCl^-–$Cl^- + H_2O$
Br_2	Br_2–$2Br^-$
IO_3^-, iodate ion	$IO_3^- + 5I^- = 3I_2 + 3H_2O$
HIO_4 or IO_4^-	$IO_4^- + 7I^- = 4I_2 + 4H_2O$
Iodine pentoxide	$I_2O_5 + 10I^- = 6I_2 + 5H_2O$
Other reductions	
Cu^{+2}, $CuSO_4$ (N.F.)	Cu^{+2}–$CuI(s)$
As(V)	As(V)–As(III)
Arsanilic acid, $C_6H_8AsNO_3$	As(V)–As(III)
Drocarbil, $C_8H_{10}AsNO_5$	As(V)–As(III)
Halazone tablets, $C_7H_5Cl_2NO_4S$	$R-Cl + 2I^- = RH + I_2 + Cl^-$

[a] Taken in part from G. L. Jenkins, A. Knevel, and F. E. DiGangi, *Quantitative Pharmaceutical Chemistry*, 6th ed., McGraw-Hill, New York, 1967, pp. 191–98.

analysis is the determination of copper or copper sulfate (National Formulary method). If copper is present in metallic form, it is first dissolved in nitric acid as follows:

$$3Cu°(s) + 2NO_3^- + 8H^+ = 3Cu^{+2} + 2NO + 4H_2O$$

Then the copper(II) ion, or copper sulfate, is treated with excess iodide ion.

$$2Cu^{+2} + 4I^- = 2CuI(s) + I_2$$

The reaction produces insoluble copper(I) iodide and iodine, which is titrated with standard sodium thiosulfate.

Other Reductions. Methods based on other reducing agents are also used, but they are fairly specialized or are not used as frequently as iodide reduction because the reducing agents are oxidized by oxygen in the air. An example of a specialized type of reducing agent is sodium nitrite, $NaNO_2$. It is used for the analysis of sulfa drugs and other organic compounds that contain an amino group (—NH_2) bonded to an aromatic ring. The reaction is conducted in acid which converts the sodium nitrite to nitrous acid, HNO_2. The product is a diazonium salt, $RN_2{}^+Cl^-$. For example, with sulfanilamide, the reaction is

$$H_2NSO_2C_6H_4\text{—}NH_2 + HNO_2 + HCl = H_2NSO_2C_6H_4\text{—}N_2{}^+Cl^- + 2H_2O$$

(sulfanilamide)

The end point may be detected by dipping starch-iodide paper into the solution, giving a blue color when excess nitrous acid is present to oxidize iodide to iodine.

11-5 SIMPLE POTENTIOMETRIC ANALYSIS

Introduction

Titrations are not the only types of techniques that can utilize oxidation-reductions for analysis. It is also possible to use a potentiometer to measure ion concentrations with a single measurement. Potentiometric measurements were described in general in Section 11-3. A potentiometric measurement requires the general equipment pictured in Figure 11-1—a reference electrode and an indicating electrode in addition to the potentiometer. Equation **11-18** gives the relationship between the voltage measured on the potentiometer, E_{pot}, and the potential, or voltage, at each of the two electrodes. If the standard hydrogen electrode is used as the reference electrode, then its potential is zero and **11-18** simplifies to

$$E_{pot} = E_{ind\ el} \tag{11-26}$$

Since the potential at the indicating electrode is equal to the potential of the half reaction, which is in equilibrium with this electrode, an equation can be developed from this for analysis. The potential at the indicating electrode must be equal to E, the potential of the half reaction. E is defined by the Nernst equation (**11-20**), where n is the number of electrons in the half reaction.

$$E_{ind\ el} = E = E° + \frac{0.059}{n} \log \frac{[Ox]}{[Red]} \tag{11-27}$$

Substituting **11-27** into **11-26** gives

$$E_{pot} = E° + \frac{0.059}{n} \log \frac{[Ox]}{[Red]} \tag{11-28}$$

Equation **11-28** can be rearranged to give

$$\log \frac{[Ox]}{[Red]} = \frac{n(E_{pot} - E°)}{0.059} \tag{11-29}$$

Equation **11-29** shows that the ratio of the concentration of the oxidized form to the concentration of the reduced form in a half reaction can be measured simply by measuring the voltage, on a potentiometer, of a solution containing these forms. If one or the other of these forms is insoluble, then the concentration of the other form can be measured directly.

Potentiometric Measurement of Chloride Ion

The classical type of indicating electrode half reaction that can be used to measure the chloride ion concentration is

$$AgCl(s) + e^- = Ag°(s) + Cl^-$$

The Nernst equation expression for this half reaction is

$$E = E° + 0.059 \log 1/[Cl^-] \tag{11-30}$$

Note that the only concentration that affects the potential of the half reaction is the chloride ion concentration. If silver/silver chloride indicating electrode is used in a measurement setup, such as shown in Figure 11-1, and a hydrogen reference electrode is also used, then **11-28** becomes for this system

$$E_{pot} = E° + 0.059 \log 1/[Cl^-] \tag{11-31}$$

Equation **11-31** can be rearranged to give

$$-\log[Cl^-] = \frac{(E_{pot} - E°)}{0.059} \tag{11-32}$$

Thus the concentration of chloride ion can be found by a single measurement of the potential of the solution. Unfortunately, the measurements give only two significant figures under routine conditions (three under carefully controlled conditions), so the molarity of the chloride ion is known to only one significant figure (see Ch. 2, Sec. 1).

More modern types of electrodes have been devised to replace the classical silver/silver chloride electrode; these are the so-called *specific ion* electrodes. They function somewhat differently but give the same type of result (Ch. 16).

A more complete discussion of potentiometric measurements, including pH measurement, is given in Chapter 16.

QUESTIONS AND PROBLEMS

(Answers to most even-numbered problems are in Appendix 5.)

Balancing with the Ion-Electron Method
1. Using the ion-electron method, complete and balance the following inorganic half reactions involving the oxidation of the reducing agents to the species at the right. Assume an acid solution unless otherwise indicated.
   a. $H_3AsO_3 \rightarrow H_3AsO_4$
   b. $Mn^{+2} \rightarrow MnO_4^-$
   c. $I^- \rightarrow I_2$
   d. $NH_2OH \rightarrow N_2(g) + H_2O$
   e. $H_2O_2 \rightarrow O_2(g)$

Simple Potentiometric Analysis

2. Using the ion-electron method, complete and balance the following inorganic half reactions involving the reduction of the oxidizing agents to the species at the right. Assume an acid solution unless otherwise indicated.
   a. $Cr_2O_7^{-2} \rightarrow 2Cr^{+2}$
   b. $IO_4^- \rightarrow IO_3^-$
   c. $IO_3^- \rightarrow I_2$
   d. $Fe(CN)_6^{-3} \rightarrow Fe(CN)_6^{-4}$
   e. $MnO_4^- \rightarrow Mn^{+2}$
3. Using the ion-electron method, complete and balance the following organic half reactions involving the oxidation of the reducing agents to the species at the right. Assume an acid solution unless otherwise indicated.
   a. $C_2H_5OH \rightarrow CH_3CO_2H$
   b. $CH_2OHCHOHCH_2OH \rightarrow 2H_2CO + HCO_2H$
   c. $C_4H_6O_4(OH)C=COH \rightarrow C_4H_6O_4C(=O)-C=O$
   d. $[C_{19}H_{27}]CHO \rightarrow [C_{19}H_{27}]CO_2H$
   e. $H_2C_2O_4 \rightarrow 2CO_2(g)$
4. Combine the balanced half reactions of the inorganic reducing agents in Problem 1 with those of the oxidizing agents in Problem 2 to obtain five balanced equations. Combine part *a* of Problem 1 with part *a* of Problem 2, etc.
5. Combine the balanced half reactions of the organic reducing agents in Problem 3 with those of the oxidizing agents in Problem 2 to obtain five balanced equations. Combine part *a* of Problem 3 with part *a* of Problem 2, etc.

Quantitative Reaction
6. Calculate the E_{rxn} of each of the following reactions and decide whether the reaction is quantitative or not (the oxidizing agent is the first species).
   a. $Ce^{IV} + Fe^{+2} \rightarrow Ce^{III} + Fe^{+3}$ (H_2SO_4)
   b. $Ag^+ + Fe^{+2} \rightarrow Ag°(s) + Fe^{+3}$
   c. $I_2 + H_3AsO_3 + H_2O \xrightarrow{1MH^+} 2I^- + H_3AsO_4 + 2H^+$
   d. $I_2 + H_3AsO_3 + H_2O \xrightarrow{pH\ 7} 2I^- + H_3AsO_4 + 2H^+$
   e. $I_2 + $ ascorbic acid $\xrightarrow{pH\ 7} 2I^- + $ dehydroascorbic acid $+ 2H^+$
   f. $2Cu^{+2} + 4I^- \rightarrow 2CuI(s) + I_2$
   g. $I_2 + 2S_2O_3^{-2} \rightarrow 2I^- + S_4O_6^{-2}$
   h. $2Fe^{+3} + Sn^{+2} \rightarrow 2Fe^{+2} + Sn^{+4}$
7. Periodic acid (H_5IO_6) or potassium periodate are frequently used to oxidize Mn^{+2} to MnO_4^- to determine manganese in steel spectrophotometrically. Answer the following questions involving this method.
   a. Assuming that the minimum difference in standard potentials $(E^°_{ox} - E^°_{red})$ for this reaction is $+0.124$ V, decide whether an *equivalent* amount of either periodate compound will oxidize Mn^{+2} quantitatively $(>99.9\%)$ to MnO_4^-.
   b. In the laboratory, either compound can be used for quantitative oxidation of Mn^{+2} to MnO_4^-. How can you rationalize this fact with your answer to *a*?
8. Calculate the equilibrium constant for the following reaction carried out in $1M$ acid.

$$VO_2^+ + Fe^{+2} + 2H^+ \rightleftharpoons VO^{+2} + Fe^{+3} + H_2O$$

Is the oxidation of Fe^{+2} to Fe^{+3} quantitative $(>99.9\%)$? How might conditions be adjusted to make the oxidation of iron more complete?

9. Calculate the equilibrium constant for the following oxidation-reduction reaction.

$$2Cu^{+2} + 4I^- \rightarrow 2CuI(s) + I_2$$

Calculation of Iodine Titration Results
10. A 0.5000 g sample of impure ascorbic acid (form wt = 176.12) in tablet form is oxidized to dehydroascorbic acid by 45.10 ml of $0.1000N$ ($0.0500M$) I_2. Calculate the per cent purity of the tablets.
11. A 1.2500 g sample of pure As_2O_3 is weighed into a 250 ml volumetric flask, dissolved, and diluted to volume. A 25.00 ml aliquot of this solution requires exactly 26.00 ml of iodine solution for titration to the starch end point. Calculate the percentage of As in a 1.0000 g ore sample which requires 15.00 ml of this same iodine solution for titration.
12. Calculate the percentage of copper in 1000 mg of a copper ore which requires 12.10 ml of $0.1000M$ sodium thiosulfate to titrate the iodine produced by the reaction of copper(II) with KI.
13. A 1.0000 g sample containing As_2O_3, As_2O_5, and an inert salt is dissolved and titrated in neutral solution to the starch end point with 20.00 ml of $0.2000N$ iodine. The resulting solution is made strongly acid and excess KI is added, releasing iodine. The iodine is titrated with 40.00 ml of $0.1500N$ sodium thiosulfate. Calculate the percentage of As_2O_3 and the percentage of As_2O_5 (form wt = 229.84) in the sample.

Calculation of Strong Oxidizing Agent Titration Results
14. A 1.0000 g sample of impure vitamin E (tocopherol, form wt of $C_{29}H_{50}O_2 = 430.7$) is titrated with 41.10 ml of $0.1000M$ cerium(IV) titrant. Calculate the per cent purity of the vitamin E, assuming the electron change is the same as that involved in the oxidation of hydroquinone to quinone (Sec. 11-4).
15. The iron in a 1.0000 g rock sample is dissolved, reduced to Fe^{+2}, and titrated with dichromate. If the titration requires 12.40 ml of $0.0500N$ dichromate, what is the percentage of Fe_2O_3 in the sample?
16. A 1.0000 g sample of limonite iron ore ($2Fe_2O_3 \cdot 3H_2O$) is dissolved, reduced to Fe^{+2}, and titrated with 20.00 ml of $0.2000N$ cerium(IV). Calculate the percentage of $2Fe_2O_3 \cdot 3H_2O$ (form wt = 373.38) in the sample.
17. A solution of potassium permanganate is standardized with pure iron metal as primary standard. The iron is dissolved in acid, reduced to Fe^{+2}, and titrated with permanganate as follows:

$$MnO_4^- + 5Fe^{+2} + 8H^+ \rightarrow 5Fe^{+3} + Mn^{+2} + 4H_2O$$

If 33.00 ml of permanganate is needed to titrate a 0.5585 g sample of iron, what is the normality of the permanganate?

18. A cerium(IV) solution is standardized with As_2O_3 as primary standard according to the following reactions.

$$As_2O_3 + 6NaOH \rightarrow 6Na^+ + 2AsO_3^{-3} + 3H_2O$$

$$AsO_3^{-3} + 3H^+ \rightarrow H_3AsO_3$$

$$2Ce^{IV} + H_3AsO_3 + H_2O \rightarrow H_3AsO_4 + 2Ce^{III} + 2H^+$$

A 0.1980 g sample of As_2O_3, when dissolved, requires 20.00 ml of cerium(IV) for titration. Calculate the normality of the cerium(IV) solution.

SEC. 11-5   Simple Potentiometric Analysis

19. A 1.0000 g sample of impure hydrogen peroxide is analyzed by titration with 28.10 ml of 0.1000M cerium(IV) titrant (Table 11-16). Calculate the per cent purity of the hydrogen peroxide.

Devising New Methods
20. Hairwaving preparations contain organic mercaptans (thiols) as the active ingredient. Suggest an oxidation-reduction method for the determination of the mercaptan (RSH).
21. Outline an oxidation-reduction method for each of the following determinations.
    a. Cu in a CuS ore containing some Fe_2O_3 and CaO
    b. Ti in ferrotitanium (Fe-Ti)
    c. Cr in a stainless steel (Fe-Cr-Ni)
    d. Sodium hypochlorite (NaClO) in a laundry bleach
    e. U in a uranium-aluminum alloy (The alloy dissolves in HNO_3, forming UO_2^{+2} and Al^{+3}.)
22. Many modern laundry bleaches are powdered oxidants containing positive halogen atoms. Suggest a way to determine the relative amounts of oxidizing power of competing brands of laundry products.
23. It is desired to titrate Fe^{+2} using one of the titrants listed below. Assume that in each case, the second compound listed is present with the Fe^{+2} and that it must be decided whether it will cause an error by reacting with the titrant. (Recall that the reaction need only be *spontaneous* to cause an error.)
    a. $KMnO^4$ titrant and HCl
    b. $K_2Cr_2O_7$ titrant and HCl
    c. Ce(IV) titrant and $Ce_2(SO_4)_3$
    d. $K_2Cr_2O_7$ titrant and $SnCl_2$
    e. Ce(IV) titrant and H_3AsO_3

Challenging Problems
24. An iron ore is mistakenly calculated to contain 10.00% Fe_2O_3 (form wt = 159.7) when it should have been calculated as % Fe_3O_4 (form wt = 231.5). Calculate the per cent Fe_3O_4 without knowing the sample weight or any other data from the analysis.
25. A silver indicating electrode and a standard hydrogen reference electrode are dipped into a saturated aqueous solution of silver bromide. The potential measured on the potentiometer (E_{meas}) is $+0.434$ V. From the standard potential of the Ag°(s)/Ag$^+$ electrode given below, calculate
    a. The [Ag$^+$] using the correct number of significant figures, and
    b. The solubility product constant for silver bromide.

$$Ag^+ + e^- = Ag°(s), \quad E° = +0.800 \text{ V}$$

26. Suppose that a new buret has been invented which has a relative accuracy of 0.1 ppt (1 part in 10,000), so that a *quantitative* reaction must now be defined as one in which no more than one part in ten thousand of A_{ox} and B_{red} remains when the titration reaction below is complete.

$$A_{ox} + B_{red} \rightleftharpoons A_{red} + B_{ox}$$

Assuming that both reactants undergo a one-electron change, calculate the minimum difference in standard potentials ($E°_{ox} - E°_{red}$) for a *quantitative* reaction.

27. The minimum difference in standard potentials $(E_A^\circ - E_B^\circ)$ for a quantitative reaction in which both reactants undergo a one-electron change is $+0.354$ V at 25°C (298°K). What would this minimum difference be at 83°C?
28. Show that the 0.059 term in the Nernst equation has the unit of volts, given that at 25°C, $0.059 = 2.3\ RT/F$.

PART II

Some Instrumental Methods of Analysis

This part of the book is an elementary treatment of some selected instrumental methods of analysis. We are not attempting to cover every method because little useful information could be included. This part can be used at two levels—the level of the better student and the level of the average student.

For the better student we recommend a very careful reading of Chapter 12, which is an introduction to instrumental methods. This chapter delineates the common features of methods based on the measurement of light and other radiation as well as those based on electrochemical reactions. It emphasizes the transducer as a common feature so that the better student can gain a unified viewpoint of instrumental measurements. If he or she is interested in the construction of a phototube, it can be read about in this chapter rather than in Chapter 13 on spectrophotometers. In addition, a complete discussion is given of all types of radiation, rather than just ultraviolet (UV), visible, and infrared (IR) radiation.

For the average student we recommend a quick reading, or no initial reading at all, of Chapter 12. Instead, this student is probably better off spending time on Chapters 13 and 14, which cover absorption spectrophotometry. These chapters are written to stand on their own, although references are occasionally made to Chapter

12. *Chapters 13 and 14 specifically cover just UV, visible, and IR absorption so that the average student is not overwhelmed by having to comprehend all types of radiation, just the most interesting types. He or she also is not exposed to other nonabsorption processes such as fluorescence, refraction, and light scattering.*

12. Introduction to Instrumental Methods

> *"Are they blood-stains, or mud stains, or rust stains, or fruit stains, or what are they? That is a question which has puzzled many an expert, and why? Because there was no reliable test,"* (Holmes said).
>
> ARTHUR CONAN DOYLE
> A Study in Scarlet, Chapter 1

For our purposes, an instrumental method of analysis is one performed by some instrumental technique in which the final measurement does not involve weighing or a visual titration end point. Instrumental methods came into being because of the need for the "reliable test" for which even Sherlock Holmes yearned. Instrumental measurements remove much of the uncertainty of the vague titration end point and obviate coprecipitation errors that may accompany a gravimetric determination.

Nearly any physical property can be made the basis for an instrumental method of analysis. Thus Chapters 13–16 have to do with the absorption, emission, refraction, or scattering of some form of *radiation*, or *light*. Chapters 17 and 18 are concerned with *electrical* voltage, potential, or conductance measurements. Even the *thermal conductivity* of gases is important to the subject of gas chromatography discussed in Chapters 20 and 21.

The fundamental operation of the analytical instrument is to *translate* chemical or physical information into a form that the analyst can observe and use. The heart of most instruments is a *transducer*, which transfers energy from the chemical domain to another domain, such as the electrical domain [1]. Other electronic components of the instrument then isolate the electrical signal, amplify it, compare it with a standard, and enable it to be read out on some form of meter.

[1] G. W. Ewing, *Instrumental Methods of Chemical Analysis*, McGraw-Hill, New York, 1975, pp. 1–6.

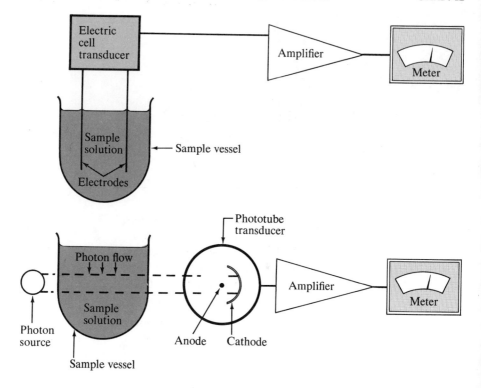

FIGURE 12-1. Two types of transducers. Above: an electric cell type of transducer in which the electrodes of the transducer are immersed in the solution. Below: a phototube type of transducer in which the electrodes of the transducer are outside the solution and respond to photons flowing through the solution.

There are several types of transducers, two of which are shown in Figure 12-1. One type, exemplified by the phototube transducer shown in the lower part of Figure 12-1, *produces a current*. This type of transducer typically is located outside of the sample solution and responds to a beam of radiation (photons) which has been partially absorbed by the solution. This type of transducer, like most transducers, has two electrodes (the anode and the cathode in Fig. 12-1). It will be discussed in more detail in Section 12-1.

The other type of transducer in Figure 12-1 *produces a potential*. The electric cell shown in the upper half of Figure 12-1 may be used for pH measurements, potentiometric titrations, etc. This type typically is located in the sample solution and uses the sample solution to complete the circuit of the cell. It also has two electrodes, which frequently are called an indicating electrode and a reference electrode. This type will be discussed in more detail in Section 12-2.

A third type of transducer *acts as a variable resistance*. It includes photoconductive cells, thermistors, and cells for electrolytical conductivity [1]. It will not be discussed further.

12-1 TRANSDUCER MEASUREMENT OF LIGHT AND OTHER RADIATION

As shown in Figure 12-1, the phototube transducer responds to a beam of electromagnetic radiation. Before going into this in more detail, we should discuss the nature of electromagnetic radiation to discover just what types of radiation are involved.

The Nature of Electromagnetic Radiation

Light is but one form of electromagnetic radiation. Other types are ultraviolet (UV) radiation, infrared (IR) radiation, and X rays. As long as a beam of radiation is characterized by a wavelike motion and consists only of photons, it is considered electromagnetic radiation. Let us give special attention to the wave and photon characteristics of such radiation.

Waves and Photons. Light and other forms of radiation *appear* to have a dual nature. On one hand, it behaves as if it consisted of particles, which we call photons. It also travels through space as if it consisted of waves. It has been said [2] that light *does not really* possess a dual nature because under no circumstances does it behave in both ways at the same time. In other words, when it is absorbed or emitted by matter, it behaves only as a photon; when it travels through space or through a prism, it behaves only as a wave and does not travel in a straight line as a particle does.

The connection between these two natures is given by the following equation.

$$E = h\nu = \frac{hc}{\lambda} \tag{12-1}$$

where E = energy in joules (J)
 ν = frequency in number of waves per second, or s^{-1}
 h = Planck's constant, 6.62×10^{-34} J-s
 c = speed of light, 3.0×10^{17} nanometers/second, or nms^{-1}
 λ = wavelength of radiation in nanometers (nm)

Equation **12-1** specifies that radiation which passes a given point at a rate of a certain number of waves per second must have a particular energy. Thus each photon in that radiation also must have that particular energy.

Note that the *nanometer* (nm) is the unit of wavelength most commonly used by chemists. It is equal to 10^{-9} meter. (It is essentially the same as the millimicron (mμ) unit used before 1968.) We will characterize all radiation in terms of nanometers; some texts use wavenumbers ($1/\lambda$), which are symbolized as $\bar{\nu}$. Both the cm^{-1} and the μm^{-1} (reciprocal micrometer) are common wavenumber units.

[2] G. H. Begbie, *Seeing and the Eye*, Anchor Press/Doubleday, Garden City, 1973.

The Electromagnetic Spectrum. We can now discuss the various regions of the electromagnetic spectrum in terms of wavelength and energy. (Recall from **12-1** that wavelength is inversely proportional to energy.) The various regions are summarized in Figure 12-2. At the top are microwaves and radio waves which

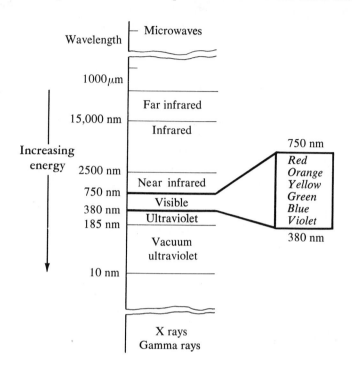

FIGURE 12-2. Types of radiation in the electromagnetic spectrum. Note that as the wavelength decreases, the energy of the radiation increases. (The micrometer (μm) wavelength unit is equal to 1000 nm.) At the right is shown the division of the visible region into its respective colors from 380–750 nm.

have very large wavelengths and very small energies. In the middle is the very useful infrared-visible-ultraviolet radiation region.

The infrared region consists of two subdivisions—the middle infrared (or simply, infrared) region from 2500–15,000 nm, and the near infrared from 750–2500 nm. The near infrared borders the visible region, from 380–750 nm. It is in this region that we perceive color. The various colors will be discussed in more detail in Section 13-1. Ultraviolet radiation borders the other side of the visible region and extends from 185–380 nm. The next chapter discusses the measurement of the absorption of all three of these types of radiation using an instrument called a spectrophotometer.

At the bottom of Figure 12-2 we find the vacuum ultraviolet region (10–185 nm) and other high energy radiation such as X rays and gamma rays.

When a beam of radiation is focused on a sample solution, various things can happen (Fig. 12-3). One is absorption (A) of some of the photons. Three of the interactions pictured are such that the resulting photons cannot be measured by a phototube transducer set at the wavelength of the original photons. These

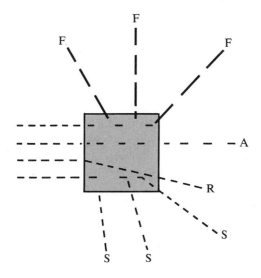

FIGURE 12-3. Hypothetical representation of four different interactions of photons with sample solution as follows:
F = Photons are absorbed and emitted as longer wavelength fluorescence photons (note longer lines representing photons) in all directions.
A = Photons are partially absorbed; the remaining photons travel straight through the solution.
R = Photons are bent or refracted in a different direction but none are absorbed, and their wavelength remains the same.
S = Photons are scattered in all directions; none are absorbed and their wavelength is the same as that before scattering.

are fluorescence (F) which produces photons of longer wavelengths, refraction (R) in which the photons are bent away from their original pathway, and scattering (S) in which the photons are scattered away in all directions at the same wavelength as the original photons. Not shown is the special case of emission of photons by gaseous atoms (Ch. 16).

Absorption of Photons

The absorption of electromagnetic radiation can be explained in terms of the interaction of photons with electrons of a chemical species, or the bond vibrations of this species. Although such processes are complicated, they can be described in general terms as an introduction. First, it is necessary to define the ground state and the excited state or states of chemical species, as the states refer to the electrons and to the bond vibrations. An electronic ground state is one in which all of the electrons are in their most stable orbitals. A vibrational ground state is one in which all of the bonds possess the smallest possible amount of vibrational energy. An electronic excited state is one in which one electron occupies a higher energy orbital than it does in the ground state. A vibrational excited state is one in which one or more bonds has a larger amount of vibrational energy than it does in the ground state. Although there is only one electronic ground state and one vibrational ground state, there may be a number of excited states.

To illustrate the absorption of photons by electrons, let us consider the usual one-photon absorption by the gaseous sodium atom. The ground state of gaseous sodium, Na_0, absorbs appreciable amounts of radiation at both 330 nm and 589 nm. Since 589 nm radiation is the lowest energy radiation absorbed, this must promote the atom to the first, or lowest energy, excited state, Na_1. Absorption at 330 nm will be assumed to promote the atom to the second excited state, Na_2. These processes can be represented as

$$Na_0 + photon_{589\,nm}(10^{-15}\text{ s}) \rightarrow Na_1$$

$$Na_0 + photon_{330\,nm}(10^{-15}\text{ s}) \rightarrow Na_2$$

The 10^{-15} second time for either process is the time required for a sodium electron to "jump" from its most stable orbital to a higher energy orbital. In terms of orbitals the Na_1 and Na_2 excited states differ from one another in that the excited electron occupies a $3p$ orbital in the Na_1 state and a $4p$ orbital in the Na_2 excited state.

As an example of absorption of photons by bond vibrations, let us consider the absorption of infrared photons by the O—H bond of tertiary butyl alcohol. The ground state, GV_0, absorbs considerable amounts of radiation at both 2740 nm and 1380 nm. Since 2740 nm is the lowest energy radiation absorbed, this must promote the alcohol molecule to its lowest energy OH vibrational excited state, EV_1. Absorption at 1380 nm will be assumed to promote the alcohol molecule to the second OH vibrational excited state, EV_2. These processes can be represented as

$$GV_0 + photon_{2740\,nm}(10^{-12}\text{ s}) \rightarrow EV_1$$

$$GV_0 + photon_{1380\,nm}(10^{-12}\text{ s}) \rightarrow EV_2$$

Note that it takes longer to reach a vibrational excited state than an electronic excited state (10^{-15} s) because bonds vibrate more slowly than an electron can jump.

The Fate of an Excited Molecule or Atom. After reaching an excited state, a molecule or atom does not remain in this state for more than 10^{-11} s (vibrational excitation) or 10^{-8} s (electronic excitation). It returns to the ground state by one or both of the following processes.

$$S_1 \text{ (or } EV_1) \xrightarrow{\text{emission}} S_0 \text{ (or } GV_0) + photon \quad (12\text{-}2)$$

$$S_1 \text{ (or } EV_1) \xrightarrow{\text{relaxation}} S_0 \text{ (or } GV_0) + heat \quad (12\text{-}3)$$

In Step **12-2** the excited state loses its excess electronic or vibrational energy by emitting a photon. This may occur via emission in the gaseous state (Ch. 16) or via fluorescence in solution (Ch. 15). Competing with Step **12-2** is Step **12-3**, which is usually faster in water. Part of the reason Step **12-3** is so fast in water is that water bonds so strongly, via hydrogen bonding, etc., to molecules that the excess electronic or vibrational energy is transferred to excess vibrational energy in the water bonds. These bonds eventually lose this energy as heat. The entire process is then repeated over and over.

The absorption of radiation is measured on a *spectrophotometer*. In Chapter 13 the measurement of UV, visible, and IR absorption on the spectrophotometer is introduced, followed by applications in Chapter 14.

Nonabsorption Processes

Recall from Figure 12-3 that there are three interactions by which photons are diverted from their original pathway to the phototube transducer. Fluorescence (F) occurs as a result of Step **12-2**. Photons of longer wavelengths are emitted in *all directions*, not just in the original direction. A small amount of these will strike the phototube if the optical system will accept photons of longer wavelength. However, fluorescence is usually measured at right angles to the original direction of the absorbed radiation, using an instrument called a *fluorometer* (Ch. 15). The fluorometer, like the spectrophotometer, employs a phototube as a transducer.

Refraction occurs when a beam of photons is bent away from its original direction (see the R beam in Fig. 12-3). Its effect is compensated for automatically in spectrophotometric measurement of absorption, but refraction itself is not measured on a spectrophotometer. An instrument called a *refractometer* is used to measure the amount of bending (Ch. 15).

Scattering occurs when a beam of photons is scattered in all directions from the original direction of the beam. It differs from fluorescence in that all of the scattered photons are of the same wavelength as the original radiation. Its effect is also compensated for in spectrophotometric measurement of absorption. Scattering may be measured on a spectrophotometer or on an instrument like a fluorometer (see Ch. 15).

Response of the Phototube (Transducer) to Photons

You now should fully understand that a solution of a chemical species absorbs only *some* of the photons of radiation striking it. The remaining photons pass through unabsorbed and strike the phototube (Fig. 12-1). You may now wonder just how the phototube acts as a transducer; that is, how does it convert chemical information (the number of unabsorbed photons) into electrical information.

Let us begin with the common components of most transducers—the two electrodes. In the phototube they are called the cathode and the anode (Fig. 12-4).

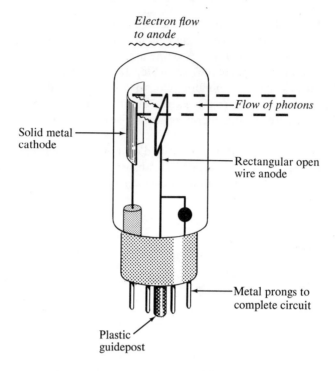

FIGURE 12-4. *A schematic diagram of a phototube. Photons of light flow past the open wire anode, strike the cathode, and are converted to electrons which are attracted to the positive anode. (In some phototubes the anode is a single wire.)*

The semicylindrical cathode shown in Figure 12-4 is coated with a metal-metal oxide mixture that releases electrons as photons of radiation strike it. It is only capable of doing this when a high voltage is impressed across it. By means of a resistance, the anode is maintained at a positive voltage relative to the cathode. Thus as electrons are ejected from the surface of the cathode, they are attracted to the positive anode. Ordinarily these electrons would react with oxygen in the air, but the phototube is evacuated so that the electrons travel unimpeded through a vacuum to the anode. The result is an electrical current that flows from the anode. (Recall that a phototube is a type of transducer that produces a current.)

Amplification and Readout. The electrical current produced in a phototube is so small that it cannot give a readout on the galvanometer type of readout device used on a spectrophotometer. It must be fed through an *amplifier* (Fig. 12-1); the amplified current is then measured on an ammeter whose current scale is calibrated to read in units of relative photon absorption, called absorbance. Section 13-3 describes the entire measuring process, including the phototube as a detector, for the measurement of UV, visible, and IR absorption.

12-2 TRANSDUCER MEASUREMENT OF pH AND ELECTRICAL POTENTIAL

As shown in Figure 12-1, the electric cell transducer responds to various concentrations of chemical species in solution. The two electrodes of the transducer are in direct contact with the sample solution, in contrast to the phototube type of transducer. A common way of measuring the signal from the transducer is by means of a potentiometer, which we will discuss first, before showing how it is used to measure pH.

The Potentiometer

The concentrations of the chemical species in an electric, or galvanic, cell are what determines the voltage of the cell. Unfortunately, if we try to measure this voltage, the voltmeter draws enough current from the electrical cell to change the initial concentrations of the ions. Thus we do not obtain an accurate measure of the initial concentrations. What is needed is a *potentiometric measurement*, not a measurement with a voltmeter.

The difference between these two types of measurements is shown in Figure 12-5. The reaction between the chemical species in the cell is

$$H_2(g) + Ox\ agt = 2H^+ + reductant$$
$$\underset{2e^-}{\underbrace{\qquad\qquad\qquad}}$$

In this reaction the hydrogen gas gives up two electrons to some oxidizing agent, becoming two hydrogen ions in the process. The upper half of Figure 12-5 shows the setup of the cell that would be used with a voltmeter (not shown) for measurement of cell voltage. Note that electrons released from the hydrogen gas flow through the electrode, as indicated by the current-measuring galvanometer. The electrons flow through the wire and combine with the oxidizing agent at the other electrode. (The oxidizing agent and the hydrogen gas parts of the cell are divided from one another.) Thus as soon as the cell is hooked up, the concentrations change and the voltage measured is an inaccurate measure of the initial concentrations.

The lower half of Figure 12-5 shows the cell setup for potentiometric measurement. An *external* cell whose voltage can be varied to known values is connected to the external circuit. When the circuit is completed, a few electrons released from the hydrogen gas begin to flow through the electrode. However, the voltage of the external cell is quickly varied until it exactly equals the voltage of the hydrogen-oxidizing agent cell; the circuit is arranged so that the external voltage *opposes* the voltage of the latter cell. Thus no current flows, as indicated by the current-measuring galvanometer. However, the voltage of the external cell is known and gives the value of the voltage of the hydrogen gas-oxidizing agent cell at the *initial* concentrations of all species. This value is labeled a *potential* rather

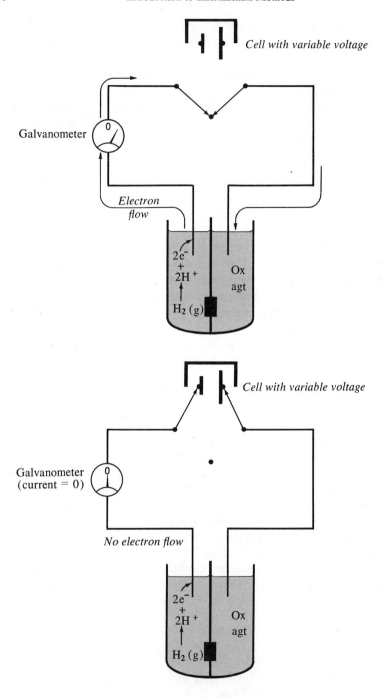

FIGURE 12-5. *Above: measurement of cell voltage with a galvanometer. Note that current is flowing. Below: potentiometric measurement of cell voltage as potential. Note that no current is flowing through the galvanometer.*

than a voltage because it can be viewed as a measure of *potential energy*. It is the value of the voltage that would be measured if a voltage-producing reaction were to occur.

Because a few electrons flow before the external voltage can be adjusted to be equal to that of the electric cell, no potentiometric device will insure that the current will be exactly zero, but for the accuracy needed for most measurements, there will be no appreciable error.

Indicating and Reference Electrodes. Modern potentiometric measurements are made with an electrochemical cell consisting of two electrodes called the indicating and reference electrodes (see also Sec. 11-3). The indicating electrode responds to changes in the concentration of the chemical species to be measured in the electrochemical cell according to the Nernst equation **(11-20)**. The reference electrode is a self-contained system whose potential remains constant; it is isolated from reaction with the solution in the cell, although it is constructed to allow electrical contact.

By using potentiometric measurement in connection with an indicating and a reference electrode, the potentials of reactions can be measured to find the concentrations of species in the solution. This will be discussed in detail in Chapter 17. A specific example is the measurement of pH, which will be discussed.

Measurement of pH

The pH of a solution can be measured by using an indicating electrode, which responds to the [H^+], and a reference electrode, which will maintain a constant potential in the solutions to be measured. The classical example of an indicating electrode is the hydrogen electrode. This is a platinized platinum electrode over which hydrogen gas is bubbled. The oxidation-reduction half reaction for this electrode is

$$2H^+ + 2e^- = H_2(g)$$

If the pressure of the hydrogen gas is one atmosphere, the Nernst equation **(11-20)** for the potential measured at this electrode is

$$E = E° + 0.059 \log a_{H^+}$$

Because $E°$ for a standard hydrogen electrode is zero,

$$-E = -0.059 \log a_{H^+} = 0.059 \text{ pH}$$

Note that this measurement is based on the exact definition of pH as the negative log of the *activity* of H^+ (see Sec. 8-1). A reference electrode such as a standard calomel electrode (Fig. 12-6) is used to complete the circuit.

Because the hydrogen electrode is very awkward to use and not always resistant to chemical attack, a glass electrode (Fig. 12-6) is used as the indicator

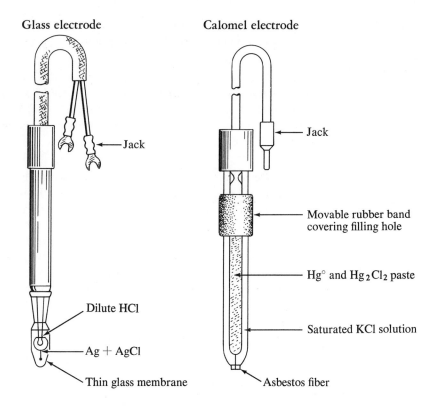

FIGURE 12-6. The glass electrode (left) and the calomel electrode (right).

electrode for measuring pH. It consists of a special silver-silver chloride electrode immersed inside of a thin membrane of a special glass. A potential difference develops across the glass membrane; the potential is a function of the difference of the hydrogen ion activity in the sample solution and the solution inside the glass.

A pH meter in many ways is simply a potentiometer used to measure pH. The glass electrode and the calomel electrode are connected through jacks (Fig. 12-6) to the pH meter to establish an electrical circuit. Details of the ultimate pH measurement are given in Chapter 17.

QUESTIONS AND PROBLEMS

(Answers to most even-numbered problems are in Appendix 5.)

Radiation and Light
1. Explain how a phototube acts as a transducer.
2. Define a photon.
3. Define a nanometer.

SEC. 12-2   Transducer Measurement of pH and Electrical Potential   283

4. Give the wavelength regions for the following types of radiation.
   a. Ultraviolet radiation
   b. Near infrared radiation
   c. Visible radiation
   d. Infrared radiation
5. Explain the difference between
   a. Absorption of light and fluorescence.
   b. Fluorescence and scattering.
   c. Refraction and scattering.
   d. Refraction and fluorescence.
6. Define a ground state and an excited state in terms of electrons and in terms of bond vibrations.
7. Explain the function of the cathode and the anode in a phototube.
8. Why does the electrical current from a phototube need to be amplified?

Potentiometry and pH
9. Explain how a potentiometer acts as a transducer.
10. Why can't a voltmeter be used to measure concentrations accurately?
11. Why is the value read from a potentiometer called a potential and not a voltage?
12. Explain the function of an indicating electrode and a reference electrode.
13. Compare the hydrogen electrode and the glass electrode for the measurement of pH.
14. Explain what a calomel electrode is and how it is constructed.

13. Introduction to Spectrophotometry and Spectrophotometric Instruments

> *Holmes shut the slide across the front of his lantern and left us in* pitch darkness. . . . *The smell of hot metal remained to assure us that the* light *was still there ready to flash out at a moment's notice.*
>
> ARTHUR CONAN DOYLE
> The Red-Headed League

In modern laboratories spectrophotometry is the most frequently used method of analysis; hence two chapters will be devoted to it. This chapter discusses the various types of electromagnetic radiation used in absorption spectrophotometry, the law governing the absorption of these types of radiation, and the components of simple spectrophotometers and colorimeters. The following chapter will cover primarily the applications of ultraviolet, visible, and infrared spectrophotometry, as well as automated spectrophotometric analysis.

Absorption spectrophotometry, frequently called just *spectrophotometry*, is the science of the measurement of the amount of electromagnetic radiation absorbed at a particular wavelength or set of wavelengths. If only visible radiation, or light, is measured, it may be called colorimetry. Instruments used to measure the absorption of visible and/or ultraviolet and infrared radiation are called *spectrophotometers*; instruments that are capable of measuring only visible radiation or light are called *colorimeters* or visible region spectrophotometers. They are operated much as Sherlock Holmes operated his lantern in the case of *The Red-Headed League*; they are initially adjusted to a zero reading with the detector in "pitch darkness" with a slide in front of the source which remains "ready to flash out at a moment's notice."

Because we are concerned only with ultraviolet-visible-infrared spectrophotometry, we will begin by discussing these three types of radiation. If necessary, you should review the general discussion of electromagnetic radiation in Section 12-1.

13-1 ULTRAVIOLET, VISIBLE, AND INFRARED RADIATION

Wavelength and Energy

Ultraviolet, visible, and infrared radiation are all characterized by *wavelength* and *energy*. Recall (Sec. 12-1) that the most common unit of wavelength is the nanometer, nm, which is equal to 10^{-9} meter. Wavelength is related to energy through the speed of light, c, which is about 3.0×10^{17} nm sec^{-1}:

$$E = hc/\lambda \qquad (13\text{-}1)$$

where h is Planck's constant (6.62×10^{-34} J-s), λ is the wavelength of the radiation in nm, and E is the energy in Joules.

Ultraviolet radiation extends from 185 nm to 380 nm and is the highest energy radiation of the three types to be discussed (Table 13-1). Its energy is such that wavelengths below 280 nm can seriously harm the eye as well as cause electronic

TABLE 13-1. *Wavelength and Energy Ranges for Ultraviolet, Visible, and Infrared Radiation*

Type of radiation	λ range, nm	Energy range, Joules[a]
Ultraviolet (UV)	185[b]–380	1.07×10^{-18}–5.23×10^{-19}
Visible (light)	380–750	5.23×10^{-19}–2.65×10^{-19}
Near infrared	750–2500	2.65×10^{-9} –7.94×10^{-10}
Infrared (IR)	2500–15,000	7.94×10^{-10}–1.32×10^{-10}

[a] Energy calculated from **13-1** (1 Joule = 0.24 calorie).
[b] This is a limit imposed by optical materials and atmospheric absorption.

energy charges. Visible radiation extends from 380–750 nm; it causes electronic energy changes responsible for the perception of color. (Visible radiation will be discussed in detail below.) Useful infrared radiation is divided into two regions—the near infrared, and the infrared (Table 13-1). This radiation is of the lowest energy; its energy is so low that it cannot cause electronic changes and so is invisible to the eye.

The visible region range is somewhat arbitrary since it depends on the *average* eye response. At the ultraviolet-violet borderline the response of the eye is limited by the absorption of radiation by the lens [1]. At the infrared-red

[1] R. K. Clayton, *Light and Living Matter*, Vol. 2, The Biological Part, McGraw-Hill, New York, 1971, pp. 93–127.

borderline the eye's response is limited by the absorption of radiation by water in the aqueous humor and other parts of the eye.

The response of the eye in the visible region also varies with wavelength, as shown in Table 13-2. This table lists the response of the eye relative to a value

TABLE 13-2. *Response of the Eye to Various Colors*

Color	Wavelength	Response of eye in middle of range[a]
Ultraviolet	185–380 nm	0.0000 (at 380 nm)
Violet	380–450 nm	0.0022
Blue	450–495 nm	0.10
Green	495–550 nm	0.83
Yellow-green	550–570 nm	0.995
Yellow	570–590 nm	0.87
Orange	590–620 nm	0.57
Red	620–750 nm	0.10
Near infrared	750–2500 nm	0.0001 (at 750 nm)

[a] Relative to a maximum response of 1.0000 at 555 nm.

of 1.0000, or 100%, for yellow-green radiation at 555 nm. It can be seen from the table that the eye also has a large response to green and yellow light. The color that a solution appears to the eye depends first of all on whether yellow and/or green (as well as yellow-green) are transmitted or not. If light of 495–590 nm is transmitted, the solution will appear yellow-green. If light of 495–570 nm is absorbed, the solution will appear yellow; if light of 550–590 nm is absorbed, the solution will appear primarily green, and so on. This response is that of the cones of the eye, not of the rods [1].

The responses listed in Table 13-2 are for the *average* eye, assuming a *normal* flux, or flow, of photons striking the eye. It should not be assumed that one cannot see radiation below 380 nm just because the response is 0.0000. The eye does respond, but the magnitude of the response will have a value in the fifth or sixth decimal so the eye will only see radiation with a higher than normal flux. Color has been observed as low as 313 nm, and even X rays (~ 1 nm) have been reported to appear bluish to monkeys in certain experiments. The eye can also see high fluxes of red photons above 750 nm; for example, the eye can observe the "red" flame emission of potassium at 766–69 nm [2].

Although Table 13-2 implies that the response of the eye varies smoothly with wavelength, this is not true. The eye behaves as though it has three independent color receptors. There are two major receptors—one in the yellow-green at 540 nm and the other in the yellow at about 580 nm. There is also a minor receptor at about 450 nm in the blue region.

[2] F. C. Strong III, *J. Chem. Educ.* **45**, 178 (1969).

Color blindness is apparently the absence of one of these receptors. In one class of color-blind subjects, the 540 nm receptor is missing as a result of a deficiency of a certain pigment. In another class of color-blind subjects, the 580 nm receptor is missing as a result of a deficiency of still another pigment in the eye [1].

Absorption of Photons

The *absorption* of electromagnetic radiation has been explained in terms of the interaction of photons with the electrons of a substance (Sec. 12-1). For example, a gaseous sodium atom in its ground state, Na_0, absorbs light of 589 nm and is promoted to its first excited state, Na_1 (Sec. 12-1).

$$Na_0 + photon_{589\ nm} \xrightarrow{10^{-15}\ s} Na_1$$

The Na_1 state differs from the Na_0 state in that one electron occupies a *higher energy* orbital in this state than it does in the ground state.

The absorption of photons by molecules or ions in solution is more complicated than that by atoms in the gaseous state. There is still just one ground state, but each electronic excited state is split into many (n) sublevels, each of slightly different energy. Each sublevel will be populated by absorption of photons of slightly different energy. For example, the absorption of light by oxyhemoglobin in the ground state, symbolized as $[Hb(O_2)_x]_0$, can be represented as follows:

$$[Hb(O_2)_x]_0 + n\ photons_{560\text{-}620\ nm} \longrightarrow n[Hb(O_2)_x]_1 \qquad (13\text{-}1)$$

where the symbol $[Hb(O_2)_x]_1$ represents the many sublevels of the first excited state of the oxyhemoglobin molecule. Of the molecules in this first excited state, there will be molecules whose excess energies range from a low, corresponding to the energy of 620 nm light, to a high, corresponding to the energy of 560 nm light.

To rationalize the well-known red color of oxyhemoglobin with the previous absorption process, we must examine the *absorption spectrum* of oxyhemoglobin. This is a plot of the amount of light absorbed, or absorbance, versus wavelength and is shown in Figure 13-1. It can be seen that oxyhemoglobin absorbs nearly all visible radiation until 620 nm; above this wavelength it transmits most, but not all, of red light (plus a small amount of orange light).

We also see that oxyhemoglobin has three absorption *maxima*, or *bands*, in Figure 13-1. The highest energy band is not completely shown but is close to 450 nm. The most intense absorption band is at 540 nm; the lowest energy absorption band is at 575 nm. Since absorption at 575 nm gives the first excited state of oxyhemoglobin, absorption at 540 nm would appear to give the second excited state, and absorption below 450 nm would then give the third excited state of oxyhemoglobin. The energy differences between these states may be described in the following manner. As long as any two molecules are in the same excited state, they must possess the same amount of excess electronic energy and can vary only in the amount

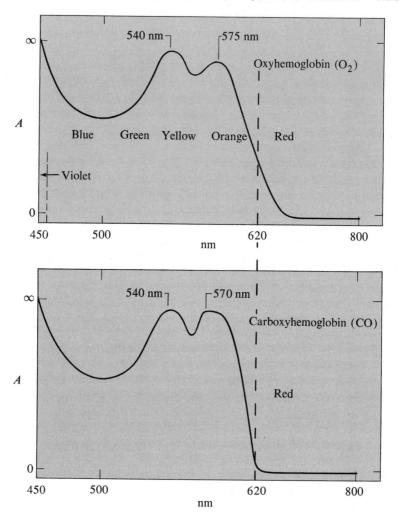

FIGURE 13-1. The absorption spectra of oxyhemoglobin (top) and carboxyhemoglobin (bottom). The left axis is absorbance, A, or the amount of light absorbed. Note that when hemoglobin is combined with oxygen it absorbs almost all colors except most of the red, so it appears red. When hemoglobin is combined with carbon monoxide it does not absorb any of the red, so it appears redder or pinker than oxyhemoglobin.

of excess vibrational energy possessed by their bonds. If any two molecules are in different excited states, they will possess different amounts of excess electronic energy and probably also different amounts of excess bond vibrational energy.

It should be emphasized that although 13-1 describes the interaction of n photons with hemoglobin, the equation is still describing a one-photon absorption process. No more than one photon is absorbed by a single hemoglobin molecule at a time. The reason for this is that most light (photon) sources are not very intense, so only one photon can be absorbed from the source during the 10^{-15} second needed to reach the excited state. If a laser, such as a ruby laser, is used as a source, the very concentrated beam of photons emitted by the laser can be used to achieve a two-photon absorption process.

A two-photon absorption process has also been postulated to show how photosynthesis can occur via absorption of relatively low energy red light [3]. Red photons have such low energies that scientists have been unable to explain satisfactorily how a relatively high energy process such as photosynthesis could utilize a single red photon per chlorophyll. Absorption of two red photons would of course double the energy available.

It should be mentioned in conclusion that a molecule such as hemoglobin does not remain in the excited state very long. It returns to the ground state in less than 10^{-9} second by one of two different processes (Sec. 12-1). Once it has reached the ground state it again can absorb a photon and become excited.

13-2 MEASUREMENT OF CONCENTRATION: THE BEER-LAMBERT LAW

The spectrophotometer and the colorimeter are instruments which can be used to measure the amount of absorption of UV, visible, or IR radiation by solutions of chemical substances. The basic law governing such measurements is the combined Beer-Lambert law. Before describing this law, we should try to visualize the flow of photons through the spectrophotometer cell which contains such a solution.

To begin with, a solution containing a chemical substance is poured into a spectrophotometer cell and inserted into the spectrophotometer. Various wavelengths of radiation are directed at the solution in the cell. Figure 13-2 pictures how beams of photons of two different wavelengths pass through the solution. At a wavelength such as wavelength x, the solution will absorb some of the photons striking it. As shown in the figure, about half of the photons are absorbed and half pass through. The number of photons in a beam passing a given point per unit time (one second, for example) is called the *radiant power* of the beam. The radiant power of the beam striking the solution is P_0, and that transmitted by the solution is P. At wavelength x (Fig. 13-2), P_0 is about twice P. Radiation at other wavelengths, such as wavelength y, will not be absorbed by the solution. Note that P_0 is the same as P at this wavelength.

If you understand the difference between P_0 and P, you are ready to read about absorbance and transmittance, the two sets of units used on the scale of the spectrophotometer and the colorimeter.

[3] *Chemical & Engineering News* **52**, 23 (Apr. 8, 1974).

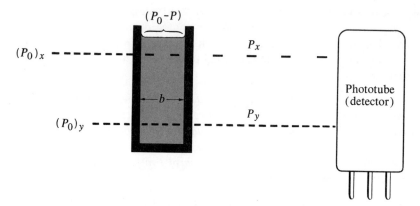

FIGURE 13-2. *A schematic diagram of the cell of a spectrophotometer containing a solution that absorbs radiation at wavelength x, but not at wavelength y. The internal cell length is symbolized as b. The radiant power of either beam striking the solution is P_0; the radiant power of either beam transmitted by the solution (and striking the phototube detector) is P; the radiant power of the absorbed radiation is (P_0-P).*

Absorbance, Transmittance, and Per Cent Transmittance

The terms to be discussed in this section are absorbance, A, and transmittance, T, and per cent transmittance, $\%T$. The amount of radiation absorbed by a solution is measured in *absorbance units*, ranging from 0.00 to 1.0 or 2.0. The fraction of radiation transmitted is measured in units of transmittance, or per cent transmittance, ranging from $100\%T$ ($T = 1.00$) to $0\%T$ ($T = 0.00$). Both appear on the spectrophotometer scale; the key scale values are as follows:

$\%T =$	0%	1%	10%	50%	90%	100%
$A =$	∞	2.0	1.0	0.30	0.05	0.00

Both the absorbance and per cent transmittance readings may be understood and related by using the concept of the radiant power of the beam of photons striking a solution and being transmitted by a solution.

First, consider the solution of an absorbing species shown in a spectrophotometer cell in Figure 13-2. Let the radiant power of the beam in photons/s striking the cell be P_0. The beam then passes through the solution and part of its radiant power is absorbed by ions or molecules in the solution. Let the radiant power of the beam leaving the cell be symbolized as P. (P will be less than or equal to P_0.) Absorbance may then be defined as

$$A = \log P_0 - \log P = \log \frac{P_0}{P} \tag{13-2}$$

Note that absorbance is a *logarithmic difference* and is best understood in the mathematical form of the middle terms. For mathematical manipulations the right-hand term is more convenient, as will now be illustrated.

Mathematically stated, transmittance is defined as a fraction.

$$T = \frac{P}{P_0}$$

This is termed *transmittance*. From **13-2** it is easily shown that

$$A = -\log T \tag{13-3}$$

Since per cent transmittance is simply $100 \times T$, absorbance may be calculated from $\%T$ in a manner similar to **13-3**.

$$A = -\log(\%T/100) \tag{13-4}$$

Significant Figures and Spectrophotometer Readout. The number of justifiable significant figures for a spectrophotometer reading depends on whether the spectrophotometer has a *scale readout* or a *digital readout*. Record the former as follows:

Absorbance, scale readout
0.00–0.99 units: Two significant figures to the right of the decimal. Since this is a log term, *all* zeroes to the right of the decimal are significant. Thus readings of 0.05 and 0.00 both have two significant figures.
1.0–2.0 units: One significant figure to the right of the decimal.

% Transmittance, scale readout
Two significant figures to the left of the decimal above 9%. One decimal to the right is sometimes estimated but is uncertain by ± 0.3 to $0.4\%T$.

To calculate absorbance from per cent transmittance or vice versa using **13-3** or **13-4** the rules for log terms in Section 2-1 must be used. Thus absorbance must be expressed with the same number of significant figures to the right of the decimal as the total number of significant figures in the transmittance or per cent transmittance readings.

Many modern spectrophotometers have a *digital* readout where all of the digits displayed are considered significant. Record as follows:

Absorbance, digital readout
0.0–3.0+ units: All digits are significant. Usually three significant figures to the right of the decimal are displayed. As above, *all* zeroes to the right of the decimal are significant.

% Transmittance, digital readout
All digits displayed are significant. Usually two significant figures to the left and one to the right of the decimal are displayed. Thus, the zero in $9.0\%T$ on a digital spectrophotometer would be significant even though it would not be on an ordinary spectrophotometer *scale*.

EXAMPLE 13-1. Calculate the absorbance corresponding to 9.0%T reading if the reading is from a) a digital readout and b) a scale readout.

Solution to a: Using **13-4**,

$$A = -\log(9.0\%/100) = -\log 9.0 - \log 10^{-2}$$

$$A = -0.954 + 2 = 1.046$$

Solution to b: The zero in 9.0%T is not significant on a scale readout; therefore $A = 1.0_{46}$ or 1.0 (one significant figure to the right of the decimal).

Self-Test 13-1. Absorbance, Transmittance, and Radiant Power

Directions: Check answers given at the end of the test. If necessary, review significant figures for logs in Section 2-1.

A. Yellow (575 nm) light with a radiant power of 2.00×10^{14} photons/s strikes a solution of hemoglobin; the solution transmits yellow light with a radiant power of 4.0×10^{13} photons/s. Calculate the
   a. Transmittance   b. % Transmittance   c. Absorbance
B. The absorption of 575 nm light by various solutions of hemoglobin is read as per cent transmittance and transmittance from a spectrophotometer with a *scale readout*. Decide whether all the digits are significant and then calculate the absorbance to the correct number of significant figures.
   a. %T = 40%; A = _____   b. %T = 5%; A = _____
   c. T = 0.010 (1.0%T); A = _____   d. 0.080 (8.0%T); A = _____
   e. %T = 91%; A = _____
C. The same solutions above are read in a spectrophotometer with a *digital readout*. Calculate the absorbance to the correct number of significant figures if the per cent transmittance readings are
   a. %T = 40.0%; A = _____   b. %T = 5.0%; A = _____
   c. %T = 1.0%; A = _____   d. %T = 8.0%; A = _____
   e. %T = 91.0%; A = _____
D. The following absorbance values were read from a spectrophotometer with a *scale readout*. Calculate the transmittance, T, to the correct number of significant figures.
   a. A = 1.7; T = _____   b. A = 0.05; T = _____   c. A = 0.60; T = _____

Answers to Self-Test 13-1

A. a. T = 0.20   b. %T = 20%   c. A = 0.70
B. a. $A = 0.39_8$ b. A = 1.3   c. A = 2.0 (last zero in 0.010 not significant)   d. A = 1.1 (last zero in 0.080 not significant)
   e. A = 0.04 (two sig. figs.)
C. a. A = 0.398   b. A = 1.30   c. A = 2.00   d. A = 1.09
   e. A = 0.041
D. a. T = 0.02   b. T = 0.89 (0.05 has two sig. figs.)   c. T = 0.25

The Beer-Lambert Law

The law governing the relation between concentration and the amount of light or radiation absorbed is the Beer-Lambert law (also inaccurately called Beer's law). It relates the absorbance to the concentration of the absorbing species and to the path length of the solution through which radiation must pass. This law states that if the concentration of the absorbing species increases, the absorbance must increase. Similarly, it also states that if the path length of the solution increases, the absorbance must increase.

The usual symbol used for the path length of a solution is b. As shown in Figure 13-2, b is also the internal cell *length*, or internal tube *diameter* if a test tube is used instead of a square cell. The usual symbol used for concentration is c.

If the concentration, c, of the compound is stated in terms of molarity, then the Beer-Lambert law gives the relation between A and c as

$$A = \varepsilon b c$$

where b is given in cm and ε is a proportionality constant known as the *molar absorptivity*. The dimensions of ε are always $cm^{-1}M^{-1}$ or $liter \cdot mole^{-1} cm^{-1}$. Hypothetically speaking, ε is the absorbance of a $1M$ solution in a 1 cm cell.

If the concentration is given in units other than molarity, then the Beer-Lambert law is stated in the form

$$A = abc$$

where the proportionality constant is symbolized as a and is called the *absorptivity*. The dimensions of a depend on the concentration unit used. If c is in g/liter, then the dimensions of a are $liter \cdot g^{-1} cm^{-1}$.

If A is plotted against c, a straight line passing through the origin is the usual result (Fig. 13-3). Such a system is said to obey the Beer-Lambert law. If the line curves upward at some point, it is said to exhibit a *positive* deviation from this law. If it curves downward at some point, it is said to show a *negative* deviation from the law. (Both situations are also shown in Figure 13-3.) The slope of the straight line portion of the curve in Figure 13-3 will be ε if a 1 cm cell is used.

It should also be stressed that ε is a constant for a *specific* wavelength only. For example, the $Co(H_2O)_6^{+2}$ ion has an ε of 10 at 530 nm. If the wavelength were changed to 500 nm it would have a different ε. The ε at 500 nm would in fact be lower than that at 530 nm, because the peak of the absorption band of $Co(H_2O)_6^{+2}$ is at 530 nm.

Frequently literature values of ε are reported in logarithmic form. To evaluate or compare such values, they should be converted to semiexponential form.

EXAMPLE 13-2. The amino acid phenylalanine has a $\log \varepsilon$ of 2.30 at 258 nm. What is the semiexponential value of ε?

Solution: Convert the characteristic and mantissa separately to antilogs (numbers).

$$\varepsilon_{258\,nm} = \text{antilog of } 2.30 = (\text{antilog of } 0.30) \times (\text{antilog of } 2) = 2.0 \times 10^2$$

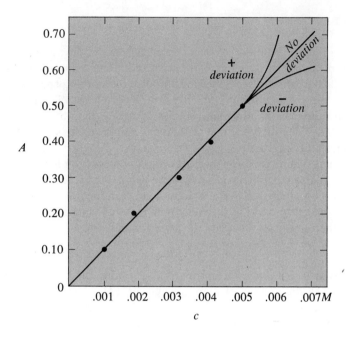

FIGURE 13-3. *A typical Beer-Lambert law plot. Below 0.005M the plot of A vs. c obeys the Beer-Lambert law.*

Beer-Lambert Law Calculations. The simplest situations involve the calculation of one variable, such as A or c, given the other three variables.

EXAMPLE 13-3. The amino acid phenylalanine, $C_6H_5CH(NH_2)CO_2H$, has a $\log \varepsilon$ of 2.30 at 258 nm. Calculate the absorbance of a $2.00 \times 10^{-2} M$ solution in a 1.00 cm cell.

Solution: Use the Beer-Lambert law.

$$A = \varepsilon bc = (2.0 \times 10^2 M^{-1} \text{ cm}^{-1})(1.00 \text{ cm})(2.00 \times 10^{-2} M)$$
$$A = 0.40 \text{ (2 sig. figs.)}$$

Somewhat more complicated calculations arise if the concentration of an unknown is to be found by comparing the absorbance of the unknown (u) with the absorbance of a standard (s) solution. The general equation can be derived by writing a ratio of the Beer-Lambert law for each solution.

$$A_u = \varepsilon b_u c_u$$
$$A_s = \varepsilon b_s c_s$$

assuming $\quad b_u = b_s$

$$c_u = c_s \frac{A_u}{A_s} \tag{13-5}$$

SEC. 13-2   Measurement of Concentration: The Beer-Lambert Law   295

EXAMPLE 13-4. The absorbance of a $1.0 \times 10^{-4} M$ solution of the amino acid tyrosine at 275 nm is 0.200. The absorbance of an unknown solution at 275 nm is 0.500. Calculate the concentration of tyrosine in the unknown.

Solution: Use **13-5** to calculate c.

$$c_u = 1.0 \times 10^{-4} M \frac{(0.500)}{(0.200)} = 2.5 \times 10^{-4} M$$

A Beer-Lambert plot such as in Figure 13-3 can also be used for finding the concentration of an unknown. This will be discussed in Chapter 14.

Self-Test 13-2.   Beer-Lambert Law Calculations

Directions: Work one problem at a time and check answers before proceeding to the next problem. If necessary, review significant figures in Section 2-1.

A. Given the value for molar absorptivity, calculate the log ε in each case.
   a. $\varepsilon = 2.0 \times 10^3$ b. $\varepsilon = 0.021$ c. $\varepsilon = 10$
B. Given the value for log ε, calculate ε in each case.
   a. log $\varepsilon = 0.4$ b. log $\varepsilon = 1.004$ c. log $\varepsilon = -0.4$
C. Calculate the designated variable using the Beer-Lambert law. Assume that $b = 1.000$ cm. (Use only Rule 2 in Sec. 2-1.)
   a. Given $\varepsilon = 1.0 \times 10^4$ and $c = 3.00 \times 10^{-6} M$, calculate A.
   b. Given $\varepsilon = 10^4$ and $c = 2.00 \times 10^{-6} M$, calculate A.
   c. Given log $\varepsilon = 4.30$ and $c = 3.00 \times 10^{-6} M$, calculate A.
   d. Given $A = 0.400$ and $c = 2.0 \times 10^{-5} M$, calculate ε and log ε.
D. The aromatic amino acid phenylalanine is to be determined by comparing the absorbance of the unknown with that of a single standard solution of phenylalanine. In each case calculate the concentration of phenylalanine in the unknown, assuming all absorbance measurements are at 259 nm. Assume $b_u = b_s$ unless otherwise stated.
   a. The standard concentration is $1.0 \times 10^{-2} M$ and its absorbance is 0.200. The absorbance of the unknown is 0.300.
   b. The standard concentration is $2.00 \times 10^{-2} M$ and its absorbance is 0.400. The absorbance of the unknown is 0.300.
   c. The standard concentration is $1.0 \times 10^{-2} M$ and its absorbance in a 2.00 cm cell is 0.400. The absorbance of the unknown in a 1.00 cm cell is 0.360.

Answers to Self-Test 13-2

A. a. 3.30   b. −1.68   c. 1.00
B. a. $2._5$ b. 1.01×10^1 c. 4×10^{-1}
C. a. 0.030   b. 0.0_2 c. 0.060   d. $\varepsilon = 2.0 \times 10^4$, log $\varepsilon = 4.30$
D. a. $1.5 \times 10^{-2} M$ b. $1.50 \times 10^{-2} M$ c. $1.8 \times 10^{-2} M$

13-3 SPECTROPHOTOMETERS AND COLORIMETERS

The spectrophotometer, you recall, is an instrument that can measure the amount of visible, UV, and/or IR radiation absorbed by a solution at a given wavelength. Likewise, the colorimeter is an instrument which essentially measures only light, using a filter to select wavelength. There are other less important differences between many of each of these two types of instruments that will be discussed in this section. First, we will describe the major components of both types of instruments and then some specific examples of these instruments.

The Major Components of a Spectrophotometer

The four major components of a spectrophotometer are the source, the monochromator, the cell, and the detector (Fig. 13-4). The amplifier and scale readout have been discussed in Section 12-1. Let us begin with the first component, the source.

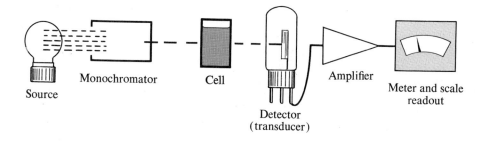

FIGURE 13-4. *A schematic diagram of a simple spectrophotometer. The dashes represent numbers of photons emitted by the source.*

The Source. In general, the sources used are: in the visible region the tungsten filament bulb; in the ultraviolet region the hydrogen, or deuterium, discharge tube; and in the infrared region the globar or Nernst glower. The output of the tungsten filament bulb is shown in Figure 13-5. It can be seen that this bulb has very low intensity below 330 nm, so it can be used in the ultraviolet from 380–330 nm, but not below 330 nm. Its intensity increases to a maximum in the near infrared region near 1000 nm, so it is an excellent source for visible region spectrophotometers, or colorimeters.

For ultraviolet measurements the hydrogen, or deuterium, discharge tube is used from below 330 nm to 200 nm or below. This is a glass tube with a quartz window that transmits ultraviolet radiation (glass absorbs radiation below 310 nm).

For infrared measurements either a Nernst glower or globar is used from 1000–15,000 nm. Both are electrically heated rods rather than bulbs. The globar has an intensity maximum at 1900 nm and the Nernst glower at 1500 nm. Each has enough intensity to be used at longer wavelengths in the infrared.

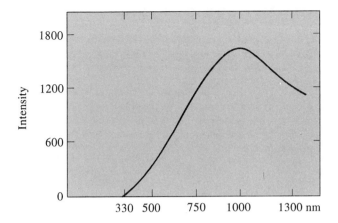

FIGURE 13-5. The intensity of a tungsten filament source at various wavelengths. The intensity units on the left axis are relative.

The Monochromator. After passing through an entrance slit which serves to define its path, radiation from the source enters some type of monochromator. This is necessary because a monochromator is used to select the wavelength(s) desired for absorption measurements from the wide range of wavelengths emitted by the source. Without a good monochromator, it would be impossible to obtain an absorption spectrum such as that of hemoglobin in Figure 13-1. In addition, the Beer-Lambert law is defined strictly for a single wavelength (monochromatic radiation); as soon as more than a 10–20 nm bandwidth of radiation is used, it is possible that deviations from this law will occur (Fig. 13-3). Therefore filters, gratings, or prisms are employed as monochromators in colorimeters and spectrophotometers. We will discuss each briefly.

The *filter* is usually a square piece of glass containing certain colored chemicals which transmit generally in the wavelength region of the color of the filter. Figure 13-6 is a plot of the transmittance of a typical violet filter at various wavelengths. This type of filter is a poor monochromator since its bandwidth of 50 nm is much larger than the desirable bandwidth of 10–20 nm. In addition, its transmittance varies with wavelength; that is, it transmits a different number of photons at each wavelength. A number of such filters are available, but not in such variety that one can obtain a filter for any desired wavelength.

Even though a filter is not suitable for obtaining an absorption spectrum, it can be used for colorimetric analysis where high accuracy is not required. Filters are also very useful in fluorescence analysis (Ch. 15). A very expensive type of filter, called an interference filter, does have 10–24 nm bandwidths and can be used in colorimetric or fluorescence analysis.

The *prism* functions as a monochromator by means of the refraction of light. Different wavelengths of light are bent to different degrees as they travel through a prism; the violet-blue wavelengths are bent more than the red wavelengths, so a dispersion of shorter wavelengths from longer wavelengths occurs. The prism is a

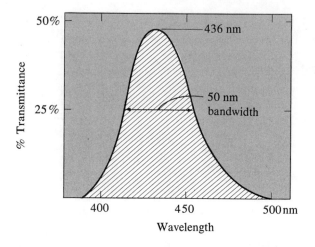

FIGURE 13-6. A plot of per cent transmittance versus wavelength for a typical violet filter (color specification 47B). The 50 nm bandwidth refers to the width of the band at half of the height of the curve at maximum transmittance.

good monochromator, and it is possible to utilize it in a spectrophotometer to obtain an absorption spectrum (Fig. 13-1). By setting the wavelength dial of the spectrophotometer to a certain selected wavelength, it is possible to obtain radiation of a narrow band of wavelength(s) in the ultraviolet to blue region. However, the separation between two wavelengths of light of relatively longer wavelength (say, red) is much poorer than that between two wavelengths in the ultraviolet-blue region. Therefore radiation selected at longer wavelengths will not be monochromatic and will have a large bandwidth. This will be especially true in the red region (620–750 nm).

The *grating*, or diffraction grating, is a piece of aluminized glass (original grating) or plastic (replica grating) which has a large series of accurately spaced lines ruled on the surface. When radiation strikes the unruled portion of its surface, most wavelengths are destroyed (via a destruction interference process) and do not reach the cell. The selected wavelength and a few wavelengths on either side of the selected wavelength undergo constructive interference giving a related series of wavelengths known as first order diffraction (same wavelength as the selected wavelength), second order diffraction (twice the wavelength of the selected wavelength), and so on.

The grating is a good monochromator, and thus it is possible to utilize it in a spectrophotometer to obtain an absorption spectrum. The desired wavelength is obtained simply by setting the wavelength dial. An advantage of the grating over the prism is that the bandwidth is constant (e.g., 20 nm) over the entire range of wavelengths. This is because the separation between any two wavelengths achieved by the grating is the same regardless of the wavelength. Because of this advantage, most spectrophotometers employ gratings instead of prisms.

The Cell. After leaving the monochromator through an exit slit, radiation only of one wavelength or a band of several wavelengths strikes an absorbing liquid sample in a cell or cuvette (Fig. 13-4). Construction of the cells will vary, depending on whether UV, visible, or IR radiation is being measured. For UV or visible region measurements, the cell is commonly a rectangular type with a 1.00 cm length and width so the internal cell length, b, is always 1.00 cm (Fig. 13-2). Cells for the IR region are different and will be discussed separately. As will be stressed, a cell used for a particular spectral region must be as transparent as possible to that region.

Cells used in the visible region can be made of optical quality borosilicate glass. You should note in Figure 13-7 that the transmittance of this material is

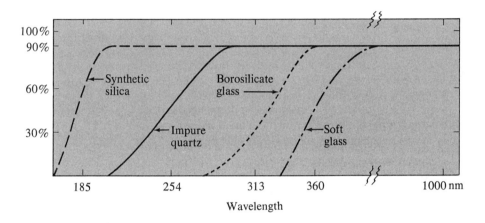

FIGURE 13-7. Plot of transmittance of various optical materials versus wavelength.

minimal below 310 nm so it cannot be used for most of the UV region. Some colorimeters or inexpensive spectrophotometers use carefully selected cylindrical cells (test tubes) which may be made of borosilicate glass or soft glass which transmits only to about 350 nm (Fig. 13-7). To insure a reproducible cell length, the cylindrical cells are usually marked so that they are inserted into the cell holder always in the same position.

Cells used in the UV region must be made of an optical "glass" that is free of the constituents in ordinary glass that absorb UV radiation. Fused standard silica may be used to give transparency from 1000–220 nm and is satisfactory for most UV measurements. Ordinary quartz is impure and is not as useful for UV work (Fig. 13-7), but pure quartz or synthetic silica (Fig. 13-7) may be used down to 180 nm.

Cells used in the IR region are a special type of sealed cell unit. The optical material is either polished sodium chloride or polished potassium bromide, both in the form of transparent "windows." Two such windows are joined to an amalgamated lead or Teflon spacer and sealed together with metal plates having apertures somewhat smaller than the windows. Glass or quartz cannot be used for IR measurements because they do not transmit over the entire IR region. Cells made from sodium chloride or potassium bromide must be handled carefully since these materials are soft, easily scratched, and easily damaged by water.

The Detector. After the unabsorbed radiation has passed through the sample cell, it impinges on some type of detector (Fig. 13-1). The type used will vary depending on the wavelength region (and the cost of the instrument), but usually the detector converts radiant energy into electrical current or voltage which can easily be measured. (If you read Ch. 12, you will recall that the detector is a type of transducer.) In single beam instruments the cell is first filled with the pure solvent and the $100\% T (A = 0)$ setting is established at the desired wavelength. Then the solvent is replaced with the sample, and the detector measures the amount of radiation absorbed compared to the pure solvent. In double beam instruments two cells (for the solvent and the sample) are used, eliminating the need to change cells. Five detectors used are the following.

1. The phototube
2. The silicon photodiode
3. The electron-multiplier phototube (photomultiplier)
4. The bolometer
5. The thermocouple

The *phototube* was discussed in Section 12-1. Briefly, it is a tube with a glass or quartz envelope containing a cathode and an anode (Fig. 12-4). Photons striking the cathode release electrons to the anode, causing an increase in electrical current proportional to the flow of photons falling on the cathode. The current is fed through an amplifier and read out on a scale calibrated in absorbance and per cent transmittance units. *Blue* phototubes cover the UV through 600 nm; *red* phototubes cover 600–1000 nm.

The *silicon photodiode* is replacing the phototube in newer spectrophotometers such as the Spectronic 21 (see below). It is a solid-state detector rather than a tube detector like the phototube and is frequently operated to produce a potential rather than a current. An advantage of this detector is that UV-sensitive photodiodes cover the 200–1100 nm region.

The *electron-multiplier phototube* (photomultiplier) incorporates several stages of amplification directly within the tube by means of dynodes. Each electron from the cathode strikes the first in a series of dynodes, ejecting two to four additional electrons. These in turn strike another dynode, and the process is repeated ten to fifteen times until the electrons finally strike the anode. The current resulting is amplified by a factor of 10^6–10^8. It is then transmitted to the meter to give an absorbance readout. This type of detector is more costly than a phototube and is used in more expensive spectrophotometers, primarily those that measure both UV and visible radiation.

The cathode in both the phototube and photomultiplier tube can be coated with various materials—those primarily sensitive to UV radiation, those sensitive to violet through orange radiation, and those sensitive to red-near IR radiation (620–1000 nm). It is not feasible to obtain a cathode coating which will emit electrons when low energy IR radiation is involved, so phototubes or photomultipliers are not used to detect IR radiation. Instead, bolometers or thermocouples are used.

The *bolometer* is simply a heat detector, that is, a temperature-sensitive resistor consisting of two thin foils or thermistor materials. The infrared radiation passing through the cell strikes only one of the foils—the other is shielded from the radiation and acts as a reference. The foils are joined electrically in a balanced Wheatstone bridge circuit which becomes unbalanced when radiation strikes the indicator foil. The resulting electrical signal is fed through an amplifier and the response is again read on a scale in absorbance units.

The *thermocouple* is also a heat detector which converts heat to an electrical signal. A blackened noble metal foil welded in two thermoelectric substances absorbs the infrared radiation reaching the detector. The foil is kept in a housing to minimize heat losses; a window in the housing transmits IR radiation to the foil. As with the bolometer, the resulting electric signal is amplified and read on an absorbance scale.

Types of Colorimeters and Spectrophotometers

Various general types of colorimeters and spectrophotometers are available; most commercially available instruments fall into one of these general types. The colorimeters are the least expensive and are, strictly speaking, limited to the visible region. Recording spectrophotometers of the UV-visible type or IR type are the most expensive. We will describe each general type in the necessary detail.

Filter Colorimeters (Photometers). A filter colorimeter, or photometer, measures light absorption by using different filters rather than a good monochromator. This type of instrument is used for routine measurements by unskilled operators. The filter is frequently inserted into the instrument before the operator ever uses it. Glass test tubes are used for cells, and a simple phototube is used for detection. The Coleman Nepho-Colorimeter and Klett-Summerson Colorimeter are examples of this type. Although they are inexpensive, they are rugged and dependable for routine work.

"Visible" Spectrophotometers. Although the spectrophotometer is unique in that it can measure outside the visible range of the colorimeter, it is also unique in that it is equipped with a good monochromator to select any desired wavelength of radiation. The so-called *visible* spectrophotometer actually covers a wavelength range much wider than the visible range and utilizes an inexpensive grating as a monochromator. For example, the Bausch & Lomb Spectronic 20 spectrophotometer (formerly called a colorimeter) covers the 330–950 nm range, the

Coleman Junior II spectrophotometer covers the 325–825 nm range, and the Turner Model 350 spectrophotometer spans the 335–1000 nm region. All of these relatively inexpensive spectrophotometers thus measure part of the UV, all of the visible, and some of the near IR regions.

The optical diagram of the Bausch & Lomb Spectronic 20 in Figure 13-8 is

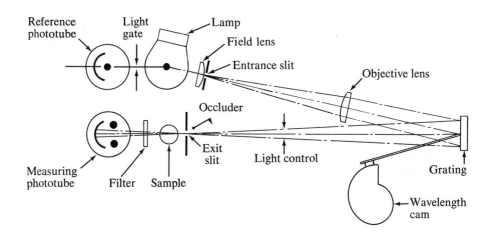

FIGURE 13-8. *Optical diagram of the Spectronic 20 spectrophotometer, top view. (Courtesy of Bausch and Lomb, Incorporated, Rochester, N.Y.)*

typical of this type of spectrophotometer. Radiation emitted by the tungsten lamp is directed by two lenses and a slit onto the diffraction grating. The radiation emitted by this lamp (Fig. 13-5) starts at about 325–330 nm and ends well above 1000 nm.

> *The tungsten lamp in Figure 13-9 is a very small version of the usual household tungsten lamp, except that it has a clear, not frosted, glass envelope, and it is a 6 volt, not 110 volt, lamp. The clear glass permits optimum transmittance of radiation in the 325–1000 nm range covered by the visible spectrophotometer.*

The grating selects radiation with a bandwidth of 20 nm centered around the nominal wavelength setting. Thus if that wavelength is 550 nm, the bandwidth of radiation actually falling on the sample is 540–560 nm. Any nominal wavelength from 330–950 nm may be selected by turning the wavelength control knob which controls the wavelength cam (Fig. 13-8) which in turn moves the grating. The selected band of radiation then is reflected off the grating through the light control and exit slit onto the sample solution. Any radiation not absorbed by the sample impinges on the measuring phototube where the light energy is converted into an electrical signal in terms of absorbance units.

SEC. 13-3   Spectrophotometers and Colorimeters   303

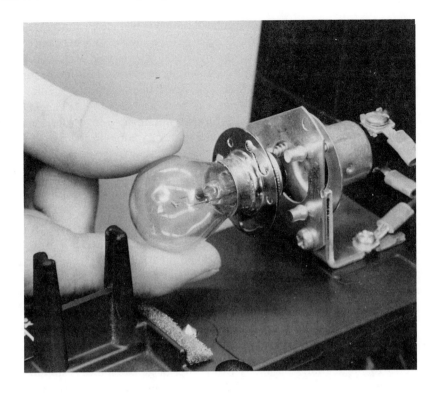

FIGURE 13-9. The 6 volt tungsten lamp of the Spectronic 20 spectrophotometer. The lamp is fitted to the housing by a clockwise turn, and the housing is connected below by leads to the power system. Directly above the lamp is the holder for the red filter used above 625 nm.

The Spectronic 20 spectrophotometer is also equipped with a reference phototube (Fig. 13-8). Since a change in the line voltage would change lamp intensity, the voltage fed to the lamp is monitored by the reference phototube so that it can be changed to keep lamp intensity, and scale readings, constant.

Ultraviolet-Visible Spectrophotometers. This type of spectrophotometer covers the UV region starting around 200 nm, the entire visible region, and part of the near IR through at least 1000 nm. Examples of this type are the Beckman DU-2 spectrophotometer, the Perkin-Elmer Model 139 spectrophotometer, and the Spectronic 21 (UV model) spectrophotometer. In addition to covering the ultraviolet region, an important use of these spectrophotometers is that many are equipped with recorders that provide a plot of absorbance vs. wavelength. For this plot to be accurate, the radiation striking the sample must be as close to monochromatic as possible. Thus these spectrophotometers use prisms or gratings that have an effective bandwidth of one nanometer or less.

Other features of this type of spectrophotometer include two interchangeable sources—the hydrogen (or deuterium) discharge tube and the tungsten lamp. Both can be operated simultaneously, so the hydrogen tube is used from 200–325 or 330 nm, after which a simple mirror adjustment directs emission from the tungsten source to the sample. The cells are usually fused silica or synthetic silica (pure quartz), so radiation from 200–1000 nm may be measured without changing cells. Finally, two interchangeable phototubes (or photomultipliers) are used in some such spectrophotometers, the changeover occurring at 625 nm where the red-near IR phototube is used. In others, a phototube containing two photosensitive surfaces (195–625 and 625–950 nm) is used. The Spectronic 21 utilizes the silicon photodiode detector (see above) which covers the 200–1100 nm range.

Infrared Spectrophotometers. This type of spectrophotometer covers the IR region from about 2 μm (5000 cm^{-1}) to about 15 μm (660 cm^{-1}). Because of the complicated infrared spectra of most molecules, automatic recording of IR spectra is essential; thus almost all such spectrophotometers are equipped with recorders. Because of the sensitivity of the instruments to many variables and because of the complicated nature of IR measurements, it is desirable to employ a *double-beam* spectrophotometer, or its equivalent, to cancel out variations when comparing the solvent to the sample.

The essential parts of the double-beam spectrophotometer are shown in Figure 13-10. Radiation from the Nernst glower or globar source is split into two beams falling simultaneously on sealed cells containing the pure solvent and the

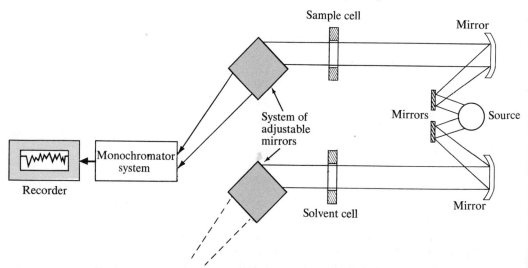

FIGURE 13-10. A schematic diagram of a double-beam infrared spectrophotometer. A single source is split by mirrors so that its beams strike the sample cell and solvent cell simultaneously. The signals are then fed alternately to the monochromator (the sample beam is shown being fed and the solvent beam is diverted) and the difference between the signals is recorded as the infrared spectrum.

sample in the same solvent. Thus in comparing the solvent and the sample, any variation in the intensity of the source will not affect the comparison. After each beam is passed through the respective cell, it is caused to travel on alternate cycles to the thermocouple or bolometer detector, automatically establishing $100\%T$.

13-4 ERRORS IN SPECTROPHOTOMETRY

A theoretical discussion of the errors in spectrophotometry is beyond the scope of this text. We will mention errors from experimental variables and measurements.

Experimental Variables

The cells are an important potential source of error whether a single-beam visible spectrophotometer or a double-beam infrared spectrophotometer is used. All cells should be carefully cleaned, especially where samples such as proteins might adsorb on cell walls. Fingerprints should be avoided; the organic compounds present may absorb ultraviolet radiation, and any aqueous residue may absorb infrared radiation in the same way water does.

The positioning of the cells in the cell compartment must be reproducible, a good reason for the line etched on test tubes used as cells in inexpensive visible spectrophotometers. (This line is matched with a line in the cell compartment.) Finally, cells whose diameters are carefully matched are important in double beam instruments for qualitative analysis and in all instruments for good quantitative work.

Other variables that must be controlled include the solvent composition, which should be the same in the standards and the unknown sample, in quantitative analysis. The pH in aqueous solutions should be controlled by a buffer if an acid-sensitive substance is measured, and the reagent concentration used to produce an absorbing species in quantitative work should likewise be controlled. In all these cases, deviations from the Beer-Lambert law (Fig. 13-3) may result.

Measurements

In general, the wavelength setting and the absorbance reading are the two instrumental settings most subject to errors and deviations. The wavelength setting may be subject to an *error* in the actual wavelength selected by the monochromator, and to *deviations* when the wavelength must be reset between two or more separate measurements. The error can be eliminated by calibrating the instrument to obtain an *accurate* wavelength; the deviations can be minimized by careful wavelength selection and checking to obtain good *precision*.

Like the eye, the phototube is subject to large errors in discerning differences between two faintly colored solutions and between two intensely colored solutions. It can be theoretically shown that the minimum concentration error will result when the absorbance reads between 0.12 and 1.0 (%T of 75 to 10%). If an error of $\pm 0.4\% T$ is made in a transmittance reading, it can also be shown that the relative error in concentration will be of the order of ± 2 pph. This assumes that the per cent transmittance scale is divided by 100 lines, each representing 1%, and that a third significant figure can be estimated between any two lines with an uncertainty of $\pm 0.4\% T$.

13-5 LOOKING BACK AND AHEAD

After reading this chapter, you should have gained a certain understanding of the absorption of ultraviolet, visible, and infrared radiation and how this is measured. Until you actually *use* the instruments for this measurement, your understanding will not be complete. Try at this point to review the concepts involved by rereading the chapter before you read, or as you read, the next chapter on applications of spectrophotometric measurements to chemistry, biochemistry, and medicine. This will help your understanding of that chapter.

To help you review, you should try working as many of the following problems as possible. Concentrate on the types of problems where you have the most difficulty. The problems are organized into different types, and each type or group is labeled. If you have a weakness in calculating absorbance, transmittance, or Beer-Lambert law variables, don't forget to go over Self-Tests 13-1 and 13-2.

At this point, you are ready for Chapter 14.

QUESTIONS AND PROBLEMS

(Answers to most even-numbered problems are in Appendix 5.)

Definitions and Concepts
1. Define the following types of radiation as to wavelength and effect on the eye.
    a. Ultraviolet radiation
    b. Infrared radiation
    c. Light
    d. Near infrared radiation
2. Define the following terms in words and by means of an equation where possible.
    a. Absorbance
    b. Radiant power
    c. Transmittance
    d. Beer-Lambert law

SEC. 13-5 Looking Back and Ahead

3. Explain the difference between each pair of terms.
   a. Absorbance and absorption
   b. Positive deviation and negative deviation from Beer-Lambert law
   c. Molar absorptivity and absorbance
   d. A colorless solution and a black solution
4. Define a nanometer. Express 500 nm (green light) in terms of meters.

The Absorption Process

5. Explain how a molecule absorbs a photon of radiant energy. Include a definition of the ground state and the excited state in the explanation.
6. In Figure 13-1 it can be seen that there are three possible excited states of oxyhemoglobin [$Hb(O_2)_x$]. Write three different equations showing the absorption of radiation by the ground state of oxyhemoglobin to give the three excited states.
7. Use Table 13-1 to answer the following questions on color and the response of the eye (recall that purple is a mixture of blue and red light).
   a. Suppose that oxyhemoglobin transmitted radiation from 450–495 nm as well as that shown in Figure 13-1. What color would it then appear?
   b. Deoxyhemoglobin (hemoglobin without oxygen) transmits radiation from 450–495 nm as well as from 620–750 nm. What color is it?
   c. Oxyhemoglobin can be oxidized to methemoglobin (contains Fe^{+3}). This molecule transmits from about 590–620 nm and has a small absorption band in the 620–700 nm region. What color should it appear?
   d. Butter absorbs all light from 380–570 nm and reflects all other light. What color should it appear? Draw an absorption spectrum for it.
   e. Lettuce absorbs all light from 550–750 nm and reflects all other light. What color should it appear? Draw an absorption spectrum for it.
8. The maximum response of the eye at night (rod vision) is shifted to 507 nm; in addition, the eye has very little response to radiation above 620 nm at night. Indicate what color(s) transmitted by the following ions the eye will respond to under nighttime illumination, but do not indicate their actual color. (See Problem 7 and Fig. 13-1 for some daytime colors.)
   a. Oxyhemoglobin
   b. A purple solution of permanganate (MnO_4^-).
   c. Deoxyhemoglobin
   d. A solution of the usual color lipstick

Absorbance and Transmittance Calculations

9. The following per cent transmittance values were read from a spectrophotometer with a scale readout. Decide whether all digits are significant before calculating A.
   a. $\%T = 20\%$ b. $\%T = 10.0\%$ c. $\%T = 4\%$ d. $\%T = 1.6\%$
10. The following per cent transmittance values were read from a spectrophotometer with a three-place per cent transmittance digital readout. Calculate the absorbance for each using the correct number of significant figures, and compare your answers to those in the previous problem.
    a. $\%T = 20.0\%$ b. $\%T = 10.0\%$ c. $\%T = 4.0\%$ d. $\%T = 1.6\%$
11. Calculate the absorbance to the correct number of significant figures for each value of

per cent transmittance given below, assuming that the values are read from a spectrophotometer scale. Note that each absorbance value you calculate should have two significant figures to the right of the decimal point, and that the first of these in each case is a zero. Explain why the first zero to the right of the decimal is significant when the general rule is that all initial zeroes are not significant.
   a. $\%T = 82\%$ **b.** $\%T = 85\%$ **c.** $\%T = 95\%$

12. The following absorbance values were read from a spectrophotometer with a *scale readout*. Decide whether all digits are significant and then calculate the per cent transmittance to the correct number of significant figures.
   a. $A = 0.10$ **b.** $A = 0.202$ **c.** $A = 2.00$

Beer-Lambert Law Calculations

13. The following solutions of potassium permanganate were measured at 540 nm in a 1.0 cm cell. Prepare a Beer-Lambert law plot of absorbance versus concentration and find the concentration of an unknown which has a per cent transmittance of 50%.

$0.000050M$	$A = 0.10$
$0.00020M$	$A = 0.40$
$0.00030M$	$A = 0.61$
$0.00040M$	$A = 0.81$

14. A $0.00010M$ solution of potassium permanganate measured at the same wavelength as in Problem 13 has a per cent transmittance of 69%. Enter this point on your plot for Problem 13 and decide whether this point obeys the Beer-Lambert law or is a positive or negative deviation from this law.

15. Given the molar absorptivity, calculate its log to the correct number of significant figures. Use a four-place log table or calculator to obtain the uncorrected $\log \varepsilon$ before rounding off to the correct number of significant figures. (See answers to a and b in the Appendix.)
   a. $\varepsilon = 2.0 \times 10^4$ **b.** $\varepsilon = 11$ **c.** $\varepsilon = 0.02$ **d.** $\varepsilon = 1.02$

16. Given the log of the molar absorptivity, calculate the molar absorptivity to the correct number of significant figures, using subscripts instead of rounding off where necessary.
   a. $\log \varepsilon = 0.3$ **b.** $\log \varepsilon = 4.0$ **c.** $\log \varepsilon = 4$ **d.** $\log \varepsilon = 3.4$

17. A solution has a $\%T$ of 40% in a 1.00 cm cell. Given the following changes in cell diameter, b, calculate the new absorbance and new $\%T$.
   a. $b = 2.00$ cm   **b.** $b = 0.500$ cm

18. Calculate the designated variable using the Beer-Lambert law. Assume $b = 1.000$ cm in each case. Use the correct number of significant figures.
   a. Given $\varepsilon = 1 \times 10^4$ and $c = 2.00 \times 10^{-6}M$, calculate A.
   b. Given $\log \varepsilon = 4.30$ and $c = 3.00 \times 10^{-6}M$, calculate A.
   c. Given $A = 0.400$ and $c = 2.0 \times 10^{-5}M$, calculate ε.

Absorption Spectra

19. In *The Invisible Witness* (W. Turner, Bobbs-Merrill, Indianapolis and New York, 1968, p. 131) the story of a man accused of robbing a woman is told. The suspect had lipstick on his hand. How might absorption spectra be used as evidence in this case?

20. Plot the absorption spectrum of acetylsalicylic acid (aspirin) from the following data. Decide whether one or two absorption bands are present and calculate the molar absorptivity for the band(s), using $c = 2.0 \times 10^{-4}M$ as the concentration of the aspirin. Also decide if the aspirin solution is colored.

Wavelength, nm	A
215	0.10
225	0.15
235	0.22
245	0.16
255	0.19
275	0.24
280	0.28
285	0.25
295	0.20
305	0.15
315	0.05
325	0.01
335	0.00
400	0.00

21. The absorption spectrum of a 0.0010M solution of p-nitrophenol, $C_6H_4(NO_2)OH$, in a 1.00 cm cell has absorption bands at 450 nm ($A = 0.80$) and at 300 nm ($A = 0.51$). Is the solution colored? Calculate the molar absorptivities at each wavelength.

22. In 1966 inorganic nitrate was reportedly found in tobacco. The nitrate ion, NO_3^-, absorbs at 203 ($\varepsilon = 1 \times 10^4$) and at 300 nm ($\varepsilon = 7.5$). What kind of spectrophotometer must be used to determine the nitrate ion in the tobacco? Calculate the minimum detectable concentration at each wavelength if the minimum detectable absorbance is 0.01.

23. The permanganate ion (MnO_4^-) absorbs at 225 nm ($\varepsilon = 3 \times 10^3$), 310 nm ($\varepsilon = 1.5 \times 10^3$), and at 525 nm ($\varepsilon = 2 \times 10^3$).
    a. What type of spectrophotometer would have to be used to measure absorption at each wavelength?
    b. Calculate the minimum detectable concentration at each wavelength for a minimum detectable A of 0.01.
    c. Comment on the feasibility of determining manganese in steel at each wavelength after dissolving in nitric acid and oxidizing to MnO_4^-. (Recall that iron(III) absorbs UV radiation.)

Challenging Problems

24. The flame test for potassium yields two lines—404 nm (relative no. of photons = 1) and 766 nm (relative no. of photons = 100). The relative response of the eye to 404 nm radiation is 0.0006 and to 677 nm radiation is 0.00004. Which color will the eye perceive predominately? (*Hint:* Calculate a ratio.)

25. Assume that in a copper(II) solution, 10^3 $Cu(H_2O)_6^{+2}$ ions are excited every 10^{-15} second and that a liter of 0.010M copper(II) is being excited for 10 hours without any excited copper(II) being able to return to the ground state. Show by calculation what the concentration of the excited state will be after this time. (Use Avogadro's number.) Will the solution be colored after 10 hours?

26. The dark-adapted molecules in the retina of the eye are excited and transmit a signal to the optic nerve when the rate of the incident radiation from a point source on a black background reaches 2×10^{-15} Watt.
    a. What is the minimum number of photons of 550 nm wavelength which must reach the retina *per second* to produce vision? (*Hint:* Use 1 J = 1 W s.)
    b. If vision persists for about 1/30th s, what number of photons are required in this time to produce vision?

14. Applications of Spectrophotometry

"Now I add this small quantity of blood to a litre of water," Holmes said. "The proportion of blood cannot be more than one in a million. I have no doubt, however, that we shall be able to obtain the characteristic (color) reaction."

ARTHUR CONAN DOYLE
A Study in Scarlet, Chapter 1

In this chapter we will discuss primarily the applications of ultraviolet, visible, and infrared spectrophotometry, as well as automated spectrophotometric analysis. Many of the applications will fall into the area of trace analysis, in concentration range of $10^{-6} M$, where Sherlock Holmes himself dabbled.

You will find that the applications divide into two categories—qualitative analysis and quantitative analysis. Qualitative analysis involves identification of compounds whose identity is unknown and screening for the presence or absence of a suspected compound. Quantitative analysis involves both the "one-shot" analysis for a known constituent and the continuous monitoring or analysis of a changing sample, such as city air, for one or more constituents.

Before learning the principles of analysis, you should understand something about the types of groups within molecules, or ions, that may absorb radiation; these groups are called *chromophores*. Molecules with the same chromophore usually absorb in the same wavelength region, though not exactly at the same wavelength.

14-1 ABSORPTION BY SOME COMMON CHROMOPHORES

The two general classes of chromophores are (1) *electronic*, in which certain electrons absorb ultraviolet or visible radiation, and (2) *vibrational*, in which certain bonds absorb infrared radiation. We will discuss the former first.

Electronic Chromophores

The electronic ground state and excited state were defined in Section 12-1. After an electron of a ground state species absorbs a photon, the latter undergoes a transition to an electronic excited state, as shown in Figure 14-1. In that state the

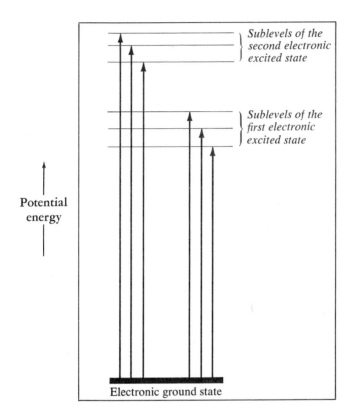

FIGURE 14-1. Two typical electronic transitions from the ground state to various sublevels of the first excited state, and from the ground state to various sublevels of the second excited state. The energy absorbed in going to the second excited state may be larger because a different electron is jumping to the same orbital, or because the same electron is jumping to a higher energy orbital.

electron occupies a different type of orbital. It helps to think of chromophores in terms of what type of *electron* will absorb radiation and also what type of *orbital* the electron will occupy in the excited state. Remember this when reading the following detailed discussion; it will help you see the forest in spite of the trees.

Unsaturated molecules or ions with double- or triple-bond chromophores form a broad class of compounds having π electrons which will absorb UV or visible radiation. One of the bonds in a double bond is a stable sigma (σ) bond; the other bond, the π bond, is weaker, and either of its two electrons can easily be excited by a UV or visible photon to a higher energy empty orbital.

Whether the double- or triple-bond chromophore is present in an organic molecule or in an inorganic ion or molecule does not matter. The radiant energy absorbed may differ somewhat, but in most cases a π electron absorbs in the UV-visible region. For example, one resonance structure of the purple MnO_4^- ion can be written as

$$O=\underset{\underset{O}{\|}}{\overset{\overset{O^-}{|}}{Mn}}=O$$

The permanganate ion should therefore absorb in the ultraviolet-visible region, and it does, both at 525 nm and at 227 nm as a result of absorption by π electrons [1].

Ozone (Table 14-1) *is another example. It can be represented by a resonance structure having a double bond.*

$$\overset{..}{\underset{..}{O}}=O-\overset{..}{\underset{..}{O}}:$$

O_3 *in the sky absorbs UV radiation from the sun from below 240 nm to about 310 nm and thereby protects the eyes and skin from harmful damage by such high energy UV radiation. (It is interesting that ozone is itself formed by oxygen molecules absorbing still higher energy UV radiation below 240 nm.)*

As can be seen from Table 14-1, there are many molecules and ions which have the double-bond chromophore and a few which have the triple-bond chromophore. Most absorb UV radiation rather than the visible radiation absorbed by the permanganate ion.

One other point is important—molecules having two or more *conjugated* double bonds have one absorption band at longer wavelengths than those which have only one double bond. For example, hexatriene and benzene, each of which has three conjugated double bonds, absorb at 268 and 254, respectively; these absorption bands are at much longer wavelengths than the 160 nm band of ethylene (Table 14-1).

The substitution of certain functional groups on conjugated double-bond chromophores commonly shifts absorption bands to longer wavelengths also. For example, acetylsalicylic acid (Table 14-1), the active ingredient in aspirin tablets, absorbs at longer wavelengths than benzene. The two functional groups responsible will be discussed in the next section.

Unsaturated molecules or ions having a pair of nonbonding (*n*) electrons on one of the two atoms in a double-bond chromophore form a second class of compounds which will absorb in the UV-visible region, as seen in Table 14-2. The *n*

---

[1] A. Viste and H. B. Gray, *Inorg. Chem.* **3**, 1113 (1964); and G. H. Schenk, *Rec. Chem. Progr.* **28**, 135 (1967).

SEC. 14-1 Absorption by Some Common Chromophores

TABLE 14-1. *Ultraviolet-Visible Absorption for π Electron Chromophores*

Name of chromophore	Formula of chromophore	Example	λ_{max}, nm	ε
Oxygen	O=O	O_2	~100 to ~180	—(gas)
Ozone	O=O$^+$—O$^-$	O_3	~200 to ~270	—(gas)
Nitroso	—N=O	NO_3^-	203	1×10^4
		NO_2	~290 to ~360	—(gas)
—	Mn=O	MnO_4^-	227	1.6×10^3
			525	2.2×10^3
Carbonyl	C=O	$(CH_3)_2C=O$	166	1.6×10^4
Ethylene	C=C	Ethylene	160	2×10^4
		2-Hexene	193	10^4
Acetylene	C≡C	Acetylene	173	6×10^3
Polyene	H(HC=CH)$_n$H	Butadiene	217	2×10^4
		Hexatriene	268	3.5×10^4
Benzene		Benzene	254	2×10^2
		Salicylic acid	308	4×10^3
		Acetylsalicylic acid	280	1.5×10^3
Phenanthrene			251	5×10^4
			293	1.5×10^4
Anthracene		Anthracene	253	1.5×10^5
			340, 357, 375	all 9×10^3

TABLE 14-2. *Ultraviolet-Visible Absorption for n Electron Chromophores*

Name of chromophore	Formula of chromophore	Example	λ_{max}, nm	ε
Carbonyl	C=O	Acetone	270	18
		$(C_6H_5)_2C=O$	330	180
Carboxyl	R—C=O(OH)	Acetic acid	204	60
Nitro	$^-$O—$^+$N=O	NO_3^-	300	7.5
Nitroso	—N=O	Nitrosobutane	300	100
			665	30
Thiocarbonyl	C=S	CS_2	318	108
		$(C_6H_5)_2C=S$	620	70
Azomethine	C=N	$(C_6H_5)_2C=NH$	340	125
Azo	N=N	C_6H_5—N=N—C_6H_5	448	425

electrons that absorb in this manner are found on atoms such as oxygen, sulfur, and nitrogen, but not on carbon. Two examples of such chromophores are acetone and the nitrate ion:

$$\begin{array}{c} CH_3 \\ \diagdown \\ C\!=\!O\!: \\ \diagup \\ CH_3 \end{array} \qquad\qquad \begin{array}{c} {}^-O \\ \phantom{{}^-O}\diagdown \\ \phantom{{}^-O}\overset{+}{N}\!=\!O\!: \\ \phantom{{}^-O}\diagup \\ {}^-O \end{array}$$

In each compound the pair of n electrons on oxygen are obviously not involved in bonding and easily can be excited by UV radiation to a higher energy empty orbital. It does not matter whether the oxygen is part of an ion or an organic molecule; the process is the same.

As can be seen from Table 14-2, there are a number of different chromophores in which n electrons absorb UV or visible radiation. The nonbonding electrons are not shown, but are present on the atom at the right of the double bond. The molar absorptivity of each chromophore is quite low, of the order of 10^2, as compared to the molar absorptivities of the chromophores listed in Table 14-1. This is characteristic for absorption of radiation by n electrons as compared to π electrons.

It is obvious that all of the chromophores in Table 14-2 have n electrons on an atom which has a double bond, and you should be wondering why this is so. The reason has to do with the empty orbital to which the n electron is excited. In each case the n electron jumps to a so-called π^* orbital associated with the double bond. This orbital has four lobes, two of which are centered on each of the atoms of the double bond. For acetone the n-π^* electronic transition looks like

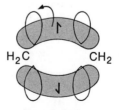

(In the excited state the electron can occupy any of the four lobes of the π^* orbital, not just the one lobe.) This explains the absorption of radiation by all of the chromophores listed in Table 14-2 at the wavelengths listed.

The π electron in the chromophores in Table 14-1 also jumps to the same type of π^* orbital as the n electron. The π-π^* transition in ethylene looks like the following from the side.

The shaded area represents the π electron density and the line represents the sigma bond. Of course, the excited electron may occupy any of the four lobes of the

π orbital, not just the one shown. The same kind of transition occurs in any of the chromophores in Table 14-1. Also note that some of these chromophores can undergo both π-π* and n-π* transitions. For example, in a solution of nitrate ions, some ions absorb at 203 nm (π-π* transition) while others absorb at 300 nm (n-π* transition). (Review Fig. 14-1 to picture this.) This is typical of small molecules or ions; the n-π* transition occurs at longer wavelengths.

Inorganic ions or chelates form the last broad class of chromophores or substances which absorb in the UV-visible region. Many of the common inorganic ions are colored because a d electron in a lower energy d orbital absorbs light and jumps to a higher energy unoccupied or half-filled d orbital. Recall that there are five d orbitals; in ions of the type $M(H_2O)_6^{+x}$ or ML_6^{+x} the lower energy d orbitals are the d_{xy}, d_{xz}, and d_{yz} orbitals, and the higher energy d orbitals are the d_{z^2} and $d_{x^2-y^2}$ orbitals.

A *typical ion* that is colored because of d-d transitions is the green $Fe(H_2O)_6^{+2}$ ion. This ion has six d electrons, so four of the d orbitals hold one electron each, and the fifth, let us assume the d_{xy} orbital, is filled. A d-d transition would look like this,

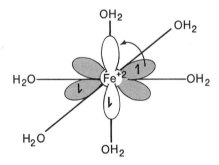

wherein an electron in the filled d_{xy} orbital (shaded) absorbs a photon of red light (690 nm) and jumps to the half-filled d_{z^2} orbital.

The d_{z^2} orbital has higher energy than the d_{xy} orbital because of greater repulsion from electron pairs on oxygen which focus directly at this orbital. An electron in the d_{xy} orbital is between the regions of greater repulsion by oxygen electron pairs; it uses the light energy to overcome the greater repulsion it encounters briefly in the excited state.

Similar types of d-d electronic transitions are responsible for the colors of the common metal ions listed in Table 14-3. As long as a metal ion possesses at least one d electron, it will absorb in the UV-visible region, although it may not always be colored. For example, the $Mn(H_2O)_6^{+2}$ ion does absorb light at 532 and 435 nm, but the molar absorptivities of these bands are so low that a $0.1M$ solution is colorless. (A $2M$ solution is a distinct pink color.) A metal ion having a filled subshell of ten d electrons will of course not exhibit any d-d transitions and frequently will be colorless.

Inorganic chelates are frequently colored because of slightly different types of electronic transitions. A good example is the chelate of iron(II) and 1,10-phenanthroline (phen), which has the formula $Fe(phen)_3^{+2}$. Each 1,10-phenanthroline molecule bonds to the iron(II) via two nitrogen donor atoms as follows:

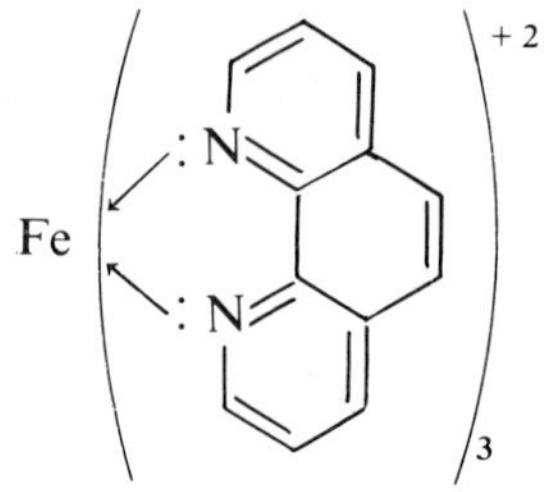

Note that a total of six electron pairs from six nitrogen donor atoms are coordinated to iron(II), so this chelate is equivalent to a ML_6^{+x} type of complex ion. A close look at the structure reveals that a five-membered *chelate ring* is formed by the iron(II), the two nitrogen donor atoms, and two carbon atoms connecting the nitrogen atoms. When light of 512 nm wavelength is absorbed by this chelate, a d electron such as a d_{xy} electron of iron(II) is excited to an empty π^* orbital on the unsaturated ring as follows:

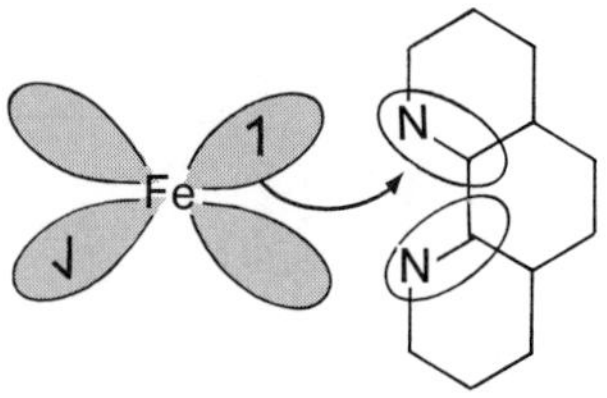

Since the two π^* orbitals shown are equivalent, the d electron may jump to either orbital, not just the one shown. Although the d electron may be excited to an unfilled d_{z^2} or $d_{x^2-y^2}$ orbital, as for $Fe(H_2O)_6^{+2}$, this requires much more energy than that of a photon of 512 nm (green light) because the 1,10-phenanthroline exerts more repulsion on these orbitals than does a water molecule.

The light absorption process for at least one of the three absorption bands of oxyhemoglobin (Fig. 13-2) is probably very similar to that for $Fe(phen)_3^{+2}$. Oxyhemoglobin is a chelate of iron(II), being bonded to the unsaturated porphyrin molecule in the heme group as follows:

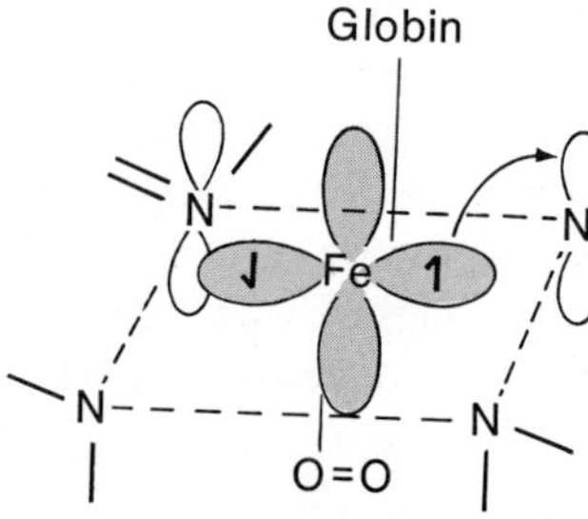

SEC. 14-1   Absorption by Some Common Chromophores

As shown above, a d electron such as a d_{xy} electron can absorb a photon and jump to one of the empty π^* orbitals of the heme nitrogens bonded to the iron(II). Other transitions are also possible.

Not all important ions of iron involve iron(II). Iron(III), or the ferric ion, forms many important colored ions. Strangely enough, $Fe(H_2O)_6^{+3}$ is essentially colorless in dilute solution. It absorbs radiation so weakly in the red-near infrared and yellow-green regions (Table 14-3) that only concentrated solutions appear

TABLE 14-3.   Ultraviolet-Visible Absorption for d Orbital Chromophores

Formula of ion	Number of d electrons	λ_{max}, nm (color)	ε
Ions with d-d transitions			
$Ti(H_2O)_6^{+3}$	1	500 (purple)	4
$Mn(H_2O)_6^{+2}$	5	435, 532 (colorless)	0.02
$Fe(H_2O)_6^{+3}$	5	794, 540 (colorless)	0.1
$Fe(H_2O)_6^{+2}$	6	450, 690, 962 (lt. green)	2
$Fe(CN)_6^{-4}$	6	420 (lt. yellow)	1
$Ni(H_2O)_6^{+2}$	8	400, 740 (green)	5
$Cu(H_2O)_6^{+2}$	9	790 (green-blue)	12
Ions and chelates with transitions involving d orbitals			
$Fe(phen)_3^{+2}$	6	512 (orange)	1.1×10^4
Oxyhemoglobin [Iron(II)]	6	575, 540, 415 (red)	—
$Co(NCS)_4^{-2}$	7	312, 620 (blue)	10^3
$Fe(SCN)^{+2}$ $[Fe(SCN)(H_2O)_5^{+2}]$	5	450 (red)	5×10^3
$FeCl^{+2}$ $[FeCl(H_2O)_5^{+2}]$	5	340 (yellow)	2×10^3
$Fe(CN)_6^{-3}$	5	415 (yellow)	1×10^3

faintly colored. As soon as ions such as the hydroxide or chloride ion (see $FeCl^{+2}$ in Table 14-3) are mixed with $Fe(H_2O)_6^{+3}$, it forms yellow complex ions. The absorption process probably involves an electron of the hydroxide or chloride being excited to an unfilled d orbital of the iron(III) ion.

Another important colored complex ion of iron(III) is that formed with the thiocyanate (SCN^-) ion. The formation of a red color when thiocyanate is added to a solution is an important qualitative analytical test for the presence of iron(III). The test is so selective that few ions interfere with it. Iron(II) does not form a colored complex with thiocyanate; hence the test is specific for iron in the $+3$ oxidation state.

The following self-test is intended as an aid mainly for the average student. The material discussed above has been rich, but the fundamentals are not difficult to understand. Once you complete the self-test, you should have a good grasp of the fundamentals and you will then recognize that the remaining material is enrichment. After mastering the fundamentals, you should proceed to the problems at the end of the chapter.

Self-Test 14-1. Electronic Absorption

Directions: Work one problem at a time and check the answer at the end of the test.

A. Recognition of chromophores. Indicate whether each of the following molecules or ions should absorb in the UV-visible region and explain why.

a. $[:O=C(-O)_2]^{-2}$, carbonate ion; $:O=C=O:$, carbon dioxide; $[(:O=)_2 Cr(-O)_2]^{-2}$, chromate ion; $CH_3C=O:$, acetic acid
$\phantom{[(:O=)_2 Cr(-O)_2]^{-2}, chrom} |$
$\phantom{[(:O=)_2 Cr(-O)_2]^{-2}, chroma} OH$

b. $:NH_3$, $CH_3-\dot{N}H_2$, $H-\ddot{C}\dot{l}:$, $[:\ddot{F}:]^-$

B. Metal ions with d orbitals. Indicate whether each of the following ions should absorb in the UV-visible region or whether they should be colorless and explain why.

a. $Cr(H_2O)_6^{+3}$, a d^3 ion; $Co(H_2O)_6^{+2}$, a d^7 ion; $Mn(H_2O)_6^{+3}$, a d^4 ion
b. $Ca(H_2O)_6^{+2}$, a d^0 ion; $Ti(H_2O)_6^{+4}$, a d^0 ion; $K(H_2O)_6^+$, a d^0 ion
c. $Zn(H_2O)_6^{+2}$, a d^{10} ion; $Cu(H_2O)_6^+$, a d^{10} ion; $Ag(H_2O)_6^+$, a d^{10} ion

C. Ions and chelates with d orbitals. Indicate whether each of the ions or chelates should absorb in the UV region and explain why.

$FeBr^{+2}$ ion; $Fe(C_2O_4)_3^{-3}$ oxalate chelate; $Fe(CN)_6^{-4}$ ion

Answers to Self-Test 14-1

A. a. All absorb because each has a double bond and because each has a pair of nonbonding electrons on an atom which is part of a double bond. (In each case the n electrons are on oxygen. Both $\pi-\pi^*$ and $n-\pi^*$ transitions can occur.

b. None should absorb because none of the atoms with n electrons is part of a double bond.

B. a. All should absorb because they have d electrons for a $d-d$ transition.
b. None should absorb because they have no d electrons; all are colorless.
c. None should absorb because their d orbitals are filled; all are colorless.

C. All should absorb in that an electron on the ligand may be excited to an unfilled d orbital on iron (III) or iron (II).

Chromophores Undergoing Vibrational Absorption

Absorption of infrared radiation by a molecule or by a certain bond in a molecule leads only to vibrational, not electronic, energy changes. Hence during infrared absorption, all the electrons remain in their most stable orbitals, and only the vibrational energies of the bonds in the molecule increase. The electronic ground state is divided into many sublevels, starting with the lowest energy sublevel, called the GV_0 sublevel, and continuing with sublevels of increasing vibrational energy—$GV_1, GV_2, ..., GV_n$. These are shown in Figure 14-2.

SEC. 14-1   Absorption by Some Common Chromophores   319

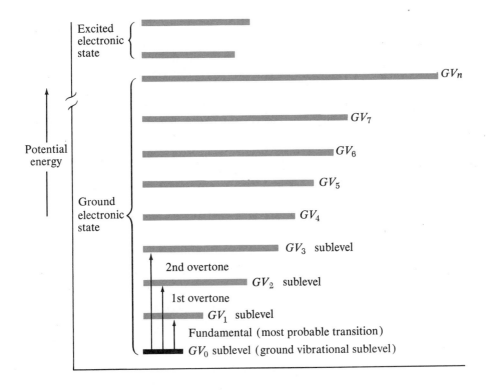

FIGURE 14-2. Subdivision of the ground electronic state into vibrational sublevels—$GV_0, GV_1, GV_2, GV_3, \ldots, GV_n$. Typical transitions occurring upon absorption of infrared radiation are shown. The first and second overtones occur with a much lower probability; they are generally found in the near infrared region at higher energies than the fundamental absorption band. (The length of each horizontal line may be taken to symbolize the internuclear distance for the bond absorbing the infrared radiation.)

Nearly all molecules exist in the GV_0 sublevel, so it is the ground vibrational sublevel. The most probable infrared absorption process, as shown in Figure 14-2, is

$$GV_0 + \text{IR photon} \rightarrow GV_1$$

in which the GV_1 sublevel is the first excited vibrational sublevel. The resulting absorption band is called the *fundamental*. The molecule does not remain in the GV_1 sublevel very long but will return to the ground vibrational sublevel.

$$GV_1 \rightarrow GV_0 + \text{Heat}(\rightarrow \text{solvent})$$

The return to the GV_0 sublevel is accompanied by the loss of the absorbed energy as heat following collision with a solvent molecule. Other, weaker absorption bands, called overtones, also occur in the near infrared region (as shown in Fig. 14-2), but they are beyond the scope of our discussion.

Although infrared absorption is quite complicated, a simple generalization can be made to help you at the beginning. The generalization is this: molecules with similar bonds, or *functional groups*, do exhibit similar fundamental infrared absorption bands. Thus all molecules with a C—O—H group will have similar C—O and O—H absorption bands, etc. The locations of the absorption bands listed in Table 14-4 illustrate this in more detail. To understand the *relative* locations of these bands, we will discuss a few similar types of bonds which absorb fairly close to one another.

Bonds consisting of a hydrogen atom and another light atom have the largest vibrational energies in the GV_0 sublevel and therefore must absorb the largest energy, smallest wavelength, infrared radiation to jump to the GV_1 sublevel. Examples of such bonds are the O—H bond, the N—H bond, the $\diagdown$C—H bond, and the =C—H bond. The essential information concerning absorption by these bonds is given in Table 14-4; only a few points will be elaborated below.

TABLE 14-4. IR Absorption Bands of Bonds Having a Hydrogen Atom

Bond or functional group	Wavelength, μm	Comment and confirming absorption
CO—H (alcohol or phenol)	2.74–2.82	Sharp O—H band only in dilute solution. Confirm by band at 7–8 and C—O band at 8.3–9.5.
	2.82–3.1	Broad O—H band from hydrogen bonding if solution is not dilute.
$\overset{O}{\underset{\|}{C}}$O—H (acid)	2.8–3.0	O—H band is sharp only in very dilute solution. Confirm by C=O band at 5.8–6.
	3.3–4.0	Broad O—H band from hydrogen bonding in most solutions.
CN—H (amide or amine)	2.82–3.0	Weak band from N—H. Confirm amide by C=O band at 5.9–6.06. Confirm amine by medium band at 6.1–6.7 or C—N band at 8.1–9.7.
C=C—H (aromatic or alkene)	3.29–3.32	Sharp C—H band of varying intensity. Confirm aromatic by strong C=C band at 6.7–6.8. Confirm alkene by variable C=C band at 6.0–6.2.
C—H (aliphatic carbon)	3.4–3.6	Usually strong C—H band because of large number of C—H bonds in organic molecules. Almost always present.
$\overset{O}{\underset{\|}{C}}$—H (aldehyde)	Two bands at 3.5 & 3.65	The 3.5 band is usually overlapped by the C—H band, but the 3.65 band is usually sharp and clearly separated from the C—H band. Confirm by C=O band at 5.75–6.0.

Absorption by the O—H bond is complicated both by the nature of the OH molecule and by the concentration of the solution measured. For alcohols, phenols, and acids, the O—H bond absorbs at different wavelengths depending on the con-

SEC. 14-1 Absorption by Some Common Chromophores

centration. In dilute solution O—H-containing molecules tend not to interact with one another and exist as *monomers*. The O—H absorption band is sharp. In moderate to concentrated solutions (0.03–1M) O—H-containing molecules hydrogen bond to one another as follows:

$$R-O\overset{H}{\underset{\cdot\cdot}{\diagup}} \longrightarrow H-O\overset{R}{\diagup}$$

This lowers slightly the vibrational energy of the O—H bond and the energy of the infrared radiation absorbed by this bond. You can see in Table 14-4 that the hydrogen bonding absorption is at longer wavelengths and is a broad, not a sharp, band. In fact, in solutions of moderate concentrations of alcohols or phenols, both absorption bands will occur in the infrared spectrum. Hydrogen bonding in acids is much stronger, so only the hydrogen bonded O—H absorption band is observed except for very dilute solutions.

Because bonds having hydrogen atoms are *all* of relatively high energy, they absorb at wavelengths quite close together in the infrared region. Consider the absorption spectra of acetaldehyde, CH_3CHO, shown in Figure 14-3. The 3.5 μm absorption band arising from hydrogen bonded to the C=O group is barely discernible because it is overlapped by the C—H absorption band of the CH_3 group. Fortunately, the 3.65 μm band is clearly separated from the intense C—H absorption band of the methyl group.

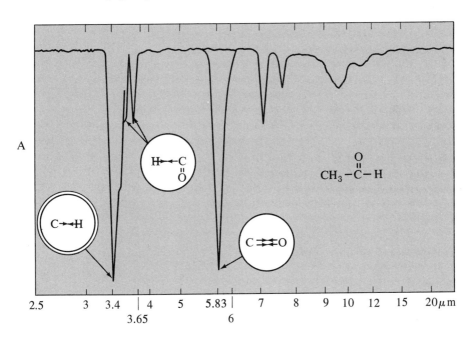

FIGURE 14-3. *The infrared absorption spectrum of acetaldehyde. Note that the 3.5 μm aldehyde C—H band is overlapped by the 3.4 μm CH_3 band, but that the 3.65 μm aldehyde C—H band is clearly separated.*

Bonds consisting of atoms other than hydrogen have smaller vibrational energies than bonds having a hydrogen atom in the GV_0 sublevel. For a jump to the GV_1 sublevel, such bonds therefore absorb smaller energy, longer wavelength infrared radiation. The essential infrared wavelengths absorbed by such bonds are given in Table 14-5, and only a few additional comments are necessary.

TABLE 14-5. IR Absorption Bands of Bonds Without a Hydrogen Atom

Bond or functional group	Wavelength, μm	Comment
Moderate energy bonds		
C≡N, nitrile	4.40–4.5	—
C≡C, acetylenic	4.42–4.76	—
C≡O, carbon monoxide	4.65–4.7	—
O=C=O, carbon dioxide	4.30	Confirm at 2.6 & 2.7 μm (weak bands)
—C=O(OR), ester	5.71–5.76	—
—CH, aldehyde ‖ O	5.75–5.95	Confirm at 3.65 μm
—COH, acid ‖ O	5.80–5.95	Confirm by broad band at 3.3–4.0 μm
\C=O, ketone	5.80–6.01	—
—CO⁻, acid anion ‖ O	6.21–6.45	—
C=C, alkene	5.95–6.17	Very weak
O=S—O, sulfur dioxide	7.25–7.45	Confirm at 8.60–8.85 μm
Low energy bonds		
C—N	7.46–9.8	—
C—O	8.7–9.5	—
C—Cl	12.5–16.6	—
C—Br	16.6–20.0	—

Triple bonds are the highest energy bonds of those listed in Table 14-5, so such bonds absorb at the smallest wavelengths listed. Note that this includes the nitrile C≡N, the acetylenic C≡C, and even the C≡O triple bond of carbon monoxide.

Double bond energies are below those of triple bonds, but about twice those of single bonds, so their absorption bands are located between those of the triple bonds and single bonds. Note the various carbonyl (C=O) absorption bands in the table. Slight differences in location indicate an aldehyde (see Fig. 14-3), a ketone, an acid, or an ester. Gaseous air pollutants like carbon dioxide and sulfur dioxide absorb in the double-bond region.

The absorption bands of C—C single bonds are not very useful because they are so common. However, other single bonds such as C—Cl, C—Br, C—N, and C—O are more significant.

No self-test has been included for infrared absorption because the discussion has been brief. However, you can check your understanding by working the problems at the end of the chapter in the infrared absorption group.

14-2 QUALITATIVE ANALYSIS

Qualitative analysis involves the *identification* of a compound whose identity is unknown or only partially known, as well as *screening* on a routine basis for the presence or absence of significant amounts of a suspected compound. Since both of these facets of qualitative analysis depend on obtaining the proper absorption spectrum or spectra, we will discuss this point first.

Obtaining the Proper Absorption Spectrum

The nature of the sample will help you decide whether to obtain ultraviolet, visible, and/or infrared spectra. For example, if the sample is an aqueous solution of primarily inorganic substances, it will be difficult to obtain an infrared spectrum because the typical cell (sodium chloride, etc.) will be partially dissolved by the water. A *colorless* aqueous solution will obviously not yield any information in the visible region but might give some information in the UV region. The 200–400 nm region should therefore be scanned on the spectrophotometer for one or more absorption bands. A *colored* solution of inorganic and/or organic compounds may absorb in the UV as well as in the visible region, so it should be scanned from 200–750 nm. Any organic sample, whether liquid, solid, or gas, should yield an infrared absorption spectrum which will give some valuable structural information. The infrared spectrum of an inorganic gaseous sample is also frequently useful.

The best and fastest way to obtain an absorption spectrum is to use a spectrophotometer that automatically scans and records the spectrum. If a visible-UV spectrophotometer is not so equipped, then the absorbance should be measured throughout the visible or UV regions in steps of 10 or 20 nm. The absorbance is then plotted on graph paper against wavelength and a curve is drawn to connect all the points. Care should be taken in the region of the absorption band maximum to make enough measurements to locate the maximum within ± 1 to 2 nm.

The exact wavelength location of the *peak* of each absorption *band*, as well as the relative absorbance of each band, is much more important than the appearance of the entire absorption spectrum. In a mixture of several compounds the entire absorption spectrum may not resemble the absorption spectra of any of the individual compounds present, so the peak wavelength(s) will be the most important evidence for the presence of one or more compounds.

Identification

Identification of an unknown compound is usually based on the measurement of *several* physical and/or chemical properties. *One* absorption band or one other physical property is seldom considered as *rigorous proof* of an identity, although it may be enough for *tentative identification*.

If a compound can be isolated in sufficient quantity, a molecular weight determination, an elemental analysis (for C, H, etc.), and at least one absorption spectrum are usually sufficient for rigorous identification. If the compound cannot be isolated in sufficient quantity for the two former measurements, then at least three other measurements may suffice. Not all three of these measurements should be taken from the same spectrum, but other confirming measurements should be obtained. (For example, the fluorescence spectra described in the next chapter may be used to confirm UV absorption spectral evidence.) Of course, the spectrum of the unknown sample should be carefully compared to the spectrum of the pure compound run on the same instrument whenever possible.

Ultraviolet absorption spectra are especially useful in identifying aromatic hydrocarbons in complex mixtures. A good example is the study of Giger and Blumer [2] of the aromatic hydrocarbons present in near-shore marine sediments. The analysis of these sediments is an important indication of the effects of industrial pollution and oil spills on the coastal environment. Because their samples were so complex, they had to separate the large number of compounds into fractions which then had to be identified. They used a *mass spectrometer* to obtain the equivalent of a molecular weight measurement and a characteristic mass spectrum having several peaks for each aromatic hydrocarbon. Two of the aromatic hydrocarbons they identified with the help of ultraviolet absorption spectra and mass spectra were phenanthrene and anthracene. The structure and peak wavelengths of these compounds are given in Table 14-1.

Their samples were especially rich in phenanthrene; the top spectrum in Figure 14-4 is a hypothetical representation of their UV absorption spectrum from their Sampling Station *A*. The identity of phenanthrene is confirmed by the sharp 293 nm peak. This peak is unique among the aromatic hydrocarbons that could possibly be present with the phenanthrene after separation. The 251 nm peak, although part of the same absorption band as the 293 nm peak, is not unique. Anthracene, for example, absorbs strongly in this same region. However, its presence does tend to support the identification in that phenanthrene would be expected to absorb at this wavelength. If the absorption were weaker at 251 nm than at 293 nm, this would lead us to suspect that phenanthrene is not present since its molar absorptivity at 251 is larger than that at 293 (Table 14-1). The phenanthrene is identified, then, *on the basis of several peaks in the mass spectrum, the two ultraviolet absorption peaks, and its molecular weight*.

The bottom spectrum in Figure 14-4 is a hypothetical representation of the ultraviolet absorption spectrum from a possible phenanthrene fraction from Station *B* of Giger and Blumer. The spectrum is "smoothed out" in the 293 nm region, so phenanthrene cannot be identified. However, anthracene is confirmed by the strong peak at 253 nm and three weaker peaks from a different band at 340, 357, and 375 nm. Although the latter peaks are quite weak, they are unique to anthracene. Anthracene is identified *on the basis of the mass spectrum, the four ultraviolet absorption peaks, and its molecular weight*.

[2] W. Giger and M. Blumer, *Anal. Chem.* **46**, 1663 (1974).

SEC. 14-2 Qualitative Analysis 325

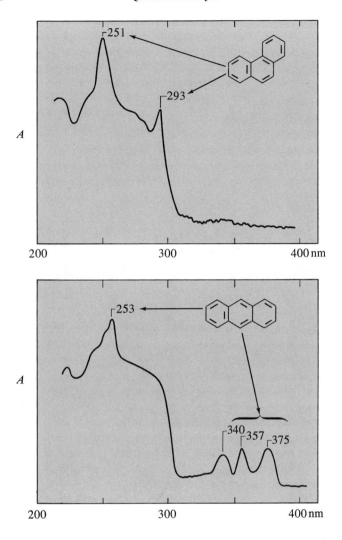

FIGURE 14-4. Hypothetical ultraviolet absorption spectra of some aromatic hydrocarbon separated fractions from near-shore marine sediments analyzed by Giger and Blumer [2]. The spectrum at the top is that of a fraction rich in phenanthrene whose identity is confirmed by the normally sharp peak at 293 nm and supported by absorption at 251 nm. The spectrum at the bottom is that of a fraction rich in anthracene, whose identity is confirmed by the intense peak at 253 nm as well as the three weak peaks at longer wavelengths.

Infrared absorption spectra are very useful for identifying solids, liquids, or gases. Gases will be discussed later in the next section. A good example of the use of infrared spectrophotometry in identifying a compound is the analysis of impure samples suspected of containing heroin. Heroin itself is a tertiary amine.

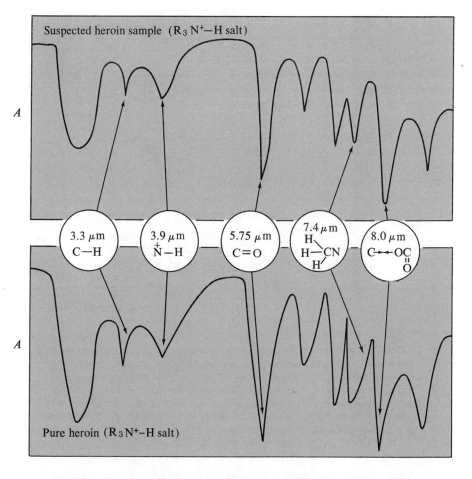

FIGURE 14-5. The infrared absorption spectra of a sample suspected to contain heroin (top) and of a sample of pure heroin (below), both in the salt form so the tertiary amine (R_3N) of heroin exists in a protonated form (R_3N^+—H) which has a characteristic absorption band for the N^+—H bond at 3.9 μm.

It also can exist as a hydrochloride salt in which the pair of electrons on the nitrogen is protonated, giving it a characteristic R_3N^+—H bond.

As stated above, the best way to identify a compound such as heroin is to compare its infrared spectrum with that of a pure sample, as is done in Figure 14-5. You observe that most of the absorption bands of the impure and pure heroin correspond, even though their absorbance values are of course different. A key band is that at 3.9 μm, indicating a protonated tertiary amine N^+—H bond. Most N—H bonds absorb in the 2.82–3.3 μm (Table 14-4). Another key band is the 5.75 μm band indicating an ester carbonyl group (Table 14-5). Heroin has two such groups, both $CH_3C{=}O$ groups. A third key band is the 7.4 μm band indicating a CH_3 group bonded to the tertiary nitrogen; the C—H bonds actually absorb at this wavelength.

Routine Screening

Screening on a routine basis is done for compounds whose identity is already known. All that is desired is that a definite positive test or a definite negative test be obtained. The following factors are important.

1. The test should be rapid to handle large numbers of samples.
2. The test should be simple and easy to interpret, for technicians.
3. The test should involve a simple instrument or a single wavelength measurement.

Screening is especially important in drug analysis and in drug abuse. For example, in the drug abuse area the identity of most drugs, such as barbiturates, are already known. It is only a matter of establishing whether a certain drug is present in a body fluid, such as blood or urine. Both barbiturates (I = acid form) and phenylethylamines (II) such as amphetamine (II, $R{=}CH_3$) can be identified by

routine screening using their ultraviolet and/or infrared absorption spectra. In alkaline solution most barbiturates form a -2 anion having a conjugated π electron system with a strong π-π^* absorption band at 250–275 nm ($\varepsilon = 10^4$). The phenylethylamines are of course substituted benzenes and have a moderate π-π^* absorption band at 247–290 nm ($\varepsilon > 300$).

Phenylethylamines, and amphetamine specifically, stimulate the central nervous system. Barbiturates have a sedative effect on the body because they depress the central nervous system.

An interesting use of screening in the area of drug analysis is the test for the presence of salicylic acid in aspirin tablets. Acetylsalicylic acid (ASA) is slowly hydrolyzed by water in the air or absorbed water to salicylic acid (SA) as follows:

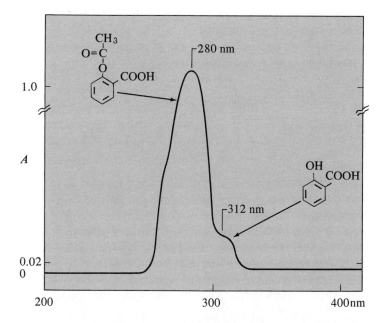

Since acetic acid is somewhat volatile, the main impurity that remains is salicylic acid. Screening for this impurity can be done using ultraviolet spectrophotometry. Although ASA and salicylic acid both have a benzene chromophore, salicylic acid has a π-π^* absorption band at 321 nm, a much longer wavelength than that of ASA (Table 14-1). A measurable absorbance at 312 nm should be a *definite positive test* for salicylic acid (Fig. 14-6).

The Food and Drug Administration (FDA) has established a *tolerance* of 0.10% salicylic acid in unbuffered aspirin. To test whether a particular brand meets this tolerance, a specified quantity of that brand of aspirin is dissolved and its absorbance is measured at 312 nm. If the absorbance exceeds 0.02 (Fig. 14-6),

FIGURE 14-6. Hypothetical ultraviolet absorption spectrum of a solution of an aspirin tablet which contains more than 0.10% of salicylic acid. The salicylic acid impurity absorbs more than the allowable 0.02 units at 312 nm.

then that brand exceeds the tolerance. It is not necessary to run the entire spectrum as shown in Figure 14-6, although it may be done to have a record.

The reason that the tolerance for salicylic acid in aspirin is so low is that salicylic acid is very irritating to the stomach lining, much more so than ASA. Once ASA passes through the stomach, it is hydrolyzed to salicylic acid, so the function of the CH_3—C=O group on ASA is simply to protect the stomach lining. The active, pain-killing form of aspirin is thus salicylic acid, not ASA.

14-3 QUANTITATIVE ANALYSIS

Quantitative analysis involves the determination of the amount of a substance present by spectrophotometric measurement. The substance may already be known to be present, or qualitative analysis using spectrophotometry, etc., may have established its presence. Before discussing the one-time analysis type of sample and the continuous monitoring of a sample whose composition changes, we will first discuss some of the important steps in a quantitative analysis.

Steps in Quantitative Analysis

A quantitative method may be based on measuring the radiant energy absorbed by a species as it exists or on measuring the radiation absorbed by a species after it has undergone a chemical reaction. In the latter case, careful control of the solvent, the reagents added, the reaction time, and the pH is necessary. If these variables are not controlled, deviations from the Beer-Lambert law may result. Each variable should be tested by obtaining a complete spectrum at several values of the quantity being varied.

The choice of wavelength is governed by the importance of measuring at the wavelength at which one can achieve the maximum absorbance of the desired constituent with the minimum error from other constituents that may be present and that absorb in the same general region. If nothing else absorbs at the peak wavelength of the absorption band, then this wavelength would be chosen for analysis.

Figure 14-7 illustrates a situation where choosing the proper wavelength is not as simple as above. Assume first that you are only determining oxyhemoglobin. If you chose either 415 nm or 540 nm as the proper wavelength, your measurement would be in error. The error at 415 nm would be large because bilirubin absorbs strongly there; the error at 540 nm would be small because it absorbs weakly at this wavelength. If there were no other choice, measurement at 540 nm could be used provided that accuracy is not needed. However, the best choice for the analysis would obviously be 575 nm where bilirubin does not absorb at all. If the situation were reversed and bilirubin had to be determined, a correction would have to be made for the absorbance of oxyhemoglobin at 461 nm. This will be discussed later.

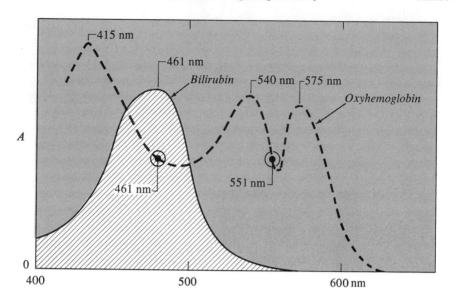

FIGURE 14-7. *Visible absorption spectra of a mixture of bilirubin (shaded area), 30 mg per 100 ml, and oxyhemoglobin, 800 mg per 100 ml, at typical blood concentrations. At the circled points of 461 nm and 551 nm, oxyhemoglobin always has the same absorbance (or same molar absorptivity).*

A plot of absorbance versus concentration should be prepared before the proposed method and wavelength are finally adopted. If such a plot is linear, then the Beer-Lambert law is said to be obeyed (Fig. 13-5). Even if the plot deviates, as shown in Figure 13-5, it still may be used if the deviation is reproducible. As long as an absorbance reading gives an accurate value corresponding to only one concentration, the plot can be used for quantitative work. Occasionally not all of the points will fall exactly on a straight line. In that case the line should be drawn so that an equal number of the points are above and below the line. Any point(s) quite far away from the line should be ignored in drawing the line, or else the point(s) should be rechecked.

In preparing the above plot, conditions should be adjusted to be the same for each solution of known concentration of the constituent to be measured. For example, if iron(III) is to be determined by adding anion X^- to form the colored complex ion FeX_2^+, enough of X^- should be added so that all of the iron(III) is present as FeX_2^+ and not as FeX^{+2}, etc.

One-Time Analysis

A one-time determination of a substance is a sort of "one-shot" analysis in that information is required about a substance whose composition does not change with time or whose composition at a particular time is all that is needed. An example of the former would be the analysis of a drug such as aspirin tablets which have just been bottled at the plant; an example of the latter would be a blood analysis performed in the morning before breakfast foods disturb the levels of chemicals in the blood. A one-time determination would be different than the continuous analysis or monitoring for a substance whose composition may change with time. The latter will be discussed separately.

Many metal ions are determined on a one-time basis because of their stability or because their concentrations do not change rapidly with time. An example of a simple one-time analysis is the colorimetric determination of iron(III) in water by forming a chelate with 1,10-phenanthroline (phen). The characteristics of this chelate were discussed in the first section and listed in Table 14-3. Because this ligand chelates strongly only to iron(II), hydroxylamine is first added to reduce the iron(III) as follows:

$$2Fe^{+3} + 2NH_2OH \rightarrow 2Fe^{+2} + N_2(g) + 2H_2O + 2H^+$$

$(-1$ N ox. st.$)$

The resulting iron(II) is then chelated by the 1,10-phenanthroline.

$$Fe^{+2} + 3 \text{ phen} \rightarrow \text{orange } Fe(phen)_3^{+2}$$

A few ions such as copper(I) also complex with 1,10-phenanthroline, but none is as intensely colored or absorbs as strongly at 512 nm, so the method is essentially free of interferences for water analysis.

The determination of total bilirubin in newborn infants is an example of a somewhat more complicated type of one-time analysis. Bilirubin is a yellow degradation product of hemoglobin in the body; it is responsible for the color of the bile, urine, and for the yellow color accompanying jaundice in newborn infants. It absorbs strongly at 461 nm, but spectrophotometric measurement is complicated by the absorption of oxyhemoglobin at this wavelength also (Fig. 14-7). Since the latter absorbs everywhere the bilirubin does, the absorbance of the oxyhemoglobin must be corrected for.

The correction for oxyhemoglobin is simplified by the fact that the molar absorptivity, and thus the absorbance, of hemoglobin is the same at 551 nm, where bilirubin does not absorb, as it is at 461 nm (Fig. 14-7). Thus, the absorbance of bilirubin at 461 nm can be found from the total absorbance at 461 nm by subtraction.

$$\text{bilirubin } A_{461 \text{ nm}} = [\text{total } A_{461 \text{ nm}} - \text{oxyhemoglobin } A_{551 \text{ nm}}]$$

The concentration of bilirubin can then be found using a plot of the absorbance of bilirubin at 461 nm versus concentration.

A special instrument made by the American Optical Corporation enables medical laboratories to perform the above analysis automatically. It splits the light transmitted by the bilirubin sample into two beams, so the absorbances at 461 and 551 nm are measured simultaneously and the readout is corrected automatically for oxyhemoglobin.

Continuous Monitoring of a Changing Sample

The composition of the air we breathe and the water we drink is constantly changing, however slightly. It is now recognized that the amounts of certain constituents in each must be determined continuously, or monitored continuously, to protect our health.

Monitoring Public Water Supplies. The continuous monitoring of public water supplies is much simpler than monitoring the air because all of the water supply from a given plant must pass through a certain pumping system. A sample of the water flowing past a certain point can be withdrawn continuously and analyzed for a desired constituent by continuous spectrophotometric measurement. Examples of constituents in the water that can be measured continuously are chlorine and the fluoride ion (added as sodium fluoride). The concentration of the chlorine must be monitored so that it does not fall below the level below which the bacteria content of the water would make it unsafe to drink. The concentration of the fluoride must be monitored so that it does not fall below the 1 ppm level needed for protection of the teeth. It also must be monitored so that it does not rise appreciably above this level and poison the water. The monitoring for both chlorine and fluoride is usually done by some type of *repetitive* colorimetric analysis. The analyses are usually done at definite intervals throughout the day; the absorbance readings from the colorimeter may be traced on a circular chart used for a given 24-hour interval.

Monitoring Air. The continuous monitoring of the air in any city is very difficult because the changing wind direction and speed make the sampling of the air prior to analysis difficult. Metropolitan areas such as Los Angeles and New York have as many as 10 to 12 stations at which levels of sulfur dioxide, carbon monoxide, particles in the air, etc., are measured spectrophotometrically and reported to a central headquarters. Air pollution *alerts* are announced on the basis of these rapid spectrophotometric measurements.

In addition to monitoring the air supply, potential sources of pollution are also monitored. *Infrared absorption spectra* are very useful for identifying and measuring harmful constituents in flue gases from plant exhaust stacks. A rapid-scan infrared spectrophotometer can give a complete spectrum in 12.5 seconds for almost continuous checks on the composition of the flue gas (Fig. 14-8).

Harmful gases such as sulfur dioxide and carbon monoxide are both unsaturated molecules and absorb in the *medium energy* infrared region (Table 14-5). Even carbon dioxide can be identified in the spectrum in Figure 14-8. The fact that the spectrum can be run many times not only gives a qualitative analysis but any increase in the absorbance would be a quantitative estimation of an increase in concentration of the gases in the flue.

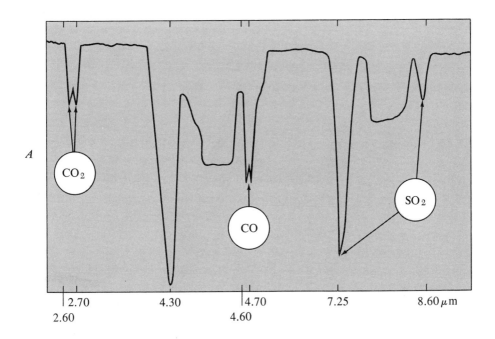

FIGURE 14-8. The hypothetical infrared spectrum of a flue gas similar to that reported by Comberiati [3]. The identity of carbon dioxide is confirmed by the intense band at 4.30 μm and supported by weaker bands at 2.60 and 2.70 μm. The identity of carbon monoxide is established by the two peaks at 4.60 and 4.70 μm of the single band. The identity of sulfur dioxide is confirmed by the intense band at 7.25 μm and supported by the weaker band at 8.60 μm.

14-4 AUTOMATED ANALYSIS

In laboratories where large numbers of different samples must be analyzed for the same constituent(s), the quantitative analytical methods have been automated to obtain the results rapidly. This is especially true in the clinical or hospital laboratory. Many types of instruments are available; we will discuss the common Technicon® AutoAnalyzer® system.

In any automated spectrophotometric measurement, the problem is how to keep the samples separated until the absorbance is measured, and then to measure one sample at a time automatically. In the above instrument, air bubbles, with or without a wash solution, are used to keep the samples separate. The bubble-separated samples are treated if necessary to form a light-absorbing species, and then pumped to a one-piece debubbler and flowcell (Fig. 14-9). The sample "falls" into an elbow as the bubble of air is pumped out through a debubbler exhaust. It then passes into a narrow flowcell which has a volume equal to that of only one sample. Pumping action is delayed long enough for light to pass through the

[3] J. R. Comberiati, *Anal. Chem.* **43**, 1497 (1971).

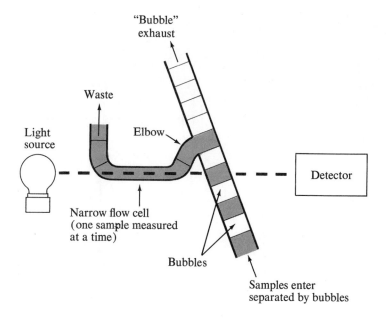

FIGURE 14-9. A typical Technicon® AutoAnalyzer® one-piece debubbler and flowcell. Bubbles leave the light path (dotted line) before the absorbance is measured by the detector. The flowcell is bent so that the light only passes through it and not the sample in the "elbow" behind it.

flowcell and for the absorbance to be measured by the detector. The sample is then pumped out through a waste exit, and the next sample enters and fills the flowcell before being measured.

Automated Determination of Glucose

One of the most common constituents determined in body fluids is glucose, $C_5H_6(OH)_5CHO$, which contains an easily oxidized aldehyde. Many of the reagents employed to determine glucose oxidize it to gluconic acid or condense with the glucose aldehyde group (Sec. 22-2).

$$\text{glucose} + o\text{-toluidine reagent} \xrightarrow{\text{8 min boiling}} \text{colored product (630 nm)}$$

$$\text{glucose} + \text{copper neocuproine} \xrightarrow{\text{heat at 90°}} \text{colored product (460 nm)}$$

In the ferricyanide method glucose reacts with yellow ferricyanide and destroys the color of the ferricyanide by forming the virtually colorless ferrocyanide ion as follows:

$$\text{glucose} + 2\text{Fe(CN)}_6^{-3} \xrightarrow{\text{heat at 95°}} 2\text{Fe(CN)}_6^{-4} + \text{gluconic acid}$$
$$\text{(colorless)}$$

SEC. 14-4 Automated Analysis 335

The decrease in the absorbance of the yellow ferricyanide ion is thus proportional to concentration of the glucose originally present.

A typical AutoAnalyzer® used to determine glucose by any of the above methods is shown in Figure 14-10.

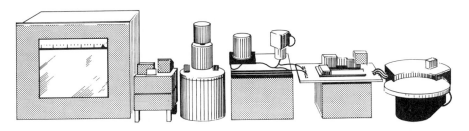

TYPICAL EXAMPLE OF A SINGLE CHANNEL
(here a Blood Urea Nitrogen Analysis)

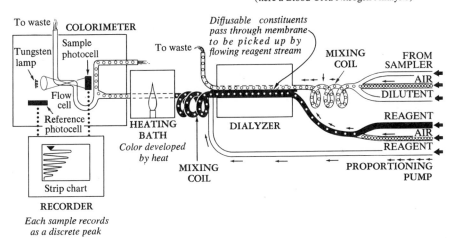

FIGURE 14-10. A single-channel Technicon® AutoAnalyzer® used for analysis of blood glucose or blood urea nitrogen. Each sample is kept separate from the others by means of an air bubble. The absorbance measured by the colorimeter is recorded for each sample on the recorder at bottom left.

Automated Determination of Protein-Bound Iodine (PBI)

Protein-bound iodine is present at such low levels in blood samples that it is measured by first decomposing the blood sample to inorganic iodide ion.

$$R-CH_2-I \xrightarrow{\text{decomp.}} I^- + CO_2$$

The aqueous part of the sample (which contains the iodide ion) is added to a mixture of yellow cerium(IV), arsenic(III), and sulfuric acid. In this acid these two reagents react extremely slowly.

$$2Ce(IV) + As(III) \xrightarrow[\text{catalyst}]{\text{slow without}} 2Ce(III) + As(V)$$

Iodide is an effective catalyst for this oxidation-reduction reaction, so as soon as the sample is added, the reaction rate increases. The decrease in the color of cerium(IV) is measured at 420 nm and this decrease is proportional to the amount of iodide present. None of the other substances present absorb at 420 nm so there are no interferences.

The AutoAnalyzer® can also be used to follow this reaction; a modification of the instrument shown in Figure 14-10 is used. Since the method has been automated it gives much more reliable results than were possible before automation.

QUESTIONS AND PROBLEMS

(Answers to most even-numbered problems are in Appendix 5.)

Definitions and Concepts
1. List and explain the two types of
   a. Qualitative spectrophotometric analysis.
   b. Quantitative spectrophotometric analysis.
2. Explain how each of the following chromophores absorbs radiation.
   a. Electronic chromophore
   b. Vibrational chromophore
3. Explain the difference between an n-π^* and a π-π^* transition.
4. Indicate which bond in each following pair of bonds has the higher vibrational energy and would therefore absorb at shorter wavelengths in the infrared.
   a. C—H and C—C   b. O—H and C—H   c. C=O and C=C
   d. C=O and C=O   e. Ester C=O and ketone C=O
5. Explain how protein-bound iodine is measured spectrophotometrically.

Electronic Chromophores
6. As discussed in the chapter, when the purple MnO_4^- absorbs light a π electron is excited to a higher energy orbital.
   a. It is known that π electron is not excited to a π^* orbital. To what other orbital could the electron be excited? (*Hint:* The manganese in MnO_4^- has how many d electrons?)
   b. Why doesn't the manganese in MnO_4^- act as a d orbital chromophore similar to those ions in Table 14-3?
7. The yellow chromate ion has the structure

$$:\overset{\overset{\displaystyle O^-}{|}}{\underset{\underset{\displaystyle O^-}{|}}{O=Cr=O}}:$$

a. What type of electron can absorb the visible radiation that results in a yellow color?
b. What two different kinds of orbitals could the excited electron occupy in the excited state? If chromate is like permanganate in the previous problem, which orbital is more likely?
c. Why doesn't the chromium(VI) in CrO_4^{-2} act as a d orbital chromophore similar to those ions in Table 14-3? (*Hint:* Find the number of d electrons in chromium(VI).)
8. The gas nitrogen dioxide can be represented by the structure O=N=O
a. Would you predict that it would absorb in the UV-visible region. Why or why not?
b. Absorption of UV radiation by this molecule is thought to be the first step in formation of smog. It decomposes to form a molecule and a reactive atomic species. Suggest what the molecule and atomic species are.
c. Some components of smog are irritating aldehydes. What two species could react to produce aldehydes?

Vibrational Chromophores

9. As a solution of ethanol is diluted from $1.2M$ to $0.025M$, the $3.0\ \mu m$ band of the $1.2M$ solutions decreases in size and a new band appears at $2.8\ \mu m$. At $0.025M$ concentration, the $3.0\ \mu m$ band disappears. Explain the appearance of the new band and the disappearance of the first band.
10. Indicate whether each of the following molecules will absorb or not in the 2.7–$3.0\ \mu m$ region. Explain which bond is primarily responsible.
   a. CH_3CO_2H b. CH_3CHO c. CH_3CH_2OH d. $CH_3-\underset{\underset{O}{\|}}{C}-CH_3$
11. Indicate whether each of the following molecules will absorb or not in the 4.4–$5.90\ \mu m$ region. Suggest which bond is primarily responsible.
   a. CH_3CN b. $CH_3CH_2NH_2$ c. $CH_3C{\equiv}CH$ d. $CH_3CH{=}CH_2$
12. The infrared spectrum of $CH_3-\underset{\underset{O}{\|}}{C}-CH_2-\underset{\underset{O}{\|}}{C}-CH_3$ has a broad absorption band at $3.3\ \mu m$. Explain why this band should be observed when both oxygens appear to be double bonded to carbons; also explain why the band is broad.

Qualitative Analysis

13. 0.400 g of regular aspirin has an absorbance of 0.012 at 312 nm and meets FDA tolerance for salicylic acid. A special 0.600 g time-release aspirin tablet has an absorbance of 0.040 at 312 nm. Does the time-release tablet meet the salicylic acid tolerance for the regular weight of aspirin tablet?
14. A sample suspected of containing heroin exhibits UV absorption in the 250–280 nm region and IR absorption at $3.9\ \mu m$. Does this
   a. Indicate heroin rigorously?
   b. Tentatively indicate heroin?
   c. Fail to indicate heroin?
15. A sample is suspected of being a barbiturate or a phenylethylamine. It exhibits UV absorption at 260 nm and IR absorption at $3.0\ \mu m$, and at 5.8–$6.0\ \mu m$. Suggest what might be present.

Quantitative Analysis

16. A mixture of bilirubin and oxyhemoglobin has an absorbance of 0.412 at 461 nm. The oxyhemoglobin alone has an absorbance of 0.103 at 551 nm. A $1.00 \times 10^{-4} M$ bilirubin standard has an absorbance of 0.400 at 461 nm. Calculate the concentration of bilirubin in the mixture using the absorbance of the standard and its concentration.

17. Suggest a suitable alternative for the ferricyanide ion for the automated determination of glucose. Bear in mind that it should be an oxidizing agent of at least equal or greater standard potential (see Appendix 2) as ferricyanide and that for the purpose of simplicity, its reduced form should be colorless or faintly colored so as not to interfere with the measurement of the unreacted oxidizing agent.

18. A mixture of iron(III) and thiocyanate ion reacts very slowly to produce iron(II). Researchers studying this reaction were able to follow the rate of the reaction by complexing the iron(III) with fluoride ion and forming a color by adding a suitable complexing agent to form a color with the iron(II). Suggest a suitable complexing agent for this purpose.

19. In 1966 German researchers found that inorganic nitrates in tobacco could be determined by UV spectrophotometry after extraction. However, impurities in the solution absorbed both in the 203–210 nm region and at 232 nm. Suggest two possible approaches for determining the nitrates without error from the impurities.

Challenging Problems

20. The molar absorptivity of the $FeSCN^{+2}$ ion is 5000, but the molar absorptivity of the $Fe(SCN)_2^+$ ion is 9800. Suggest why the molar absorptivity of the second ion might be twice that of the first. What are the implications of this for quantitative analysis for iron(III)?

21. The following compounds absorb at the listed wavelengths, with the molar absorptivity given in parenthesis: CH_3I, 257 nm (378); CH_3Br, 202 nm (80); CH_3Cl, 177 nm (small molar absorptivity).
    a. Suggest a reasonable electronic transition to account for all of the bands.
    b. Rationalize the trends in the location and molar absorptivity.

15. Fluorometry, Refractometry, and Light Scattering

Even now in the stillness of death, the huge jaws seemed to be dripping with a bluish flame.... I placed my hand upon the glowing *muzzle and as I held them up my fingers* gleamed *in the darkness. "Phosphorus," I said. "A cunning preparation of it," said Holmes.*

ARTHUR CONAN DOYLE
The Hound of the Baskervilles, Chapter 14

In the previous chapters we have described the measurement of absorption of radiation. In this chapter three other measurements of the interaction of radiation with solutions are discussed—fluorometry, refractometry, and the measurement of light scattering.

These measurements are related in that each is a measurement of radiation after it has been *affected* in some way by interacting with a species in solution. In contrast, spectrophotometry involves measurement of radiation which has *not* interacted with a species in solution.

Fluorometry is concerned with the measurement of the fluorescence emitted by a species which has absorbed incident radiation. The fluorescence is emitted in *all directions* and is of longer wavelengths and lower energies than the absorbed radiation. As shown in Figure 15-1, not all of the incident radiation is absorbed; some of it passes through the solution unchanged. Only that which is re-emitted changes in wavelength.

Refractometry involves measuring the *change in direction* of incident radiation as it passes from air into a solution or solid. As shown in Figure 15-1, the incident radiation is not absorbed; it interacts with the solution species in such a

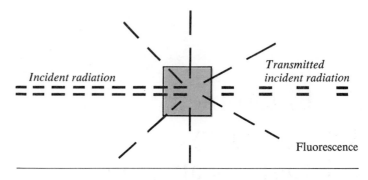
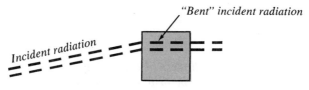
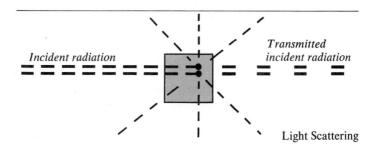

FIGURE 15-1. *Schematic representations of fluorescence (top), refraction (middle), and light scattering (bottom). The length of the line (photon) represents the wavelength. Only in the case of fluorescence (top) does the wavelength change, becoming longer. The solid dots in the bottom portion of the figure represent suspended solids which scatter the indicident radiation.*

way that only its direction, not its wavelength, is changed. If only refraction occurs, we can say that none of the photons of the incident radiation are absorbed; they merely undergo a change in direction.

Measurement of light scattering is the measurement of the scattering of photons of the incident radiation by solid particles suspended in the solution. Some of the incident radiation passes through the solution (Fig. 15-1), but much of it is scattered in all directions, *without a change in wavelength*.

We will begin our discussion of these topics with a discussion of fluorometry.

15-1 THE FLUORESCENCE OF MOLECULES AND IONS

The nature of fluorescence and the measurement of fluorescence are both complicated; in this chapter we are presenting only the essentials—a simplified picture of the energy changes occurring during fluorescence, the fluorescence-concentration relation, and a few applications out of thousands.

You should review the discussion of absorption of photons in Section 13-1 before reading the material below, and also review the components of a spectrophotometer in Section 13-3.

Energy Changes Occurring During Fluorescence

The absorption process discussed in the two previous chapters is the first step in fluorescence. In this step a molecule or ion is *excited* by a *photon* of radiation; this is strictly referred to as *photoexcitation*. The emission of photons following photoexcitation is in general called *photoluminescence*, or less accurately, luminescence. Fluorescence is one type of photoluminescence; phosphorescence, which is beyond the scope of this chapter, is another type. In a photoluminescent process such as fluorescence, the emitting species always reaches the excited state by absorbing a photon.

> *Chemiluminescence and bioluminescence are processes which differ from a photoluminescence process like fluorescence in that chemical energy is used to promote the emitting species to the excited state. For example, the firefly exhibits bioluminescence as a result of chemical reactions, producing a luminescent species.*

The exciting radiation in fluorescence is usually ultraviolet (UV) radiation, but occasionally visible radiation is used. The excitation promotes the ground state ion or molecule to the excited state. The entire process is summarized simply as

$$\text{ground st.} + \text{UV} \xrightarrow{\text{excitation}} \text{excited st.} \xrightarrow{\text{emission}} \text{ground st.} + \text{fluor.}$$

In general, the energy of fluorescence radiation is much lower than that of the exciting radiation. Since wavelength is inversely proportional to energy, fluorescence radiation is located at longer wavelengths than the exciting radiation. Usually fluorescence emission is visible to the eye, but it may occur in the UV above 300 nm also. Just as substances in solution absorb over a range of wavelengths (see the absorption *bands* of hemoglobin in Fig. 13-2), fluorescent ions or molecules in solution emit a range of wavelengths. A fluorescence spectrum can be obtained using a *spectrofluorometer* just as an absorption spectrum is obtained using a spectrophotometer.

A typical fluorescence spectrum is shown in Figure 15-2 for the organic molecule *quinine*, an alkaloid used to treat malaria. The left spectrum is a plot of the UV radiation that will excite quinine and is called an *excitation spectrum*.

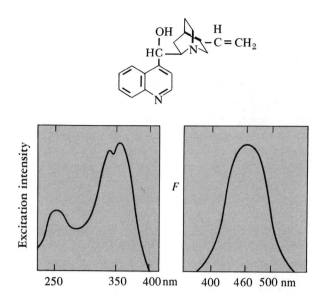

FIGURE 15-2. *The fluorescence excitation spectrum (left) and fluorescence emission spectrum (right) of the organic molecule quinine. The structure of quinine is given at the top.*

The right spectrum is a plot of the radiation emitted by excited quinine and is called a fluorescence emission spectrum or just a *fluorescence spectrum*.

To explain why quinine or other species emit fluorescence at longer wavelengths (lower energies) than the radiation they absorb, it is necessary to employ a molecular energy diagram. As shown in Figure 15-3, such a diagram displays the electronic ground state and its vibrational sublevels and the electronic excited state and its vibrational sublevels in order of increasing energy. (Recall from Fig. 14-2 and the discussion of IR absorption that an electronic state is divided into many vibrational energy sublevels.)

The Excitation Process. This process is the same as absorption; it begins with all molecules in the ground state, symbolized as S_0 in Figure 15-3. The electronic transition can be represented by one equation,

$$S_0 + \text{UV photons} \rightarrow S_1$$

where S_1 is the symbol for the first excited state. If vibrational sublevels are included in a description of this process, then *many* transitions must be written to describe excitation accurately. Just three are shown in Figure 15-3—the lowest energy transition to the EV_0 sublevel of S_1, a middle energy transition to any middle energy EV_x sublevel of S_1, and the highest energy transition to the highest energy, EV_n, sublevel of S_1. The many transitions to the many sublevels of the S_1 state thus account for the excitation *band* such as that of quinine in the left spectrum of Figure 15-2.

SEC. 15-1 The Fluorescence of Molecules and Ions

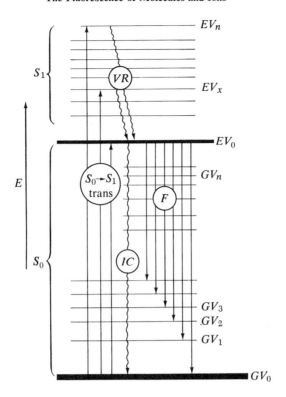

FIGURE 15-3. A molecular energy diagram showing excitation (absorption) of the ground state, S_0, to the excited state, S_1, and fluorescence (F) from the lowest vibrational sublevel, EV_0, of the excited S_1 state. After excitation to various vibrational sublevels of S_1, all molecules relax vibrationally (VR) to the EV_0 sublevel of S_1. Many of the molecules emit fluorescence and return to various ground vibrational sublevels (GV_2, GV_1, GV_0, etc.). Many molecules do not fluoresce but internally convert (IC) their energy to vibrational energy and return to the S_0 state with loss of heat.

Relaxation: Loss of Energy. What follows the excitation process largely explains why fluorescence is a lower energy process than excitation (absorption). Molecules or ions in all S_1 vibrational sublevels except EV_0 *relax* to the EV_0 sublevel, losing part of their energy as heat. Since the relaxation occurs through the vibrational motions of the bonds, it is called vibrational relaxation (*VR* in Fig. 15-3). At this point all the S_1 molecules have less energy than when they were excited, except for the few originally excited to the EV_0 sublevel. The energy lost as heat is transferred to solvent molecules during collisions.

The Fluorescence Emission Process. Any molecule in the EV_0 sublevel of S_1 potentially can fluoresce by emitting a photon; the electronic transition can be represented by one equation as follows:

$$S_1 \rightarrow S_0 + \text{photons (fluorescence)}$$

However, as can be seen in Figure 15-3, S_1 molecules undergo *many* transitions, emitting photons of varying energies. Just six are shown in Figure 15-3, and all but one emit photons of lower energy than needed to excite an S_0 molecule. This shows why fluorescence is of lower energy and longer wavelengths than absorption (Fig. 15-2) and why a fluorescence *band* is observed.

Molecules that drop to sublevels such as GV_3, GV_2, etc., do not remain in these sublevels but lose their excess vibrational energy by relaxation. This energy is lost as heat to the solvent.

Internal Conversion: A Competing Process. Not all molecules, and very few ions, fluoresce in solution. There are many reasons for this, including processes that compete with fluorescence. Internal conversion, shown as *IC* in Figure 15-3, is one such process. In this process an S_1 species interacts strongly with the solvent, converting its electronic energy to vibrational energy which is lost to the solvent.

Fluorescence Instrumentation

The two types of fluorescence instruments are the filter fluorometer and the spectrofluorometer. The chief difference between them is that the former utilizes filters as monochromators and the latter uses gratings as monochromators. Both of these were discussed in Chapter 13; Figure 13-6 shows the optical characteristics of a typical filter that can also be used in a filter fluorometer.

The basic design of both instruments is similar, and their four essential components are shown in Figure 15-4. The source is one of the various mercury arc lamps or a xenon arc lamp; the latter is used in the spectrofluorometer. The radiation from the source passes through an excitation grating or primary filter, either of which allows only a certain wavelength range to excite the sample. As the radiation is absorbed by a substance in the sample cell, it emits fluorescence in all directions. Fluorescence emitted at right angles to the exciting radiation passes through a fluorescence grating or secondary filter, either of which transmits only a certain wavelength range. The transmitted fluorescence is measured by a detector and the detector output is recorded or displayed on a meter.

Filter Fluorometer Measurements. Filter fluorometers are primarily used for quantitative analysis; because they are simple to use, they are well suited for routine work. The source is a mercury arc lamp, either an 85-watt medium-pressure lamp (366, 405, 436 nm) or a 4-watt low-pressure lamp. The latter may be of the clear quartz envelope type for emitting the *254 nm mercury line*, or of the white phosphor coated type for emitting a *300–400 nm band* of radiation. A filter fluorometer is thus limited to the above wavelengths for excitation.

Exciting radiation is selected by using a *primary filter*. The bandwidth transmitted by the band pass filters, which are used as primary filters, is similar to that of the filter in Figure 13-6—about 40–60 nm. The wide bandwidth and higher transmittance of such a filter is one of the advantages of a filter fluorometer over a spectrofluorometer—P_0, the radiant power of the radiation striking the cell, is larger than in the spectrofluorometer. (Recall that P_0 is measured in photons/s.)

The sample cell may be a glass test tube, square glass cell, optical grade quartz cell, or synthetic silica cell. Glass cells transmit ultraviolet radiation down to about 320 nm, but quartz or silica cells must be used below 320 nm.

As shown in Figure 15-4, fluorescence is emitted in all directions, but only that emitted *at right angles* to the exciting radiation is measured. This minimizes scattering of the exciting radiation through the slit and secondary filter to the detector. Since this cannot be eliminated, the secondary filter is chosen so it will transmit at longer wavelengths than the exciting radiation. Thus the secondary filter *absorbs* any scattered exciting radiation.

Secondary filters can be of two types—the *band pass* filter, also used as a primary filter, and the *sharp cut* filter. The latter transmits nearly all radiation above a given wavelength. The sharp cut filter will thus transmit a larger number of fluorescent photons/s than the band pass filter. As shown in Figure 15-4, the symbol

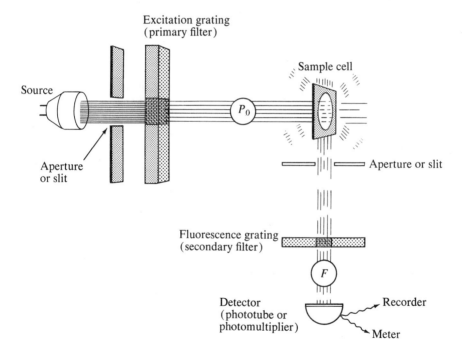

FIGURE 15-4. Schematic diagram (top view) of the components of a fluorometer (filter fluorometer or spectrofluorometer). The source is either a mercury arc or xenon arc lamp. As shown, the excitation grating or primary filter transmits only a portion of the radiation emitted by the source. Most of the exciting radiation passes through the sample cell without being absorbed. The radiation that is absorbed causes the sample to fluoresce in all directions, but only the emission that passes through the aperture or slit and through the secondary filter or fluorescence grating is measured by the phototube, or photomultiplier. The output of the detector is either measured on a meter or plotted on a recorder.

F is used for the number of fluorescent photons/s passing through the secondary filter and striking the detector. F is also used to refer to the intensity of fluorescence read from the meter or recorder. The units for F are usually arbitrarily set from 0–100.

Spectrofluorometer Measurements. A spectrofluorometer has basically the same type of design as a filter fluorometer, so only the differences will be mentioned. The source is a xenon arc lamp; this emits continuously from 200 nm to above 600 nm. It can thus be used to obtain spectra like those in Figure 15-2, because the amount of excitation at each wavelength and the intensity of fluorescence at each wavelength can be measured. This cannot be done using mercury arc excitation, because none of the mercury arcs emits continuously in the 200–600 nm range.

Both the exciting radiation and the emitted fluorescence are selected by using gratings. Usually the grating bandwidth is small and constant, such as 10 nm. Thus the radiant power, P_0, of the radiation striking the cell in the spectrofluorometer will be less than that in a filter fluorometer. The same will be true for F, the number of fluorescent photons transmitted by the emission grating. The advantage of using a grating is that you can select any given wavelength for excitation and for emission. Thus if two fluorescent molecules A and B are present in a mixture, A can be measured in the presence of B using one of the following techniques.

1. If A absorbs radiation in a region where B doesn't, select a unique excitation wavelength so that only A is excited. Only A will fluoresce and be measured. (This can sometimes be done on a filter fluorometer by choosing the correct primary filter.)
2. If both A and B absorb in the same region, but A emits fluorescence at different wavelengths than B, then select an excitation wavelength that will be more favorable for A, if possible. Then select a unique fluorescence emission wavelength (where only A emits). When the mixture is excited, only the fluorescence of A will be measured. (This can sometimes be done using a secondary filter also.)

For example, suppose you must determine molecule A in the presence of quinine (Fig. 15-2). Assume that molecule A is excited in the same regions as quinine, but that it emits a fluorescence band from 450–560 nm with a peak at 520 nm. Since quinine does not emit above 520 nm, molecule A can be measured at 520 or 530 nm without error from quinine.

As for the filter fluorometer, the fluorescence intensity readout on the spectrofluorometer is symbolized as F. The units for F also range from 0 to 100 on the recorded fluorescence spectrum or meter readout.

The Relation Between Concentration and F

The symbol F has already been defined as the intensity of fluorescence measured by an instrument and read from the meter or graphical output of the instrument. F is directly proportional to the molarity, c, of a fluorescent species even though not all of the fluorescence from that species is actually measured by any instrument.

This is because at low concentrations F is a *constant* fraction k of the total fluorescence transmitted by a specific grating or secondary filter.

$$F \cong k \text{ (total fluorescence)} \quad (15\text{-}1)$$

The total fluorescence is related approximately to c as follows:

$$\text{total fluorescence} \cong \varepsilon bc(2.3 P_0 \phi_f) \quad (15\text{-}2)$$

The ϕ_f term is the *quantum efficiency*, or *fraction* of excited states that actually emit fluorescence. Substitution of **15-2** into **15-1** gives the usual equation relating concentration and F.

$$F \cong k\varepsilon bc(2.3 P_0 \phi_f) \quad (15\text{-}3)$$

Equation **15-3** can be derived using the Beer-Lambert law by methods beyond the scope of this text [1].

Equation **15-3** shows mathematically how the various experimental variables affect values of F read from an instrument. If concentration is varied and all other variables are held constant, then a plot of F vs. c should give a straight line (Fig.

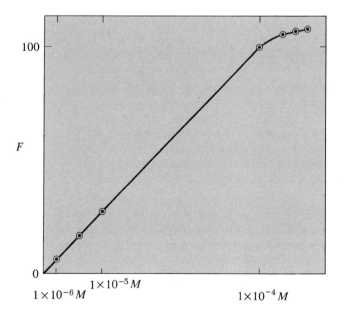

FIGURE 15-5. *A typical plot of F, fluorescence intensity vs. concentration. The plot is linear for most compounds until the solution becomes too concentrated; it bends downward typically above $10^{-4}M$ for weak fluorescers, but not for all compounds. Concentrations much lower than $10^{-6}M$ can be measured for many intensely fluorescent compounds.*

[1] G. H. Schenk, *Absorption of Light and Ultraviolet Radiation: Fluorescence and Phosphorescence Emission*, Allyn & Bacon, Boston, 1973, pp. 159–78, 219–42, 267–68.

15-5). As is the case for plots of absorbance *vs.* c (Fig. 13-3), deviations from linearity do occur at higher concentrations in plots of F *vs.* c. The reason for the downward curvature in Figure 15-5 is that the approximation in **15-1** is no longer valid. What happens is that the solution is so concentrated ($\geqslant 10^{-4} M$) that a disproportionate amount of excitation occurs at the front portion of the cell where the fluorescence cannot reach the detector. This effect is called the *inner filter effect*; it is discussed in detail in more advanced texts [1, 2].

It can be seen from **15-3** that *increasing* any of the variables on the right side of the equation will increase F. For example, k can be increased by using a wider bandwidth grating or secondary filter, ε can be optimized by exciting at a wavelength where ε is at a maximum, b can be increased by using a somewhat larger cell, and P_0 can be increased by using a wider bandwidth grating or primary filter or by using a source with a higher radiant power. The effect of these variables is best explored by working the following self-test. This self-test is *not* recommended for the below-average student; this student should concentrate on mastering the general details of fluorescence first before going into the specific details covered in the self-test.

Self-Test 15-1. Fluorescence-Concentration Relation

Directions: Work one problem at a time and check the answers given below.

A. Predict the effect of b, the cell path length on F, if all other variables in **15-3** are held constant and
   a. The cell path length is increased from 1.0 cm to 2.0 cm.
   b. The cell path length is decreased from 1.0 cm to 1.0 mm.
B. Predict the effect of ε on F if all other variables in **15-3** are held constant and
   a. Excitation is changed to a wavelength at which $\varepsilon = 1 \times 10^4$ from excitation at a wavelength at which $\varepsilon = 1 \times 10^3$.
   b. Excitation of anthracene (Table 15-1) is changed to 375 nm ($\varepsilon = 9 \times 10^3$) from excitation at 253 nm ($\varepsilon = 1.5 \times 10^5$).
   c. Excitation of phenanthrene (Table 15-1) is changed to 251 nm from excitation at 293 nm.
C. Predict the effect of k on F if all other variables in **15-3** are held constant and
   a. Fluorescence emission is measured using a grating instead of a secondary filter.
   b. Fluorescence emission is measured using a sharp cut secondary filter instead of a band pass secondary filter.

[2] C. A. Parker, *Photoluminescence of Solutions*, Elsevier, New York, 1968, pp. 220–34; A. L. Conrad, *Treatise on Analytical Chemistry*, Part I, Vol. 5, Wiley Interscience, New York, 1964, pp. 3057–78.

SEC. 15-1   The Fluorescence of Molecules and Ions   349

D. Predict the effect of P_0 on F if all other variables in **15-3** are held constant and
   a. Excitation is done using a primary filter instead of a grating.
   b. Excitation is done using a spectrofluorometer instead of a filter fluorometer.
   c. A sample is excited at 254 nm with a mercury arc source instead of a xenon arc source, which is very weak in the 200–270 nm region.

Answers to Self-Test 15-1

A. a. F is doubled.   b. F is reduced to 0.1 of that in a 1 cm cell.
B. a. F is increased tenfold.   b. F is reduced to 0.06 of that at 253 nm.
   c. F is increased by a factor of 3.3.
C. a. F is decreased.   b. F is increased.
D. a. F is increased.   b. F is decreased.   c. F is increased, since P_0 of the mercury arc at 254 nm is very large compared to P_0 of the xenon arc.

Types of Molecules and Ions That Fluoresce

A number of organic molecules and inorganic ions fluoresce. In most cases it is the species which are *rigid to some degree* that appear to fluoresce. This will be emphasized as each of these two classes are discussed.

Organic Molecules. Most of the organic molecules that fluoresce are aromatic hydrocarbons or molecules that contain aromatic rings. Even molecules with rings having one or two double bonds such as cyclohexadiene do not fluoresce. Whenever at least one benzene ring is present in a molecule, it has the potential to fluoresce. Apparently, internal conversion (Fig. 15-3) to the ground state is too fast for fluorescence to be significant until the carbon-carbon bonds in the conjugated π system are *all rigid*. Then when the π system is excited, it can lose *some* of its energy by fluorescence instead of by bond vibrational energy increases in the internal conversion process.

Aromatic hydrocarbons, being rigid molecules in general, are usually intensely fluorescent. Benzene is the least rigid because it has only one ring; its quantum efficiency (Table 15-1) is only 0.01 in a solvent exposed to air. This means that only 1% of excited benzene molecules ever emit fluorescence. Because of this, benzene is rarely determined by fluorescence; however, the larger, more rigid aromatic hydrocarbons are commonly determined by fluorescence. These include the others listed in Table 15-1—anthracene, phenanthrene, and benzo[a]pyrene.

TABLE 15-1. Some Fluorescent Organic Molecules (Solvent = Ethanol)

Organic molecule	ϕ_f, room temp.	λ_{ex}, nm	λ_{em}, nm
Aromatic hydrocarbons			
Benzene	0.01 (in air)	250	270–310
Anthracene	0.3	254, 375	380–450
Phenanthrene	0.1	251, 265, 293	340–420
Benzo[a]pyrene	>0.4	366 (520[a])	400–500 (540–550[a])
Aromatic amino acids (H_2O solvent)			
Phenylalanine	<0.04	254	270–310
Phenylalanine	Chem. rxn[b]	365	500–530
Tyrosine	0.2	225, 280	280–340
Tryptophan	0.2	220, 280	310–420
Substituted benzenes			
Aniline	0.1	280	310–405
Phenol	0.2	270	285–365
Salicylic acid	≥0.1	310	400–550[c]
Acetylsalicylic acid	0.02[d]	280[d]	310–390[d]
Miscellaneous			
Quinine	0.55	366	400–500
Morphine	Chem. rxn[e]	254	390–420

[a] Sulfuric acid solvent.
[b] Ninhydrin reaction produces a more fluorescent product.
[c] Solvent is 1% acetic acid-chloroform.
[d] Only fluorescent in acetic acid-chloroform type solvents.
[e] Oxidation produces fluorescent pseudomorphine.

The three aromatic amino acids listed in Table 15-1 are all determined by fluorescence also. Phenylalanine, whose structure and determination will be discussed in detail later, is a very weakly fluorescent molecule like benzene. Tyrosine and tryptophan have the following structures.

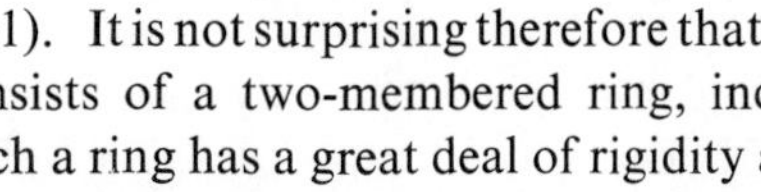

Tyrosine 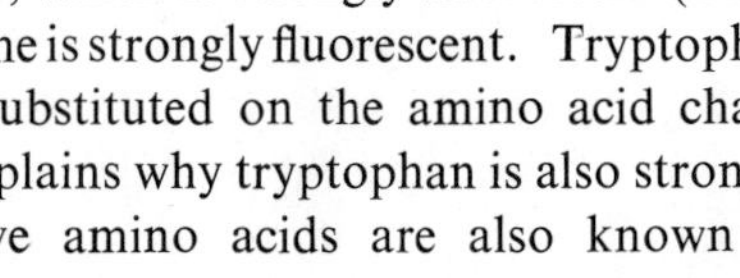   Tryprophan

Tyrosine is very similar structually to phenol, which is strongly fluorescent (Table 15-1). It is not surprising therefore that tyrosine is strongly fluorescent. Tryptophan consists of a two-membered ring, indole, substituted on the amino acid chain. Such a ring has a great deal of rigidity and explains why tryptophan is also strongly fluorescent. Proteins containing the above amino acids are also known to fluoresce [3].

Although benzene itself is very weakly fluorescent at room temperature, substitution of electron-donating groups on the benzene ring increases the fluor-

[3] F. W. J. Teal, *Biochem. J.* **76**, 381 (1960).

escence, probably as a result of increased rigidity, among other factors. Molecules such as aniline, phenol, and salicylic acid fluoresce intensely. In contrast, acetylsalicylic acid (ASA) does not fluoresce under most conditions, so salicylic acid impurities in aspirin tablets may be measured fluorometrically as well as spectrophotometrically (Ch. 14). When a small percentage of acetic acid, such as 1%, is added to a solvent such as chloroform, the ASA also fluoresces [4].

Many molecules that are nonfluorescent or weakly fluorescent can be converted to fluorescent species by adding a reagent to change their structure. Morphine is very weakly fluorescent, but can be oxidized to a highly fluorescent molecule called *pseudomorphine*. This will be discussed in detail later.

Inorganic Ions. Very few unchelated metal ions fluoresce in solution because they lose their absorbed energy to water molecules that bond strongly to them. The O—H bonds "soak up" the absorbed electronic energy by converting it to vibrational energy. The inorganic ions that do fluoresce in solution at room temperature are UO_2^{+2}, the uranyl ion; Ce^{+3}, the cerium(III) ion; and Tl^+, the thallium(I) ion. Boric acid, H_3BO_3, also reacts with benzoin to form a rigid, green, fluorescent, addition compound.

The largest class of fluorescent inorganic ions is made up of those that form relatively rigid chelates with organic ligands. The electrons on the organic ligand are excited, not the electrons on the metal ions. However, the metal ion chelated by the organic ligand enables the system to be rigid enough to fluoresce rather than to undergo internal conversion. Metal ions that form such fluorescent chelates are those in Group IIA—Mg(II), Ca(II), and Sr(II)—and those in Group IIIA—Al(III), Ga(III), and In(III). Details of the determination of such metal ions may be found in advanced references [1, 5].

Al(III) can be determined using many different organic ligands. One of these is 8-hydroxyquinoline, which forms the following fluorescent chelate.

$$\left[\begin{array}{c} \text{quinoline-O:} \\ \text{N:} \end{array} \right]_3 Al^{+3}$$

Such chelates have been used for the determination of from 10^{-6}–$10^{-8} M$ Al(III) in drinking water, boiler water, and other water samples.

Qualitative Fluorometric Analysis

As we pointed out in the previous chapter, qualitative spectral analysis consists of both *identification* and *screening*. *Identification* of a specific compound usually

[4] C. I. Miles and G. H. Schenk, *Anal. Chem.* **42**, 656 (1970).
[5] C. E. White and R. J. Argauer, *Fluorescence Analysis—A Practical Approach*, Dekker, New York, 1970.

involves obtaining the excitation and fluorescence emission spectra on a spectrofluorometer and comparing them with that of a pure sample of the compound. For example, if a compound is suspected of being the alkaloid quinine, then the excitation and fluorescence emission spectra can be obtained and compared with spectra similar to those in Figure 15-2. Excitation bands at 250 nm and 350 nm, as well as a fluorescence emission band at 460 nm, would be good evidence that the compound is indeed quinine.

Identification in some cases may be accomplished by the observation of fluorescence without obtaining both excitation and fluorescence spectra. The detection of oil spills on the oceans or lakes is an example. All oils contain many fluorescent aromatic hydrocarbons. Since natural water does not fluoresce, an oil spill can be located by observing fluorescence with the proper excitation. Investigations have been made of the use of lasers operated from airplanes to excite oil spills, and also of the use of buoys containing a low-pressure mercury arc source [6]. In the case of the buoys, the mercury arc emitted 254 nm UV radiation which excited fuel oil spilled during unloading at a tanker dock. A detector in the buoy was adjusted to give an alarm only when the fluorescence emission resembled that of the fuel oil, not when fluorescence from outboard motor oil, etc., was detected.

Screening for specific compounds can in many cases be done very simply by using a filter fluorometer rather than an expensive spectrofluorometer. The screening for morphine in body fluids such as urine is a good example. Morphine has the following structure.

(Note that its structure is similar to that of heroin; see the discussion of IR identification in Sec. 14-2.)

Morphine contains a *phenol chromophore*; that is, the left ring is a substituted phenol. It would therefore be expected to fluoresce like phenol (Table 15-1). Thus it could be excited with 254 nm radiation from a mercury arc source in a filter fluorometer. The difficulty is that body fluids contain other compounds that also fluoresce with 254 nm excitation. Because morphine is a metabolite of heroin, its concentration is not always such that its fluorescence intensity will be significant compared to that of the other compounds in the urine. This difficulty is overcome by oxidizing the morphine with ferricyanide ion to the intensely fluorescent pseudomorphine. The probable reaction is

$$2\text{Fe}(\text{CN})_6^{-3} + 2 \text{ Morphine} \rightarrow \text{Pseudomorphine} + 2\text{Fe}(\text{CN})_6^{-4}$$

[6] *Chem. & Eng. News* **52**, 30 (Mar. 25, 1974).

The pseudomorphine is a dimer of morphine, because ferricyanide is known to oxidize phenols (R—OH) to dimers of the type R(OH)—R(OH).

The actual screening test is based on filter fluorometric measurement of the pseudomorphine after obtaining a blank reading. The blank reading is made before adding ferricyanide, thus giving the fluorescence morphine itself plus other compounds present in the urine. The ferricyanide ion reagent is added and the fluorescence intensity is again measured after two minutes. Since pseudomorphine is twenty times as fluorescent as morphine, a significant increase in fluorescence intensity definitely indicates morphine. No increase in fluorescence intensity indicates the absence of morphine.

A simple morphine analyzer is commercially available which utilizes disposable glass cells rather than the expensive quartz cells that are normally used with 254 nm excitation. This analyzer is a filter fluorometer which utilizes a mercury source positioned above the glass cell so that the 254 nm radiation passes through the open top of the cell. Thus although glass absorbs 254 nm radiation in the usual fluorometer, it does not in this case. The fluorescence emission, being in the 400 nm range, is transmitted through the glass and is measured by the detector.

Quantitative Fluorometric Analysis

The quantitative analysis of a fluorescent compound in a mixture is accomplished by using either one of the two techniques discussed in connection with fluorescence instrumentation. Either a unique excitation band or a unique fluorescence emission band must be used for accurate analysis. A good example is the determination of the carcinogenic aromatic hydrocarbon, benzo[a]pyrene in samples of particles from polluted air. The structure of benzo[a]pyrene is

It is a very rigid, highly fluorescent aromatic hydrocarbon (Table 15-1). Unfortunately, polluted air contains many other aromatic hydrocarbons that emit fluorescence in the same region as the benzo[a]pyrene. To determine it without separation, a solvent of sulfuric acid is used. In this acid benzo[a]pyrene is excited by light of 520 nm rather than by the ultraviolet (Table 15-1). Although a few other aromatics in polluted air absorb very weakly at this wavelength, none emits at 540–550 nm [7]. It was determined in the presence of 40 other compounds by the technique of measuring a *unique fluorescence band.*

[7] E. Sawicki et al., *Int. J. Air. Poll.* **2**, 273 (1960).

Quantitative analysis may also be carried out using a unique excitation band. An example of this is the determination of phenylalanine, whose structure is

$$\langle\!\!\!\bigcirc\!\!\!\rangle\text{---}CH_2\text{---}\underset{NH_2}{\overset{H}{\underset{|}{C}}}\text{---}CO_2H$$

Fluorometric measurement of phenylalanine is useful in testing for phenylketonuria, a hereditary metabolic disorder that can result in mental retardation. Phenylketonuriac babies cannot convert their phenylalanine efficiently to tyrosine, so a high phenylalanine blood level indicates this disorder. Since phenylalanine is not fluorescent enough (Table 15-1), it is treated with ninhydrin, copper(II) ion, and L-leucyl-L-alanine [8] to give a fluorescent species which is excited at 365 nm. This is a *unique excitation band* because tyrosine and tryptophan are not excited at this wavelength (Table 15-1). The fluorescence from the species is at 515 nm, in the green region.

For other examples of quantitative analysis, you should work the following self-test.

Self-Test 15-2. Quantitative Fluorometric Analysis Problems

Directions: Suggest a fluorometric *technique*, specifying *both* excitation wavelength and range of fluorescence emission wavelengths for the determination of the stipulated compound. One of the two must be unique (see preceding and p. 313).

A. Aspirin tablets contain mainly acetylsalicylic acid (ASA) and salicylic acid (SA). After the tablets have been dissolved in 1% acetic acid-chloroform, suggest a fluorometric technique for the determination of
   a. ASA (see Table 15-1).
   b. SA (see Table 15-1).
B. A mixture of aromatic hydrocarbons contains both benzene and anthracene (see Table 15-1). Suggest a fluorometric technique for the determination of anthracene by means of
   a. A unique excitation wavelength.
   b. A unique fluorescence wavelength range only.
C. A mixture of aromatic hydrocarbons contains both benzene and phenanthrene (see Table 15-1). Suggest a fluorometric technique for the determination of phenanthrene by means of
   a. A unique excitation wavelength.
   b. A unique fluorescence wavelength range only.

[8] P. K. Wong, *Clin. Chem.* **10**, 1098 (1964).

D. A mixture of aromatic hydrocarbons contains benzene, anthracene, phenanthrene, and benzo[a]pyrene. Suggest fluorometric techniques for the determination of as many of these compounds as possible without separation.

Answers to Self-Test 15-2

A. a. Excite both at 280 nm; measure ASA only at 310–390 nm.
    b. Excite only SA at 310 nm; measure SA only at 400–550 nm.
B. a. Excite at 375 nm where benzene does not absorb; measure at 380–450 nm.
    b. Excite both at 250–254 nm where both absorb; measure anthracene only at 380–450 nm.
C. a. Excite at 293 nm where benzene does not absorb; measure at 340–420 nm.
    b. Excite both at 250–251 nm where both absorb; measure phenanthrene only at 340–420 nm.
D. Determine benzo[a]pyrene by dissolving the mixture in sulfuric acid, exciting at 520 nm and measuring it only at 540–550 nm. Determine anthracene by exciting it plus benzo[a]pyrene at 375 nm and measuring it only at 380–395 nm.

15-2 THE MEASUREMENT OF REFRACTION AND REFRACTIVE INDEX

Refraction of radiation occurs when a beam of photons changes in velocity as it passes from air into a solution. As a result of the change in velocity, the direction of the beam also changes. Refractometry is the measurement of this change in direction; it is mainly concerned with the measurement of the refractive index, n_i, of a solution or pure solvent. This is defined as

$$n_i = \frac{c}{v_i} \tag{15-4}$$

where c is the velocity of radiation in air (strictly speaking, in a vacuum) and is constant, and v_i is the velocity of the beam of photons in the solution or pure solvent.

This subject is a large one, like fluorescence, and we are presenting only the essentials—a simplified picture of what occurs during refraction, a brief introduction to the refractometer, and a few applications out of many.

Changes Occurring During Refraction

The refraction of a beam of photons does not involve a change in energy of the photons or of the solution through which the photons pass. As shown in Figure

15-1, the photons in the beam are not absorbed. The interaction of the photons with the molecules in solution is such that only the direction and the velocity of the photons are changed, not the wavelength. Since the interaction slows down the photons traveling through a solution, the index of refraction of the solution will be larger than one according to **15-4**.

Since the refractive index depends on the wavelength of radiation passing through a solution and on the temperature of that solution, the wavelength is specified with a subscript and the temperature with a superscript. The yellow D line (589 nm) from a sodium vapor lamp is most often used as a source of photons and measurements are commonly taken at 20° or 25°C, so n_D^{20} is the common term for refractive index.

The refraction of radiation by molecules in the liquid state depends on the number of atoms per molecule, their atomic weights, and the structure of the molecule. The larger the atomic weight of an atom or the larger the number of atoms in a molecule, the larger the refraction and the refractive index. Evidently a beam of photons interacts more with a larger electronic system, thus reducing its velocity to a smaller value, which in turn results in a larger index of refraction **(15-4)**. Thus in a series of *pure* liquids such as CCl_4, $CHCl_3$, and CH_2Cl_2, the larger chlorine atoms, rather than the smaller hydrogen atoms, have the greatest effect on the index of refraction. As you can see in Table 15-2, the order of decreasing

TABLE 15-2. Refractive Indices of Some Pure Liquids

Compounds	Formula	n_D^{20}
Chloroalkanes		
Carbon tetrachloride	CCl_4	1.4573
Chloroform	$CHCl_3$	1.4426
Methylene chloride	CH_2Cl_2	1.4237
Alcohols		
Ethyl alcohol	C_2H_5OH	1.3590
Ethylene glycol	$C_2H_4(OH)_2$	1.4318
Glycerine (glycerol)	$C_3H_5(OH)_3$	1.4740
Others		
Acetic acid	CH_3CO_2H	1.3715
Propronic acid	$C_2H_5CO_2H$	1.3874
Butyric acid	$C_3H_7CO_2H$	1.3979
Water	H_2O	1.3328

refractive index is n_D^{20} CCl_4 > n_D^{20} $CHCl_3$ > n_D^{20} CH_2Cl_2. The same is true for the series of alcohols listed in the table; again, the relatively large oxygen atoms cause the index of refraction of glycerine, $C_3H_5(OH)_3$ to be larger than that of the other two alcohols.

It follows from the preceding discussion that a mixture of two liquids will have a different refractive index than either of the pure liquids. The actual refractive index will depend on the relative concentration of the two liquids. In a more general sense, a solution will have a different refractive index than the pure solvent because the solute molecules will refract the beam of photons to a different degree than the solvent.

Since water is the most common solvent, let us consider the refraction of radiation in aqueous solutions briefly. Note that water, being a small molecule, has a rather small refractive index compared to the organic molecules in Table 15-2. Aqueous solutions of any of the organic molecules in Table 15-2 would be expected to have larger refractive indices than that of pure water. The same should be true for most inorganic solutes since they are composed of larger atoms than water. As the concentration of most solutes in water is increased, we would expect that the refractive indices of such solutions would increase also. This is true, as can be seen for the plot of refractive index *vs.* concentration of glycerine in Figure 15-6. In fact, the refractive index in this case increases in direct proportion to the concentration of glycerine, as evidenced from the straight line.

It is not always true that the refractive index *vs.* concentration plot will be a straight line as it is in Figure 15-6. Because of interactions between water and solute molecules, such plots will frequently be curved. However, the refractive index generally increases as the concentration of the solute increases, even in these cases.

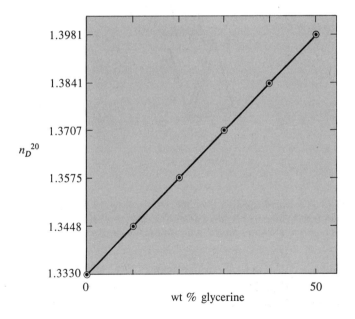

FIGURE 15-6. *A plot of n_D^{20} vs. weight per cent glycerine for aqueous solutions of glycerine. Under the conditions shown water is the solvent. Above 50 wt %, glycerine would be in excess and would become the solvent.*

The Refractometer

The refractometer is used to measure the refractive index of liquids. Most refractometers are based on the use of so-called *Amici prisms* in which the light beam is bent to give a dark and a light field separated by a sharp boundary as viewed through a telescopelike eyepiece. The most common refractometer is the Abbe refractometer shown in Figure 15-7. It is equipped with hollow prism casings

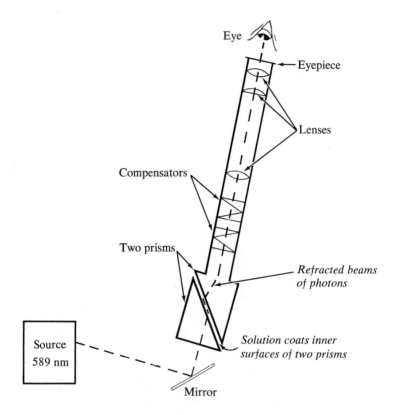

FIGURE 15-7. *A schematic diagram of an Abbe refractometer. The liquid whose refractive index is to be measured is added dropwise to the lower prism in the open position; the two prisms are then adjusted to the closed position seen above.*

through which water at a constant temperature, such as 20°C, is pumped to maintain a constant temperature. (Recall that the refractive index depends on the temperature.) Instead of having a yellow sodium source of 589 nm radiation, the Abbe refractometer has a white light source and a special prism arrangement which gives a beam with the same path as that of 589 nm radiation.

To measure the refractive index of a solution or pure liquid, one puts a drop of it on the lower prism and adjusts the prisms to the closed position shown in

Figure 15-7. A beam of photons is reflected off the mirror through the prisms and liquid sample and thence to the eye. After an adjustment of the viewer, the eye sees a dark and a light field with a sharp boundary. The prism angle is then changed until the sharp boundary coincides with the intersection of the cross hairs on the eyepiece. The refractive index is then read from a scale (not shown) to four decimal places.

Other types of refractometers are also important. A *dipping refractometer* is used for large amounts of samples in a beaker or other large container. For the continuous measurement of fractions from a liquid chromatographic column (Ch. 22), a *differential* type of refractometer is employed. This compares the refractive index of a solvent containing a solute leaving the column with that of the pure solvent. The difference in the two refractive indices is plotted by a recorder to indicate the relative concentrations of various different solutes leaving the column.

Qualitative Analysis

The refractive index of a *pure* unknown liquid is very useful for the identification of the compound when used with other data such as boiling point, elemental analysis, and absorption spectra. Since the presence of another compound can change the refractive index appreciably, the compound should be carefully purified before taking the measurement. If the presence of a functional group can be verified, it can make the analysis much simpler. For example, suppose that it has been established that the unknown contains one carboxylic acid group, $-CO_2H$, per molecule and that the refractive index is 1.3698. It is very possible that the compound is acetic acid (Table 15-2). Even though the measured refractive index is lower than that of pure acetic acid, it can be seen from the table that the higher molecular weight acids have higher, not lower, refractive indices. Since water has a lower refractive index than that of acetic acid, it is possible that a trace of water is present, causing the refractive index of the unknown to be lower than that of pure acetic acid.

Quantitative Analysis

The measurement of the refractive index of two-component mixtures is often the simplest and most rapid method of analysis. If more than two components are present, analysis is probably not accurate by this means, however. All that is needed is a plot of refractive index *vs.* concentration of the solute component, such as that shown in Figure 15-6. After the refractive index of the unknown sample has been carefully measured, its concentration is read from the plot.

Since larger molecules in solution refract light to a greater degree than small molecules, it is not surprising that many high molecular weight compounds have been determined in solution by refractometry. The classic example is the analysis of aqueous sugar solutions for dissolved sucrose, $C_{12}H_{22}O_{11}$. The sucrose content of many commercial preparations, such as syrups, may be found routinely by measuring their refractive indices. The presence of other disaccharides such as lactose or maltose will not cause a significant error because they have the same formulas and about the same refractive index.

Refractometry can be used in clinical chemistry to estimate the specific gravity of urine. The refractive index of the urine is approximately linearly related to the total solids per unit volume, and is therefore related to the specific gravity [9].

15-3 MEASUREMENT OF LIGHT SCATTERING

Scattering of radiation occurs when a beam of photons changes its direction after colliding with solid particles in solution (Fig. 15-1). If there are a large number of particles, more of the photons will be scattered and less of the photons will pass through the solution in the original direction. If there are relatively few particles, less of the photons will be scattered and more of them will pass through the solution. In any case there is *no change* in the wavelength of the scattered radiation as occurs in the fluorescence process (Fig. 15-1).

Two methods of analysis are based on scattering of light. *Turbidimetric* analysis is based on the measurement of the decrease in radiant power $(P_0 - P)$ of the incident radiation. *Nephelometry* is based on the measurement of the radiant power of the radiation scattered at right angles to the incident radiation. Both methods will be discussed briefly.

Measurement of Changes Resulting from Scattering

Light scattering occurs in a *reproducible manner* when a beam of photons of constant radiant power, P_0, passes into a *stable* suspension of solid particles in solution. As shown in Figure 15-8, a certain fraction of the photons, of radiant power P, pass through the solution without being scattered. A detector placed in the path of these photons will be able to measure both P_0 and P; such an instrument will function as a *turbidimeter*. The loss in radiant power $(P_0 - P)$ is related to the "concentration" c of the suspended particles as follows:

$$\log(P_0 - P) = \log\frac{P_0}{P} = kcb \tag{15-5}$$

in which b is the path length in cm of the cell, and k is a proportionality term which is hopefully constant over the concentration range of interest. A plot of the log term in **15-5** *vs.* c will be linear if k is constant, but will be curved if k varies with concentration.

Equation **10-5** is utilized in turbidimetric analysis in the same way as the Beer-Lambert law is utilized in spectrophotometric analysis. Values of $\log(P_0/P)$

[9] R. J. Henry et al. *Clinical Chemistry*, Harper and Row, New York, 1974, p. 1545.

SEC. 15-3   Measurement of Light Scattering   361

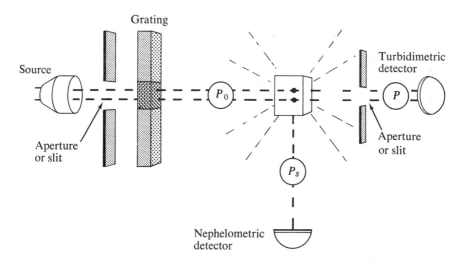

FIGURE 15-8. *A schematic diagram of a turbidimeter-nephelometer. When the detector measures the transmitted radiation P, the instrument functions as a turbidimeter. When the detector measures the scattered radiation, P_s, the instrument functions as a nephelometer.*

are measured for several standard solutions of varying concentrations of suspended particles, and a plot is made of $\log(P_0/P)$ vs. c. The unknown sample's $\log(P_0/P)$ is then measured and the concentration of the suspended particles in the unknown sample is read from the plot.

The relationship between concentration and the radiant power, P_s, of the scattered radiation for measurements in a *nephelometer* is different than that in 15-5. In this type of measurement (Fig. 15-8) the radiant power reaching the detector is directly proportional to the concentration.

$$P_s = k'bc \qquad (15\text{-}6)$$

where k' is a proportionality term which is constant over a limited range of concentration. This expression is similar to **15-3** for fluorescence in that the measured radiant power in each case is directly proportional to the concentration of the species involved. A plot of P_s vs. c will be linear if k' remains constant over the range of concentrations involved.

Analytical Applications

The relative magnitudes of P, P_0, and P_s will determine whether a suspension is analyzed by nephelometric or turbidimetric means. If relatively few particles are present such that the difference between P and P_0 is very small, turbidimetric analysis will be inaccurate and a nephelometric measurement should be used. If a relatively large number of particles are present, then the difference between P and P_0 will be large enough to be measured accurately using a turbidimetric method.

Both methods are useful for the determination of traces of sulfate ion, for which there are few convenient spectrophotometric methods. Barium(II) is added to produce a suspension of barium sulfate. Nephelometry is used to measure sulfate as low as a few parts per million and turbidimetry is utilized to measure higher concentrations. The lead(II) ion interferes by forming insoluble lead sulfate.

Turbidimetric methods are important in pollution analysis. Water, whether polluted or used for drinking, contains many suspended particles which may be measured by turbidimetry.

QUESTIONS AND PROBLEMS

(Answers to most even-numbered problems are in Appendix 5.)

Definitions and Concepts
1. Explain the difference between
    a. Absorption and fluorescence.
    b. Fluorescence and light scattering.
    c. Light scattering and refraction.
    d. Absorption and light scattering.
2. Explain the difference between a spectrofluorometer and a filter fluorometer.
3. What is the inner filter effect in fluorescence? Does it have anything to do with an optical filter?
4. Explain how morphine can be determined in a glass fluorescence cell, even though it is excited by 254 nm radiation which is absorbed by glass.
5. Which compound in each following group should have the smallest index of refraction, and why?
    a. CH_3CH_2Cl, $CHCl_2CH_3$, CCl_3CH_3
    b. $CHBr_3$, CBr_4, CH_2Br_2
6. Explain the difference between a turbidimetric and a nephelometric method of analysis.

Fluorometry
7. Which instrument will give the most efficient excitation of anthracene (absorbs as in Table 15-1)—a filter fluorometer with a clear envelope low pressure mercury arc lamp, or a spectrofluorometer?
8. A standard solution of phenylalanine containing 3.0 mg/100 ml has an F reading of 25 units. On the same fluorometer, an unknown solution containing phenylalanine gives an F reading of 21 units. Calculate the concentration of phenylalanine in the unknown.
9. When a solution of anthracene is excited at 375 nm using a xenon arc in a spectrofluorometer, it gives a low F reading. Name several ways to increase that reading.
10. Calculate the ratio of F of anthracene to F of the same concentration of phenanthrene at 252 nm where their respective molar absorptivities are 1.5×10^5 and 5×10^4.

16. Emission and Absorption of Radiation by Gaseous Atoms

> *"Really, Watson, you excel yourself,"* said Holmes.... *"It may be that you are not yourself* luminous, *but you are a conductor of* light."
>
> ARTHUR CONAN DOYLE
> The Hound of the Baskervilles, Chapter 1

In this chapter we will see how the interaction of *light* or ultraviolet radiation with gaseous atoms can be used for chemical analysis. We will present the two most common of a number of analytical methods based on the excitation of atoms in the gaseous state. Since radiation is measured in both methods, both are types of *spectrometry*. (We will use this term whenever we refer to measurement of radiation involving atoms in the *gaseous state*, reserving the term *spectrophotometry* for the measurement of absorption of radiation by species *in solution*.)

The two spectrometric methods are flame emission spectrometry (an atomic emission method) and atomic absorption spectrometry. Flame emission spectrometry was formerly called flame photometry and is still referred to by that name in much of the literature. It is based on the measurement of the amount of radiation emitted by atoms in a flame. It is much used in clinical laboratories for the determination of sodium and potassium ions in body fluids.

Atomic absorption spectrometry is a more recent method, and it is also a more general method for the determination of metal ions than flame emission spectrometry. It is not as applicable to the determination of alkali metal ions such as sodium and potassium. It is based on the measurement of the amount of radiation absorbed by gaseous atoms usually, but not always, in a flame.

16-1 EXCITATION AND EMISSION PROCESSES IN ATOMS

Before we consider flame emission spectrometry and atomic absorption spectrometry, we should discuss briefly the nature of excitation and emission processes in atoms. Because the determination of sodium ions is so important and is such a common analysis, we will describe the excitation and emission of sodium atoms. In our discussion we'll assume that you are familiar with the terms *electronic ground state* and *electronic excited state* as introduced in Section 12-1.

Events Occurring in a Flame

Since both flame emission and atomic absorption spectrometry usually involve spraying or atomizing an aqueous sample into a flame prior to analysis, we will describe the entire sequence of events in a flame for both methods. Briefly, once a solution of sodium chloride is placed into an atomizer below the flame, the following events take place.

1. The sodium chloride solution is sprayed into the flame in the form of small droplets.
2. The water is vaporized, leaving a fine suspension of solid sodium chloride.
3. The solid sodium chloride is partially vaporized to gaseous sodium atoms and gaseous chlorine atoms.
4. Some of the sodium atoms ionize and form oxides.
5. a. Some sodium atoms are thermally excited to the excited state.
   b. In atomic absorption spectrometry, some sodium atoms absorb photons to reach the excited state.
6. The excited sodium atoms emit photons or heat and return to the ground state.

A more detailed description of these events follows. At stage 1 (Fig. 16-1) sodium chloride solution passes through the capillary of a *nebulizer* located inside the burner. The burner gas helps sweep the sample into the flame, and the solution is dispersed in the form of fine droplets, so a fog or aerosol is produced. In each droplet, each sodium chloride is of course surrounded by water molecules.

At stage 2 (Fig. 16-1) the droplets enter the flame and the water is completely vaporized, momentarily leaving gaseous sodium chloride in ionic form. This is quickly dissociated (stage 3) to gaseous atomic chlorine and gaseous atomic sodium in its ground state (Na_0).

$$NaCl(g) \rightarrow Cl(g) + Na_0$$

It is important to recognize that it is atomic sodium, with its outer $3s$ electron intact, that is produced rather than ionic sodium, which has no outer $3s$ electron.

Some of the atomic sodium may ionize and possibly form oxides, but these reactions are incidental at this point. At stage 4 (Fig. 16-1) atomic sodium in its electronic ground state is excited to its first excited state (Na_1). In flame emission

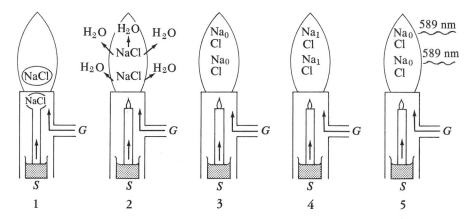

FIGURE 16-1. *Schematic diagram of the sequence of events in a flame for sodium chloride sample, S.*
1. *Sodium chloride sample, S, transported into flame by burner gas, G.*
2. *Water evaporates from around sodium chloride.*
3. *Atomic ground state sodium and chlorine produced.*
4. *Atomic excited state sodium produced.*
5. *Atomic sodium returns to ground state, emitting radiation.*

spectrometry this is accomplished thermally; i.e., heat from the flame excites the sodium.

$$\text{Na}_0 + \text{heat} \rightarrow \text{Na}_1 \qquad (16\text{-}1)$$

For most atoms this is not very efficient, and most of the sodium atoms remain in the ground state. In atomic absorption spectrometry the excitation is accomplished more efficiently in general by exciting the sodium with a beam of photons, some of which are absorbed.

$$\text{Na}_0 + 589 \text{ nm photons} \rightarrow \text{Na}_1 \qquad (16\text{-}2)$$

In atomic absorption the measurement occurs at this point; the amount of photons absorbed are measured and related to concentration. Light of 589 nm wavelength is used because this wavelength has the exact amount of energy needed to promote ground state sodium to its first excited state. (Some thermal excitation occurs **(16-1)**, but this is not measured.)

At stage 5 (Fig. 16-1) atomic sodium returns to its electronic ground state. The excess electronic energy of the excited state is dissipated in two ways. Some of the sodium atoms emit photons, primarily at 589 nm, and some lose energy thermally, i.e., by transferring heat to the surroundings.

$$\text{Na}_1 \rightarrow \text{Na}_0 + 589 \text{ nm photons} \qquad (16\text{-}3)$$

$$\text{Na}_1 \rightarrow \text{Na}_0 + \text{heat} \qquad (16\text{-}4)$$

In flame emission spectrometry the measurement is performed at this point

(16-3). The 589 nm emission is classified as a *line emission* because a single wavelength is emitted, rather than a *band*. This is sometimes called the *principal* line of sodium since it is the most intense line of all those emitted by sodium. (It is actually split into two very close lines, at 589.0 and 589.6 nm, but for convenience we refer to the two lines as one line.)

In atomic absorption spectrometry the measurement has already been made at this point, and it is immaterial what happens to the excited sodium atoms. Of course, they return to the ground state via the reactions in **16-3** and **16-4**.

In flame emission spectrometry the temperature of the flame is critical, more so than in atomic absorption spectrometry. The flame must be hot enough not only to form atoms but also to excite the atoms thermally to an excited state. In a given type of flame, the relative amounts of ground state and excited state atoms will vary widely from element to element. Whether flame emission or atomic absorption is the better method to use will depend in part on the element.

Other Sodium Lines

Atoms can exist in more than one electronic excited state. This makes it possible for atoms such as sodium to exhibit other lines in addition to the principal (most intense) line. Some of the other sodium lines involve both the ground state and one excited state. These are termed *resonance* lines since the atom can *resonate* back and forth directly between the ground state and the excited state by absorbing or

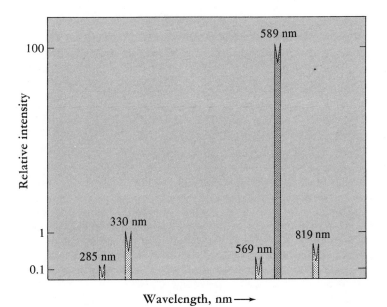

FIGURE 16-2. *Hypothetical flame emission spectrum of the sodium atom in a hot flame.*

emitting at this line. Other sodium lines involve two different excited states and are rather weak lines. They can only be used in flame emission spectrometry because atomic absorption spectrometry involves absorption by the ground state atoms only.

A hypothetical sodium emission in a hot flame is shown in Figure 16-2. The principal, or first, resonance line at 589 nm is of course the most intense, giving the flame a yellow, or yellow-orange color. Other resonance lines are at 330 and 285 nm. Weaker, nonresonance lines are at 819 and 569 nm. Many of the weaker lines are not observed in a cooler flame, such as the Bunsen flame, because of the

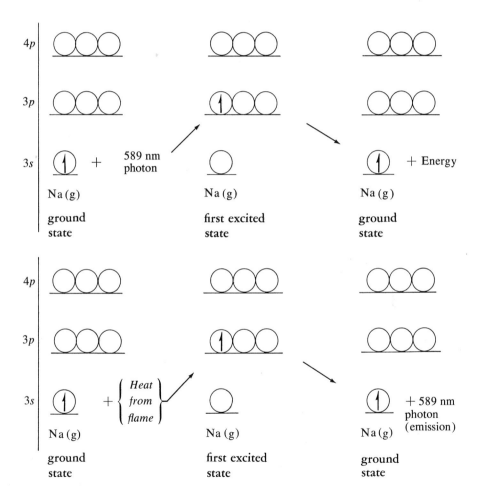

FIGURE 16-3. Electronic orbital diagram of the excitation of sodium. Above: excitation by 589 nm light (as in atomic absorption spectrometry). Below: thermal excitation of sodium atoms (as in flame emission spectrometry) with emission of 589 nm light.

smaller thermal energy available for excitation in such a flame. (Most of the lines are split like the 589 nm line into two very close lines.) Atomic absorption measurements are generally made only at 589 and 300 nm, these being the two most intense resonance lines.

Electronic Orbital Picture

Although it is not essential, you may be interested in a simple electronic orbital picture of the 589 nm line of sodium. This is shown in Figure 16-3.

The electron that is excited thermally by the flame or by absorbing 589 nm radiation is the $3s$ electron of atomic sodium. (Note that the sodium ion does not possess a $3s$ electron and thus cannot undergo such an electronic excitation.) Of the various higher energy orbitals available to it, the excited electron can most easily occupy a $3p$ orbital. This represents the electronic configuration of the first excited state—$1s^2 2s^2 2p^6 3s^0 3p^1$. The electron can occupy higher energy orbitals, such as the $4p$, if more heat can be absorbed or if higher energy radiation can be absorbed. Because the excited state is unstable, the electron loses its excess electronic energy quickly and again occupies the $3s$ orbital in the ground state.

16-2 FLAME EMISSION SPECTROMETRY (FLAME PHOTOMETRY)

Flame emission spectrometry (formerly known as flame photometry) is the instrumental measurement of the amount of radiation emitted by excited atoms in a flame. It is now one of the most general methods of analysis, because flames have been found to excite atoms of essentially all the metals, including the rare earths. At the same time, it can be used with very simple instrumentation for thousands of rapid, routine determinations of sodium and potassium ions in body fluids every day in clinical laboratories.

The instrument used to measure the amount of emission is generally known as a *flame atomic emission spectrometer*. (Some older models are still called flame photometers.) The spectrometer (Fig. 16-4) consists of a nebulizer, a burner and two gases, a monochromator, a detector system, and a readout device. The sample is sprayed by the nebulizer into the burner, carried into the flame, atomized, and excited. The emission from the excited atoms passes into the monochromator where the appropriate wavelength is selected for measurement. The intensity of that emitted wavelength is measured by the detection system, and the units of intensity indicated on the readout meter. We will discuss each component in a little more detail.

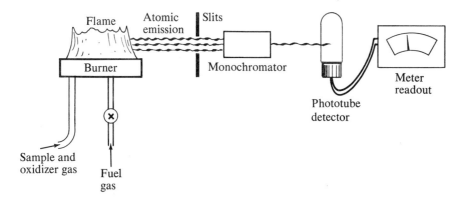

FIGURE 16-4. *Schematic diagram of a flame emission spectrometer.*

Instrumentation

We begin our description of each component with a discussion of the nebulizer and burner system.

The Nebulizer-Burner System. The nebulizer and burner are operated essentially as one unit. The purpose of the nebulizer is to spray a sample into the gas stream and nebulize (fragment) it into a fine mist or aerosol. The burner serves to mix the *fuel gas* and the *oxidizer gas* in proper proportions to produce a stable flame hot enough to atomize and excite the sample.

There are two broad types of burners, each of which utilizes the nebulizer in a slightly different manner. These are the *premixed burner* and the *turbulent burner* (or sprayer-burner) [1]. A schematic diagram of the premixed burner and its nebulizer is shown in Figure 16-5. The sample is sucked into the nebulizer and sprayed into a mixing chamber where it is mixed with the oxidizer gas. The larger drops are trapped and removed in this chamber; only the fine drops are swept into the burner by the oxidizer gas. Before entering the burner, the sample-oxidizer gas mixture is premixed with the fuel gas. The mixture then ignites as it passes into the flame. The flame produced is quite stable but has the disadvantage that not all of the sample enters the flame.

In the turbulent burner the fuel gas and the oxidizer gas are mixed at the point they enter the flame. The sample passes through its own channel via a capillary tip into the flame also. Because all of the sample enters the flame, the burner is a *total consumption* burner. However, nebulization is less efficient than in the premixed burner and large drops of sample are carried into the flame.

The type of fuel gas-oxidizer gas mixture used in either burner is important for achieving proper flame temperature. The fuel gas must be a compound that

[1] R. D. Dresser, R. A. Mooney, E. M. Heithmar, and F. W. Plankey, *J. Chem. Educ.* **52**, A403 (Sept., 1975).

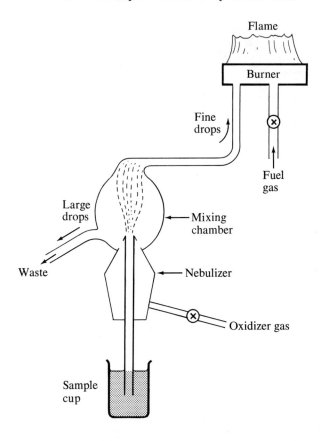

FIGURE 16-5. *Schematic diagram of premixed burner with nebulizer.*

can be burned in a controllable manner to give a reproducible flame, as well as a hot flame. Table 16-1 lists a few typical mixtures.

The natural gas used in Bunsen burners has such a low flame temperature that it is capable of exciting primarily the alkali metals (Group IA) and some of the alkaline earth metals (Group IIA). For that reason, hydrogen and acetylene have replaced the natural gas as fuel gas. The acetylene-nitrous oxide flame is used for

TABLE 16-1. *Typical Gases for Flames [1]*

Fuel gas	Oxidizer gas	Maximum flame temp., °K
Natural gas	air	1700°
Acetylene (C_2H_2)	air	2500°
H_2	O_2	2940°
Acetylene	N_2O	2970°
Acetylene	O_2	3400°

the determination of most metals because it prevents oxide formation and other interfering reactions. This gas mixture is not used for alkali metals because it produces a low fraction of atomic metal in the flame (only 0.3 of all sodium species exists as atomic sodium for example). Instead an acetylene-air flame is used for the alkali metals (over 0.6 of all sodium species exists as atomic sodium).

The Monochromator. There are two types of monochromators used—the optical filter and the grating (Sec. 13-3). The optical filter is used in simpler instruments such as those made for the clinical determination of sodium and potassium in body fluids. The grating is used in the more sophisticated instruments for better resolution of emission lines in more complicated mixtures. The latter instruments also have an adjustable slit width to aid in resolving lines.

The Detector System. The simpler instruments may be equipped with a photocell or a phototube (Sec. 13-3). The more sophisticated instruments are equipped with an electron multiplier phototube, or photomultiplier.

The Meter Readout. The meter readout is usually in arbitrary units such as 0 to 1000 or 0 to 100 intensity units. On flame spectrometers consisting of a nebulizer-burner attached to a spectrophotometer, the per cent transmittance scale is used for the readout. A reading of $0\%T$ corresponds to no flame emission, and that of $100\%T$ corresponds to a maximum flame emission intensity.

Flame Emission Analysis

We will first discuss the nature of the various methods of determining the concentration of metal ions in solution, and then discuss the various groups of metal ions that can be determined.

Direct Determination. In the direct determination method the emission intensity of the unknown, I_u, is directly compared with the emission intensity of one or more standards, I_s. The simplest approach is to measure the emission intensity of just one standard solution, one that is fairly close in concentration to that of the unknown. Then, the concentration of the unknown, c_u, may be calculated from the known concentration of the standard, c_s, as follows:

$$c_u = c_s \frac{I_u}{I_s}$$

This approach is generally not used because the emission intensity does not always vary in a linear manner as the concentration increases. Unless the concentration of the unknown is quite close to that of the standard, a serious error may result.

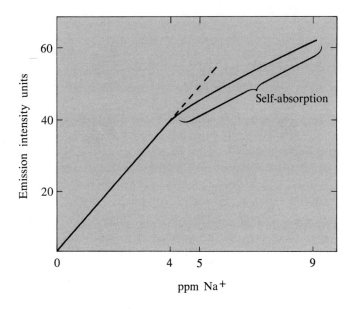

FIGURE 16-6. *Calibration plot for sodium ion, showing self-absorption deviation from linearity above 4 ppm sodium ion.*

A better approach is to plot the emission intensity of a series of standard solutions *vs.* the concentration of these solutions (Fig. 16-6). As you can see, the emission intensity of sodium at 589 nm increases in a linear manner until about 4 ppm sodium ion; then the emission intensity shows a negative deviation from linearity. Above this concentration some of the sodium atoms in the outer cone of the flame are absorbing some of the 589 nm emission from excited sodium atoms in the inner cone of the flame. No error will result if the concentration of the unknown is read from the calibration plot.

Internal Standard Method. If there are interferences present which cannot be eliminated, or whose effect on the emission cannot be corrected for, the internal standard method can be used. This involves adding to each unknown a fixed amount of another metal ion, called the internal standard. The emission wavelength of the internal standard should be fairly close to that of the metal ion being measured, and the emission process should be similar. The *ratio* of the emission intensity of the ion determined to that of the ion used as internal standard is plotted against the concentration of the metal ion being measured, using log-log graph paper. This usually cancels out the variations in flame and compositions of the standards and unknown.

Standard Addition Method. In this method intensity readings are obtained for two solutions—a solution containing an aliquot of the unknown only, and a solution containing the same aliquot of the unknown as well as a measured volume of a standard solution of the same metal ion. The concentration of the metal ion in each solution is then read from a calibration plot such as that in Figure 16-6. If there is no interference, then subtracting the concentration found for the unknown from the concentration should give the concentration of the standard added. If there is an interference, then the foregoing will not be true. A correction factor of the ratio of the standard concentration added divided by the standard concentration found can be multiplied times the concentration of the unknown found, to give the actual concentration of the unknown.

Metal Ions Commonly Determined

Flame emission spectrometry can be used to determine essentially any of the metallic elements, but not nonmetals such as chloride, bromide, sulfide, etc. There are a number of metal ions for which flame emission is more sensitive [2] than atomic absorption spectrometry (Sec. 16-3). These are

Group IA: Li, Na, K, Rb
Group IIA: Ca, Sr, Ba
Group IIIA: Al, Ga, In, Tl
Transition metals: Ru, W, Re
Rare earths: Eu, Ho, La, Lu, Nd, Pr, Sm, Tb, Tm, Yb

A few of these metal ions are deserving of further comment.

Determination of Sodium and Potassium. The flame emission determination of sodium and potassium ions in the clinical laboratory is by far the most important flame measurement. These ions are determined in body fluids such as blood serum and urine. Their concentrations in urine vary widely, but the concentration ranges of these ions in the blood serum of normal adults is

$$Na^+: 1.38 \times 10^{-3}\text{--}1.46 \times 10^{-3} M \text{ (138--146 meq/100 ml)}$$

$$K^+: 0.038 \times 10^{-3}\text{--}0.050 \times 10^{-3} M \text{ (3.8--5.0 meq/100 ml)}$$

The ratio of serum (extracellular) sodium to potassium is thus about 30:1. The sodium line from such a sample will therefore be much more intense than the potassium line (Fig. 16-7).

Since the sodium and potassium principal emission lines are so far apart, only a simple optical filter is necessary to enable one line to be measured in the presence of the other. In some situations the sodium and potassium are determined by the direct determination method. A separate calibration plot (Fig. 16-6) is made up for each metal ion.

Other samples are analyzed by the internal standard method in which the lithium ion is used as the internal standard ion. The ratio of intensity of sodium

[2] E. E. Pickett and S. R. Koirtyohann, *Anal. Chem.* **41**, 28A (Dec. 1969).

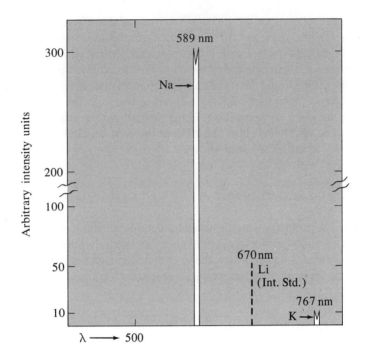

FIGURE 16-7. *Hypothetical flame emission spectrum of a mixture of sodium and potassium ions in blood serum. The internal standard line of lithium ion is shown with a dotted line.*

to the intensity of lithium is measured and plotted *vs.* the concentration of sodium, and the same is done for potassium. This minimizes variations in the flame and the effects of other substances in the blood samples that might affect the sodium in the blood sample. Lithium is suitable for an internal standard because it behaves nearly the same as sodium and potassium, and isn't found in the blood.

Large concentrations of sodium also enhance the emission intensity of potassium in certain flames. This causes a serious error in acetylene-air flames, but a smaller error in a natural gas-air flame. The errors in both flames decrease as the concentrations of both ions decrease.

Determination of Other Metal Ions. Whereas the alkali metals are generally measured in an air-acetylene flame, or an air-natural gas flame, most other metals are measured in a nitrous oxide-acetylene flame. This flame minimizes oxide formation and other interferences, and makes it possible to determine some metals that cannot be measured in an air-acetylene flame. For example, in Table 16-2, you can see that aluminum(III), barium(II), and magnesium(II) are all readily measured in the nitrous oxide-acetylene flame, but not as readily in the air-acetylene flame.

TABLE 16-2. *Detection Limits in PPM of Metal Ions in a Flame* [2]

Metal	Wavelength, nm	Detection limit, ppm	
		N_2O—C_2H_2	Air—C_2H_2
Al	396.2	0.005[a]	—
Ba	553.6	0.001[a]	—
Ca	422.7	0.0001[a]	0.005[a]
Mg	285.2	0.005	—
K	766.5	—	0.0005
Na	589.0	—	0.0005

[a] Determined in the presence of a high concentration of potassium chloride.

16-3 ATOMIC ABSORPTION SPECTROMETRY

Atomic absorption spectrometry is the instrumental measurement of the amount of radiation absorbed by unexcited atoms in the gaseous state. It, like flame emission spectrometry, is one of the most general methods of analysis because atoms of all the metals, plus some nonmetals, absorb radiation in the gaseous state. It also can be used with special sample handling for the determination of traces of mercury in fish and in other foods.

The instrument used to measure the amount of absorption is known as the atomic absorption spectrometer. There are two broad types of spectrometers—

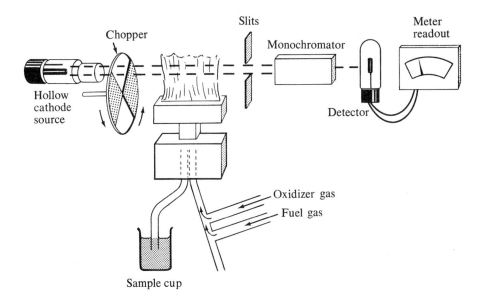

FIGURE 16-8. *Schematic diagram of atomic absorption spectrometer.*

flame and nonflame. We will discuss mainly the flame spectrometer at this point, and later describe the flameless spectrometer, where atoms are vaporized without a flame.

The flame atomic absorption spectrometer (Fig. 16-8) consists of a premixed burner system, a source of radiation, a monochromator, a detector system, and a readout device. The sample is sprayed in fine drops by the nebulizer into the burner and carried into the flame where it absorbs some of the radiation from the source. The remaining radiation from the source passes into the monochromator where the appropriate wavelength is selected for measurement. The radiant power of the transmitted radiation is measured by the detector system, and the amount of radiation absorbed indicated in absorbance units on the readout meter. (The Beer-Lambert law, Sec. 13-2, is used to treat atomic absorption measurements.) We will discuss each component in a little more detail below.

Instrumentation

We begin our description of each component with a discussion of the nebulizer and burner system.

The Nebulizer-Burner System. Recall from the Beer-Lambert law that absorption depends on b, the path length traveled by the radiation through the absorbing medium. For this reason, a long flame path length is needed to increase the absorbance. The *slot burner* shown in Figure 16-8 is used for this purpose. It is much like the premixed burner (Sec. 16-2) used in flame emission spectrometry, except for the longer flame path length. Although the slot burner is used on most commercial instruments, the turbulent burner is also usually available as optional equipment. The premixed slot burner gives a quieter flame but is generally limited to fairly low-burning velocity flames. The nebulizer systems for these burners are of course fairly similar to those described in Section 16-2.

Insofar as flames are concerned, the air-acetylene flame (Table 16-1) is most commonly used in atomic absorption spectrometry.

The Source. The most common source is the hollow cathode lamp (Fig. 16-9). Most often, the cathode is made of the metal ion to be determined. The emission from the lamp is an emission line, in contrast to the continuum emission of sources used in spectrophotometry (Sec. 13-3). This of course means that a separate lamp is needed for each metal ion to be determined.

The name of the source derives from the lamp's cylindrical hollow cathode, which emits the line of the metal from which it is made. (In the case of metals such as sodium, an alloy of sodium with a less malleable metal is used.) A high voltage current (240 to 400 V) flows between the hollow cathode and the anode. This strikes atoms of an inert filler gas such as neon, forming positive ions. The ions bombard the negative cathode, causing "sputtering" of cathode metal atoms in vapor form into the inner hollow of the cathode (the outer surface of the cathode

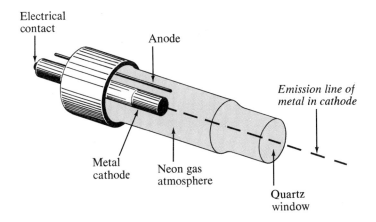

FIGURE 16-9. Schematic diagram of a hollow cathode tube used as a source for an atomic absorption spectrometer.

is covered with glass to confine emission to the inner hollow). The gaseous metal atoms become excited and emit the characteristic line(s) of that metal. The line(s) passes through the window at the end of the tube. For general purposes, the window may be made of quartz, but for elements such as sodium, a pyrex glass window can be used since pyrex transmits visible lines such as 589 nm.

Although a number of lines may be emitted by a given source, the *principal* (or first resonance line) is generally used (Sec. 16-1). This line is the most intense line because it is the result of a transition between the first excited state and the ground state. The reason that a line source is preferred over other types is that the width of the absorption line is quite narrow, similar to the emission lines in Figure 16-7. Thus most of the photons in the line emitted by the source will be absorbed by the atoms in the flame, rather than just a few photons emitted by a continuum source.

Choppers and Modulation. Recall that after atoms have been excited via atomic absorption, some of them return to the ground state via emission of photons of the same wavelength used for excitation (16-3). This emission will cause an error in the measurement of absorption by yielding an apparent increase in the number of photons reaching the detector. This will cause an apparent decrease in the absorbance read from the meter.

The above error can be eliminated by using a chopper, or by employing modulation. The chopper is shown in Figure 16-8. It alternately transmits and then blocks the hollow cathode beam during its rotation. The detector is synchronized and tuned to the intermittent source radiation and does not respond to the constant emission of radiation from excited atoms in the flame. Instead of a chopper, modulation can be used. The hollow cathode can be run from a modulated power supply. The transmitted photons from the source are measured by a detector system employing an a.c. amplifier at the same frequency as the modulated power supply. Emission from excited atoms in the flame is of course not modulated and therefore cannot affect the output of the a.c. amplifier.

Monochromator and Detector. These components are generally the same as for a sophisticated fluorescence emission spectrometer. A grating is used as the monochromator, and a photomultiplier tube is used as the detector. An a.c. amplifier is employed as previously mentioned.

Atomic Absorption Analysis

We will first discuss the direct determination method of analysis, and then some examples of the determination of various metal ions.

The Direct Determination Method. The basic law relating the absorbance of radiation, A, to the concentration of the metal ion in solution, c, is the Beer-Lambert law (Sec. 13-2).

$$A = abc \qquad (16\text{-}5)$$

where b is the path length over which absorption occurs in the flame and a is a proportionality constant. The path length is obviously difficult to measure, so a cannot be readily calculated and reported in the literature as the molar absorptivity is (Sec. 13-2). It is necessary to employ a proportionality constant in the above equation to reflect the proportionality between the number of absorbing atoms in the flame and the concentration of the ions in the solution sprayed into the flame. Thus, a could be defined as

$$a = k \left[\frac{\text{no. of atoms in flame}}{c} \right]$$

where k is a constant depending on the type of flame and other conditions. This definition of a keeps **16-5** "honest" in that the left side represents an absorbance in a flame, and the right side also represents an identity that is valid for the flame.

As **16-5** predicts, a plot of absorbance *vs.* concentration should be linear, as long as there are no interactions that cause deviations from the Beer-Lambert law. Two such plots are shown for atomic absorption by strontium in Figure 16-10. Since it is known that the potassium ion represses the ionization of strontium atoms in a nitrous oxide-acetylene flame, it is obvious that the concentration of the strontium atoms *in the flame* is higher when potassium ions are present. Interpreting this in terms of **16-5**, we would say that a for the plot where potassium was present is greater than a where potassium is absent.

Although they will not be discussed here, the internal standard method and the standard addition method described in flame emission spectrometry (Sec. 16-2) are applicable to some degree in atomic absorption, particularly if the direct determination method cannot be made to give a useful Beer-Lambert plot.

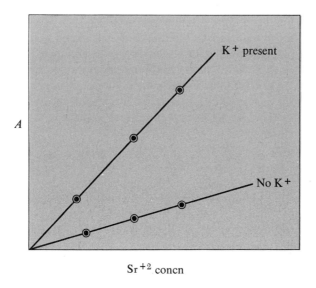

FIGURE 16-10. Beer-Lambert law plot of atomic absorption by strontium in a nitrous oxide-acetylene flame.

Metal Ions Commonly Determined

Atomic absorption spectrometry can be used to determine essentially any of the metallic elements, given the proper flame. One important limitation of *flame* atomic absorption is the determination of mercury; however, this can be determined by flameless atomic absorption, which we will discuss later. Atomic absorption can also be used for certain nonmetals which cannot be determined at all, or with difficulty, by flame emission spectrometry. These nonmetals include selenium, tellurium, arsenic, and silicon.

In addition, there are a number of metal ions for which atomic absorption spectrometry is more sensitive [2] than flame emission spectrometry. These are

Group IB: Ag, Au
Group IIA: Be, Mg
Group IIB: Zn, Cd, Hg
Group IVA: Si, Pb
Group VA: As, Sb, Bi
Transition metals: Co, Fe, Pt

Determination of Lead. One toxic metal that has been prominent in the news is lead. Not only does it cause lead poisoning in children who eat paint from older buildings, but it also endangers anyone who drinks liquids or eats food stored in pottery and ceramic ware coated with lead-based paints. It had been previously assumed that the protective glazes used on these vessels prevented any dissolution of lead into liquids or food stored in them. However, this was found not to be true.

The U.S. Food and Drug Administration (FDA) then developed a procedure to simulate the leaching of lead(II) ion into materials stored in these vessels. The FDA procedure involves soaking the glazed ware with a dilute 4% solution of acetic acid (similar to vinegar) for 24 hours at room temperature. The solution is then tested for lead(II) ion to see if it exceeds the allowable maximum of 0.007 mg of lead(II) ion per ml of test solution. Atomic absorption measurement of lead at 217 nm is used for the analysis since it can detect as little as 1×10^{-5} mg of lead per ml (0.01 ppm lead).

Determination of Toxic Metals in Biological Samples. Flame atomic absorption spectrometry is sensitive enough in most cases to detect the very low concentrations of toxic metals found in biological samples. Table 16-3 lists some of these metals

TABLE 16-3. AA Detection Limits and Biological Threshold Limits for Toxic Metals

Toxic metal	AA det. limit, ppm	BTL[a], ppm	Biological sample
Arsenic	0.1	0.3	Blood
		0.4 mg/g	Hair
Cadmium	0.002	0.05	Blood
Lead	0.01	0.8	Blood
		0.024 mg/g	Hair
Mercury	0.5	0.06	Blood
Selenium	0.1	0.07	Urine
Tellurium	0.1	0.02	Urine

[a] BTL is the biological threshold limit in mg/liter (=ppm).

with the flame atomic absorption detection limit in ppm (mg/liter), as reported by Moffitt [3], and the biological threshold limit (BTL). BTL indicates excessive toxic *exposure* to the toxic metal but not necessarily toxicity.

The detection limit for mercury listed in Table 16-3 is fairly high compared to BTL. However, the detection limit for flameless atomic absorption is much lower (see following) so flameless atomic absorption is preferred for the determination of mercury.

Flameless Atomic Absorption

Although the flame is a good method of atomizing samples, it is not the only method. For metal-containing samples, a *carbon-rod* atomizer has been used with excellent results. This is an electrically heated graphite tube, open at both ends, and placed axially in the optical pathway of the hollow cathode lamp. Liquid or solid samples are heated gently in a cavity in the rod to volatilize water and organic compounds;

[3] A. E. Moffitt Jr., et al., *Amer. Lab.* **3**, 8 (1971).

this is followed by strong heating to vaporize the remaining metals. The beam from the lamp passes through the vapor of the metal atoms just above the sample cavity, and absorption is measured as for flame atomic absorption. Most metals can be determined with as good or better detection limits as flame atomic absorption. Mercury, however, is determined by a different type of flameless atomic absorption instrument.

Determination of Mercury. Mercury is a unique metal in that it is so toxic that sensitive methods are needed to detect the low levels where toxicity starts. For example, the FDA tolerance for mercury in fish is 0.5 ppm (0.305 mg/g fish). Flame atomic absorption is usually not as sensitive as is needed for all samples needing testing.

In 1968 Hatch and Ott [4] devised a highly sensitive flameless atomic absorption method for mercury that is based on a simple volatilization of free mercury by sweeping air through the liquid sample. The actual method varies according to the type of sample, but in general the sample is treated with an oxidizing acid to convert mercury to its highest oxidation state, the mercury(II) ion. Next, an acidic solution of tin(II) ion is added to reduce the mercury(II) ion as follows:

$$Hg^{+2} + Sn^{+2} \rightarrow Hg^{\circ}(soln.) + Sn^{+4}$$

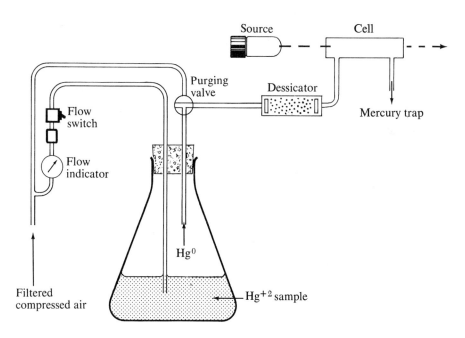

FIGURE 16-11. *Schematic diagram of apparatus for the flameless atomic absorption determination of mercury.*

[4] W. R. Hatch and W. L. Ott, *Anal. Chem.* **40**, 2025 (1968).

At this point, the free mercury is dispersed in the solution. A flow of air is then passed through the solution, sweeping the free mercury in vapor form out of the reaction flask (Fig. 16-11). The air stream containing the mercury is dried by passing it through a desiccator, and then it passes through a 15 cm long optical cell. The intense 254 nm mercury line from the source passes through the optical cell, and the absorbance at 254 nm is measured as in flame atomic absorption.

Note that in contrast to flame atomic absorption, the flameless determination of mercury employs an optical cell to confine the mercury vapors. Because the concentration of gaseous mercury atoms is quite low in the air, the cell is 15 cm long rather than the 1 cm used in solution spectrophotometry. This increases the absorbance (**16-5**) and permits analysis of samples containing as low as 0.005–0.05 ppm mercury.

QUESTIONS AND PROBLEMS

(Answers to most even-numbered problems are in Appendix 5.)

Concepts and Definitions
1. Explain the role of the flame in both flame emission and atomic absorption spectrometry.
2. Contrast and compare flame emission and atomic absorption spectrometry as to the following steps.
   a. Flame excitation
   b. Loss of excitation energy
   c. Measurement step
3. Explain the difference between flame and flameless atomic absorption spectrometry.
4. Explain the difference between the following flame emission spectrometric methods of analysis.
   a. The direct determination method
   b. The standard addition method
   c. The internal standard method
5. State whether flame emission or atomic absorption spectrometry is the preferred method for the determination of each element in each following group.
   a. Li, Na, Ag, Au   b. Be, Mg, Ca, Sr   c. Al, As, Sb, Tl
   d. Co, Ru, Fe, Re   e. Zn, Ga, Cd, Tl
6. Explain why flame emission is preferred to atomic absorption spectrometry for the determination of sodium(I) and potassium(I) ions in most clinical laboratories.
7. Draw orbital diagrams to explain the excitation and emission processes for the following elements in a flame (see Fig. 16-3).
   a. Li   b. K

Quantitative Analysis Problems
8. A standard solution containing $1.50 \times 10^{-3} M$ sodium(I) ion gives an emission intensity of 50 units at 589 nm in a flame emission spectrometer. Calculate the sodium(I) ion

concentration in each blood serum sample below, and state whether the concentration falls in the normal adult range.
   a. Blood with emission intensity of 70 units
   b. Blood with emission intensity of 46 units
   c. Blood with emission intensity of 42 units
9. A standard solution containing $1.50 \times 10^{-3} M$ sodium(I) ion and $5.0 \times 10^{-5} M$ potassium(I) ion gives an emission intensity of 100 units at 589 nm and 10 units at 766 nm in a flame emission spectrometer. Calculate the concentrations of these ions in the following blood serum samples and indicate if possible whether the concentration falls in the normal adult range.
   a. Sample with 97 units at 589 nm and 9.0 units at 766 nm
   b. Sample with 80 units at 589 nm and 9.5 units at 766 nm
   c. Sample with 96 units at 589 nm and 6.0 units at 766 nm
   d. Sample with 82 units at 589 nm and 6.5 units at 766 nm
10. A sample is to be analyzed for $\%Na_2O$. The following readings are obtained on the flame emission spectrometer at 589 nm.

mg/liter Na_2O	Emission intensity units
0	3
10	22
25	46
50	70
Unknown sample (10 mg/ml)	28

Calculate the $\%Na_2O$ in the sample after constructing a calibration plot for intensity vs. $\%Na_2O$.

11. The following absorbance readings were obtained for cadmium(II) solutions on an atomic absorption spectrometer. Plot absorbance vs. concentration and determine the concentration of an unknown having an absorbance of 0.205.

ppm Cd^{+2}	A
0	0.000
3.80	0.104
5.80	0.160
8.00	0.220
11.20	0.310

17. Direct Potentiometric and pH Measurements

> "... *the triggers* were wired together so that, if you pulled on the hinder one, both barrels were discharged." (*he said to Holmes*).
>
> ARTHUR CONAN DOYLE
> The Valley of Fear, Chapter 4

You may recall from earlier chapters that a potential is the electrically measured tendency of a chemical species to lose or gain electrons. Such a measurement can be used as the basis for an instrumental determination of concentration (more accurately, activity). Briefly, this is done by using an appropriate *potentiometer* (Sec. 12-1) connected to a pair of electrodes in the sample solution to measure the tendency, in volts, for an energy-producing reaction to occur. The concentration is then calculated from the voltage.

Direct potentiometric and pH measurement is one of the truly fascinating methods of chemical analysis. Just by inserting two appropriate electrodes *wired* to a potentiometer into a solution, a positive result is obtained (in contrast to the negative result achieved by Sherlock Holmes' protagonist in *wiring* two guns together!). Even more interesting is the fact that the analytical result is obtained without any appreciable reaction; there isn't even the *temporary* formation of an excited state as in light absorption.

How is it possible to make a measurement without using a chemical reaction or forming an excited state? The answer lies in an understanding of the potentiometric measurement introduced in Chapter 12. Briefly, the potentiometer voltage (Fig. 12-5) is adjusted to oppose the voltage developed in the electric cell of which the species to be measured is a part. The potentiometer voltage opposes the electrochemical reaction also. Because this voltage can be accurately measured, its value is a direct measure of the concentration of the species present. Thus, the potential energy of the species is measured rather than the kinetic energy associated with a reaction.

17-1 GENERAL APPROACH TO POTENTIOMETRIC MEASUREMENTS

Before describing specific potentiometric measurements, such as the measurement of pH, and pCl, we will first outline a general approach to making potentiometric measurements of any ion. We will begin with a general description of the apparatus.

Apparatus for Potentiometric Measurements

Although a somewhat schematic diagram of potentiometric apparatus was given in Figure 12-5, Figure 17-1 is a more specific illustration of the apparatus and will be referred to in what follows. Essentially Figure 17-1 depicts an electric cell; the basic circuit consists of the potentiometer connected to two electrodes immersed in the sample solution.

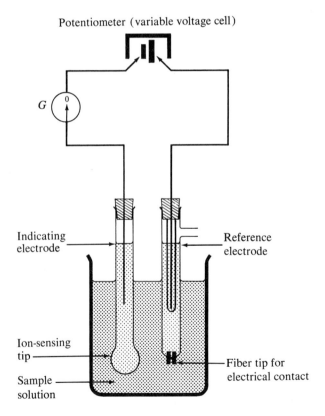

FIGURE 17-1. Apparatus for direct potentiometric or pH measurement. The ion-sensing tip of the indicating electrode responds to changes in the activity (corrected concentration) of the ionic sample. The galvanometer, G, indicates zero current flow.

The Reference Electrode. The reference electrode is a self-contained system whose potential is constant and independent of concentrations of ions in the sample solution. It is equipped with a fiber tip (Fig. 17-1) which provides contact with the sample solution; diffusion of the sample solution into the electrode is negligible so the sample does not affect the composition of the fluid inside the electrode.

The most common reference electrode is the saturated calomel electrode (SCE); its construction was shown in Figure 12-6. It is composed of metallic mercury and insoluble mercury(I) chloride (calomel) in equilibrium with a saturated solution of potassium chloride. It has a standard reduction potential of $+0.2458$ V at 25°C. The half reaction that governs its potential is

$$Hg_2Cl_2(s) + 2e^- = 2Hg(l) + 2Cl^-$$

The Nernst equation (Sec. 11-5) for the reaction is

$$E = E° + \frac{0.059}{2} \log \frac{1}{[Cl^-]^2}$$

Since the potential of this half reaction depends only on the concentration of the chloride ion, and since this is constant in the SCE, the potential of this electrode should theoretically be constant.

The Indicating Electrode. The indicating, or indicator, electrode must respond rapidly to the activity (corrected concentration) of the chemical species to be measured. There are various types of indicating electrodes—metallic, glass membrane, liquid membrane, and solid-state. Each type responds in a somewhat different manner at its ion-sensing tip (Fig. 17-1). They all, in general, measure E, the potential of the half reaction involving the chemical species to be measured. The potential of the indicating electrode, $E_{ind\ el}$, is related to E and the activity of the chemical species through the Nernst equation (25°C) as follows:

$$E_{ind\ el} = E = E° + \frac{0.0590}{n} \log \frac{a_{ox}}{a_{red}} \quad (17\text{-}1)$$

where a_{ox} and a_{red} are the activities of the oxidized form and reduced form, respectively, of the chemical species to be measured. If the total ionic content of the solution is below $10^{-4} M$, the activities are essentially the same as the concentrations of the two forms, but otherwise they are somewhat smaller. Of course, only one of the two forms usually can be measured; the activity of the other form is assumed to remain constant for all samples measured.

The *metallic indicating electrode* consists of an active metal such as silver, mercury, or copper, or an inert metal such as platinum. Most active metal metallic electrodes consist of a wire made of the active metal dipping into solution containing the metal ion. For the active silver metal electrode, the half reaction is

$$Ag^+ + e^- = Ag°(s) \qquad E° = +0.800 \text{ V}$$

Some metallic electrodes consist of a wire made of the active metal coated with a layer of an insoluble salt containing the metal ion and surrounded by a solution of constant concentration of the anion of the salt. A silver-silver chloride electrode can be made this way; its half reaction is

$$AgCl(s) + e^- = Ag°(s) + Cl^- \qquad E° = +0.222 \text{ V}$$

The silver electrode can be used to measure pAg directly and pCl, pBr, etc., indirectly via the solubility equilibria of the corresponding insoluble silver(I) salt. The silver-silver chloride electrode can be made to measure pCl directly.

Inert metal electrodes consist of a wire such as platinum which dips directly into the sample solution without being enclosed in a tube. The function of the inert electrode is simply to transfer electrons; it does not take part in the reaction itself. It can thus be used to measure activity ratios of soluble ions such as Fe^{+3}/Fe^{+2} which cannot form a solid metal electrode.

The *glass membrane electrode* has a thin, fragile, glass ion-sensing tip. A typical glass electrode was pictured in Figure 12-6. Such an electrode usually has an internal silver-silver chloride reference electrode to serve as a medium for the current flow. A solution of the ion to be measured is placed inside the thin ion-sensing tip. A potential difference develops across the glass membrane that is a function of the difference of the activities of the ion inside and outside the tip in the sample solution.

This potential difference arises because of an ion-exchange reaction that occurs at the surface of the glass tip. For example, with a glass electrode used for pH measurement, the ion-exchange site is either $-SiO^-Na^+$ (older electrodes) or $-SiO^-Li^+$ (newer electrodes). In the sample solution whose pH is to be measured, an equilibrium ion-exchange reaction is established; for the older electrodes, this is

$$-SiO^-Na^+ + H^+ = -SiO^-H^+ + Na^+ \qquad (17\text{-}2)$$

$$\text{(glass)} \quad \text{(solution)} \quad \text{(glass)} \quad \text{(solution)}$$

The establishment of an equilibrium between these sites on the glass is thought to be responsible for the potential difference that results. Different glasses with different types of sites can be used for the measurement of monovalent cations such as Na^+ and K^+.

The *liquid membrane electrode* has a thin, porous polymer membrane tip instead of a glass tip like the glass membrane electrode. It is similar to the latter in that it has an internal silver-silver chloride reference electrode to serve as an element for current flow. A solution of the ion to be measured is placed inside the ion-sensing tip (Fig. 17-1). A potential difference develops across the porous polymer membrane that is a function of the difference of the activities of the ion inside and outside the tip, in the sample solution.

The porous polymer membrane is saturated with a solution of a liquid ion exchange resin dissolved in a water-insoluble organic solvent. One such ion exchange resin has phosphate groups [=P(O)O$^-$H$^+$] which can react via an ion-exchange equilibrium with ions such as calcium(II) as follows:

$$2{=}P(O)O^-H^+ + Ca^{+2} = Ca^{+2}[{=}P(O)O^-]_2 + 2H^+$$

(polymer)   (solution)   (polymer)   (solution)

The change in the ratio of the two different sites on the ion-exchange resin is thought to be responsible for the potential difference observed. Different ion exchange resins with different types of sites can be used for the measurement of divalent cations such as Ca^{+2}, Mg^{+2}, Cu^{+2}, etc.

The *solid state electrode* has a single crystal or pellet membrane tip instead of a glass tip or porous polymer tip. It may have an internal silver-silver chloride reference electrode or merely a simple electrical contact to the inner surface of the membrane. A solution of a compound containing the anion (or cation) to be measured is placed inside the ion-sensing tip (Fig. 17-1). The membrane itself is a conducting solid and contains as a lattice ion the anion (or cation) to be measured. Electrical conduction occurs by a lattice defect mechanism in which only the lattice ion (ion to be measured) moves into vacant positions in the lattice. This type of electrode is used mainly for the measurement of anions such as F^-, Cl^-, Br^-, SCN^-, and S^{-2}.

The Potentiometer. One type of potentiometer is the classical potentiometer (Fig. 12-5) which utilizes an external cell having a variable voltage which can be read to ± 0.001 volt (± 1 millivolt). Such a type of potentiometer is rarely used in modern laboratories. A more popular type is the *electronic voltmeter* with a special amplifier designed to have a high input resistance. This design has the same effect as the classical potentiometer in that it minimizes the current flow to an insignificant value. (This arrangement cannot be shown in a simple schematic fashion and thus has been omitted from Fig. 17-1). Thus the concentration of the ionic species interacting with the indicating electrode does not change (Sec. 12-2).

The potentiometer reading, E_{pot}, is now an accurate indication of the potential of the indicating electrode, $E_{ind\ el}$. Once the constant potential of the reference electrode, E_{ref}, is subtracted from $E_{ind\ el}$, the difference is equal to E_{pot}.

$$E_{ind\ el} - E_{ref} = E_{pot} \tag{17-3}$$

Since E_{ref} and E_{pot} are the known parameters in any potentiometric measurement, 17-3 is better rearranged for calculation purposes to

$$E_{ind\ el} = E_{pot} + E_{ref} \tag{17-4}$$

Once $E_{ind\ el}$ is known, **17-1** can be used to calculate the activity of the species to be measured. Frequently, however, the scale of the potentiometer is precalibrated to read directly in units such as pH, pAg, pCl, etc.

Since the voltage scale of the potentiometer is uncertain to ± 0.001 V, only three significant figures are justified for reporting $E_{ind\ el}$ (or pH, pAg, pCl, etc.). When $E_{ind\ el}$ is used with **17-1** to calculate an activity, this usually limits the activity to two significant figures because of the log values involved. See the following example.

EXAMPLE 17-1. A silver metal indicating electrode and a saturated calomel reference electrode are used to measure pAg potentiometrically. The potentiometer reading is $+0.419$ V; the potential of the SCE is $+0.246$ V. Calculate pAg and [Ag$^+$] to the correct number of significant figures.

Solution: Use **17-4** to calculate the potential at the indicating electrode.

$$E_{\text{ind el}} = +0.419 \text{ V} + 0.246 \text{ V} = +0.665 \text{ V}$$

Next, use the $E°$ of $+0.800$ V for the reduction of Ag$^+$ to Ag°(s) in conjunction with the Nernst equation **(17-1)** to calculate [Ag$^+$].

$$E_{\text{ind el}} = +0.665 \text{ V} = +0.800 \text{ V} + 0.0590 \log a_{\text{Ag}^+}$$

(Since the activity of silver metal is unity, its activity does not affect the above expression.)

Combining the indicating electrode potential and the $E°$ gives

$$+0.665 \text{ V} - 0.800 \text{ V} = -0.135 \text{ V} = 0.0590 \log a_{\text{Ag}^+}$$

Solving for the activity of the Ag$^+$ ion, we obtain

$$\log a_{\text{Ag}^+} = \frac{-0.135 \text{ V}}{0.0590} = -2.28_8 = +0.71_2 - 3$$

$$a_{\text{Ag}^+} = 5.1_5 \times 10^{-3} M \text{ (2 sig. figs.)}$$

Since pAg is the negative log of the activity of the silver(I) ion,

$$\text{pAg} = 2.28_8, \text{ or } 2.29 \text{ (3 sig. figs.)}$$

Note that the activity of the silver(I) ion has one significant figure less than the log value because the first significant figure of the log represents the exponent.

At this concentration the activity of the silver(I) ion is about 5% lower than the molarity, so [Ag$^+$] = $5.4 \times 10^{-3} M$.

17-2 POTENTIOMETRIC MEASUREMENT OF pH

The classical measurement of pH, using the hydrogen indicating electrode, was discussed in Section 12-2. Here we will describe the modern measurement of pH, using the glass indicating electrode. Recall that pH is defined as

$$\text{pH} = -\log a_{\text{H}^+}$$

and that the activity, a_{H^+}, of the hydrogen ion is the *product* of the activity coefficient and [H$^+$]. At an ionic strength of 0.001, the activity coefficient of hydrogen ion is 0.97, so below $10^{-3} M$ H$^+$, the activity and molarity of the hydrogen ion are about the same (as long as the ionic strength is $<10^{-3}$).

Electrodes for pH Measurement

The usual reference electrode is the saturated calomel electrode (Fig. 12-6; Sec. 17-1). It has a constant potential of $+0.2458$ V and does not interact with the sample solution. It is connected to the back of the pH meter (Fig. 17-2) and held upright in a rigid holder.

The glass electrode shown in Figure 17-2 is used as the indicating electrode for most of the pH measurements made presently. The H^+ in a sample solution equilibrates with the $-SiO^-Na^+$ groups on the electrode tip's surface, and affects the potential of the electrode. Strictly speaking, the activity of the sodium(I) ion in the sample solution can also affect the potential of the glass electrode because it is involved in the equilibria in **17-2** (see following).

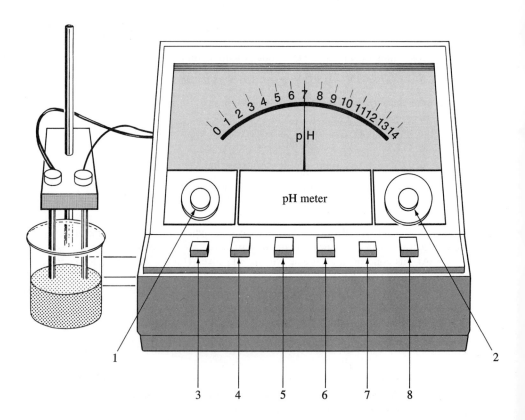

FIGURE 17-2. *A schematic diagram of a modern pH meter with scale readout. The controls are (1) standardization control for measurements; (2) temperature compensator control; (3) manual pushbutton to connect temperature compensator (usually kept "on"); (4) automatic pushbutton to disconnect temperature compensator for automatic temperature compensation; (5) ± 700 mV pushbutton; (6) 1400 mV pushbutton; (7) standby pushbutton for removing electrodes from solution; (8) read pushbutton for reading pH from scale.*

The relation between $E_{\text{ind el}}$, the potential at the glass electrode, and the activity of the hydrogen ion in a sample containing sodium(I) ion is

$$E_{\text{ind el}} = C + 0.0590 \log[a_{H^+} + (K_{H^+, Na^+})a_{Na^+}] \tag{17-5}$$

where C is a constant that includes a number of sources of potential—such as the potential of the SCE, a liquid junction potential, and an asymmetry potential—and K_{H^+, Na^+} is a type of selectivity coefficient of the glass electrode for H^+ in the presence of Na^+.

Equation **17-5** can be simplified for lithium oxide glass electrodes used at pHs below 13. Below this pH, the second term in the brackets is negligible compared to the a_{H^+} and we obtain

$$E_{\text{ind el}} = C + 0.0590 \log a_{H^+} = C - 0.0590 \text{ pH} \tag{17-6}$$

Unfortunately, the value of C in **17-6** can never be known accurately. The value of the asymmetry potential varies from day to day, and the value of the liquid junction potential is uncertain. Therefore, most laboratory pH measurements must involve calibration of the pH meter against a standard buffer solution, which is assumed to have the same value of C. The National Bureau of Standards has determined the pH values of the buffers listed below very accurately.

Buffer	pH (25°C)
0.05M potassium tetroxalate	1.679
Saturated potassium hydrogen tartrate	3.557
0.05M potassium acid phthalate	4.008
0.025M $H_2PO_4^-$, 0.025M HPO_4^{-2}	6.865
0.01M Borax	9.180

It is recommended that buffers whose pHs bracket the pH of the sample be used to standardize the pH meter to reduce the error from variation in liquid junction potentials.

Sodium Error. In alkaline media glass membrane electrodes exhibit a sodium error, usually referred to as an alkaline error. The error arises from high activities of sodium ion interacting with $-\text{AlOSi}^-\text{H}^+$ sites via equilibria like that in **17-2**, causing a change in potential. The magnitude of this effect depends on the type of glass in the electrode and on the ratio of the sodium ion to the hydrogen ion. The $-\text{AlOSi}^-\text{H}^+$ sites in the older soda-lime-aluminum glass electrodes have a relatively small selectivity coefficient for H^+ over Na^+ compared to that of the sites in the new lithium oxide glass electrodes.

With the older soda-lime glass electrodes, the presence of 1M sodium ion caused pH values to be low by 0.2 pH unit at pH 10 and by 1.0 unit at pH 12. When the newer lithium oxide glass electrodes are used, the presence of 1M sodium ion causes an error of less than 0.2 pH unit at pH 13. Almost all commercially available glass pH electrodes are of this type.

Last, but not least, glass electrodes should be stored upright in distilled water when not in use. New glass electrodes should also be presoaked in water before use.

The reason for both of these requirements is that the outer layers of the glass electrode must become hydrated before their ion exchange sites can function effectively. During the soaking several layers of the glass surface of an electrode become hydrated and the ion exchange sites exchange metal ions for hydrogen ions. Without this soaking a steady potential reading would not be attained. Of course, if a glass electrode is not to be used for a long time, it is best stored dry since the glass slowly dissolves, like any glass, in water.

The Use of the pH Meter

There are three basic types of pH meters, two of which are based on the potentiometric principle of opposing the voltage of the sample solution. These are the classic potentiometric null-point indicator type and the new, direct-reading, feedback circuit type (Fig. 17-2). With the former, calibration and adjustment of the null-point indicator require time and careful adjustment for reading. The reading obtained, however, is generally very accurate and precise. With the latter, the direct-reading feature gives a more rapid but possibly less accurate and precise reading of the pH. The third type is simply an electronic voltmeter.

> Because of the high electrical resistance of the glass electrode, all types require special design. Electronic amplification of the small current flowing into the galvanometer is required in the potentiometric type. The direct-reading types require a high input resistance or a corrector amplifier-capacitor combination to correct for zero drift (Beckman Zeromatic® pH meters).

Operation. To standardize a particular pH meter with a particular set of electrodes, two standard buffer solutions are used. The meter is put in the standby mode (Control 7, Fig. 17-2) to protect the circuits when removing electrodes from a solution. The electrodes are then removed from solution, rinsed of the previous sample solution, and excess water is blotted off. They then are immersed in a standard buffer solution and the circuits reconnected (Control 8, Fig. 17-2). The temperature compensator (Control 2, Fig. 17-2) is adjusted to the temperature of the buffer solution, and the standardization control (Control 1, Fig. 17-2) is turned until the meter reading corresponds to the pH of the first buffer. The above is repeated for the second buffer.

After the electrodes have been removed from the buffer they are dipped into the sample solution. A temperature adjustment is made if necessary (Control 2) and the pH of the sample is read. Generally one standardization is enough for several samples.

pH Readout. There are two types of readout for pH meters—scale readout (Fig. 17-2) and digital readout. In the usual scale readout, only two significant figures to the right of the decimal point are justified. The first is read directly from the scale and the second is interpolated. Since pH is a log term, *all* zeroes to the right of the decimal are significant. Thus pH readings of 0.05 and 0.00 both have two significant figures to the right of the decimal point.

EXAMPLE 17-2. A scale readout pH meter gives a pH reading of 5.96 for a sample of water which has been purified in a water plant. Calculate the activity of the hydrogen ion and [H$^+$].

Solution: Insert the pH value into the definition of pH.

$$\text{pH} = -\log a_{\text{H}^+} = 5.96$$

$$\log a_{\text{H}^+} = -5.96 = +0.04 - 6$$

$$a_{\text{H}^+} = 1.1 \times 10^{-6}$$

Note that the 0.04 term represents two significant figures since it is a log term (Sec. 2-1), so the activity should be expressed with two significant figures.

Since this is a purified water sample, we may tentatively assume that the activity and the molarity are the same; thus [H$^+$] = $1.1 \times 10^{-6} M$.

Digital readout pH meters give a direct readout of pH to two or three digits to the right of the decimal point. There is some question that a third digit is significant because of stability problems and activity coefficients. Some expensive pH meters do possess the stability necessary for a third digit when they are used with special low resistance glass electrodes. In general, it is best to check the manufacturer's literature and the performance of the pH meter before assuming an uncertainty of ± 0.001 for pH readings.

17-3 POTENTIOMETRIC MEASUREMENT OF SPECIFIC IONS

We have already described indicating electrodes useful for direct potentiometric measurement of other ions (Sec. 17-1). We will now discuss the measurement of a few important ions using specific ion-indicating electrodes. As for pH measurement, this measurement gives the *activity* of the ion, not its molarity. Thus pNa is

$$\text{pNa} = -\log a_{\text{Na}^+}$$

where a_{Na^+} is the product of the activity coefficient and [Na$^+$]. There is quite a difference in body fluids. In normal human serum, the activity coefficient of sodium(I) ion is 0.78, for example. If analytical results in terms of concentration are desired, a plot of potential in millivolts against molarity, or ppm, is constructed.

Measurement of Chloride Ion

The measurement of chloride is very important. For example, in food processing the chloride content of milk and butter can be measured potentiometrically using a solid-state electrode (Sec. 17-1). An increase of the amount of chloride in milk is an indication of mastitis in cows; this is easily detected with the chloride electrode.

In biomedical research, increased or decreased chloride levels can be quickly measured by potentiometric measurement. Measurement of chloride is important because of its contribution to the electrolyte balance of the blood (Sec. 22-1); in addition, chloride is necessary for the manufacture of hydrochloric acid in the stomach.

Sweat Chloride. The measurement of chloride in body perspiration is a useful clinical measurement. Formerly, a thirty-minute sweat collection procedure was needed before measurement. Now, sweating is induced by rubbing the palm or wrist with a compound called pilocarpine and applying a weak electric current to induce passage of chloride ions through the skin (iontophoresis).

Pilocarpine is an alkaloid, $C_{11}H_{16}O_2N_2$. The two nitrogens are part of a five-membered heterocyclic ring. One nitrogen is basic enough to form a hydrochloride salt.

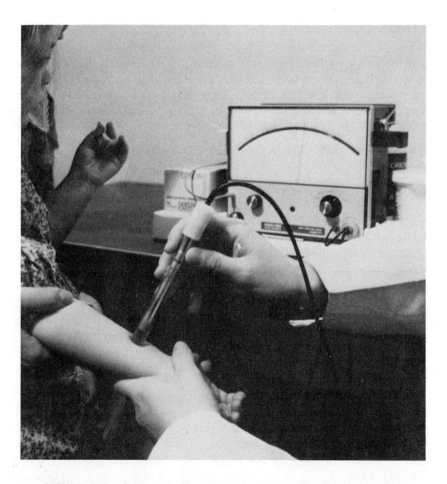

FIGURE 17-3. Orion cystic fibrosis screening system with a flat-headed chloride specific ion electrode for pCl measurement. (Courtesy of Orion Research, Inc.)

The chloride in the sweat is measured directly on the unbroken skin using a *flat-headed* chloride specific ion electrode (Fig. 17-3). The entire procedure takes five minutes using a commercial chloride screening system.

Since most chloride abnormalities result in decreased chloride levels, the occurrence of increased chloride levels is of great significance in diagnosing *cystic fibrosis*, a childhood lung disease. Cystic fibrosis affects one in every 600–2000 babies born, and is characterized by high sweat levels of chloride and other ions. The flat-headed chloride specific ion electrode (Fig. 17-3) is used for rapid screening of babies in hospitals. The results are interpreted with reference to a mean infant chloride level of 20 meq/liter ($2.0 \times 10^{-2} M$) and a safe upper level of 40 meq/liter ($4.0 \times 10^{-2} M$). If the level of chloride exceeds 60 meq/liter ($6.0 \times 10^{-2} M$), then cystic fibrosis is indicated. It has been known to run as high as 140 meq/liter [1]. If the chloride level is between 40 and 60 meq/liter, cystic fibrosis is not decisively indicated. Further observation would probably be carried out.

Measurement of Fluoride Ion

The measurement of fluoride ion is very important in many areas related to health as well. Fluoride is measured with a solid state, single crystal, specific ion electrode (Sec. 17-1). Probably the most important application is the analysis of drinking water to which fluoride ion has been added for the protection of teeth. The single crystal used in the electrode is a lanthanum fluoride crystal doped with divalent europium to increase the electrical conductivity.

The fluoride ion electrode is highly specific for fluoride ion down to an activity of about 10^{-6}. The only known anion interference is the hydroxide ion. In spite of this, there are definite pH limitations to the method. As the level of fluoride decreases, the pH *window*, or *range*, over which it may be accurately measured shrinks (Fig. 17-4). Below pH 3.5 fluoride forms hydrofluoric acid, which is not measured by the electrode.

To analyze drinking water for 1 ppm ($5 \times 10^{-5} M$) of fluoride, the pH must be adjusted to be within a pH window of 5–8. In addition, something must be done to prevent iron(III) in the water from forming fluoride complex ions since these are not measured by the electrode. A buffer is added to keep the solution slightly acidic; a solution of sodium citrate is also added to complex the iron(III) and prevent it from complexing the fluoride ion. Once this is done, a calibration plot (Figure 17-4) can be used for the analysis of ppm fluoride.

Measurement of Cations

Glass electrodes have been modified for the measurement of monovalent cations such as Na^+, K^+, Ag^+, Li^+, and even NH_4^+. Since the hydrogen ion interacts also

[1] R. L. Searcy, *Diagnostic Biochemistry*, McGraw-Hill, New York, 1969, p. 159.

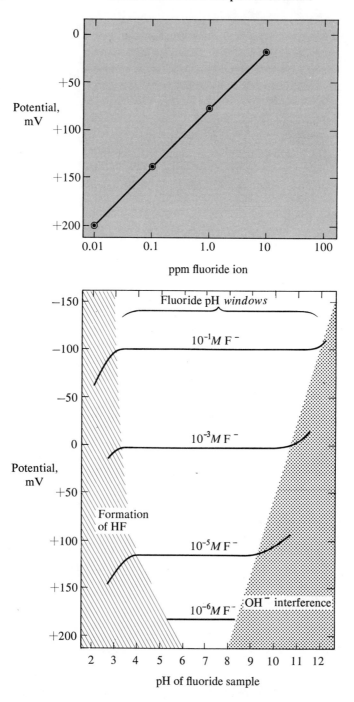

FIGURE 17-4. *Above: typical calibration plot for fluoride analysis. Below: behavior of fluoride electrode as a function of pH.*

with the ion exchange sites on such electrodes, most measurements are best done in neutral or basic solution. In neutral solution, for example, a typical sodium electrode has a selectivity coefficient (**17-5**) for sodium over potassium of about 300. This means sodium(I) can be measured in nearly a hundred times as much potassium(I) without appreciable error (10%).

A glass electrode has been proposed for the measurement of sodium(I) in body fluids, such as blood, but a preliminary separation of interfering materials is necessary. Flame emission (photometry) analysis is still preferred by many clinical and hospital laboratories [2].

QUESTIONS AND PROBLEMS

(Answers to most even-numbered problems are in Appendix 5.)

Concepts and Definitions
1. Name the items of general equipment that are necessary for the direct potentiometric measurement of a particular ion in solution.
2. Explain the difference between an indicating electrode and a reference electrode.
3. Explain why a measured potential is a form of potential energy, not kinetic energy.
4. Describe a calomel electrode.
5. List the four types of indicating electrodes and state how the ion-sensing tip of each differs.
6. Explain what is meant by a sodium error for the glass electrode.
7. Modern glass electrodes (lithium oxide) have a very low sodium error. Would you predict that they would also have a very low lithium error? If so, why is this not as serious a problem for most measurement as the sodium error?
8. Explain how a potentiometer differs from a voltmeter.
9. Why does a pH measurement give only the approximate $[H^+]$?
10. Explain how a sweat chloride measurement is made to detect cystic fibrosis.

pH Measurements
11. Calculate the activity of the hydrogen ion corresponding to each of the following pH values (assume each represents the correct number of significant figures).
    a. pH = 1.96   b. pH = 0.939
    c. pH = −0.04   d. pH = 0.00
    e. pH = 4.996   f. pH = 6.9
12. Assuming that no other ionic species are present other than the acid ionizing to hydrogen ion in Problem 11, state whether the $[H^+]$ will be larger, smaller, or the same as the activity.
13. If K_w for water is 1.00×10^{-14}, calculate the pOH corresponding to the following pH values, using the correct number of significant figures.
    a. pH = 1.96   b. pH = 0.939
    c. pH = −0.51   d. pH = 14.00
14. The following pH measurements were made with the older soda-lime glass electrode.

[2] B. Zak, Wayne State University, Detroit, personal communication, 1976.

State whether there will be an error in the pH and, if so, how large the error might be.
    a. pH = 12.1 (1M Na$^+$)   b. pH = 10.2 (1M Na$^+$)
15. Choose the correct National Bureau of Standards approved buffer(s) for samples whose [H$^+$] will fall in the range of [H$^+$] given below.
    a. $5 \times 10^{-4} M$–$9 \times 10^{-4} M$ H$^+$
    b. $1.1 \times 10^{-4} M$–$2.3 \times 10^{-4} M$ H$^+$
    c. $8.1 \times 10^{-6} M$–$2.9 \times 10^{-5} M$ H$^+$

Specific Ion Electrodes
16. Calculate the activity of each metal ion below from the pMetal reading obtained from potentiometric measurement using a specific ion indicating electrode.
    a. pNa = 1.23   b. pK = 10.96
    c. pF = 0.20   d. pCa = -0.301
17. An unknown sample of calcium(II) ion has a potentiometer reading of +10 millivolts using a calcium specific ion electrode. By making a semilog plot of millivolt readings versus concentration using the following calcium(II) ion standards, calculate the concentration in molarity of the calcium(II) ion in the unknown.

M of Ca^{+2}	mV
1.0×10^{-4}	-26
8.0×10^{-4}	0.0
6.0×10^{-3}	$+25$
3.0×10^{-2}	$+47$

18. Electrochemical Methods of Analysis

"You know my methods. Apply them,"
said Holmes.

ARTHUR CONAN DOYLE
The Hound of the Baskervilles, Chapter I

The previous chapter dealt entirely with one electrical method of analysis. This was the direct measurement of the potential energy (or simply the potential) of a particular reaction. No actual electrochemical reaction occurred to an appreciable extent. In this chapter we deal with methods of analysis which are truly *electrochemical*; i.e., there is a conversion of electrical energy into chemical energy, or vice versa (recall the transducer mentioned in Ch. 12). It should be your goal for this chapter to "know" these methods and be able to "apply them," as Holmes was wont to admonish Dr. Watson!

We have chosen to start this chapter with a short review of the principles of electrolytic and galvanic cells. These principles will then be applied to the succeeding topics of electrogravimetry (electrodeposition), coulometry, and voltammetry, including polarography. Don't become discouraged as you begin to understand that these methods are complicated. Try to grasp the essentials first—later, master the details.

These methods are important because there are several important health-science oriented applications. For one thing, coulometry can be used for very accurate measurement of the chloride ion, an important constituent of body fluids and an ion not easily measured by spectrophotometry. For another, the amount of oxygen in body fluids such as the blood can be measured by the use of two different types of electrodes. One of these is based on a galvanic type of electrical measurement, and the other is based on a polarographic type of measurement.

18-1 A REVIEW OF GALVANIC AND ELECTROLYTIC CELLS

Electrolytic and Galvanic Cell Reactions

There are two different types of electrochemical cells—the galvanic cell, and the electrolytic cell. In the galvanic cell a chemical reaction occurs *spontaneously* and produces electrical energy. Such chemical reactions were discussed in Chapter 11, and the calculation of the standard potential was outlined in Section 11-3. In the electrolytic cell there is no spontaneous reaction; electrical energy is needed to cause a cell reaction, which is the opposite of the spontaneous reaction.

To illustrate the difference between the two types of cells, let us consider the spontaneous reaction and the nonspontaneous reaction that can occur from the combination of the same two half reactions. The half reactions are

$$Cu^{+2} + 2e^- = Cu°(s) \qquad E° = +0.337 \text{ V}$$

$$\frac{1}{2}O_2(g) + 2H^+ + 2e^- = H_2O \qquad E° = +1.229 \text{ V}$$

The spontaneous reaction is that which results from the subtraction of the first half reaction and its $E°$ from the second half reaction as follows:

$$\frac{1}{2}O_2(g) + 2H^+ + Cu°(s) = H_2O + Cu^{+2} \qquad E°_{cell} = +0.892 \text{ V} \qquad (18\text{-}1)$$

We deduce that the reaction is spontaneous because the $E°_{cell}$ is positive. The nonspontaneous reaction is that which results from the subtraction of the second half reaction and its $E°$ from the first half reaction as follows:

$$H_2O + Cu^{+2} = \tfrac{1}{2}O_2(g) + 2H^+ + Cu°(s) \qquad E°_{cell} = -0.892 \text{ V} \qquad (18\text{-}2)$$

This reaction will not occur spontaneously because $E°_{cell}$ is negative, but it will occur if enough electrical energy is applied to overcome the voltage of the spontaneous reaction and thus reverse the direction of the equilibrium.

In the galvanic cell, measurement of the potential, or the flow of current, may be used for analytical purposes. But how can an electrolytic cell be used for analytical purposes? In an electrolytic cell involving a reaction such as that in **18-2**, the applied voltage forces electrons from the anode to the cathode. At the anode electrons are taken up from the species that is oxidized (in **18-2** water is oxidized to oxygen gas). At the cathode electrons combine with metal ions to form the free metal which deposits on the cathode. If the deposition is quantitative, the gain in weight of the cathode can be measured, indirectly giving the amount of the metal ion originally in the sample solution. This process is called *electrogravimetry*, or *electrodeposition*. Because not all metal ions deposit quantitatively, a more general approach is to measure the quantity of electrical current required to reduce or oxidize an ion in solution. This is known as *coulometry*. Frequently, a so-called coulometric titration is performed as a variation of direct coulometry.

In addition to electrogravimetry and coulometry, a third approach is the *polarographic* reduction of metal ions at a dropping mercury electrode. Measurement of the so-called diffusion current gives a measure of the ionic concentration.

Electrode Reactions

Electrode reactions in a galvanic cell are simply the spontaneous half reactions that would ordinarily occur as a part of a complete oxidation-reduction reaction. The oxidizing agent is reduced at the cathode, and the reducing agent is oxidized at the anode. In an electrolytic cell a number of reactions are possible, depending on the applied voltage. Whatever reaction occurs first obviously can be used most easily for analysis. The half reaction that occurs first at the cathode is that which has the most positive, or least negative, potential. For example, in a mixture of Ag^+ ($E° = +0.800$ V) and Cu^{+2} ($E° = +0.337$ V) silver(I) will be deposited first, as silver metal.

Since most metal ions are soluble only in acid solution, electrolysis in acid solution involves a separation of the metal ion from the hydrogen ion.

The only metal ions commonly reduced before hydrogen ion in acid solution are Ag^+, Cu^{+2}, Hg^{+2}, and BiO^+; the $E°$ values of these are all more positive than the $E°$ of 0.0 V for the reduction of $1M$ hydrogen ion. If the pH is raised to 7, the $E°$ for the reduction of hydrogen ion is lowered to -0.43 V. Any metal ions in solution at that pH and that are reduced at a potential more positive than this, such as $Zn(NH_3)_4^{+2}$, can be deposited at the cathode.

Another way of preventing the interference of hydrogen ion is to use a mercury cathode, rather than a solid inert cathode, for deposition. This has the advantage of shifting the point of equilibrium to the right for the reduction process as follows:

$$M^{+n} + ne^- + Hg(l) = M[Hg(l)]$$

Because the metal is dissolved in the mercury, it has less tendency to dissolve into the aqueous sample solution. This has the effect of making the $E°$ value for the reduction of the metal less negative, or more positive. Many metal ions that would normally be reduced after hydrogen ion, using an inert cathode, are reduced before hydrogen ion. Thus polarography can be used for the determination of about one-third of all the elements.

Overvoltage. An effect that is of practical importance, but cannot be predicted theoretically, is the overvoltage effect for a cell. This is most significant in cases where a gas is liberated at any electrode; in such a case, additional voltage is necessary to overcome the slow rate of some step. Typical overvoltages are listed in Table 18-1. Note that the overvoltage for a specific gas, such as hydrogen, varies with the kind of electrodes used. There is a wide variation in the overvoltage necessary for a smooth platinum (inert) electrode and a liquid mercury electrode which forms an amalgam with metals. The high overvoltage for the mercury electrode is

TABLE 18-1. *Overvoltages for Formation of $O_2(g)$ and $H_2(g)$*

Current density A/cm^2	Hydrogen overvoltage			Oxygen overvoltage	
	Smooth Pt	Platinized Pt	Hg	Smooth Pt	Platinized Pt
0.001	0.024	0.015	0.9	0.72	0.40
0.01	0.068	0.030	1.04	0.85	0.52
0.1	0.29	0.041	1.07	1.28	0.64
1.0	0.68	0.048	1.11	1.49	0.77

an important factor in allowing many metal ions having $E°$ values more negative than that for hydrogen ion, to be reduced and amalgamated into the mercury electrode before hydrogen ion is reduced to hydrogen gas.

In addition to cell overvoltage, there is also an *electrode overpotential*, or *concentration overpotential*. This effect arises when metal ions (or anions) at the surface of the electrode are removed from solution at a faster rate than they diffuse to the surface, resulting in a lower concentration at the surface than in the bulk of the solution. It is the concentration at the surface, rather than that in the bulk of the solution, that should be used in the theoretical (Nernst equation) calculation of the potential required for an electrochemical reaction. Since the surface concentration is not known, it cannot be used; the difference between the theoretical potential and the observed potential (excluding other effects such as overvoltage) is simply said to be the electrode overpotential.

18-2 ELECTROGRAVIMETRY

Electrogravimetric (Electrodeposition) Apparatus

In electrogravimetry, or electrodeposition, the species determined are limited to a few metal ions which are deposited on the cathode. Usually only one metal ion from a sample is electrodeposited. The metal ion to be determined is reduced at a preweighed platinum cathode (Fig. 18-1) and deposited quantitatively on that cathode. After completion of the electrodeposition, the cathode is dried and weighed. The weight of the metal is calculated by subtracting the weight of the cathode before deposition from that after deposition. This can then be used to calculate the concentration of the metal ion in the sample solution.

In the electrogravimetric apparatus (Fig. 18-1) the current for the electrolysis is supplied by a battery or other source of direct current. The voltmeter, V, measures the applied voltage, and an ammeter, A, measures the current flow. Since the voltage must be controlled to deposit a particular metal, the variable resistor, R, is used to adjust the applied voltage. (As the amount of metal ion in solution decreases near the end of the electrodeposition, the variable resistor is adjusted to supply the voltage required at each point.)

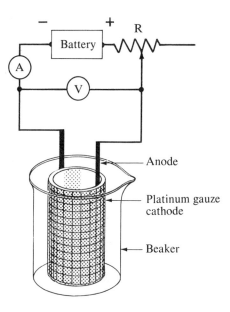

FIGURE 18-1. *Apparatus for electrodeposition.*

Metal Ions Commonly Determined. The copper(II) ion is the most common metal ion determined by electrogravimetry. The reaction at the cathode is

$$Cu^{+2} + 2e^- = Cu°(s)$$

and the reaction at the anode is

$$H_2O = \frac{1}{2}O_2(g) + 2H^+ + 2e^-$$

(The overall cell reaction is the sum of these and was given in **18-2**). The silver(I) ion is the most likely interference and should be removed by precipitation before analysis. Even small amounts of silver(I) will cause problems. More than 0.2% of silver(I) causes a rough deposit of copper metal, making handling difficult; more than 0.4% will also affect the accuracy of the method. Some nitric acid must be added to the electrolyte to prevent the reduction of hydrogen ions to hydrogen gas during the latter stages of the deposition. If too much nitric acid is used, and more than 0.5% of iron(III) is present, quantitative deposition of copper may also be inhibited. The separation of copper(II) from other metal ions by this method is also possible; this is discussed in Section 19-1.

The sequential determination of copper(II) and then nickel(II) is possible in mixtures of the two metal ions. The copper(II) is deposited from acid solution as above, leaving nickel(II) in solution since its $E°$ is -0.25 V. The electrodeposition is temporarily stopped after electrodeposition of copper is complete, and the cathode is weighed to determine the weight of the copper deposit. The cathode is then reinserted into the solution, and the solution is neutralized with ammonia to form the $Ni(NH_3)_6^{+2}$ ion. The pH at this point is such that hydrogen ion cannot be reduced to hydrogen gas before $Ni(NH_3)_6^{+2}$ is reduced to nickel on the cathode. The electrodeposition of nickel is then performed, and the cathode plated with both copper and nickel is again weighed. The increase in weight over the previous weighing gives the weight of nickel.

The silver(I) ion has the most positive $E°$ in acid solution and could theoretically be determined in the presence of any other metal ion. However, it is not deposited quantitatively in acid solution. It can be deposited quantitatively from an ammonical solution or from an alkaline cyanide solution. The reaction at the cathode for the former complex ion is

$$Ag(NH_3)_2^+ + e^- = Ag°(s) + 2NH_3 \qquad E° = +0.375 \text{ V}$$

Since the $E°$ for the electrodeposition of the $Cu(NH_3)_4^{+2}$ ion is -0.06 V, the diammine silver(I) ion can still be electrodeposited in the presence of the tetrammine copper(II) ion in ammonia solution.

Lead(II) ion is always oxidized at the anode to lead(IV) ion and is deposited as lead(IV) oxide. The reaction at the anode is

$$Pb^{+2} + 2H_2O = PbO_2(s) + 4H^+ + 2e^- \qquad E° = +1.455$$

(The $E°$ value is for the *reduction* of lead(IV) oxide.) Lead(II) and copper(II) can be deposited simultaneously, the lead(II) at the anode, and the copper(II) at the cathode, even in the presence of other ions, except of course silver(I).

Potential Requirements for Electrolysis. The potential, E, required to start electrodeposition depends on the potential at the cathode, E_c, the potential at the anode, E_a, the overpotential at the cathode, E_{oc}, the overpotential at the anode, E_{oa}, and the cell resistance, R_c. The equation relating these is

$$E = (E_c + E_{oc}) - (E_a + E_{oa}) + R_c$$

Since metal ions are generally determined by electrolysis, the potential at the cathode, E_c, is the most important variable in the equation above. This variable depends on the concentration of the metal ion to be electrodeposited, and it can be calculated using the Nernst equation.

$$E_c = E° + 0.059/n \log [M^{+n}] \qquad (18\text{-}3)$$

where $E°$ is the standard reduction potential of metal ion M^{+n} and n is the number of electrons required to reduce it to metal $M°(s)$.

For example, if $0.010M$ copper(II) ion is to be electrodeposited at the cathode as copper metal, then E_c is calculated using **18-3** as follows:

$$E_c = +0.337 \text{ V} + \frac{0.059}{2} \log [0.010M \text{ Cu}^{+2}]$$

$$E_c = +0.337 \text{ V} - 0.059 \text{ V} = +0.278 \text{ V}$$

Of all the parameters contributing to E, E_c varies most significantly during electrodeposition. From the calculation above, you can see that E_c will become less positive as the concentration of Cu^{+2} decreases during deposition. For example, if $[Cu^{+2}]$ were $1.0 \times 10^{-6} M$ at the end of the electrodeposition, then E_c would be $+0.160$ V.

18-3 COULOMETRIC ANALYSIS

Faraday's Law and the Coulomb

Coulometry, or coulometric analysis, depends on the measurement of the amount of electrical current required to reduce or oxidize an ion in solution. It is a more general method than electrogravimetry because not many ions are quantitatively deposited on an electrode, and because it can be used for reactions producing soluble products or gases as well as metals.

The key to understanding coulometry is knowing Faraday's law. This law is based on two electrical units—the *coulomb* (C) and the faraday (F). The coulomb is defined in terms of amperes of electrical current and the total time it flows in seconds.

$$C = A \cdot s \qquad (18\text{-}4)$$

The faraday may be defined in terms of the coulomb and in terms of moles of electrons. The faraday is frequently defined to three significant figures as 9.65×10^4 C, but for analytical work requiring more than three significant figures, it is defined as

$$F = 96{,}487 \pm 1 \text{ C} \qquad (18\text{-}5)$$

In terms of moles of electrons or equivalents, it is defined as

$$F = 1 \text{ mole of electrons} = 1 \text{ eq of a chemical species} \qquad (18\text{-}6)$$

(Recall that one mole of electrons will reduce one equivalent of a species by definition.) The relation between the number of equivalents, the weight in grams of a species that is oxidized or reduced, and n, the number of electrons involved per formula weight is

$$\text{eq} = \frac{g}{\text{form wt}/n} \qquad (18\text{-}7)$$

The number of equivalents of any species reduced or oxidized at a constant current for a given time in seconds may be obtained by combining **18-4** through **18-7**:

$$\text{eq} = \frac{C}{96{,}487 \text{ C/F}} = \frac{A \cdot s}{96{,}487 \text{ C/F}} \qquad (18\text{-}8)$$

Once the number of equivalents have been calculated, **18-7** may be used to calculate the number of grams involved.

EXAMPLE 18-1. A 100.0 ml volume of a copper(II) sulfate solution of unknown concentration is analyzed coulometrically by reducing the copper(II) to copper metal using a constant current of 2.000 A. (a) If 20.00 min are required for the electrolysis, calculate the equivalents of copper(II) in the sample, and the molarity. (b) Calculate the weight of the copper deposited at the cathode.

Solution to a: Using **18-8**, calculate the eq of copper(II).

$$\text{eq} = \frac{(2.000 \text{ A})(1200 \text{ s})}{96{,}487 \text{ C/F}} = 0.02487_4 \text{ eq}$$

The molarity of the copper(II) in the sample is

$$0.02487_4 \text{ eq}/2 \text{ eq per mole} = 0.01243_7 \text{ moles}$$

$$\frac{0.01243_7 \text{ moles}}{0.1000 \text{ liter}} = 0.1243_7 M$$

Solution to b: The weight of the copper deposited in grams is found by rearranging **18-7**.

$$g = (\text{eq})(\text{form wt}/n)$$

$$g = (0.02487_4 \text{ eq})(63.54/2) = 0.7902_5 \text{ g}$$

Coulometric Methods and Apparatus

There are two experimental approaches to coulometric measurement—constant potential (applied voltage) coulometry and constant current coulometry.

In constant potential coulometry the potential is set at the desired potential and the electrolysis is allowed to run until the reaction is complete, at which point the current is essentially zero. The coulombs of current used can be calculated in one of two ways. Classically, a separate electrolysis cell, or *coulometer*, is connected in series with the working cell. For example, silver(I) may be deposited in the coulometer, and the silver metal weighed to give the coulombs of current used. The modern way of measuring coulombs is to record a plot of current in amperes *vs.* time in seconds and to calculate the area under this curve to give the total coulombs used. Various mechanical and electronic devices have been used to integrate the area under the curve automatically [1].

In the constant current approach to coulometry the time necessary to complete the electrolysis, rather than the current, is measured. A digital electric timer which can be turned off automatically is used to record the exact time. The coulombs are calculated from the magnitude of the current and the time.

$$C = A \cdot s$$

The number of coulombs are then converted to equivalents using Faraday's law **(18-8)**. In order to tell when the reaction has ceased, one of the various indicating techniques used in conventional titrations is used. Visual indicators can be used, or potentiometric detection of the end point can be used. There is also an

[1] J. J. Lingane, *Electroanalytical Chemistry*, 2nd ed., Wiley-Interscience, 1958, p. 340–50.

amperometric method employing two electrodes; a diffusion current that results from excess titrant or excess sample ion is used to signal the end point (Sec. 18-4).

Coulometric methods may be *direct* or *indirect*. In the direct method the species being determined is reduced at the cathode or oxidized at the anode. In the indirect method the species being determined does not react at either electrode but reacts with a second species generated at one of the electrodes. This second species is called a *titrant* even though it is not delivered from a buret, and the indirect method is also frequently called a *coulometric titration*.

A schematic diagram of a coulometric apparatus for an indirect coulometric method is shown in Figure 18-2. The generating electrode is shown at the left. If it is to generate a metal ion, it will be an anode, and the other electrode will be the cathode. (No method of indicating the end point is shown.)

The constant current source is either a battery in series with a large resistance, or a regulated constant current device automatically controlled with an electromechanical or electronic regulator.

The timer may be a precision electric stopwatch that can be turned on and off in perfect synchronization with the current. A better, but more expensive, timer is a clock with a solenoid-operated brake, which has a deviation of ± 0.01 s per complete cycle.

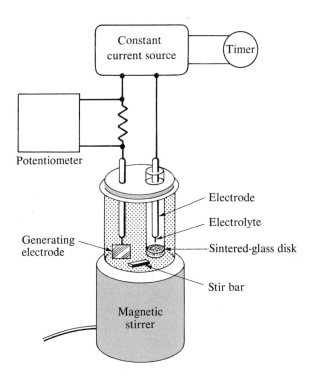

FIGURE 18-2. Schematic diagram of a constant-current coulometer.

Ions Commonly Determined

Coulometric methods are especially useful for the determination of small amounts of reducible or oxidizable species. This is because an amount of electricity as small as 0.1000 C can be measured with a relative deviation of 0.1 pph. From **18-8**, this can be calculated to be

$$0.1000 \text{ C} = 0.000001036_4 \text{ eq} = 1.036_4 \times 10^{-6} \text{ eq}$$

If this amount is dissolved in 10 ml of solution, it will be roughly a $10^{-5} M$ solution, a fairly low concentration.

Direct coulometric methods are not used for many determinations because side reactions occur in many cases near the end of the electrolysis. Indirect coulometry, or coulometric titration, is used much more because the *titrant* generated at one of the electrodes is used up immediately in a chemical reaction before any side reactions can occur. Coulometric titrations also have the advantage of being able to utilize unstable reagents, such as silver(II) and copper(I), which are difficult to use in conventional titrations.

Determination of Chloride. From a medical viewpoint, the most important coulometric method is the determination of chloride ion in body fluids. The chloride ion is colorless and is not readily determined by colorimetry, so the determination of small amounts of chloride by coulometric titration is widely used.

The basic coulometric reactions for the determination of chloride are, first of all, coulometric generation of silver(I) from a silver wire *anode* immersed in the sample.

$$Ag°(\text{anode}) \rightarrow Ag^+ + e^-$$

As it is generated, the silver(I) ion reacts with the chloride ion.

$$Ag^+ + Cl^- \rightarrow AgCl(s)$$

At the point at which the silver(I) is no longer able to react with the chloride ion (equivalence point), it is then reduced at the *cathode* as follows:

$$Ag^+ + e^- \rightarrow Ag°(\text{cathode})$$

This reduction is detected amperometrically, wherein the sudden increase in diffusion current (Sec. 18-4) signals the equivalence point.

A commercially available instrument called the Cotlove Chloridometer has been designed for the coulometric determination of chloride. This is an automatic constant-current coulometer and utilizes the above reactions. The titration is performed automatically and uses the amperometric end point. The sudden rise in the amperometric current at the end point activates a microswitch that automatically terminates the generation of silver(I) ion from the cathode, and stops the timer. The chloridometer is standardized against a known chloride solution before analysis, so the time needed for the titration can be related to the concentration of the chloride directly. The results also may be printed out on paper tape by the instrument. Protein does not interfere with the titration because no excess silver(I) ion is generated. (If an excess of silver(I) were present, it would react with the protein and give high results.) This is an advantage because protein must be precipitated in many clinical methods before the determination.

Determination of Other Ions. Almost any ion that can be titrated by conventional methods can be determined by coulometric titration. As an example, consider the coulometric titration of arsenic(III). As a titrant, iodine is generated at a platinum anode by oxidation of the iodide ion.

$$2I^- \rightarrow I_2 + 2e^-$$

The iodine reacts with arsenic(III) as it is generated.

$$I_2 + As(III) \rightarrow 2I^- + As(V)$$

As soon as the first excess of iodine is generated past the equivalence point, it reacts with the starch indicator giving a blue-purple color.

$$I_2 + starch + I^- \rightarrow I_3(starch)^-$$

Iron(II) may be similarly determined by generating cerium(IV) titrant from a solution of cerium(III). The same approach can be used for many other ions.

18-4 POLAROGRAPHY AND AMPEROMETRIC TITRATIONS

Polarography: Voltammetry at a Mercury Cathode

Voltammetry involves the measurement of small amounts of a chemical species by recording the current at various applied voltages. A voltammogram, or curve of current *vs.* applied voltage, is used for the determination of the species. The *working electrode*, or electrode at which the species to be determined reacts, can be either the anode or the cathode. The first reproducible voltammetric apparatus was devised by the Nobel laureate Jaroslav Heyrovsky in 1922 [2]. He used a dropping mercury cathode as his working electrode. He called his technique polarography, and the plot of current *vs.* voltage a *polarogram*. Although polarography is really just one specialized form of voltammetry, it has come to be used so much for analysis that we will confine our discussion to it.

Let us begin our consideration of polarography with the polarogram rather than with the apparatus. The polarogram is really a plot of the *diffusion current*, i_d, *vs.* the applied voltage, E. One reason for this is that the solution is not stirred during polarography, so the current passing through the polarographic cell depends on the rate of diffusion of ions to the dropping mercury cathode.

[2] J. Heyrovsky, *Chem. Listy* **16**, 256 (1922).

Since polarography is used for the *reduction* of chemical species, let us consider the hypothetical reduction of two metal ions, M_1 and M_2, in a polarographic cell. We will stipulate that M_1 is more easily reduced (has a larger positive $E°$) than M_2. Figure 18-3 is a *hypothetical* polarogram for this hypothetical reduction.

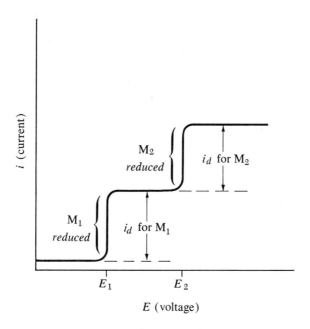

FIGURE 18-3. *Hypothetical current-voltage curve obtained for stepwise reduction of metal ion M_1 and metal ion M_2.*

We will start the polarographic reduction with the applied voltage at zero; we note that no current flows at this voltage ($i_d = 0$). We then increase the voltage until it reaches a value corresponding to E_1, the voltage at which M_1 is reduced. At this point a current begins to flow. The current increases until it reaches a limit governed by the rate at which ions of M_1 diffuse to the mercury cathode. The greater the concentration of M_1, the greater the current. The distance i_d for M_1 is called the diffusion current for M_1. The relation between i_d and the concentration of M_1 is

$$i_d = k_1 [M_1] \qquad (18\text{-}9)$$

where k_1 is a proportionality constant and $[M_1]$ is the molarity of M_1 in the solution. It should be noted that the amount of M_1 reduced during polarographic measurement is negligible, so the $[M_1]$ term is constant during the recording of the diffusion current.

If the voltage is increased above E_1, the current will remain constant for a time because M_1 is still being reduced. When the voltage reaches a value corresponding to E_2, the current begins to increase again until it again reaches a limit governed by the rate at which ions of M_2 diffuse to the mercury cathode. The total diffusion current at this point is the sum of that from M_1 and that from M_2. However, the diffusion current from M_1 can be subtracted graphically (Fig. 18-3) from the total to give the i_d for M_2. The relation between i_d and the concentration of M_2 is similar to **18-9**.

$$i_d = k_2[M_2] \qquad (18\text{-}10)$$

Because i_d in **18-10** is smaller than i_d in **18-9**, the concentration of M_2 is very likely smaller than that of M_1. However, the values of k_1 and k_2 would have to be determined before we can be sure that this is so.

The hypothetical polarogram in Figure 18-3 is said to exhibit two waves, or steps, the first starting at point E_1 and continuing to point E_2 with a *height* corresponding to i_d for M_1. The second wave starts at point E_2 and continues to the end of the curve; it has a *height* corresponding to i_d for M_2.

An Experimental Polarogram. Let us consider an actual experimental polarographic reduction of metal ions M_1 and M_2. Since many mercury drops form and fall off the dropping mercury cathode during one polarogram, it is necessary to

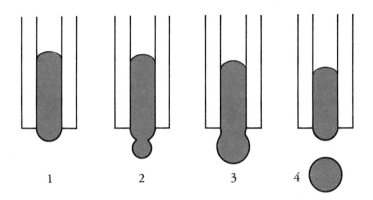

FIGURE 18-4. Stages of formation of a mercury drop from capillary in dropping mercury cathode. 1, drop starts to form and current is at a minimum. 2, drop grows and current increases slightly. 3, drop reaches maximum size and current increases slightly to maximum. 4, drop falls off, new drop starts to form, and current decreases to same magnitude as in stage 1.

visualize the stages of growth of a mercury drop and their effect on the diffusion current. As shown in Figure 18-4, the drop starts out as a small *bubble* (stage 1) with only a moderate surface available for reduction of the ions diffusing to it. The diffusion current is at a minimum at this stage. The drop grows through stages 2 and 3 to a maximum and so does the surface available for reduction of the ions diffusing to it. The diffusion current also increases to a maximum. Then the

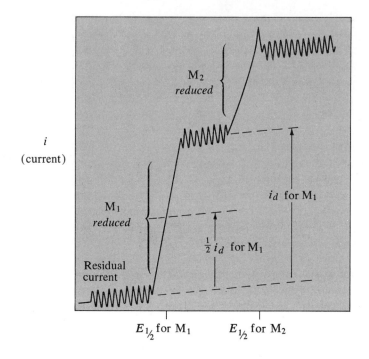

FIGURE 18-5. *Actual current-voltage curve obtained for stepwise reduction of metal ion M_1 and metal ion M_2.*

drop falls off (stage 4) and the surface and diffusion current decrease somewhat to a minimum again. The whole process repeats itself many times during the course of one polarogram.

We can now proceed to the actual polarogram of the reduction of ions M_1 and M_2 (Fig. 18-5). As the applied voltage is increased from zero to E_1 we observe a *residual current* rather than none at all. This is partly the result of the so-called *double layer* of negative and positive charges around the mercury drop. We note that here, as well as elsewhere, the curve is not a straight line, but has a sawtooth shape. This is the result of the changing size of the mercury drop.

The curve does not exhibit a sharp perpendicular break for the reduction of M_1, but instead it increases at an angle less than 90°. Instead of all the ions of M_1 being reduced at one voltage, they are reduced at slightly different voltages. This is because the ions themselves have slightly different energies. The average voltage at which the M_1 ions are reduced is symbolized as $E_{1/2}$, the *half-wave potential*. It is the value of the applied voltage at which the diffusion current is one-half the value of i_d for M_1. The half-wave potential, like the standard potential, is a constant that is characteristic of a redox half reaction such as $M^{+n} + ne^- = M°(s)$. It does not depend on concentration, and it can be used for identifying a particular ion in a simple mixture. (Too many species have similar values of $E_{1/2}$ for this approach to be used for complex mixtures.)

The wave for the reduction of M_2 is similar to that for M_1, except that it exhibits a *maximum* at the beginning. The reasons for such a maximum are not completely understood, but it is thought that the maximum is an effect arising from adsorption onto the mercury surface. Surface-active agents such as gelatin can be added to eliminate such a maximum.

The polarogram in Figure 18-5 can only be obtained if the solution is kept free of oxygen by bubbling nitrogen gas through it for at least five minutes. If oxygen is not removed, it is easily reduced at the dropping mercury cathode in two steps (giving two waves) as follows:

$$O_2 + 2H^+ + 2e^- = H_2O_2 \qquad E_{1/2} = -0.05 \text{ V (V SCE)} \qquad \text{(18-11)}$$

$$H_2O_2 + 2H^+ + 2e^- = 2H_2O \qquad E_{1/2} = -0.9 \text{ V (V SCE)} \qquad \text{(18-12)}$$

The half-wave potentials given for these reactions are given with reference to the saturated calomel electrode (SCE).

Polarographic Apparatus and Methods

A schematic diagram of a polarographic cell is given in Figure 18-6. Nitrogen gas is bubbled into the solution through a separate side arm to remove oxygen and to prevent the reactions given in **18-11** and **18-12**. The anode is usually a saturated calomel electrode which serves to complete the electrical cell as well as to act as the reference electrode. The anode is kept in electrical contact with the sample solution via a bridge plugged with agar.

The mercury is kept in a mercury reservoir and is forced through the capillary tip (Fig. 18-4) into the sample solution. Let us assume that we are going to reduce a solution of Cd^{+2} in a $0.1M$ potassium chloride *supporting electrolyte* (such an electrolyte eliminates electrical migration of Cd^{+2} so that it reaches the cathode only by diffusion). The cadmium(II) is then reduced at the cathode as follows:

$$Cd^{+2} + 2e^- + Hg(l) = Cd[Hg](l) \qquad E_{1/2} = -0.60 \text{ V (V SCE)}$$

As cadmium(II) is reduced, mercury in the calomel electrode is oxidized to mercury(I) and forms mercury(I) chloride.

$$2Hg(l) + 2Cl^- = Hg_2Cl_2(s)$$

The overall cell reaction is thus

$$Cd^{+2} + 2Hg(l) + 2Cl^- = Cd[Hg](l) + Hg_2Cl_2(s)$$

Fortunately, the amount of reduction of cadmium(II) ion is very small, so there is not much change in concentration of the chloride ions or loss of mercury metal in the calomel anode.

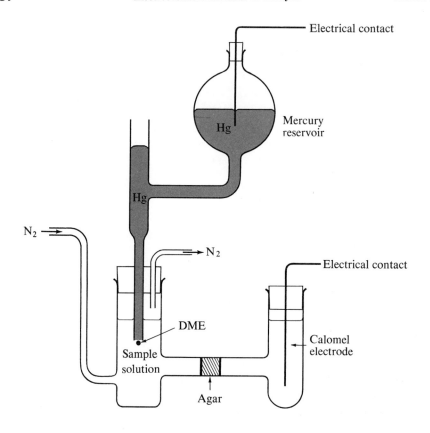

FIGURE 18-6. Schematic diagram of a polarographic cell, showing the dropping mercury electrode (DME). The anode is the calomel electrode.

The Ilkovic Equation. Under controlled conditions the relation between the diffusion current, i_d, in microamps, and the concentration, C, in mmoles/liter of a reducible species is given by the Ilkovic equation:

$$i_d = 607nCD^{1/2}m^{2/3}t^{1/6} \qquad (18\text{-}13)$$

where: 607 = a constant which is a combination of constants including the farad
n = number of electrons in the half reaction at the mercury cathode
D = diffusion coefficient of the species reduced in cm^2/s
m = mass of mercury passing through the capillary in mg/s
t = mercury drop time in s

Temperature does not appear in **18-13**, but D, the diffusion coefficient, generally undergoes a 1–2% change per degree Celsius at room temperature, so temperature should be controlled for careful work.

Quantitative Methods. There are several general methods available for quantitative analysis using polarography. The *absolute method* involves measuring all the parameters except C in **18-13** and then solving for C. This is difficult experimentally because of the effects of temperature and the nature of the solution on D, m, and t.

The *calibration plot method* is the best method for the analysis of large numbers of samples. The values of i_d for four to five solutions of varying concentrations of the sample species are measured and a plot of i_d vs. C of the species is made. This is usually a straight line, much like a Beer-Lambert law plot. The i_d of the unknown sample is measured, entered on the plot, and the concentration of the unknown read from the plot.

The *standard addition method* is recommended when only one to three samples are to be analyzed because it is faster, though somewhat less accurate. A polarogram of the unknown sample is recorded, and i_d of the sample species is measured. Then a known amount of a standard solution of the same species is added to the unknown sample, and a second polarogram is recorded, giving a value of i_d corresponding to the total of the sample species in the unknown and in the standard solution. If the volume of the standard solution is small enough compared to the volume of the unknown sample to be neglected, then the increase in the diffusion current, $i_{d(s)}$, the standard concentration, C_s, and the diffusion current from the unknown alone, $i_{d(u)}$, can be used to calculate the concentration of the sample species in the unknown, C_u, as follows:

$$C_u = C_s \frac{i_{d(u)}}{i_{d(s)}} \tag{18-14}$$

Species Commonly Determined. The useful applied voltage varied from $+0.2$ to -1.9 V *vs.* SCE on a polarograph with a DME. Any soluble species that can be reproducibly reduced within this range can be determined, in theory, by polarography. This includes many inorganic metal ions such as cadmium(II), lead(II), and others that can form mercury amalgams. Alkali metal ions such as sodium(I) and potassium(I) are not determined polarographically, unless a special electrolyte containing tetralkylammonium hydroxide is used. Inorganic anions, as long as they are reducible, can also be determined. For example, sulfate should be determinable, but sulfide is not because sulfur is in its lowest oxidation state in the sulfide anion.

Reducible organic and biochemical compounds can also be determined if they can be dissolved in a suitable polarographic solvent. For example, aldehydes and ketones are reduced to alcohols.

$$RCHO + 2e^- + 2H^+ \rightarrow RCH_2OH$$

In addition, sugars which contain aldehyde or ketone functional groups should also be amenable to polarographic determination. Glucose contains an aldehyde functional group and should undergo a two-electron reduction similar to that above.

$$C_5H_{11}O_5CHO + 2e^- + 2H^+ \rightarrow C_5H_{11}O_5CH_2OH$$
(glucose) (glucitol, or sorbitol)

Amperometric Methods

An amperometric method differs from a polarographic method primarily in that the applied voltage is kept *constant* in the amperometric method; in the polarographic method the voltage is *continuously varied*. There are at least two general amperometric methods. In one, the determination is done by comparing the diffusion current of the unknown sample to that of a standard sample and calculating the unknown concentration using an equation such as **18-14**. This approach is best for the determination of the same species in a large number of samples, for example, the determination of oxygen in aqueous solutions such as body fluids. In the second general method an amperometric detection of the end point is used.

Determination of Oxygen. Recall that oxygen is readily reduced to hydrogen peroxide (**18-11**) and then to water (**18-12**), using a mercury working electrode. It is also possible to reduce oxygen with an inert working electrode such as a gold cathode. The determination of oxygen using the commercially available oxygen meter is based on the use of such a gold cathode.

The probe holding both the gold cathode and a silver anode is inserted into the aqueous solution to be measured (Fig. 18-7). A constant voltage is applied to cause the reduction of oxygen. The oxygen diffuses through a membrane and is reduced at the gold cathode. The diffusion current that results depends on the rate of diffusion of oxygen through the membrane, which in turn depends on the pressure of the oxygen dissolved in the aqueous solution.

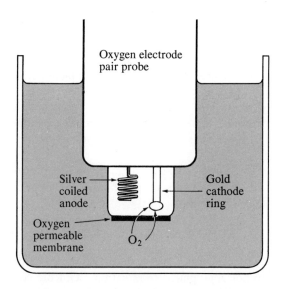

FIGURE 18-7. Oxygen electrode pair used in connection with oxygen meter (not shown) to reduce oxygen and give a diffusion current for amperometric measurement of oxygen.

The magnitude of the diffusion current reading on the oxygen meter (not shown) is calibrated by using an oxygen standard, such as fresh water at a given temperature. The fresh water has a known oxygen content at any given temperature.

Amperometric Titrations. An amperometric titration is one in which the equivalence point is located by means of measuring the change in the diffusion current throughout the titration, holding the applied voltage constant. A plot of i_d vs. the volume of titrant will show a change in slope at the equivalence point (Fig. 18-8).

To conduct an amperometric titration, the applied voltage is selected to cause reduction of the titrant, the sample species, or both. One example is the

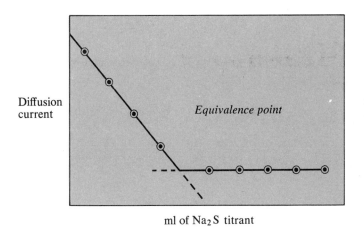

ml of Na_2S titrant

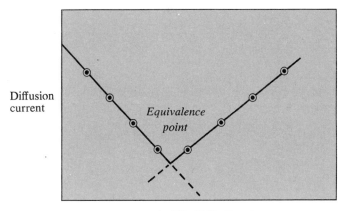

ml of K_2CrO_4 titrant

FIGURE 18-8. Two amperometric titration curves of lead(II) ion. Above: titration of lead(II) with sodium sulfide titrant. Below: titration of lead(II) with potassium chromate titrant.

titration of lead(II) by sodium sulfide, forming insoluble lead(II) sulfide. The potential is set so the lead(II) is reduced; sulfide ion of course is in its most reduced form and cannot be reduced further. The shape of the titration curve is shown in the top half of Figure 18-8. During the titration the concentration of lead(II) is continually decreasing, and so is the diffusion current. At the equivalence point the diffusion current is zero, and it remains zero as an excess of the titrant is added.

Another example is the titration of lead(II) by potassium chromate, forming insoluble lead(II) chromate as follows:

$$Pb^{+2} + K_2CrO_4 \rightarrow PbCrO_4(s) + 2K^+$$

Both of these species are reduced, but the chromate ion is not present to be reduced until after the equivalence point. As shown in the bottom half of Figure 18-8, the diffusion current increases beyond the equivalence point as the result of the excess of chromate ion added to the solution.

The scope of amperometric titrations is of course wider than these examples; any ion that is polarographically reducible can in principle be used in an amperometric titration.

QUESTIONS AND PROBLEMS

(Answers to most even-numbered problems are in Appendix 5.)

Concepts and Definitions
1. Define what is meant by an electrogravimetric method of analysis. How does it differ from a gravimetric method of analysis?
2. Explain the difference between
   a. An electrogravimetric method and a direct coulometric method of analysis.
   b. Electrogravimetry and polarography.
   c. A polarographic method and an amperometric titration method.
   d. Direct coulometry and indirect coulometry.
3. Explain why coulometry can be used to determine more elements than electrogravimetry.
4. Explain how oxygen can be determined amperometrically. Of what importance might this be for blood analysis?
5. If a metal does not form a mercury amalgam, can it be determined polarographically?
6. Indicate whether each of these anions should be able to be determined polarographically. (*Hint:* Are they reducible?)
   a. Sulfate ion
   b. Sulfide ion
   c. Iodate (IO_3^-) ion
   d. Iodide ion
   e. Triiodide (I_3^-) ion
   f. Iodine (I_2)
7. Define a diffusion current and tell how it can be used for quantitative polarographic analysis.

SEC. 18-4   Polarography and Amperometric Titrations

8. What is the nature of the titrant in
   a. An amperometric titration?
   b. A coulometric titration?

Quantitative Analysis Calculations

9. A solution containing an unknown concentration of lead(II) chloride gave a polarographic diffusion current of 1.74 microamperes. To 50 ml of this solution was added a 1.0 ml volume of lead(II) chloride such that the concentration of lead(II) chloride in the 50 ml solution from the 1.0 ml addition alone would have been $2.0 \times 10^{-4} M$. The polarographic diffusion current of the resulting mixture is 2.61 microamperes. Calculate the concentration of lead(II) chloride originally present in the 50 ml solution.

10. A solution of copper(II) nitrate is analyzed by electrogravimetry. Calculate the molarity of this compound if 0.3000 g of pure copper is deposited from
    a. A 25.00 ml aliquot.
    b. A 75.0 ml aliquot.

11. A solution contains $0.010M$ Cu^{+2} and $0.010M$ Ag^+, and it is planned to electrodeposit all of the Ag^+ before the Cu^{+2} begins to deposit. Assuming that all potential requirements for electrolysis at the cathode are the same except for the potentials calculated from the Nernst equation, calculate the percentage of Ag^+ remaining when Cu^{+2} begins to deposit.

12. A 25.00 ml aliquot of a 100 ml sample containing 1.100 g of an arsenic(III) sample is analyzed by coulometric titration, generating iodine from iodide. A constant current of 2.50 A requires 125 s to reach the starch end point, at which point all of the arsenic(III) is oxidized to arsenic(V). Calculate the percentage of arsenic in the sample.

13. In a coulometric determination of permanganate, iron(III) is reduced to iron(II), which then reduces the permanganate to manganese(II) ion. Titration is complete when a constant current of 0.00200 A runs for 11.0 min. Calculate either the molarity or normality of the permanganate solution, assuming a 25.00 ml volume.

PART III

Separations and Clinical Analysis

This part of the book is an elementary treatment of some selected separation techniques plus an introduction to clinical analysis. We are not attempting to cover every separation method because little useful information could then be included. This part can be used for two purposes—a brief survey for short courses, and a longer treatment for long courses.

For courses which require just an introduction to separation techniques we recommend a careful reading of Chapter 19. This chapter surveys both the single stage and multiple stage separation techniques, so the student will become somewhat familiar with techniques such as thin-layer chromatography and gas chromatography. The chapter also discusses liquid-liquid extraction at length, so the student will have an in-depth exposure to at least one separation technique which can be both single stage and multiple stage.

For longer courses we recommend a careful reading of Chapters 19, 20, and 21. After being exposed to the survey of separations in Chapter 19, the student can then master the theory of chromatography in Chapter 20 and the applications of that theory to separations in Chapter 21.

In both cases we recommend that students read Chapter 22 to see how clinical analysis is done.

19. Introduction to Separations: Liquid-Liquid Extraction

> *"Sherlock Holmes put the sopping bundle upon the table.... From within he extracted a* dumb-bell, *which he tossed down to its fellow in the corner."*
>
> ARTHUR CONAN DOYLE
> The Valley of Fear, Chapter 7

A separation is physically removing interferences from a species as completely as possible before measuring it. In the above quote Sherlock Holmes was actually performing a separation of sorts by "extracting" the dumbbell by hand. Although the separation of solid crystals has occasionally been done by hand, the chemist generally performs separations with the help of an instrument and/or specialized lab apparatus.

Some separations, such as liquid-liquid extraction, involve a simple transfer of a species from one phase to an immiscible phase of the same type. For a mixture of species A, B, and C in liquid phase 1, the separation of species A by removal into phase 2 may be symbolized as

$$(A, B, C)_{\text{ph 1(l)}} + (\ )_{\text{ph 2(l)}} \xrightarrow{\text{extraction}} (A)_{\text{ph 2(l)}} + (B, C)_{\text{ph 1(l)}}$$

Many more separations involve transferring a species from one phase to an immiscible phase of a different type. For a mixture of species A, B, and C in phase 1 (gas, liquid, or solid), the separation of species A by removal into phase 2 (a different phase than phase 1) may be symbolized as

$$(A, B, C)_{\text{ph 1}} + (\ )_{\text{ph 2}} \xrightarrow{\text{separation}} (A)_{\text{ph 2}} + (B, C)_{\text{ph 1}}$$

We will first survey the various types of separation methods with two purposes in mind—a brief survey of separation methods for shorter courses, and an introduction to separations to prepare you for Chapters 20 and 21. Then we will discuss liquid-liquid extraction as a detailed example of a useful separation technique.

19-1 SURVEY OF SEPARATION TECHNIQUES

Introduction

In this section we will survey various separation techniques rather than discuss them in detail. The key to understanding these techniques is to recognize that each involves a separation of two phases. Some separations involve a single stage or step, but since quantitative separation in a single stage is not common, most separations are based on multiple stages or steps. Some examples of each type of separation follow.

Single stage separation techniques
  Single step extraction
  Precipitation
  Electrodeposition

Multiple stage separation techniques
  Multiple step extraction
  Liquid chromatography: liquid-liquid and liquid-solid
    Column chromatography (atmospheric pressure)
    High-performance (pressure) liquid chromatography (HPLC)
    Thin-layer chromatography and paper chromatography
    Exclusion chromatography
  Ion-exchange chromatography
  Gas-liquid and gas-solid chromatography

We will begin with a discussion of the single stage separation techniques to establish the basic principles. Then we will introduce some of the multiple stage separation techniques.

Single Stage Separation Techniques

Since liquid-liquid extraction is to be discussed separately in the next section, we will confine our discussion to separations by precipitation and by electrodeposition. In both of these techniques there is a separation of phases that requires some time to accomplish.

Separation by Precipitation. In a separation done by precipitation, quantitative separation in a single step is often achieved because an excess of the precipitating reagent can be added without causing serious coprecipitation. Since a separation by precipitation is a necessary part of any gravimetric method of analysis, all of the gravimetric methods discussed in Chapter 5 may be considered as simple examples of this type of separation.

A separation by precipitation may of course be used in connection with methods other than gravimetric analysis. Frequently, the desired species is precipitated to separate it from interferences, and then it is measured by a rapid method of analysis such as spectrophotometric measurement. An example of this is the separation of cholesterol from interferences before measurement. In body fluids cholesterol occurs in the free form and in the esterified form.

$$C_{27}H_{45}OH \qquad\qquad C_{27}H_{45}O-\overset{\overset{\displaystyle O}{\|}}{C}-R$$

Free cholesterol   Esterified cholesterol
(a steroid alcohol)   (a steroid ester)

The Zlatkis-Zak [1] colorimetric method for cholesterol measures the sum of both the free cholesterol and esterified cholesterol. If free cholesterol alone is to be measured, then it is commonly separated by precipitating it with digitonin, $C_{55}H_{90}O_{29}$ as follows:

$$C_{27}H_{45}OH + C_{55}H_{90}O_{29} + H_2O \rightarrow C_{27}H_{45}OH[C_{55}H_{90}O_{29} \cdot H_2O](s)$$

The digitonin precipitation is quite specific for cholesterol; it will not precipitate esterified cholesterol or other sterols (steroid alcohols). Unfortunately, the precipitation is slow, requiring an overnight reaction. After the precipitate has been isolated, it is washed, dried, and dissolved before measuring the free cholesterol colorimetrically.

Separation by Electrodeposition. In a separation done by electrodeposition, quantitative separation in a single operation can be achieved as long as the voltage applied during the deposition does not reach a value where other metal ions are deposited at the cathode. Recall from Section 18-1 that electrodeposition, or electrogravimetry, involves the quantitative deposition of a *single metal ion* at the cathode. (See Fig. 18-1 for the electric cell and apparatus used for electrodeposition.) In this chapter then, we are concerned with a quantitative separation of one metal ion from others, as well as a quantitative deposition on the cathode.

The key to separating one metal ion from others is controlling the voltage applied to the cathode. By applying a voltage just large enough to start electrodeposition of a metal ion with the most positive $E°$ value, we can avoid deposition of metal ions with less positive $E°$s, like the metal ions listed in Table 19-1. Of the soluble ions listed, the metal ion most readily separated by electrodeposition at a cathode is silver(I) because it has the most positive $E°$. (Lead(IV) dioxide would of course not exist in solution.) The metal ion *most commonly* separated by electrodeposition at the cathode is the copper(II) ion. This is because copper is found in so many substances.

By looking at Table 19-1, you can see that copper(II) can easily be separated from alkali metal ion such as sodium(I), etc. However, unless the BiO^+ ion is absent, bismuth will be electrodeposited along with copper. Whether ions with an $E°$ similar to zinc(II) will interfere will depend on the voltage at the *end* of the electrodeposition of copper. In practice, this may be as low as -0.4 V; if electrodeposition were terminated at this point, neither zinc(II) nor alkali metals such as

[1] A. Zlatkis, B. Zak, and A. J. Boyle, *J. Lab. Clin. Med.* **41**, 486 (1953).

TABLE 19-1. *Potentials of Reducible Metal Ions*

Reduction half reaction	$E°$, volts
$PbO_2(s) + 2e^- = Pb^{+2}$	+1.455
$Ag^+ + e^- = Ag°(s)$	+0.800
$Cu^{+2} + 2e^- = Cu°(s)$	+0.337
$BiO^+ + 3e^- = Bi°(s)$	+0.32
$2H^+ + 2e^- = H_2(g)$	0.00
$Zn^{+2} + 2e^- = Zn°(s)$	−0.76
$Na^+ + e^- = Na°(s)$	−2.71

sodium(I) would interfere. They would be left in solution. (Silver of course would interfere.)

Another metal that can be separated readily is lead(II), which is oxidized at the *anode* to lead(IV) dioxide. It can be deposited at the anode frequently at the same time that copper is deposited at the cathode. Silver(I) can be separated from any of the metal ions listed in Table 19-1; the only interference would be some species that would have a similar or more positive $E°$ value.

Multiple Stage Separation Techniques

Since multiple step solvent extraction is to be covered in the next section, our discussion here will be concerned only with various types of chromatography. The term chromatography has come to include so many physically different separation techniques that it is difficult to define. The following are four probable, common characteristics of a chromatographic technique.

1. A chromatographic separation involves a distribution of the sample components between a stationary phase and an eluent (carrier) phase, or mobile phase.
2. The stationary phase is a bed of large surface area.
3. The eluent (carrier) is a fluid (liquid or gas) that percolates through or along the bed.
4. The sample components are eluted at different rates through the bed and emerge separated at different times.

A schematic diagram of a chromatographic separation is shown in Figure 19-1. A two component mixture of A and B is added to the bed contained in a column, and initially A and B remain evenly distributed at the top of the column (left). As eluent continues to be added, the faster moving A migrates ahead of B to some extent but still remains partially mixed with B (middle). Finally, a complete separation of A and B is achieved as essentially all of A migrates ahead of B (right). Ultimately, A is eluted out of the column before B reaches the bottom of the bed.

SEC. 19-1 — Survey of Separation Techniques

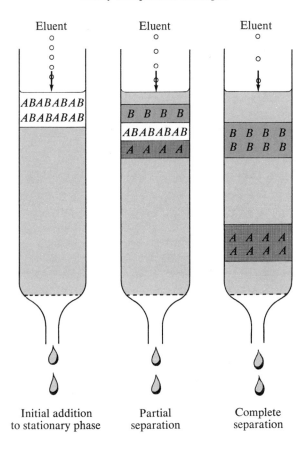

FIGURE 19-1. Schematic diagram of a chromatographic separation of species A, a fast-moving species, and species B, a slow-moving species, down a chromatographic column.

Liquid Chromatography. Liquid chromatography covers many different techniques. In addition, it is divided into two *mechanistic areas*—liquid-liquid (partition) chromatography, and liquid-solid (adsorption) chromatography. In *partition chromatography* a stationary bed is coated with a stationary aqueous or nonaqueous liquid. The eluent used is an immiscible liquid of the opposite type (nonaqueous or aqueous). The separation works much on the principle of a multiple step extraction with components dissolving in the stationary liquid to different degrees. In *adsorption chromatography*, the stationary bed itself interacts with the sample components; each component is adsorbed from the liquid eluent to a certain degree onto the stationary bed. In order for the bed material to interact with the sample, it must be a finely divided pure solid such as activated alumina, silica, nylon, or starch.

Since we are not as interested in the mechanism of liquid chromatography as in the various techniques themselves, we will proceed to a discussion of column chromatography at atmospheric pressure, high-performance (pressure) liquid chromatography, thin-layer chromatography, and paper chromatography.

Column chromatography is carried out at atmospheric pressure using fairly wide diameter columns (1–5 cm) packed with an adsorbent. The separation is done by hand; the sample is poured onto the column and elution proceeds as shown in Figure 19-1. Because the adsorbent is finely divided, elution is slow, requiring from thirty minutes to several hours. The components may be detected automatically as they emerge from the column if the components absorb visible or ultraviolet radiation. Similarly, an automatic fraction collector may be set up so that uniform-

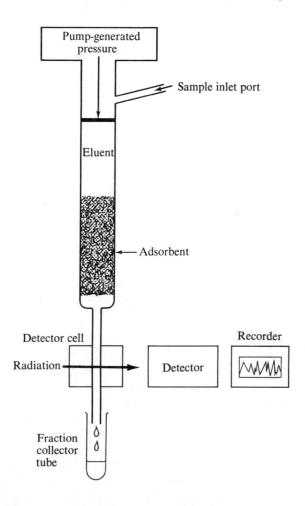

FIGURE 19-2. *Schematic diagram of a high-performance (pressure) liquid chromatographic column. Pump-generated pressure forces the eluent and sample through the bed of adsorbent, into the detector cell, and finally into a fraction collector tube.*

size fractions may be collected. This type of column chromatography is suitable for some qualitative and quantitative uses, but is *not efficient* enough for large numbers of samples involving more than a few components.

High-performance (pressure) liquid chromatography (HPLC) permits efficient, usually fast, column separation of even complex mixtures (Fig. 19-2). It employs a pump and operates at pressures high enough to achieve separations in minutes using micro-sized stationary bed particles. The column is closed to the atmosphere and constructed of metal to withstand the pressures of up to 10,000 psi that are used. The column is fairly thin, being of 2–10 mm inside diameter.

After a sample is introduced through a valve or inlet port onto such a column, liquid eluent is added to the column. The high pressure forces the eluent through the adsorbent at a high rate. The component that interacts the least with the adsorbent or stationary liquid phase is eluted first; its exit from the column is indicated by the detector system located just below the exit.

As each component leaves the column, it enters a very narrow tube which is part of the detector cell. Some type of radiation, such as UV radiation, passes through the cell and is partially absorbed by the component. (Alternatively, refractive index changes may be measured.) The absorption is measured by the detector and its output is displayed on the record graph. The component may then be collected in a fraction collector tube as part of an automatic fraction collector system. A fresh collector tube will be attached to the detector cell to collect the next increment of the eluent.

Exclusion column chromatography involves the use of a column packed with porous beads. The pores, or interiors, of the beads are held to within a certain size range; these pores hold a majority of the liquid mobile phase. An exclusion chromatographic separation is based on a slight, partial, or complete exclusion of different size molecules from the liquid mobile phase inside the stationary porous beads. As shown in Figure 19-3, the largest molecules are too large to penetrate the pores of even the largest beads. Intermediate size molecules can penetrate some of the pores of the beads, and small molecules can penetrate all of the pores of all beads. As a result, the largest molecules pass through the column first, followed by intermediate size molecules, and finally small molecules.

Exclusion chromatography is generally termed *gel permeation chromatography* when the system is composed of an organic solvent and a gel such as cross-linked styrene or porous silica beads. Because the amount of swelling and the resultant pore size are a function of the specific organic solvent used, the type of separation required can be achieved by picking the correct organic solvent.

Exclusion chromatography is termed *gel filtration chromatography* when the system is composed of an aqueous solvent and a hydrophilic solid such as dextran, acrylamide, or agarose gel. The aqueous solvent may be an unbuffered saline solution, a buffered saline solution, or a solution of a salt other than sodium chloride. Gel materials with as many as ten different ranges of exclusion size are available.

Thin-layer chromatography (TLC) involves chromatography on a thin (0.1–2 mm thickness) layer of adsorbent cemented to a rigid plate of thin glass or

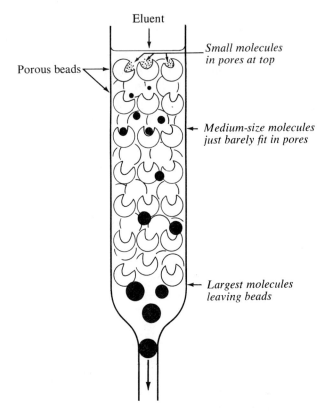

FIGURE 19-3. Schematic diagram of an exclusion chromatographic column separation showing the largest molecules leaving the column first and the smallest molecules traveling the slowest.

plastic. The adsorbent is usually a finely divided material such as silica gel or alumina. The samples are applied as uniform spots at a short distance from one end of the plate. Then the plate is placed vertically in a closed glass developing chamber filled to about a one cm depth with the eluting solvent (Fig. 19-4). The eluting solvent migrates up the plate by capillary action, eluting first the component least attracted to the adsorbent. Eventually all the components are eluted to some degree as the solvent front is allowed to reach a premarked point on the plate. The entire process is rapid, requiring from 15 minutes to an hour.

After separation each component may exist as an invisible or a colored spot on the plate. If it is invisible, it is visualized as dark spots on fluorescent plates or by spraying with a color-forming reagent like iodine in methyl alcohol or sulfuric acid. The iodine forms brown complexes with many organic compounds, and sulfuric acid forms a black carbon spot after heating. The visualization of the spot is adequate for qualitative analysis. Quantitative analysis requires special techniques discussed in Chapter 21.

Paper chromatography involves chromatography on a fairly thick sheet of

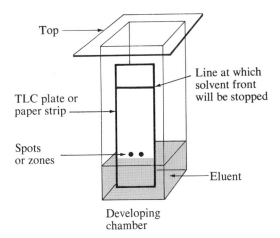

FIGURE 19-4. Block diagram of a developing chamber for a thin-layer chromatogram or a paper chromatogram.

uniform, pure paper, rather than on the thin, adsorbent layer used in TLC. (The paper is similar to that used for some filter papers.) The samples are applied as spots a short distance from one end of the paper, and the sheet is suspended in a glass developing chamber filled to a 1 cm depth with the eluting solvent (Fig. 19-4). Elution occurs as in TLC except that components partition themselves between the layer of solvent (usually water) adsorbed on the paper and the eluting solvent. Elution is usually slower than in TLC. After separation visualization is done by many of the same methods as in TLC, except that fluorescent paper is not available for use. If the compounds themselves fluoresce, this can be seen on the paper.

Ion-Exchange Chromatography. Ion-exchange chromatography is different than the previous types of chromatography because it is restricted to ions rather than molecules and because the interaction of the sample components is based on a chemical *reaction* rather than on a chemical or physical *interaction*, or attraction.

Ion-exchange resins are usually one of two types—cation-exchange or anion-exchange resins. The most common type of cation-exchange resin is a sulfonated polymer, which contains sulfonic acid groups, $-SO_3^-H^+$. The proton in the sulfonic acid groups can be exchanged for $+1$, $+2$, or $+3$ cations; for example,

$$2\text{Res-SO}_3^-H^+ + Ca^{+2} \rightarrow (\text{Res-SO}_3^-)_2 Ca^{+2} + 2H^+ \qquad (19\text{-}1)$$

The most common type of anion-exchange resin is that which contains a quaternary ammonium group, $-NR_3^+Cl^-$. The chloride ion in this group can be exchanged for -1, -2, or -3 anions; for example,

$$2\text{Res-NR}_3^+Cl^- + CO_3^{-2} \rightarrow (\text{Res-NR}_3^+)_2 CO_3^{-2} + 2Cl^-$$

To conduct an ion-exchange chromatographic separation of two cations, a column similar to that in Figure 19-1 is packed with a cation-exchange resin mixed with a solvent, frequently the eluting solvent. Then a solution of the sample containing, say, cations such as Ca^{+2} and Mg^{+2}, is poured onto the column. Both cations react with the resin (**19-1**) and are held on the resin. An eluent such as hydrochloric acid is then poured down the column, gradually displacing first the Ca^{+2} and then the Mg^{+2}. Each ion passes down the column and is detected as it leaves. An automatic fraction collector may be used to collect increments of the eluting solvent containing each ion.

Gas Chromatography. Gas chromatography involves both gas-solid and gas-liquid chromatography. We will discuss the latter rather than differentiating them in

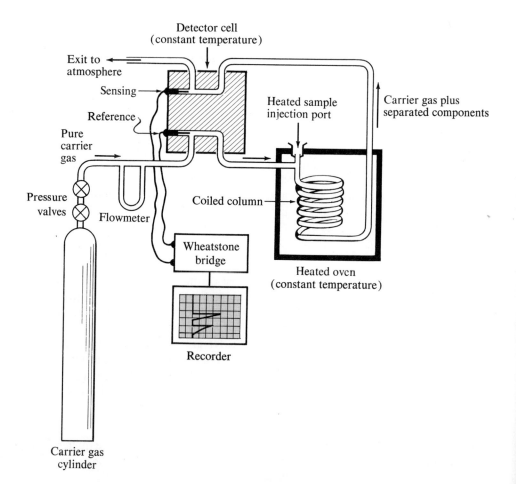

FIGURE 19-5. Schematic diagram of a gas chromatograph having a thermal conductivity detector cell. (Note that the detector measures the difference in thermal conductivity of pure carrier gas and gas plus components.)

this chapter. Gas chromatography involves chromatography on a solid support, or packing, in a coiled column. The solid support is coated with thin film of a nonvolatile organic liquid, called the stationary phase. The column is heated to maintain the sample in the gaseous state.

To separate the sample into its components, it is injected through the heated sample injection port (Fig. 19-5) into the column through which the carrier gas is flowing at a constant rate. (Helium is a typical carrier gas.) The temperature of the injection port causes the sample to volatilize if it is not already a gas. The carrier gas transports the sample to the column packing where it interacts with the liquid coating and is partially absorbed by it. The sample components are partitioned between the moving carrier gas and the liquid coating. The component that interacts least with the liquid coating is the first to be eluted, followed by others, all mixed with the carrier gas.

As each component leaves the column, it enters the detector cell (Fig. 19-5). Various types of detectors can be used to sense the emergence of each component; a thermal conductivity detector is shown in Figure 19-5. In this type of detector the thermal conductivity of the carrier gas mixed with a gaseous sample component is different from that of the pure carrier gas. This difference is measured electrically and displayed on a recorder to give an elution curve.

The detector indicates the presence of, and measures the quantity of, each component in the sample. Thus it is useful for qualitative analysis because it will indicate the number of components in a sample; however, it does not automatically indicate what the chemical identity of each is. Identification must be done by comparison with known compounds and their *retention times*, which is the number of minutes between the injection of the sample and its emergence from the column. Quantitative analysis is done on the basis of measurement of peak areas, or sometimes peak heights.

Individual components may be collected after being separated on some chromatographic columns, and further analysis may be performed to help identify the components.

19-2 LIQUID-LIQUID (SOLVENT) EXTRACTION

Principles

Liquid-liquid extraction involves the separation of two or more components in a particular solvent by preferentially dissolving one of them in a second solvent immiscible with the first. Since water is generally one of the solvents, solvent extraction is based on the fact that many organic liquids are not miscible with water, and are either heavier or lighter than water.

A *one-step* liquid-liquid extraction can be understood in terms of the following discussion. Suppose that an immiscible organic liquid is added to an

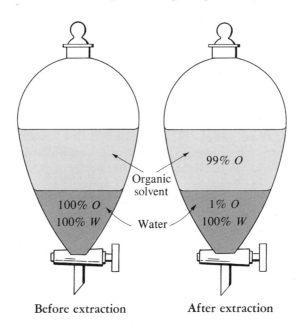

FIGURE 19-6. *Solvent extraction of organic-soluble substance O from organic-insoluble substance W into a lighter organic solvent.*

aqueous solution containing components O (organic-soluble) and W (organic-insoluble) in a separatory funnel (left, Fig. 19-6). Let us assume that the organic liquid has a smaller specific gravity (density) than water (sp gr = 1.0) so that it remains in a layer on top of the aqueous layer. We then shake the stoppered separatory funnel vigorously for a sufficient time for equilibrium to be attained. Since O is soluble in the organic layer, most of it passes into the upper layer. However, since O is also soluble in water, a small amount of it will remain in the lower aqueous layer (right, Fig. 19-6). This may be an amount such as 1%, or more, or less. Since W is insoluble in the organic solvent, it remains in the aqueous phase. In some cases a small amount of W will also be extracted into the upper layer, making the separation poorer than that shown in Figure 19-5.

If substance O is not extracted at least 99% into the organic layer, then the two layers will be separated by drawing off the heavier layer and extracting the aqueous layer with fresh organic solvent. This is a *multiple-step* extraction. More elaborate versions employing the principle of countercurrent extraction and the Craig Countercurrent Distribution Apparatus can be used for less favorable cases.

Choosing the Organic Solvent. Not just any organic solvent can be used to extract chemical substances from water, or vice versa. Obviously, it must be a good solvent for the substance being extracted, but it should also coalesce quickly and not form emulsions with water. Finally, it should also be significantly heavier or lighter than water so that two definite layers are formed. Some organic solvents that are suitable for extraction are listed in decreasing order of specific gravity as follows:

Chloroform, $CHCl_3$	1.49
Dichloromethane, CH_2Cl_2	1.31
Tributyl phosphate, $(BuO)_3PO$	0.98
Benzene, C_6H_6	0.88
Methyl isobutyl ketone, $CH_3COCH_2CH(CH_3)_2$	0.80
Ether, $(C_2H_5)_2O$	0.71

All of these are significantly different than water with respect to specific gravity, except tributyl phosphate. A lighter solvent such as benzene is frequently added to it to reduce the specific gravity and help layers separate more readily.

It is difficult to predict just what chemical species will be soluble in water and which will be soluble in organic solvents. In general, organic compounds with an oxygen atom or an —OH group will be soluble in water to a great extent if they have fewer than six carbon atoms per one oxygen (or —OH). When the ratio of carbon atoms to oxygens is more than six to one, the solubility decreases steadily as the ratio increases. Ionic substances of course tend to dissolve in water; if they have a large carbon chain on them, it is possible that they will dissolve in organic solvents if they can form an ion pair. The best way to approach the problem is to test the solubility in the laboratory.

Per Cent Extraction

As analytical chemists, we are interested in the *per cent extraction* of any component extracted into an organic solvent, or vice versa. This, however, varies with the relative volumes of the two solvents used and the initial amount of the extractable component present. We must therefore define some type of equilibrium constant to enable us to calculate the per cent extraction under any conditions.

If we are dealing with species which exist in only one form (does not ionize, etc.), we can define a partition coefficient, K_p.

$$K_p = \frac{[S]_o}{[S]_w}$$

where $[S]_o$ is the concentration of the species in the organic liquid phase, and $[S]_w$ is the concentration of the species in the aqueous phase. However, since many species exist in more than one form, most authors prefer to define a *distribution ratio*, D, as follows:

$$D = \frac{[C_s]_o}{[C_s]_w} = \frac{\text{(moles of all forms of } S)_o/V_o}{\text{(moles of all forms of } S)_w/V_w} \tag{19-2}$$

where $[C_s]_o$ is the concentration of all forms of species S in the organic phase, $[C_s]_w$ is the concentration of all forms of species S in the aqueous phase, and V_o and V_w are the respective volumes of organic and aqueous phases.

As an example of a real situation, the distribution ratio for the extraction of benzoic acid from water into the organic solvent benzene may be defined as

$$D = \frac{[C_{bz}]_o}{[C_{bz}]_w} = \frac{[HBz]_o + [HBz\ dimer]_o}{[HBz]_w + [Bz^-]_w}$$

where $[C_{bz}]_o$ and $[C_{bz}]_w$ are the total concentrations of benzoic acid in each phase. $[C_{bz}]_o$ is the sum of un-ionized benzoic acid $[HBz]_o$ and benzoic acid dimer $[HBz\ dimer]_o$; and $[C_{bz}]_w$ is the sum of un-ionized benzoic acid $[HBz]_w$ and ionized benzoic acid $[Bz^-]_w$. It is clear that defining a partition coefficient for such a situation would be much more difficult than defining a distribution ratio and dealing with the sums of the species. The disadvantage of using D is that it will vary with pH, so the pH will have to be constant for experiments where D is used.

Per Cent Extraction and D. The per cent extraction can be related to D, so you can calculate the per cent extraction corresponding to various theoretical or real values of D. Let us assume we will always be extracting a chemical species from water to an organic phase. Then % Ex, the per cent of species S extracted into the organic phase, will be

$$\% \text{ Ex} = \left[\frac{(\text{moles of all forms of } S)_o}{(\text{moles of all forms of } S)_o + (\text{moles of all forms of } S)_w}\right](100) \qquad (19\text{-}3)$$

By suitable manipulation of this equation and substitution of **19-2**, we obtain the relation between the per cent extraction and D as follows:

$$\% \text{ Ex} = \left[\frac{D}{D + (V_w/V_o)}\right] 100 \qquad (19\text{-}4)$$

where again V_w and V_o are the respective volumes of the aqueous phase and the organic phase. In the special case where V_w and V_o are equal, **19-4** simplifies to

$$\% \text{ Ex} = \left[\frac{D}{D + 1}\right] 100 \qquad (19\text{-}5)$$

Equations **19-3** and **19-4** apply only to a *single* extraction of an aqueous phase with an organic phase. The per cent extraction that is obtained for n repeated extractions with a fresh volume of organic solvent always equal to that of the aqueous phase can be derived to give

$$\% \text{ Ex} = 100\% - \frac{100\%}{(D+1)^n} \qquad (19\text{-}6)$$

In general, the per cent extraction for a given value of D will be larger for repeated extractions than if the volume of the organic phase is increased, as the following example will demonstrate.

SEC. 19-2   Liquid-Liquid (Solvent) Extraction   437

EXAMPLE 19-1. A hypothetical organic compound B has a partition coefficient of 10.0 for extraction into ether from water. (a) It is desired to extract B at least 99.0% into the ether. Will one extraction using equal volumes of ether and water be adequate? (b) If not, will it be better to increase the volume of ether relative to that of water, or use two or more repeated extractions?

Solution to a: Use **19-5** to calculate the per cent extraction.

$$\% \text{ Ex} = \left[\frac{10.0}{10.0+1}\right] 100 = 90.9\%$$

Obviously it is not adequate to use one extraction with $V_o = V_w$. We must either increase the volume of ether, or use more than one extraction.

Solution to b: First, let us consider using twice the volume of ether compared to the volume of water, and calculate the per cent extracted using **19-4**.

$$\% \text{ Ex} = \left[\frac{10}{10+2/1}\right] 100 = 95.2\%$$

The per cent extraction has increased from 90.9 to 95.2%, but it is still not a 99.0% extraction. Before considering further increases in the ratio of organic solvent to water for one extraction, let us consider two extractions with the same volume of ether as water. Using **19-6**,

$$\% \text{ Ex} = 100\% - \frac{100\%}{(10.0+1)^2} = 99.2\%$$

Obviously two extractions are more satisfactory than increasing the volume of the organic solvent.

The results of Example 19-1 are quite general for other values of *D*. Table 19-1 contains the results of calculations of the per cent extraction for all three of the above situations for various values of *D*. Several conclusions can be made from the table, depending on how we define a quantitative extraction. If we are willing to accept a 99.0% extraction as quantitative, we see that for a one-step extraction involving equal volumes of solvent, *D* must be above 90.0. If we double the volume

TABLE 19-1.   Relation of Per Cent Extraction to D

	$V_o = V_w$		$V_o = 2V_w$
D	% Ex for 1 extraction	% Ex for 2 extractions	% Ex for 1 extraction
1.00	50.0	75.0	66.7
10.0	90.9	99.2	95.2
20.0	95.2	99.8	97.6
50.0	98.0	99.9$_6$	99.0
90.0	98.9	99.9$_9$	99.4
500.0	99.8	99.9$_{994}$	99.9
900.0	99.9	99.9$_{999}$	99.9$_4$

of organic solvent used to extract, then D can be as low as 50.0, instead of greater than 90.0. Further, if we use two extractions with equal volumes instead of one, D can be as low as 10.0, instead of 50.0 or 90.0, for a quantitative extraction. If we desire a 99.9% extraction for quantitative purposes, the minimum values of D are much larger. For one step with equal solvent volumes, D must be above 900, a rather uncommon value. Even if two extractions with equal volumes are used, D must be 500.0 or larger, also a rather large value as extractions go. In such cases manual extractions involving one or more steps may be discarded in favor of a continuous automatic extraction procedure such as that involving the Craig Countercurrent Distribution Apparatus to be discussed.

Analytical Applications of Manual Extractions

Extractions are important in the analytical laboratory for isolating manually a chemical species to be determined, or a species that might interfere in the determination of another species. In some cases values for D have been reported; in other cases the per cent extraction is all that is reported; but in some instances only the fact that the extraction permits a "quantitative" separation is all that is given.

The classic inorganic example of extraction is the extraction of iron(III) from an aqueous $6M$ hydrochloric acid solution into ether, or diethyl ether as it is more accurately called. This extraction can be used to isolate iron(III) before its determination, but more often it is used to remove iron(III) as an interfering species. In $6M$ hydrochloric acid iron(III) forms the tetrachloroferrate(III) ion, $FeCl_4^-$, and it is extracted as the $H^+FeCl_4^-$ ion pair into ether. The distribution ratio for this extraction using $6M$ hydrochloric acid as the aqueous phase is 140. (It has a different value for other concentrations of hydrochloric acid.) The per cent of iron(III) extracted using equal volumes of $6M$ acid and ether is calculated using **19-5** as follows:

$$\% \text{ Ex} = \left[\frac{140}{140+1} \right] 100 = 99.3\%$$

If a more quantitative extraction such as 99.9% is desired, a larger ratio of ether to $6M$ acid can be used. The ratio needed can be calculated by rearranging **19-4** and solving for V_w/V_o.

$$\frac{V_w}{V_o} = 140 \left[\frac{100}{99.9\%} \right] - 140 = 0.140 = \frac{1}{7.14}$$

Thus about 7.14 volumes of ether per one volume of $6M$ acid will give a theoretical value of 99.9% extraction.

Certain other species are also extracted into the ether. Arsenic(III) chloride has a D value of 2 and is about 68% extracted into an equal volume of ether in one extraction step. In contrast, arsenic(V) chloride has a D value of 0.02 and is only about 2% extracted under the same conditions. Thus, the interference of arsenic(III) chloride can be eliminated by oxidizing the sample to produce arsenic(V) chloride before extracting. Antimony(V) interferes greatly, however, being 81% extracted, so a simple one-step manual extraction is not always feasible.

A classic organic example of extraction is the extraction of phenols from samples such as oil refinery waste water [2]. Phenols are weakly acidic and are converted to the phenoxide anion only in strongly basic solutions above pH 11.

$$C_6H_5OH + OH^- \rightarrow C_6H_6O^- + H_2O$$

If the refinery waste water is extracted at its normal pH, a number of other different types of organic compounds are extracted with phenol into carbon tetrachloride, the organic solvent used. First, however, the waste water is adjusted to pH 12 which converts the phenols to anions that are only soluble in water, not in carbon tetrachloride. The other types of organic compounds are then extracted away from the phenoxide anions into the carbon tetrachloride. The pH of the waste water is then adjusted back to 5, and the resulting phenols are extracted into carbon tetrachloride. The phenols can then be measured using spectrophotometry without interference from the other organic compounds.

A good clinical example of extraction is the extraction of cholesterol and cholesterol esters away from the protein in blood serum. Since the blood serum is primarily an aqueous medium, organic solvents such as ether-alcohol, or acetone, can be added to extract the cholesterol and its esters away from the protein which might interfere with the colorimetric determination of cholesterol. In Section 19-1 the further separation of cholesterol and cholesterol esters by precipitation was also discussed.

Countercurrent Extraction: the Craig Apparatus

So far we have discussed separations where the differences in the distribution ratio are fairly large and separations can be achieved by one to three manual extractions. Where the differences are small, countercurrent extraction may be used; the apparatus for this is known as the Craig Countercurrent Distribution Apparatus. The apparatus itself is rather complicated and an accurate description of it is rather involved. Essentially, it consists of a series of glass tubes mounted in a rack so that a whole set of stepwise extractions can be conducted simultaneously. When the solvent layers have separated themselves, the rack is tilted so that the upper, lighter phase in every tube is completely transferred to the upper part of the next tube.

Let us consider a simple example of a separation between two species, S and T. Suppose that they are present in equal quantities in a solvent which is heavier than the extracting solvent. Further, assume that T has a D value of 9.00 and that S has a D value of 1.00 for the extraction of each into the lighter extracting solvent. At the start each tube, except the first, in the Craig Apparatus will be filled with an equal volume of the same solvent as is in the sample. The first tube will be

[2] J. Schmauch and H. M. Grubb, *Anal. Chem.* **26**, 308 (1954).

440 Introduction to Separations: Liquid-Liquid Extraction CHAP. 19

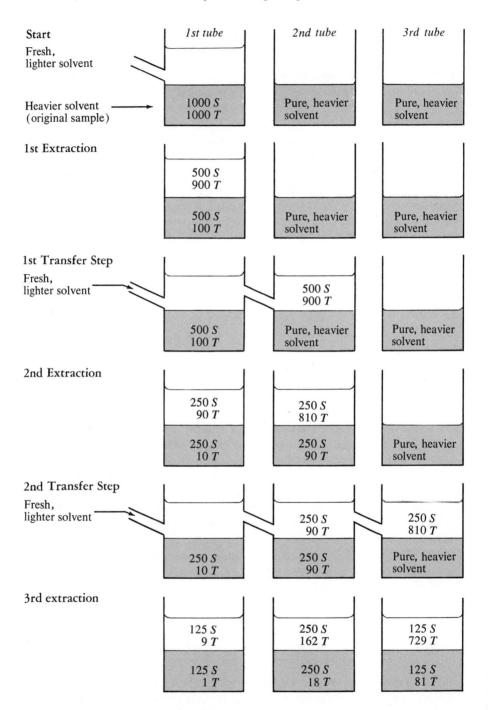

FIGURE 19-7. Schematic diagram of the first three extraction steps of the Craig Countercurrent extraction process. For species T, $D = 9.0$; for species S, $D = 1.0$.

filled with the sample. Then an equal volume of the lighter extracting solvent will be poured into the first tube (see *Start*, Fig. 19-7).

To simplify the example let us consider that 1000 molecules of S and 1000 molecules of T are present in the sample. After the first extraction (Fig. 19-7), half of the molecules of S will be found in each phase since D is 1.00. Exactly 90% of the original 1000 molecules of T, or 900 molecules, will have been extracted into the upper phase. Next, the first transfer step is carried out. The upper layer containing 500 S molecules and 900 T molecules is transferred into the second tube, and fresh extracting solvent (lighter) is poured into the first tube.

In the second extraction two extractions occur—one in the first tube and one in the second tube. In each tube 250 molecules of S end up in each phase because its distribution ratio is 1.00. Note, however, that only 10 molecules of T remain in the original sample phase in the first tube. This is because its larger D of 9.00 enables 90% of the T molecules in each lower phase to be extracted into the upper phase. Then the second transfer step is executed. The upper layer in each of the two tubes is transferred to the next tube, and fresh extracting solvent (lighter) is poured into the first tube.

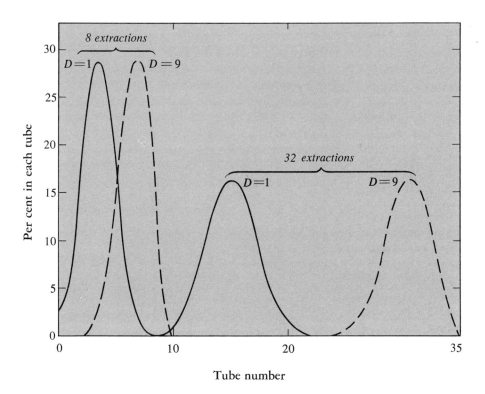

FIGURE 19-8. Theoretical distribution of molecules having D values of 1.00 and 9.00 after varying numbers of transfers in a Craig Countercurrent Distribution Apparatus.

In the third extraction three extractions occur—one in each of the three tubes shown in Figure 19-7 (in the actual Craig Apparatus there are many more than three tubes). Note that all but one molecule of T is now in the third tube, whereas most of S is in the second tube. It may seem odd that there is already no change in the number of molecules of S in each phase in the second tube, but this will change as soon as the third transfer occurs.

Of course, there is no separation at this point, but you should have now grasped the concept of the Craig Countercurrent extraction. Eventually, all of the T molecules will be found in the tubes toward the end of the apparatus, whereas the S molecules will tend to "bunch" up toward the center because D is 1.00. No matter how many tubes are used, all of T will be found farther along than all of S. This is illustrated in Figure 19-8 for two different numbers of extractions. After eight extractions, there is not a complete separation of T and S. When the number of extractions is increased to 32, there is a fairly good separation of the two species.

QUESTIONS AND PROBLEMS

(Answers to most even-numbered problems are in Appendix 5.)

Concepts and Definitions
1. Differentiate between the following pairs of separation techniques insofar as the number of stages or steps involved.
   a. Liquid chromatography and single step extraction
   b. Precipitation and paper chromatography
   c. Gas chromatography and electrodeposition
2. What is the difference between a separation by precipitation and a gravimetric method of analysis?
3. Give one point of similarity and one point of difference between a separation by precipitation and a separation by electrodeposition.
4. Why must cholesterol be separated by precipitation from cholesterol esters if the Zlatkis-Zak colorimetric method is to be used to measure cholesterol in body fluids?
5. Indicate whether the first metal ion in each of the following pairs can be separated without interference from the second metal ion by electrodeposition.
   a. Cu^{+2} and Ag^+ b. Cu^{+2} and Na^+ c. Ag^+ and Cu^{+2} d. Cu^{+2} and BiO^+
6. List the four common characteristics of a chromatographic technique.
7. Differentiate between liquid-liquid chromatography and liquid-solid chromatography.
8. Differentiate between thin-layer chromatography and paper chromatography.
9. List three advantages of thin-layer chromatography over paper chromatography including at least one advantage regarding visualization.
10. Describe the carrier and the stationary phase used in gas chromatography.
11. What is the difference between ion-exchange chromatography and liquid chromatography of organic compounds?

Solvent Extraction Concepts

12. Describe the nature of the two phases in a solvent extraction.
13. Indicate why organic solvents such as ethyl alcohol and acetic acid are not used to extract chemical species from water.
14. What is the difference between the partition coefficient and the distribution ratio?
15. Derive the relation between the per cent extraction and D, as expressed in **19-4**. (*Hint:* Start with **19-3**).
16. Why should two extractions using equal volumes be more efficient than a single extraction using a volume of the extracting solvent that is double that of the solvent containing the sample?
17. The distribution ratio for extracting iron(III) into ether from $4M$ hydrochloric acid is less than that for $6M$ hydrochloric acid. Explain.
18. Differentiate between a manual extraction involving thirty repeated extractions and an extraction involving a thirty-tube Craig Countercurrent Distribution Apparatus.

Solvent Extraction Problems

19. Calculate the per cent extraction for a single extraction into an organic phase from an aqueous sample phase of equal volume for each following value of D.
    a. $D = 0.100$ b. $D = 5.00$ c. $D = 30.0$ d. $D = 200.0$
20. Calculate the per cent extraction for a single extraction into an organic phase which is double (2.00:1.00) in volume that of the aqueous sample phase for each value of D given in the previous problem.
21. Calculate the per cent extraction for two extractions using a volume of organic phase that is each time the same as that of the aqueous sample phase for each value of D given in Problem 19.
22. A certain species has $D = 5.00$ for extraction into an organic solvent. Which will give the higher per cent extraction: a single extraction using a volume of organic solvent ten times higher than that of the aqueous phase? or two extractions of the aqueous phase with the same volume of fresh organic solvent each time?

20. Theory of Chromatography

> "*You have a* theory?"
> "*Yes, a* provisional one. *But I shall be surprised if it does not turn out to be correct.*" (*said Holmes.*)
>
> ARTHUR CONAN DOYLE
> The Yellow Face

Theories are proposed to be confirmed or disproved at some future time. Even Sherlock Holmes' theories were not always correct, as he found to his chagrin in the above story. Chromatographic theory is still in a partially unsettled state, although it does not appear that any of the present theory will be disproved in the near future. Two theoretical approaches, plate theory and rate theory, have been used by classical chromatographers to guide their research work. The classical treatment is a little like looking at the same situation twice through different glasses.

In the following treatment we have tried to approach chromatography from as *unified* a viewpoint as possible. We have tried to avoid classical phrases like the height equivalent of a theoretical plate and use the concept of plate height in a practical, real sense. We start our treatment with the concept of resolution, since, after all, the resolution of two or more sample constituents is the desired result. We then follow this with equilibrium and rate considerations, but without emphasizing two different theories and setting up artificial barriers to a unified understanding of chromatography. Thus the words *plate theory* and *rate theory* do not appear in the chapter. We hope that our treatment is at least partially successful and that you will understand enough chromatographic theory to read the following chapter with a greater depth of meaning.

20-1 CHROMATOGRAPHIC RESOLUTION

In this discussion we will focus on chromatographic separations in which, one by one, sample constituents on the stationary phase dissolve in or are eluted by the mobile phase as it moves by them. (This is in contrast to displacement chromatography where the mobile phase *displaces* sample constituents from the stationary phase.) If the stationary phase is packed in a column, the term *elution* is used to describe the process. As shown in Figure 19-1, the mobile phase passes over the stationary phase continuously until all the sample constituents are eluted from the column. If the stationary phase is an open sheet (thin layer or paper), the term *development* is used to describe the elution chromatography [1]. Because the mobile phase is stopped at a line short of the end of the sheet (Fig. 19-4), the sample constituents cannot be eluted from the stationary phase.

Because elution chromatography on a column *cannot be visualized*, it is generally *conceptualized* in most texts in terms of the detector response curve. To help you understand this, we will constantly compare it with developmental elution chromatography on thin layer plates, something that *can be visualized*.

Resolution

The quality of a chromatographic separation depends on the efficiency and the selectivity of the separation. Let us begin by considering efficiency in terms of first a single sample constituent, and then two constituents.

Efficiency. Each sample constituent can be viewed as existing and traveling in a *zone* on a thin-layer plate or in a column. On a thin-layer plate the constituent actually can be seen as a circular zone, or spot, on the stationary phase (Fig. 20-1, upper). In a column of the stationary phase the zone must be conceptualized in terms of the detector response curve plotted by a recorder (Fig. 20-1, lower). As the sample constituent leaves the column, its concentration will increase gradually to some maximum value, at V_r, the *retention volume*. (Note that the curve in the lower half of Fig. 20-1 is a plot of response against *volume* of mobile phase. The retention volume is the volume at which the constituent's concentration is at a maximum and at which half of the constituent ideally has been eluted.) The concentration of the constituent will then decrease until it is completely eluted from the column.

In qualitative terms the efficiency of a stationary phase may be thought of as how well it functions in keeping the constituent zone or the conceptual zone from spreading apart. In quantitative terms efficiency may be evaluated by first defining the quantity W (Fig. 20-1) as follows:

$$W = \text{diameter of constituent zone (TLC), or}$$
$$= \text{width of conceptual zone at its base}$$

For symmetrical zones the usual practice is to use the quantity σ (Fig. 20-1):

$$\sigma = \frac{W}{4} \tag{20-1}$$

[1] J. M. Miller, *Separation Methods in Chemical Analysis*, Wiley-Interscience, New York, 1975.

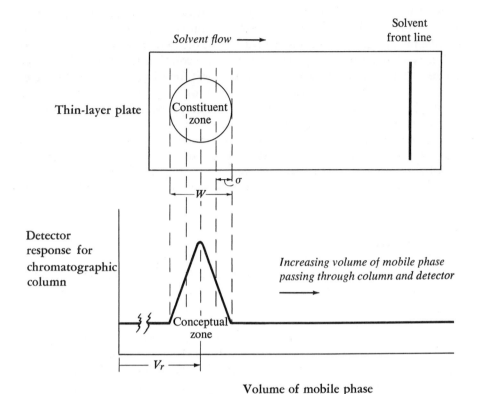

FIGURE 20-1. Constituent zone on a thin-layer plate (upper) and conceptual zone of a sample constituent eluted from a column (lower). The "W" term is the diameter of the thin-layer zone or the width of the conceptual zone at its base.

The σ term is sometimes called the quarter width of the zone. Column efficiency can then be defined in terms of L, the length of the column, and either W or σ.

$$\text{Efficiency} = \frac{W^2}{16 L}, \text{ or} \qquad (20\text{-}2)$$

$$\text{Efficiency} = \frac{\sigma^2}{L} \qquad (20\text{-}3)$$

In principle, the same definitions can be applied to thin-layer plates.

Selectivity. In addition to efficiency, the selectivity of a stationary phase is important in influencing the resolution of two or more chromatographic zones. In qualitative terms, selectivity is the separation of the *centers* of each of the constituent zones. Selectivity may be expressed quantitatively by symbolizing the distance between the zone centers as d. Note that in the upper half of Figure 20-2, d is readily seen to be the distance between the zone centers on a thin-layer plate.

SEC. 20-1 Chromatographic Resolution 447

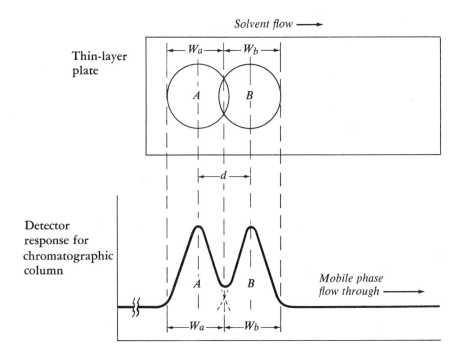

FIGURE 20-2. *Two nearly resolved sample constituent zones.*

However, for a chromatographic column, d is conceptually the distance between the maxima of the conceptual zones (Fig. 20-2, lower). We must bear in mind that the maxima *represent* the centers of the zones leaving the column, so d is also the distance between zone centers on a column.

Observe in Figure 20-2 (lower) that if we continued the "inner" half of each conceptual zone along the dotted lines shown, the dotted lines would intersect close to a point at the baseline. This represents a very small amount of overlap of zones A and B on a column.

Resolution. We are now ready to discuss the concept of resolution in terms of the efficiency and selectivity of the separation shown in Figure 20-2. It is easily seen that *resolution is directly proportional to selectivity* (distance between zone centers) and *inversely proportional to efficiency* (width of each zone). In other words, resolution is a measure of the overlap of any two adjacent zones.

Qualitatively speaking, we can also define resolution, R, as follows:

$$R = \frac{\text{zone separation}}{\text{zone width}} \tag{20-4}$$

Equation **20-4** clearly indicates that not only is resolution dependent on the separation of the zones, but that this separation is limited by how wide the zones are.

In quantitative terms, R can be defined in two slightly different ways using the symbols in Figure 20-2 as

$$R = \frac{d}{(W_a+W_b)/2} = \frac{2d}{(W_a+W_b)}$$

If the A and B zones have the same width or diameter, **20-4** becomes

$$R = \frac{d}{4\sigma} \tag{20-5}$$

(Recall from **20-1** that $W = 4\sigma$.)

Values for R can be calculated for various situations by assuming that σ for zones A and B is the same as the standard deviation from the normal distribution curve (Fig. 2-1). Note that a distance of $\pm 2\sigma$ from the center of the curve in Figure 2-1 includes 95.46% of the population. By subtraction we find that 4.54%/2 of the population is on one side or the other of the portion of the curve having a base with a total length of 4σ.

Let us now consider the separation of constituents A and B with zones as

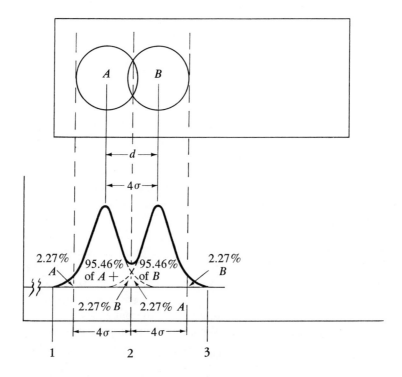

FIGURE 20-3. *The percentages of A and B in two nearly resolved constituent zones. Constituent A is collected from point 1 to point 2; constituent B is collected from point 2 to point 3.*

shown in Figure 20-3. At point 1 we begin to collect constituent A, at point 2 we begin to collect constituent B, and at point 3 we stop collecting. Assume that the distance from point 2 to the maxima of either zone is the same, 2σ. Then the distance between the maxima is 4σ. The width of each zone at the base will then be 4σ also. Using **20-4** or **20-5**, R is calculated as

$$R = \frac{4\sigma}{4\sigma} = 1.0$$

Next, let us develop an interpretation of a value of unity for R, always assuming two zones of equal width. To collect constituent A we can begin at point 1, at a distance greater than 2σ from the center of zone A because there are no other constituents on that side of the zone. We must stop collecting A at point 2 because the contamination from B will become too large. The percentages of A and B collected at point 2 are

$$\%A \text{ collected} = 2.27\% + 95.46\% = 97.7\%A$$

$$\%B \text{ collected} = 0\% + 2.27\% = 2.27\%B \text{ (contamination)}$$

To collect constituent B we begin at point 2, where the collection of A was terminated. Since there is no other constituent on the other side of the zone of B, we can collect B until point 3, where it has stopped coming off the column. The percentages of B and A collected from point 2 to point 3 are:

$$\%B \text{ collected} = 95.46\% + 2.27\% = 97.7\%B$$

$$\%A \text{ collected} = 2.27\% + 0\% = 2.27\%A \text{ (contamination)}$$

We see that a numerical value of 1.0 for R implies that the separation is a borderline one; the amount of contamination of A or B by the other is just beginning to be significant. Similar calculations for other values of R for two zones of equal size give the following interpretations.

d	R	%A collected	% contamination by B
2σ	0.5	84.13	15.87
4σ	1.0	97.7	2.27
6σ	1.5	99.87	0.13

Since a quantitative process usually implies 99.9%, a strictly quantitative separation would be achieved with an R value of 1.5.

20-2 EQUILIBRIUM CONSIDERATIONS

As noted in Section 19-1, chromatography is a continuous multiple stage separation technique. You might wonder whether it is possible for equilibrium to be achieved anywhere in such a multiple stage process. Consideration of the concept of a sample zone would seem to make it likely that equilibrium would be achieved in the *center* of the zone. In the center molecules of a particular constituent are effectively in complete equilibrium between the stationary phase and the mobile phase. This equilibrium can be described by the same D, or distribution ratio, as is used in liquid-liquid extraction (Sec. 19-2).

For example, in a liquid chromatographic system involving water as the mobile phase and benzene as the stationary phase, the distribution ratio for the chromatographic process has the same value as D for the liquid-liquid extraction process (**19-2**). Although this is well and good, values of D are unfortunately not among the fundamental parameters measured in chromatography. After discussing some of these parameters, we will describe how D is related to them.

Measured Chromatographic Parameters

Both development (planar) chromatography and column elution chromatography are characterized by a retardation parameter, R_r, called the retention ratio or retardation factor. First, let's consider R_r in general terms.

General Definition of R_r. R_r may be defined in terms of time as

$$R_r = \frac{t_m}{t_m + t_s} \tag{20-6}$$

where t_m is the time the average sample constituent molecule spends in the mobile phase, and t_s is the time that the same molecule spends in the stationary phase. R_r thus indicates the fraction of time that the average molecule spends in the mobile phase.

R_r may also be defined in terms of $\bar{u}$, the average velocity of the mobile phase, and $\bar{v}$, the average velocity of the sample constituent molecule as follows:

$$R_r = \frac{\bar{v}}{\bar{u}} \tag{20-7}$$

If R_r is 0.40, for example, this means that the average constituent molecule spends 40% of its time in the mobile phase, or travels 40% as fast along the stationary phase as the mobile phase. Similarly, if R_r is 0.0, this indicates that the constituent is not eluted or moved by the mobile phase at all. An R_r of 1.0 indicates that the constituent moves along as fast as the mobile phase and is not retained at all by the stationary phase.

Retardation Factor in Development Chromatography. In thin-layer and paper chromatography, both forms of development chromatography, R_r is not used. Instead a retardation factor, R_f, measured in terms of distance is employed. Recall that the sample is placed in a spot, or zone, at the bottom of a vertically aligned plate or sheet (Fig. 19-4), and each constituent is eluted from the spot, or zone, by the mobile phase traveling up the plate or sheet.

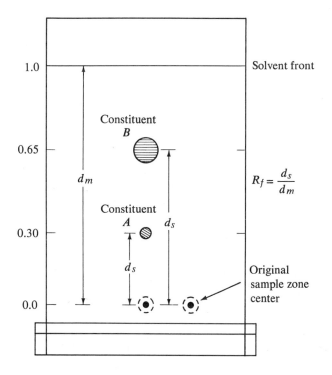

FIGURE 20-4. *Schematic diagram of the measurement of R_f on a thin-layer plate or paper chromatogram.*

As shown in Figure 20-4, an R_f is calculated for each constituent zone in terms of the distance traveled from the original sample zone.

$$R_f = \frac{d_s}{d_m}$$

In this equation, d_s is the distance the average constituent molecule travels along the planar stationary phase during the time the front of the mobile phase (solvent) travels the distance d_m. Note that d_s is measured from the *center* of the original sample zone to the *center* of the sample constituent zone. The measurement of d_m should also utilize the location of the bulk of the mobile phase, rather than the front, but this is a practical impossibility.

Note in Figure 20-4 that R_f values fall in the region of zero to one. Slower moving constituents, such as constituent A, have R_f values closer to zero, and faster moving constituents, such as constituent B, have R_f values closer to one.

Since the average velocity, $\bar{v}$, of each constituent should be proportional to d_s and since it *appears* that $\bar{u}$, the average velocity of the mobile phase, should be proportional to d_m, it appears that R_f can be exactly defined in an equation similar to 20-7. However, this is only approximately true since the *front* of the mobile phase or solvent creeps ahead a bit faster than the bulk of the mobile phase. Thus since d_m is only approximately proportional to $\bar{u}$, we can say

$$R_f \cong \frac{\bar{v}}{\bar{u}} \cong R_r \tag{20-8}$$

Strictly speaking, R_r is about 10% larger than R_f on the average, but we will neglect this difference in the rest of the discussion.

Column Elution Chromatography. In column elution chromatography we may measure elution or retention volumes rather than times (or distances). These volumes must be related to V_m, the volume of the mobile phase held in the column outside the stationary phase. (We have already defined the retention volume V_r as the volume of the mobile phase at which the concentration of the constituent has its maximum value.) Before any constituent can appear at the bottom of a column, a volume of mobile phase equivalent to V_m must pass through the column.

Now suppose that of two constituents being separated on a column, a volume of mobile phase equivalent to $2V_m$ is needed to elute constituent A to its maximum concentration at point V_r (Fig. 20-5). Further suppose that a volume of mobile phase equivalent to $4V_m$ is necessary for elution of constituent B to its maximum concentration at point V_r for B (Fig. 20-5). From this information it is possible to

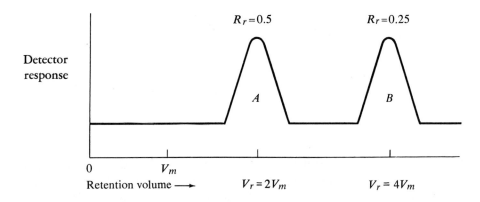

FIGURE 20-5. *Column elution separation of two constituents with different retention volumes.*

calculate the retardation ratio, R_r, as follows:

$$R_r = \frac{t_m}{t_r} = \frac{V_m}{V_r} \tag{20-9}$$

Using this equation, R_r for A is 0.5, and R_r for B is 0.25.

SEC. 20-2   Equilibrium Considerations

Equation 20-9 is derived as follows. Recall that t_m is the time the average molecule spends in the mobile phase. We can then define t_r as the longer time that corresponds to the time for the average molecule to be eluted to its maximum concentration at volume V_r (Fig. 20-5). Going back to 20-6, we see that t_s, the time the average molecule spends in the stationary phase, must be the difference between t_r and t_m

$$t_s = t_r - t_m \tag{20-10}$$

Substituting 20-10 into 20-6, we obtain

$$R_r = \frac{t_m}{t_m + (t_r - t_m)} = \frac{t_m}{t_r} \tag{20-11}$$

This gives the first half of 20-9. The second half is obtained by assuming that at constant flow rate, F, the following equations are true.

$$t_m = \frac{V_m}{F} \tag{20-12}$$

$$t_r = \frac{V_r}{F} \tag{20-13}$$

Substituting 20-12 and 20-13 into 20-11 gives

$$R_r = \frac{V_m/F}{V_r/F} = \frac{V_m}{V_r} \tag{20-14}$$

Relation between R_r and D

Recall that in Section 19-2 we defined a partition coefficient, K_p, and a distribution ratio, D (19-2), for liquid-liquid extraction. We can similarly define such constants for chromatography as

$$K = \frac{[S]_s}{[S]_m} \tag{20-15}$$

$$D = \frac{[C_s]_s}{[C_s]_m} \tag{20-16}$$

where $[S]$ is the concentration of a particular form of a solute in either the mobile or stationary phases, and $[C_s]$ is the concentration of all forms of the solute in either the mobile or the stationary phase.

We can also define a mass ratio constant, k' as

$$k' = \frac{(\text{moles})_s}{(\text{moles})_m}$$

For a constituent that exists in *only one equilibrium form*,

$$k' = K\left(\frac{V_s}{V_m}\right) = D\left(\frac{V_s}{V_m}\right) \qquad (20\text{-}17)$$

where V_s is the volume of the stationary phase, and V_m is the volume of the mobile phase. Equation **20-17** indicates that the partition coefficient and distribution ratio for a given separation system, such as liquid-liquid extraction or liquid-liquid chromatography, are directly proportional to the ratio of masses on the stationary and mobile phases.

Now how do we relate K or D to R_r, the retention ratio? First we define the mass ratio constant in terms of times as follows:

$$k' = \frac{t_s}{t_m} \qquad (20\text{-}18)$$

Rearranging,

$$t_s = (k')(t_m) \qquad (20\text{-}19)$$

and substituting **20-19** into **20-6**, we obtain

$$R_r = \frac{t_m}{t_m + (k')(t_m)} = \frac{1}{1+k'} \qquad (20\text{-}20)$$

Since R_r has the sort of inverse relationship to k' given in **20-20**, it also has the same relationship to K or D since they are proportional to k'.

Equation **20-20** is relevant to development chromatography since **20-8** indicates that R_f is approximately the same as R_r. Thus for thin-layer or paper chromatography,

$$R_f \cong \frac{1}{1+k'} \qquad (20\text{-}21)$$

Equation **20-20** can also be used to relate volumes in column elution chromatography as well. Combining the right-hand side of **20-20** and the right-hand side of **20-11**, we obtain

$$R_r = \frac{t_m}{t_r} = \frac{1}{1+k'}$$

Rearranging the above gives

$$t_r = t_m(1+k') \qquad (20\text{-}22)$$

We can now substitute volumes for times using **20-12** and **20-13** at a constant flow rate F:

$$\frac{V_r}{F} = \frac{V_m}{F}(1+k')$$

SEC. 20-2 Equilibrium Considerations

We then eliminate F and rearrange to obtain

$$V_r = V_m + k'V_m \tag{20-23}$$

Substituting the middle term of **20-17** for k', we can also obtain

$$V_r = V_m + \left(K\frac{V_s}{V_m}\right)V_m$$

which simplifies to

$$V_r = V_m + KV_s \tag{20-24}$$

again at constant flow rate of the mobile phase.

Applications. It is possible to apply some of the above equations in a general way to improve chromatographic separations. For example, in thin-layer chromatography, using a certain mobile phase may give R_f values in the 0.85–0.95 range, which means the constituents are moving too fast to be resolved effectively. By inspecting **20-21**, we see that R_f is affected only by the reciprocal of $(1+k')$. If values of k' are known or can be deduced through **20-17**, then a mobile phase with a *larger* value of k' should be substituted for the original mobile phase. Since k' for the original mobile phase must have been of the order of 0.1 to give an R_f of the order of 0.9, it is clear that a mobile phase with a k' of the order of 2 must be chosen to achieve an R_f value in the 0.5 range. If values of k' are not available, it may be possible to estimate k' from values of K_p or D from a liquid-liquid extraction system involving the same phases as used in the thin-layer separation. (This assumes that the thin-layer separation occurs via partition chromatography rather than adsorption chromatography.)

As a second example, assume that a sample constituent is being eluted too soon from a column for a good separation from the first constituent eluted from the column. If the separation is being effected by partition chromatography, the liquid stationary phase should be changed so that its V_r is increased to a greater extent than V_r for the first constituent. By means of either **20-23** or **20-24**, we are guided to choose a mobile phase with a higher k', or K, for the sample constituent. In effect, we are changing to a mobile phase which interacts *to a greater degree* with the constituent.

A final example may be taken from the field of gas chromatography. Suppose that all the sample constituents are eluted from the column too rapidly to be differentiated. In other words, V_r for all constituents is too small. This time, however, we will assume that the values of K are known to be satisfactory. Equation **20-24** predicts that we should then increase V_s, the amount of the stationary phase. In effect, we are giving the sample more reactive solvent to undergo a proper separation.

Application of the Separation Quotient (Factor). It is often useful to consider the separation of any two sample constituents, A and B, in terms of α, the separation quotient, or separation factor. The α term is usually defined as a ratio numerically greater than one, rather than less than unity. For liquid-liquid extraction of A and B (which we assume exist in only one form), it is defined as

$$\alpha = \frac{(K_p)_b}{(K_p)_a}$$

where constituent B is chosen to have the larger K_p. Similarly for development or column elution chromatography, it is defined

$$\alpha = \frac{(k')_b}{(k')_a} \tag{20-25}$$

where B again has the larger k' value. For column elution chromatography **20-25** and **20-23** can be used to show that α gives a ratio of the adjusted retention volume as follows:

$$\alpha = \frac{(V_r - V_m)_b}{(V_r - V_m)_a} \tag{20-26}$$

By reference to Figure 20-5, you can deduce that the adjusted retention volume is corrected for variation in V_m, the volume of the mobile phase held in the column.

By means of α, we can transfer separation information from, say, thin-layer chromatography to column elution chromatography. For example, for a thin-layer separation of A and B performed by partition chromatography, suppose that A and B have R_f values of 0.50 and 0.30, respectively. Then **20-21** can be used to solve for the respective mass ratio constants.

$(R_f)_a = 0.50 = \dfrac{1}{1+(k')_a}$ $\qquad$ $(R_f)_b = 0.30 = \dfrac{1}{1+(k')_b}$

$(k')_a = 1.0$ $\qquad\qquad\qquad\qquad\quad$ $(k')_b = 2.3$

Then α can be calculated using **20-25**.

$$\alpha = \frac{(k')_b}{(k')_a} = \frac{2.3}{1} = 2.3$$

This predicts that the ratio of the mass ratio constants for the column elution separation will also be 2.3 to 1. In addition, **20-26** predicts that the adjusted retention volumes for B and A will have the same ratio.

A Final Word on Column Length. The consideration of α ignores the effect of the length of a column on the resolution, or separation of two or more constituents. The resolution of two or more constituents can be improved by increasing the length of the column used. It can be shown that resolution is proportional to the square root of the length of the column. For example, if we quadruple column length, we will increase the resolution (**20-5**) by a factor of two. The limitations to this are that the time for the separation may be increased to be inconveniently long, or that the constituents may become too dilute to be isolated or to be measured accurately.

20-3 KINETIC CONSIDERATIONS

The discussion thus far has not provided any information on the relationship of the length of a column and the rate of the separation (kinetics). It is possible to establish such a relationship by defining specific parameters that control the length of a column and the separation rate, and then relating the parameters. We will begin with the length of a column.

A chromatographic column or thin-layer plate may be thought of as consisting of a number, N, of theoretical plates or stages at which equilibrium is *postulated* to occur. (The term *theoretical plates* originates from separations carried out by distillation, in which the distillation column is conceptually divided into theoretical plates.) The number of theoretical plates can be estimated empirically by

$$N = 16(d/W)^2 \qquad (20\text{-}28)$$

where d and W are defined as in Figure 20-2.

We will now indicate the relationship between N and L, the length of the column, or the migration distance in thin-layer chromatography. This is

$$L = HN \qquad (20\text{-}29)$$

where H is obviously the height of each theoretical plate, or simply the *plate height*. It turns out that H is also an efficiency parameter which characterizes zone spreading and is given by

$$H = \frac{W^2}{16L} \qquad (20\text{-}30)$$

It is H, not L, that is directly related to the rate of a chromatographic separation. Let us turn to that very point.

Rate factors Influencing H

The basic rate, or kinetic, consideration involves W, the width of a sample constituent zone (Fig. 20-1). As a constituent moves through the stationary phase, it invariably "spreads out" as it travels, so W is usually significantly larger at the end of the separation than at the start. It turns out that the rate at which the constituent moves through a column or up a plate affects the width or diameter of the constituent zone. This also affects both the efficiency and the plate height (**20-30**).

An equation that relates the plate height H, and the velocity v, of the mobile phase is the van Deemter equation:

$$H = A + B/v + Cv \qquad (20\text{-}31)$$

where A, B, and C are constants whose implications will now be described.

The A, or eddy diffusion term, is *independent* of the velocity of the mobile phase. It is a measure of the nonideal chromatographic behavior of constituent molecules traveling a large number of pathways through the stationary phase.

These pathways differ in length, so the molecules arrive at the end of the column or plate at different times.

The B/v, or longitudinal diffusion term, is *inversely proportional* to the velocity of the mobile phase. It is a measure of the molecular diffusion forces which cause migration from the center of the constituent zone toward the edges of the zone. This migration widens the zone, decreases the efficiency, and increases the plate height. As the velocity of the mobile phase is increased, longitudinal diffusion decreases.

The Cv, or mass transfer term, is *directly proportional* to the velocity of the mobile phase. It is a measure of the effect of the partitioning rate of the sample constituent between the stationary and mobile phases. As the velocity of the mobile phase increases, the time available for equilibration (partitioning) between the two phases decreases until at some point efficiency begins to decrease seriously and plate height increases significantly.

The cumulative effect of all these terms on H as velocity increases is shown in Figure 20-6 for gas chromatography and high-performance (pressure) liquid chromatography (HPLC). In gas chromatography the most desirable situation is to achieve a minimum value of H, since **20-29** predicts that for a constant L, the number of theoretical plates will increase as H decreases. At very low mobile phase velocities, the B/v term predominates and H is far too high for an efficient separation. At high velocities, the Cv term predominates and H is again too large. If the layer of stationary liquid phase can be adjusted to be fairly thin, column efficiency can be improved and the value of H will be lowered by a decrease in the

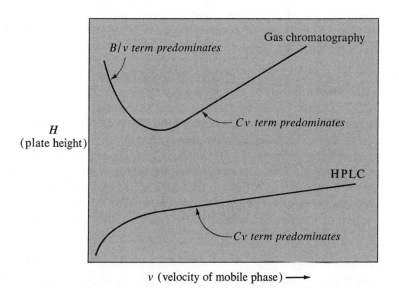

FIGURE 20-6. *Typical curve of plate height H versus velocity of the mobile phase for gas chromatography and high-performance (pressure) liquid chromatography (HPLC).*

value of C. Figure 20-6 does not predict what experimental adjustments are needed because of the A term, except that the value of A should be reduced. In fact, eddy diffusion can be reduced by careful packing of the column with small, uniformly shaped particles to avoid open channels.

In high-performance liquid chromatography (HPLC), the curve in Figure 20-6 does not exhibit a minimum because longitudinal diffusion and eddy diffusion are much slower than they are in gas chromatography. Therefore H gradually increases as a function of the Cv term. Because the increase in H is so gradual, rather high flow rates can be used to compensate for the longer columns needed to maintain a large number of theoretical plates. (Recall from **20-29** that $N = L/H$.) To achieve rapid equilibrium, it is then necessary to use a column packing consisting of small, uniformly shaped particles.

Summary

Let us now summarize very briefly the dependence of resolution on some of the key parameters. Increasing the column length, L, of course increases resolution in proportion to the square root of the length. There are limitations to this in time and constituent concentration, but we have already noted that HPLC is partly successful because of the use of long columns.

Smaller diameter columns reduce the effect of eddy diffusion and channeling of the sample constituents in directions which slow up the rate of longitudinal diffusion. Hence the need for uniformly shaped, small particles of the stationary phase or solid support for the stationary phase. Again, the use of this type of particle has made HPLC possible. Resolution is therefore increased by decreasing column diameter.

Since the sample constituent zones widen as the separation proceeds, the initial sample zone should be as narrow as possible to minimize the value of W, which in turn minimizes d (Figs. 20-2 and 20-3). Thus the sample should be as small as possible to minimize the size of the initial sample zone. Resolution is therefore increased by decreasing the sample size.

Finally, the velocity of the mobile phase affects the efficiency of the separation and H directly, rather than resolution. Since resolution is favored by greater efficiency, it is also favored by minimizing the flow rate (Fig. 20-6).

QUESTIONS AND PROBLEMS

(Answers to most even-numbered problems are in Appendix 5.)

Definitions and Concepts
1. Define the following terms.
    a. Chromatographic efficiency
    b. Chromatographic selectivity
    c. Chromatographic resolution

2. Explain the difference between chromatographic resolution, R, the retardation ratio, R_r, and the retardation factor, R_f.
3. Explain how the R_f value is measured in thin-layer or paper chromatography.
4. Explain from a theoretical point of view why R_f values are measured using the center of the zone rather than the front edge of the zone, as in the case of measuring the solvent.
5. Explain why an R_f value is not exactly the same as an R_r value for the same chromatographic system.
6. Explain the difference between t_m, the time the average constituent molecule spends in the mobile phase, and t_r, the time required for the average constituent molecule to be eluted to volume V_r (Fig. 20-5).
7. Explain why t_s, the time spent in the stationary phase for an average constituent molecule, is equal to $t_r - t_m$, the difference in the time spent in the mobile phase and the time required for elution to volume V_r (Fig. 20-5).
8. Describe the effect of resolution and column length on the separation of two constituents on a chromatographic column.
9. What effect does the flow velocity have on the separation of two constituents on a chromatographic column?

Problems
10. Consider the separation of constituents A, B, and C rather than just the separation of A and B shown in Figure 20-3. Assume that the zone of C overlaps the zone of B to the same degree ($d = 4\sigma$) that B overlaps the zone of A. Calculate
    a. The percentage of A, and the percentage of B collected at point 2.
    b. The percentage of B, and the percentage of all contaminants at point 3.
    c. The percentage of C, and the percentage of B collected at the point where C is completely eluted.
11. Repeat the previous problem using $d = 6\sigma$ for all three constituent zones.
12. For a column elution chromatographic separation of constituents A and B, calculate R_r for each constituent, given that the volume of mobile phase held in the column is 3.8 ml and
    a. Constituent A has a retention volume of 7.6 ml.
    b. Constituent B has a retention volume of 11.4 ml.
13. For a column elution chromatographic system it is known that constituent A has an R_r of 0.40 and constituent B has an R_r of 0.20. Given that the volume of the mobile phase held in the column is 4.7 ml, calculate
    a. The retention volume for constituent A.
    b. The retention volume for constituent B.
14. For a given column elution chromatographic system, it is known that the time the average sample constituent spends in the mobile phase is 1.6 min. Calculate the time for the average molecule to be eluted to its maximum concentration, given that
    a. Constituent A has an R_r of 0.50.
    b. Constituent B has an R_r of 0.20.
15. On a thin-layer partition chromatographic plate, constituents A and B have R_f values of 0.60 and 0.40, respectively. For a column elution chromatographic separation employing the same stationary and mobile phases, calculate the ratio of the mass ratio constant of B to that of A.

21. Applications of Chromatography

*"You know my methods. Apply them,
and it will be instructive to compare
results," said Holmes.*

ARTHUR CONAN DOYLE
The Sign of the Four

Sherlock Holmes was always applying his methods and judging their effectiveness by the success of the applications. In this chapter we will give you the same opportunity—to judge the effectiveness of chromatography by its applications. If necessary, you should review the introduction to and the schematic diagrams of the various chromatographic techniques and equipment in Chapter 19. Although the theory of chromatography in Chapter 20 will help you visualize how the separations occur, we will not stress the theory in the applications presented. The order of the separations discussed is the same as that in Chapter 19—liquid chromatography (column, thin-layer, paper), ion-exchange chromatography, and gas chromatography.

21-1 APPLICATIONS OF LIQUID CHROMATOGRAPHY

Recall that liquid chromatography embraces separations by column chromatography using either atmospheric pressure or high-performance (pressure) liquid chromatography (HPLC), by thin-layer chromatography, and by paper chromatography. Some of the above separations are achieved by *adsorption* on active sites of a solid adsorbent; others depend on a *partition* between a stationary liquid phase on a solid and the eluting solvent. As we discuss the adsorbents and solvents that are common to all liquid chromatographic techniques, we will refer often to these two mechanisms.

Adsorbents and Solid Supports

Adsorbents and solid supports used in column or thin-layer chromatography are listed in Table 21-1. The most common adsorbent is silica gel, which is essentially a highly purified type of *porous* sand with a uniform particle size and a controlled percentage of water (usually 0–20%), achieved through activation (heating). As long as some water is present, Si—OH *active sites* form on the silica gel surface and hydrogen bond to many organic compounds.

The next most common adsorbent is alumina, an aluminum oxide that is highly pure and hydrated from 0–15% water. The amount of water may be controlled by the addition of water to a grade of alumina containing no water. Alumina is available in its natural basic form, in a neutral form, or an acidic form. These forms possess one or more types of *active sites* such as Al^{+3}, Al—OH, Al—O$^-$, or Al—OH$^+$. Basic alumina is recommended for the separation of strong organic bases such as amines; the other grades can be used for the separation of other compounds such as aromatic hydrocarbons.

In general, we will focus on the *adsorption* of compounds on the active sites of silica gel and alumina rather than on the *partitioning* of compounds between the eluent and the alumina or silica gel adsorbent. Some adsorbents, however, function primarily by retaining a layer of solvent on their surfaces; compounds to be separated are partitioned between the adsorbed solvent and the eluting solvent. These include powdered cellulose and kieselguhr, as well as DEAE-cellulose (which also functions as an ion-exchanger).

Powdered cellulose is used mainly in thin-layer chromatography to separate polar compounds. Kieselguhr, a purified form of diatomaceous earth, is useful for the separation of various sugars, such as glucose, sucrose, etc. DEAE-cellulose is useful for the separation of large molecules such as nucleotides.

TABLE 21-1. Common Adsorbents for Liquid Chromatography

Adsorbent	Compounds separated
Silica gel (Hydrated SiO_2)	Most compounds (Important exception: strongly basic compounds such as amines.)
Alumina (Al_2O_3)	Bases, steroids, aromatic hydrocarbons, carbonyl compounds
Cellulose $(C_6H_{10}O_5)_x$	Polar compounds, ions
Kieselguhr	Sugars
Polyamides (Nylon-type polymer)	Anthocyanins, flavones, nitroanilines
DEAE-Cellulose (Diethylaminoethylcellulose)	Nucleotides

Solvents

The effectiveness of a solvent in most chromatographic elutions can be correlated to its degree of polarity. One measure of polarity is the dielectric constant, which is an electrical measurement of the effect of a solvent between two plates on the magnitude of electrical charge stored by a condenser connected to the plates.

TABLE 21-2. Polarity Series of Solvents

Solvent (increasing polarity)	Dielectric constant
n-Hexane	1.9
Cyclohexane	2.0
Carbon tetrachloride	2.2
Benzene	2.3
Trichloroethylene	3.4
Diethyl ether	4.3
Chloroform	4.8
Ethyl acetate	6.0
n-Propanol	20.1
Acetone	20.7
Ethanol	24.3
Methanol	32.6
Water	78.5 (25°C)

The solvents in Table 21-2 are arranged in order of increasing dielectric constant, so water has the largest dielectric constant and is also the most polar solvent. Fairly polar solvents are frequently needed to elute compounds from silica gel and alumina; both adsorbents will adsorb polar compounds rather strongly.

In terms of adsorption chromatography, the solvents in Table 21-2 can move compounds down a column by *eluting* or *displacing* them. In elution the compounds adsorbed on active sites dissolve in the solvent as it moves past them down the column. In displacement the solvent itself competes with the adsorbed compounds for the active sites; because the solvent is much more concentrated, it gradually displaces the compounds from the active sites. Most of the nonprotonic, low dielectric constant solvents in Table 21-2 operate primarily by elution; the high dielectric constant, protonic solvents such as alcohols operate by displacement.

Applications of Column Chromatography

Column chromatography can be carried out at atmospheric pressure using wide diameter columns and relatively little equipment. High-performance liquid chromatography (HPLC), in contrast, is carried out at high pressures in thin columns using an elaborate instrumental setup. Each technique has its definite uses.

Column Chromatography at Atmospheric Pressure. Column chromatography can be used for the separation of a synthesized compound from its reaction mixture for some qualitative analysis purposes and many quantitative purposes. (Thin-layer chromatography is preferred for most qualitative and many quantitative purposes, however.)

In general, less polar compounds should be eluted or displaced from a column before the more polar compounds, using a solvent with as low a polarity as possible. (The order of polarity of compounds parallels the order of solvent polarity given in Table 21-2.) For example, if a mixture of an ester and an alcohol are to be separated on a column by means of adsorption on silica gel or alumina, a solvent such as chloroform or ethyl acetate should be used to elute or displace the ester from the column first. Then a more polar solvent such as acetone or ethanol should be used to elute or displace the alcohol. It is not necessary in most cases to use a solvent of pure chloroform, etc.; generally a mixture of hexane containing a certain percentage of chloroform is satisfactory.

A good example of a separation by elution is the separation of vitamin A alcohol, $C_{19}H_{27}CH_2OH$, from vitamin A aldehyde, $C_{19}H_{27}CHO$ [1]. A column of alumina partially deactivated with water to separate the two vitamin A compounds and others was chosen. Since the carbonyl group of the aldehyde is less polar than the hydroxyl group of the alcohol, the vitamin A aldehyde was the obvious choice to be eluted first. An eluting solvent of 2% acetone in hexane was found to elute vitamin A aldehyde, plus another compound, and leave vitamin A alcohol on the column. An eluting solvent of 8% ethanol (ethyl alcohol) was found to elute the vitamin A alcohol. The elution could be followed by ultraviolet spectrophotometry. A hypothetical elution curve of the separation is shown in Figure 21-1.

Separation of inorganic compounds or ions is generally not performed on alumina or silica gel adsorbents; however, a few important separations based on a partition between an adsorbed stationary liquid phase and the mobile eluting phase have been reported. Fritz and Schmitt [2] have reported the separation of UO_2^{+2} from nearly all other metal ions on a silica gel column having an adsorbed $6M$ nitric acid aqueous phase. The $UO_2^{+2}(NO_3^-)_2$ ion pair is eluted from the column first using an eluting solvent of methyl isobutyl ketone. The other metal ions are not dissolved by the methyl isobutyl ketone, and they remain on the column; they can be eluted later by washing the column with an aqueous eluting solvent.

Inorganic compounds can also be separated by using inert adsorbents coated with an organic solvent; an aqueous eluting solvent is used to achieve a partition effect. An early example was the separation of UO_2^{+2}, Th^{+4}, and Pu^{+4} from other metal ions on an inert "Kel-F" resin coated with tributyl phosphate, an organic solvent [3]. The metal ions were dissolved in $5.5M$ aqueous nitric acid; the UO_2^{+2}, Th^{+4}, and Pu^{+4} were selectively partitioned into the tributyl phosphate, but the other metal ions remained in the aqueous eluting solvent and passed through the column.

[1] P. A. Plack, S. K. Kon, and S. Y. Thompson, *Biochem J.* **71**, 467 (1959).
[2] J. S. Fritz and D. H. Schmitt, *Talanta* **13**, 123 (1966).
[3] A. G. Hamlin, B. J. Roberts, W. Laughlin, and S. G. Walker, *Anal. Chem.* **33**, 1547 (1961).

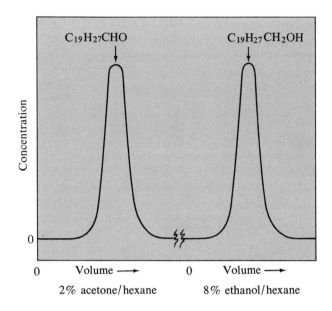

FIGURE 21-1. Hypothetical elution curve of vitamin A aldehyde and vitamin A alcohol on an alumina column.

High-Performance (Pressure) Liquid Chromatography (HPLC). HPLC partition or adsorption column chromatography can be used for the separation of preparative mixtures, but they are mainly used for rapid qualitative and quantitative analysis. The high pressures employed (Sec. 19-1) enable analyses to be conducted in a few minutes.

The stationary phase in HPLC adsorption columns may be very fine particles of silica gel or alumina, or silica gel-coated micro-size glass particles. In HPLC partition columns the stationary phase usually differs only in that it must consist of particles finely divided enough to be used in the small diameter (2–10 mm) columns used in HPLC column chromatography. In many cases, an organic coating is chemically bonded to the adsorbent particles using special techniques. A partition-like process takes place between the stationary liquid coating and the mobile eluting solvent.

Either HPLC partition or adsorption column chromatography may be used for the analysis of mixtures of pharmaceuticals, drugs of abuse, vitamins, and other mixtures. HPLC is generally preferred to gas chromatography for the analysis of polymers and other heat-sensitive materials.

In general, the composition of the eluting solvent is not changed during elution as it is in atmospheric pressure column chromatography. This is done to facilitate the separation, permit rapid separation, and avoid problems with the refractive index detector (the refractive index of each solvent differs significantly). If the high pumping pressure does not permit a rapid enough separation, a technique known as gradient elution can be used. In this technique the mixing of the components of the eluting solvent is controlled so that the proportion of the more polar component of the solvent is gradually increased. This elutes the more polar compound from the column faster.

An example of a separation by HPLC partitioning is the separation of the fat-soluble vitamin A acetate and vitamin D in vitamin tablets [4]. A preliminary treatment is done first to separate as much as possible of the inert tablet binder, or gelatin from a capsule, from the vitamins before chromatography. The separation is carried out on a chemically bonded nonpolar stationary phase consisting of n-$C_{18}H_{38}$ groups using 95% methanol-5% water as the mobile phase. Note that the stationary phase is much less polar than the mobile phase used, and was chosen because of the large nonpolar groups on the vitamins. As shown in Figure 21-2, Vitamin D (calciferol, $C_{28}H_{43}OH$) is eluted last because its steroid-type structure interacts more strongly with the n-$C_{18}H_{38}$ groups than does the nonpolar part of Vitamin A acetate ($C_{19}H_{27}CH_2O_2CCH_3$), which consists mostly of alternating single and double carbon-carbon bonds. As shown in Figure 21-2, it is eluted quite soon after the solvent peak resulting from injection of the hexane-methanol extract onto the column. The more polar vitamin D is eluted a *relatively* long time after the less polar vitamin A acetate.

The peaks, such as those shown in Figure 21-2, can be quantitated by establishing calibration plots constructed from peak heights or peak areas of the pure individual components eluted under the same conditions as the mixture.

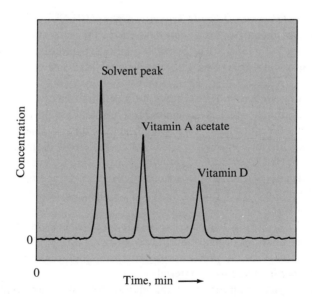

FIGURE 21-2. Hypothetical high-performance (pressure) liquid chromatographic separation of vitamins A and D on a Waters Micro Bondapak C_{18} column.

[4] "Analysis of Pharmaceutical Products," Waters Associates Bulletin AN 138, Milford, Mass., Dec., 1973.

Applications of Exclusion Column Chromatography. Exclusion chromatography is a special kind of column chromatography that can be utilized both on columns at room pressure and high-speed columns at high pressure. Separations are based on a partial or complete exclusion of molecules according to size from a mobile phase inside a stationary phase consisting of porous beads. The largest molecules are nearly completely excluded from entering the porous beads and so pass through the column first. Intermediate size molecules are able to enter only beads which have large pores, and thus pass through the column second. Small molecules can enter the pores of all of the beads and require the largest amount of solvent for elution, and therefore leave the column last.

When an aqueous mobile phase is used, exclusion chromatography is referred to as *gel filtration chromatography*; when an organic mobile phase is used, it is referred to as *gel permeation chromatography*. The latter can be carried out with high-speed columns at high pressure because the porous beads used under these conditions can withstand the high pressures used. The former must be done on columns with a low pressure drop because the stationary phases are very compressible.

Gel filtration chromatography is frequently used for *desalting* biochemical samples. After certain reactions have been carried out, biochemists need to separate biochemical compounds from small inorganic salts or buffers. The mixture is poured onto a gel filtration column and eluted with pure water; the large biochemical compounds are excluded and pass through the column first, separated from the inorganic compounds. Gel permeation chromatography is frequently used for separating proteins of different molecular weights or for determining the range of molecular weights present in an organic polymer.

Applications of Thin-Layer Chromatography

Thin-layer chromatography is carried out using the same adsorbents employed in column chromatography (Table 21-1). The adsorbents are cemented to a plate (Fig. 19-3) with a calcium sulfate binder, but function in the same manner, either by adsorption or partition chromatography. Because thin-layer chromatography is so easy to set up and requires about one-half hour on the average for a separation, it is frequently used for qualitative analysis. With a certain amount of care, it can also be used for quantitative estimation of certain compounds.

Qualitative Analysis on Thin-Layer Plates. In general, selection of an adsorbent for thin-layer chromatography should be done so that the compounds to be separated exhibit the greatest differences in R_f values. Recall from Chapter 20 that an R_f value is obtained by dividing the distance a sample spot moves by the distance the solvent front moves. For compounds of low polarity the greatest differences in R_f values are obtained by using plates of silica gel or activated alumina and solvents of low polarity (Table 21-2). The separation is achieved by adsorption

Table 21-3. Thin-Layer Chromatography of B Vitamins [5] (Silica gel adsorbent)

B vitamin (Important polar group)	R_f value in Water	R_f value in Benzene (5% acetic acid, etc.)
B_1, Thiamine (R—N^+Cl^-)	0.05	0
B_{12}, Cyanocobalamin (Co^+PO^-)	0.22	0
B_2, Riboflavin (3 OH, 2 C=O,	0.42	0.35
Pantothenic acid (COOH, —NH—C=O)	0.60	0.89
Nicotonic acid (COOH)	0.76	0.75

rather than partition chromatography. For compounds of high polarity the greatest differences in R_f values are achieved by using plates with cellulose, non-activated alumina, or silica gel [5].

Silica gel is used more extensively than any other adsorbent. It affords excellent separation of compounds of low polarity by adsorption; when coated with an appropriate stationary liquid phase, it also permits good separation of compounds of high polarity by partitioning. If acids are to be separated on silica gel, it is best to use a small amount of acetic acid in the eluting solvent. Bases are best separated on alumina, but if they are to be separated on silica gel, a small amount of ammonia or diethylamine should be added to the eluting solvent. Very polar molecules such as amino acids and carbohydrates should of course be separated on cellulose or kieselguhr rather than on silica gel.

A good example of a thin-layer separation for qualitative analysis is the separation of some of the B vitamins listed in Table 21-3 [5]. Two different solvents were used for elution—water and a mixture of 5% acetic acid, 5% acetone, and 20% methanol in benzene. Since the polarities of the B vitamins vary greatly, some good separations are possible. For example, thiamine (B_1) is a highly polar univalent cation which is strongly retained by the adsorbent close to the original spot. Although cyanocobalamin (B_{12}) is also a univalent cation, its positive charge is somewhat shielded within a complex ion of cobalt(II), so it moves somewhat more than thiamine when water is used than when the benzene mixture is used. In contrast, pantothenic acid has a higher R_f value in the benzene mixture than in water. This is probably the result of the combined effect of acetic acid, acetone, and methanol interacting with the many polar groups of this acid.

Quantitative Analysis on Thin-Layer Plates. Quantitative estimation of compounds on a thin-layer plate may be carried out on the plate (in situ) or after removing the spots from the plate. In situ quantitation may be carried out by using spectrophotometry, fluorometry, or radioisotope scanning. A special instrument called a photodensitometer, or densitometer, can be used for automatic scanning of the amount of light transmitted or reflected by colored substances on a plate. A special

---

[5] V. W. Rodwell, *Thin Layer Chromatography*, American Chemical Society (ACS Audio Course), Washington, D. C., 1975.

type of scanning fluorometer may also be used for fluorescent compounds or for compounds on a fluorescent plate. Radioisotope scanning may be used for carbon-14, phosphorus-32, or tritium (^{3}H) compounds. Quantitation after removing spots from the plate may be carried out by cutting out an appropriate section of a plate, extracting the compound from the section, and measuring it spectrophotometrically.

An example of a thin-layer separation that may have involved both qualitative and quantitative analysis is the 1975 Florida oil spill investigation in which oil samples from about 250 "suspect" ships were tested. Thin-layer chromatography, fluorometry, infrared spectrophotometry, and gas chromatography were used to analyze each oil sample [6]. The analyses conclusively proved that a Liberian tanker was the only ship that could have spilled the oil which blighted about 50 miles of Florida beaches.

Applications of Paper Chromatography

Paper chromatography is based on a partition of compounds between a stationary solvent held on the paper and the eluting solvent. It is not used as much as thin-layer chromatography because separations are much slower and spots are not as easy to detect. Paper chromatography finds its greatest uses in the separation of the same compounds best separated on cellulose thin-layer plates—amino acids, carbohydrates, and other biochemical compounds.

Selection of solvents for paper chromatography is made so that the greatest differences in R_f values are obtained for the compounds to be separated. If water is present, it is usually assumed that water is the stationary solvent held on the paper. Thus polar compounds will be adsorbed most strongly and have the lowest R_f values; compounds of low polarity will have the highest R_f values.

A good example of a paper chromatographic separation is the separation of amino acids on Whatman No. 1 paper with water-saturated phenol as the eluting solvent [7]. The R_f values for polar amino acids such as aspartic acid and glutamic acid should be quite low because they are diprotic carboxylic acids (such acids are highly soluble in water). As shown in Table 21-4, this is indeed the case. In contrast, amino acids with only one carboxylic acid group are not nearly as polar and

TABLE 21-4. Paper Chromatography of Amino Acids

Amino acid	R_f value
Aspartic acid	0.14
Glutamic acid	0.24
Alanine	0.59
Glutamine	0.59
Proline	0.90

[6] *Chem. & Eng. News* **53**, 7 (Nov. 17, 1975).
[7] G. M. Price, *Biochem J.* **80**, 420 (1961).

should be moved more readily by the water-saturated phenol-eluting solvent. This is the case for alanine and glutamine, two amino acids that have amino groups of similar basicity. Evidently the amino group of proline interacts more strongly with the phenol eluting solvent than the amino groups of alanine and glutamine since proline is eluted before either of them.

21-2 APPLICATIONS OF ION-EXCHANGE CHROMATOGRAPHY

Recall that ion-exchange chromatography is a separation based on a chemical reaction of ions with a stationary phase of ion-exchange resin. It can be performed in columns at atmospheric pressure or at high speeds using columns at high pressure (HPLC). It can be used to separate various cations from one another and various anions from one another, using cation-exchange resin and anion-exchange resin, respectively.

Ion-Exchange Resins

Cation-exchange resins consist of a backbone of an organic polymer to which is attached an anionic functional group which will exchange cations. Some of the functional groups used on cation-exchange resin are the sulfonic acid group, $-SO_3^-$, the carboxylate group, $-CO_2^-$, and the phosphonate group, $-PO_3H^-$. The most commonly used resin is that which has the sulfonic acid group. It is used in the hydrogen or sodium form; the reaction of the hydrogen form with the sodium(I) ion is

$$\text{Res}-SO_3^-H^+ + Na^+ \rightarrow \text{Res}-SO_3^-Na^+ + H^+ \qquad (21\text{-}1)$$

Anion-exchange resins consist of a backbone of an organic polymer to which is attached a cationic functional group which will exchange anions. Some of the functional groups used on anion-exchange resins are the quaternary ammonium group, $-NR_3^+$, and amine groups such as $-NHR_2^+$. The quaternary ammonium resin is the most commonly used resin. It is used in the chloride form; the reaction of the chloride form with an anion such as the nitrate ion is

$$\text{Res}-NR_3^+Cl^- + NO_3^- \rightarrow \text{Res}-NR_3^+NO_3^- + Cl^- \qquad (21\text{-}2)$$

Eluting Solvents

The effectiveness of an eluting solvent in ion-exchange chromatography does not depend on polarity, as in conventional column chromatography, but on the ions dissolved in the eluting solvent. To separate various ions, they are first exchanged onto the ion-exchange resin. Then they must be displaced by an ion of the same

TABLE 21-5 Ion Exchange Selectivities with 8% Crosslinked Resin

Ion	Relative equilibrium constant (25°C)
Li^+	1 (arbitrary standard)
H^+	1.27
Na^+	1.98
K^+	2.90
Mg^{+2}	3.29
Ca^{+2}	5.16
Sr^{+2}	6.51

charge at different rates so that each ion in the mixture is completely eluted from the column before the next ion appears. The choice of the ion used in the eluting solvent is based on the *selectivity* of a particular resin for that ion in relation to the ions being separated. For example, suppose that it is desired to separate the Li^+, Na^+, and K^+ ions on a sulfonic acid resin. The best choice for an eluting ion would be one that is preferred almost equally by the resin as compared to the ion least strongly held. Table 21-5 gives a compilation of relative equilibrium constants which make "selectivity" more quantitative [8]. You can see that the best choice for the eluting ion would be the hydrogen ion, in the form of a strong acid. This would elute lithium ion first fairly rapidly but would displace sodium ion and potassium ion more slowly since these are more strongly held.

Applications of Ion-Exchange Chromatography

Ion-exchange chromatography can be used for qualitative analysis, but is mainly used for separations prior to a quantitative analysis for each ion separated. The following generalizations are useful to predict separations [9].

1. In general, the more hydrated an ion is in solution, the smaller its selectivity on a given resin. For example, lithium ion is hydrated more than other univalent cations and has the lowest selectivity (Table 21-5).
2. Trivalent ions have higher selectivities than divalent or univalent ions, and divalent ions have higher selectivities than univalent ions.
3. The selectivity of a resin for an ion increases with increasing crosslinking of organic polymer backbone of the resin.

The simplest example of ion-exchange chromatography is the separation of interfering anions from cations using an anion-exchange column, and vice versa. For example, iron(III) ions are seriously coprecipitated with barium sulfate during the precipitation for the determination of sulfate. To avoid this, the sample con-

[8] L. Meites, *Handbook of Analytical Chemistry*, McGraw-Hill, New York, 1963, pp. 10-159.
[9] H. F. Walton, *Principles and Methods of Chemical Analysis*, Prentice-Hall, Englewood Cliffs, N.J., 1964, pp. 138–59.

taining iron(III) and sulfate ions is passed through a cation-exchange column before precipitation, removing the interfering iron(III) ion and allowing the sulfate ion to pass through the column. Another simple example of ion-exchange chromatography is the separation of divalent or trivalent cations from univalent cations using a cation-exchange column. As can be seen from Table 21-5, cations such as strontium(II) and calcium(II) have much higher selectivities than any of the univalent cations listed. When such a mixture is passed through a cation-exchange column, the univalent cations pass through rather quickly and the divalent cations are retained until all the univalent cations have been eluted.

A more difficult separation is the chromatographic separation of cations of all the same charge, such as the separation of the alkali metal cations. Fortunately, there are some fairly large differences in the hydration of these ions; lithium(I) ion is highly hydrated compared to sodium(I) and potassium(I) ions, for example. The chromatographic separation of these three alkali metal ions can be done using $0.7M$ hydrochloric acid as the eluting solvent [10]. A hypothetical elution curve for the separation is shown in Figure 21-3. Since lithium ion has the lowest selectivity (Table 21-5), it is eluted first. The sodium ion is held about twice as strongly by the cation-exchange resin, so it is eluted late enough for the separation to be quantitative; the separation of potassium is also extremely good. If a divalent cation such as magnesium(II) were present, it would be eluted much later than potassium ion without any interference.

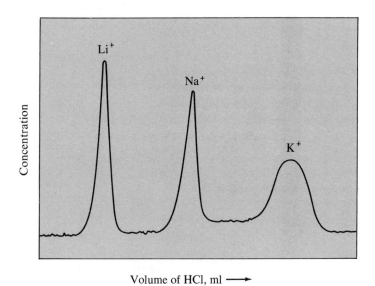

FIGURE 21-3. *Hypothetical ion-exchange elution curve for the separation of three alkali metal ions using 0.7M hydrochloric acid on a sulfonic acid cation-exchange column.*

[10] W. Riemann, *Rec. Chem. Progr.* **15**, 85 (1954).

21-3 APPLICATIONS OF GAS CHROMATOGRAPHY

Gas chromatography is a separation technique generally based on the partitioning of a compound between a mobile vapor phase and a stationary liquid phase held on a solid support. (Strictly speaking this is gas-liquid chromatography.) As you recall from Figure 19-4, the gas flows through the support (coated with a stationary phase) in a heated column and finally through a detector. By choosing the proper stationary liquid phase we can separate nearly all compounds with any significant volatility, excluding heat-sensitive compounds and most polymers.

Stationary Liquid Phases

Since the components in a sample must dissolve in the stationary phase to undergo partitioning, the effectiveness of a stationary phase depends on its ability to dissolve, and possibly interact with, each component. (It is of course also important that the stationary phase be nonvolatile and stable at the temperatures used.) A useful rule of thumb is *like dissolves like*. If a stationary phase is used that is similar to the components being separated, then this allows the volatility of the components, as indicated by boiling point, to govern the order of the separation. Thus, a more volatile component should be eluted before a second component with a higher boiling point.

Some recommended stationary phases [11] for various classes of organic compounds are listed in Table 21-6.

A simple illustration of the use of Table 21-6 is as follows. Suppose that a mixture of benzene (b.p. 80.1°C) and cyclohexane (b.p. 80.7°C) is to be separated. A suitable stationary phase might be squalane, a long chain hydrocarbon, which is saturated like cyclohexane rather than unsaturated like benzene. Cyclohexane

TABLE 21-6. Recommended Stationary Phases [11]

Organic sample component	Recommended stationary phase
Aliphatic hydrocarbons	Squalane
Aromatic hydrocarbons	Dinonyl phthalate, Phenyl methyl silicone polymer
Alcohols	Dinonyl phthalate, Phenyl methyl silicone polymer
Aldehydes or ketones	Squalane
Acids	Dimethyl silicone polymer
Ethers, esters	Dinonyl phthalate
Amines	Polyethylene glycol
Alkyl halides	Dinonyl phthalate
Sulfur compounds	Dimethyl silicone polymer
Olefins	Silver nitrate + polyethylene glycol

[11] J. M. Bobbitt, A. E. Schwarting, and R. J. Gritter, *Introduction to Chromatography*, Reinhold, New York, 1968, pp. 106–143.

should thus be more soluble in the squalane and should be eluted last (longer retention time) on the basis of its higher boiling point and higher solubility. If dinonyl phthalate were to be chosen as a stationary phase instead of squalane, then benzene should be more soluble because the dinonyl phthalate is unsaturated like benzene. There also is undoubtedly a weak interaction between the electron-rich benzene and the electron-poor ester groups of the dinonyl phthalate that further increases the solubility of benzene as compared to cyclohexane. Thus the cyclohexane will be eluted ahead of benzene because the high relative solubility of benzene will outweigh its higher volatility. In other words, benzene will have a longer retention time.

Influence of Detector on Gas Chromatography Analysis

The type of detector (Table 21-7) available is frequently important in the analysis of complex samples. For example, the thermal conductivity detector measures the difference in thermal conductivity of the carrier gas alone and the carrier gas containing the sample. It is suitable for general purposes but is not very sensitive to small amounts. Suppose that you must analyze for 10^{-3} μg amounts of an alcohol in a water solution. The thermal conductivity detector cannot detect such amounts, but the flame ionization detector can. Its output is proportional to the number of ions produced in burning the sample, which gives it a greater sensitivity than the thermal conductivity detector. In addition, it has little or no response to water, so it detects the alcohol easily without "seeing" the relatively large amount of water present with the alcohol.

The electron capture detector has additional advantages over both the thermal conductivity and flame ionization detectors. Its response is based on measurement of a flow of "slow" electrons to an anode; reducible components such as halogen, phosphorus, sulfur, and nitro-compounds react readily with these electrons and are detected at the lowest levels of all three detectors.

TABLE 21-7 Gas Chromatographic Detectors [11]

Characteristics of detectors	Thermal conductivity	Flame ionization	Electron capture
Minimum amount detected	2–5 μg	$10^{-5} \mu$g	$10^{-7} \mu$g
Responds to:	All compounds	Only combustible compounds	Only compounds with electronegative atoms
Temperature limit	450°C	400°C	225°C
Carrier gas	Helium	Helium or nitrogen	Nitrogen or argon
Special applications	Water	Water extracts	Halogen, phosphorus, sulfur, and $-NO_2$ compounds

Applications of Gas Chromatography

Qualitative Analysis Using Gas Chromatography

A gas chromatographic elution curve (Fig. 21-4) is a powerful device for qualitative analysis because the presence of, say, ten peaks generally indicates at least ten components. However, the instrument itself does not identify the components giving rise to the peaks. This can be done by using the *retention time* of each component, the retention time being the number of minutes between injection of the sample and its entrance into the detector. Comparison of the retention time of an unknown component and that of several suspected compounds on the same column under the same operating conditions will frequently identify the component. If this

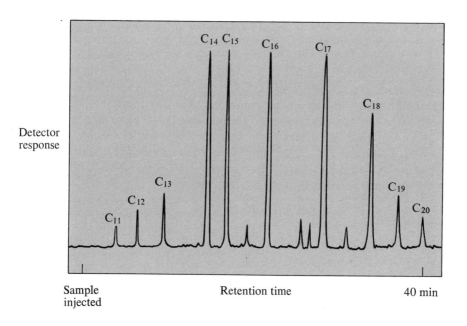

FIGURE 21-4. Gas chromatogram of a crude oil seepage sample using Apiezon L grease as stationary phase and a flame ionization detector.

does not help, then the eluent from the column can be run through a spectrophotometer, such as an infrared spectrophotometer, for identification.

An example of a gas chromatographic qualitative analysis is the 1975 Florida oil spill investigation previously discussed [6]. Crude oil samples from about 250 "suspect" ships had to be matched to the crude oil that blighted Florida beaches. Since crude oil consists of a number of hydrocarbons, gas chromatographic elution curves, similar to that in Figure 21-4, of the suspect ships were compared with the curve of the crude oil on the beaches. The gas chromatographic evidence, as well as thin-layer chromatographic evidence, was instrumental in proving that a Liberian tanker was guilty.

Quantitative Analysis

Under constant conditions (temperature, flow rate, etc.) the area under each gas chromatographic peak is proportional to the concentration of the component giving that peak. It is therefore necessary to integrate the area under the peaks accurately. One method of doing this is to use electronic digital integration [12]. In this method the peak is divided into a large number of segments or rectangles of equal width. Each segment (peak) height can be approximated by a voltage; in addition, the height of each segment is directly proportional to the area. Digital integration gives the sum of all the voltages, which is directly proportional to the peak area. Another way of integrating the area under chromatographic peaks involves using the ball and disc integrator [13]. The chromatographic output is fed into an electromechanical rotating disc. The disc rotations are plotted as up-and-down "sweeps" of a recorder pen. The number of such sweeps represents the peak area.

One method of quantitative analysis is to measure the peak areas of a number of standards of varying concentrations and to plot area versus concentration; this is the so-called *calibration curve method.* Another method is the *internal standard method*, in which the calibration curve consists of a plot of the *ratios of the peak area* of a constant amount of a standard compound to the peak areas of different known amounts of the sample *vs.* the *weight ratios.* The unknown sample is treated similarly and its weight read from the plot. This method avoids errors from unexpected experimental variations.

QUESTIONS AND PROBLEMS

(Answers to most even-numbered problems are in Appendix 5.)

Concepts and Definitions
1. Explain the difference between adsorption and partition liquid chromatography.
2. Explain the difference(s) between
    a. Silica gel and alumina adsorbents.
    b. Cellulose and silica gel adsorbents.
    c. Kieselguhr and alumina adsorbents.
3. In each solvent pair, decide which is the more polar solvent and which is the more nonpolar solvent.
    a. Ethanol and hexane
    b. Acetone and chloroform
    c. Propanol and water
    d. Cyclohexane and benzene

[12] H. L. Pardue, M. F. Burke, and J. R. Barnes, *J. Chem. Educ.* **44**, 695 (1967).
[13] J. R. Barnes and H. L. Pardue, *Anal. Chem.* **38**, 156 (1966).

4. What advantage(s) does high-speed column chromatography have over column chromatography at atmospheric (room) pressure for
   a. Qualitative analysis?
   b. Quantitative analysis?
5. In partition liquid chromatography, explain the difference between the stationary liquid phase, and the mobile liquid phase.
6. Explain the three generalizations for predicting separations using ion-exchange chromatography.
7. Recommend a suitable stationary liquid phase for gas chromatographic separation of each pair of compounds.
   a. Benzene and naphthalene
   b. Acetone and propionaldehyde
   c. Acetic acid and thiophenol
   d. Diethyl ether and ethyl acetate

Liquid Chromatographic Analysis
8. A mixture of salicylic acid and acetylsalicylic acid is separated by chromatography on a thin-layer plate. Under ultraviolet radiation a blue fluorescent spot is detected at low R_f and a violet fluorescent spot is detected at high R_f. Which compound has the low R_f, and which the high R_f? (*Hint:* Look up their structures and see Table 21-2.)
9. Calculate the R_f values for each compound in parts *a* and *b* given that the solvent front has traveled 17.5 cm.
   a. Naphthalene spot at 7.9 cm
   b. Tetraphenylethylene spot at 3.5 cm
   c. Suppose that both compounds were chromatographed on a shorter plate where the solvent front travels 8.5 cm. Calculate where the spots of the two compounds should be located (use centimeters).
10. Suggest a suitable solvent for the elution of each of the following groups of compounds from a high speed liquid chromatographic column.
    a. Phenol, *p*-bromophenol, and 2, 4-dibromophenol
    b. Acetaldehyde, formaldehyde, and butyraldehyde
    c. Benzene, chlorobenzene, and 2, 4-dichlorobenzene

Ion-Exchange Chromatographic Analysis
11. In each of the following mixtures, predict the order of elution from a sulfonic acid cation-exchange column.
    a. Ca^{+2}, Mg^{+2}, Ba^{+2} b. Rb^+, Li^+, K^+, Na^+ c. Mg^{+2}, K^+, Ca^{+2}, Li^+
12. Suggest a suitable eluting solvent for the separation of each of the following mixtures. (Do not use HCl for any of them because it will elute them too slowly.) Assume a sulfonic acid cation-exchange resin.
    a. K^+, Rb^+, Cs^+ b. Ba^{+2}, Sr^{+2}, Ca^{+2}

Gas Chromatographic Analysis
13. The following compounds are not as volatile as might be desired for gas chromatographic separation. Suggest a derivative that might be used for each class of compounds to produce a more volatile, less hydrogen-bonded type of compound.
    a. High molecular weight fatty acids
    b. High molecular weight alcohols
    c. Amino acids
    d. Carbohydrates

14. A sample is known to be a C_4H_9OH alcohol (a butanol). Write out the structures of the possible isomers (4) that the sample might be. Look up their boiling points, and predict the order of elution from a column with a dinonyl phthalate stationary liquid coating.
15. Suggest the appropriate detector for the analysis of the first component in the presence of the other components in each following mixture.
   a. CH_3OH with H_2O and $CH_3\underset{\underset{O}{\|}}{C}CH_3$

   b. $CH_3\underset{\underset{O}{\|}}{C}CH_3$ ($10^{-2}\,\mu g$) with CH_3CHO and $CH_3CH(OH)CH_3$

   c. CS_2 ($10^{-2}\,\mu g$) with CH_3OH and C_4H_9OH

22. Clinical Analysis

> "All knowledge *comes useful to the detective*," remarked Holmes. "*Even the trivial fact that in the year 1865 a picture of Greuze entitled La Jeune Fille a l'Agneau fetched... more than forty thousand pounds may start a train of reflection in your mind.*"
>
> ARTHUR CONAN DOYLE
> The Valley of Fear, Chapter 2

Even as all knowledge is useful to the detective, so all chemical knowledge is useful to the clinical chemist. Throughout this chapter are found a number of references to previous chapters to verify this statement. The clinical chemist uses all the methods of the analytical chemist plus some of his own to characterize and analyze medical samples.

Since most of us are more familiar with elements, we will begin the chapter by discussing the elements in the body. Naturally the elements exist as free or combined ions, so we will be describing the major ions or major electrolytes as well as ions present in trace concentrations. We will also try to say something about the interpretation of electrolyte levels.

The main discussion will center on clinical analysis for selected components in the blood. Both the sampling of the blood and the determination of various compounds and other substances in the blood will be discussed. Rather than discussing the determination of substances in other parts of the body, we will concentrate on the analysis for common substances in the blood. We think that the lengthy discussion of the determination of glucose will be of value and interest. More extensive presentations of clinical analysis are available [1].

---

[1] R. J. Henry, D. C. Cannon, and J. W. Winkelman, *Clinical Chemistry, Principles and Techniques*, Harper and Row, New York, 1974.

22-1 ELEMENTS IN THE BODY

Nearly all the elements exist in ionic form in the body. The major ions in body fluids are termed the major electrolytes, whereas the minor ions are frequently labeled the trace elements, or trace metal ions. We will discuss the major electrolytes first.

Electrolytes in the Serum and Urine

Typical levels of the major electrolytes in the blood serum and in urine are listed with those of the trace elements in Table 22-1. These levels are given in both ppm [2] and molarity [3]. The major serum electrolytes are Na^+, K^+, Ca^{+2}, Mg^{+2}, Cl^-, and HCO_3^-. All of these ions are important for maintaining electrical neutrality in the blood, among other things. A loss of bicarbonate ion as carbon dioxide must be balanced by the movement of an equivalent amount of one of the cations out of that part of the bloodstream. Note from Table 21-1 that the molarity ($0.14 M$) of the sodium ion is nearly balanced by the sum of the molarities ($0.13 M$) of the chloride ion and the bicarbonate ion. Unless pathological variations occur, the sum of the concentrations of the latter two ions is generally $0.01 M$ less than that of the sodium ion [3].

Analysis for the major serum electrolytes is important because it will indicate the nature of the electrolyte balance, rather than any toxic or lethal levels. (These levels will be important when the trace metals are considered.) Alkali (Group I) metal ions such as sodium and potassium are of course routinely determined by flame emission spectrometry (Sec. 16-2); they can also be measured by atomic absorption spectrometry (Sec. 16-3). On automated clinical analyzers (Sec. 14-3), sodium and potassium ion are nearly always measured by flame emission spectrometry. There is a variety of good methods for the determination of calcium and magnesium ions. They can be determined by titration with EDTA (Sec. 10-2), by atomic absorption spectrometry (Sec. 16-3), by flame emission spectrometry (Sec. 16-2), or by colorimetric analysis. On some automated clinical analyzers an organic complexing dye is added to form a colored calcium(II) chelate which is measured colorimetrically.

The chloride ion may also be determined by a variety of methods—precipitation titration (Sec. 10-1), specific ion electrode (Sec. 17-3), and coulometric measurement (Sec. 18-3). The bicarbonate ion may be titrated as a base with standard acid or it may be converted to carbon dioxide and measured as a gas.

One other measurement that is important in blood analysis is the measurement of blood pH. The pH of normal arterial blood is surprisingly constant, ranging between about 7.38 and 7.42 [2]. Blood pH is measured with the pH meter (Sec. 17-2) and is important for indicating general disorders when used with electrolyte

[2] G. D. Christian, *Anal. Chem.* **41** 24A (Jan., 1969).
[3] J. S. Annino, *Clinical Chemistry*, Little, Brown, Boston, 1964, pp. 122–31.

TABLE 22-1. Levels of the Elements in Blood Serum and Urine

Ionic form, or element	Blood serum level, ppm	Urine, mg/day
Alkali and alkaline earth metal ions [2, 3]		
Li^+	0.01	—
Na^+	3200 (0.14M)	1000–5000
K^+	120–214 (0.005M)	—
Mg^{+2}	43 (0.001M)	60–120
Ca^{+2}	90–110 (0.003M)	96–800
Sr^{+2}	—	0.4
Other metal ions [2]		
Al^{+3}	0.13–.17	0.05
As(III) & (V)	0.04–.2	$\geq$ 0.1 ppm
Cd^{+2}	0.0033	0.002–0.02 ppm
Cu^{+2}	1.05	—
Fe(II) & (III)	1.25 (men) / 0.90 (women)	0.1–0.3
Pb^{+2}	0.3–0.4 (whole blood)	0.01–.07
Anions [3]		
Cl^-	0.10M	—
HCO_3^-	0.03M	—
HPO_4^{-2}	0.002M	—
SO_4^{-2}	0.001M	—

analyses. Table 22-2 gives a summary of some electrolyte patterns and their diagnostic indications.

Although the electrolyte analyses in Table 22-2 are interrelated, sometimes a single result will rule out a certain physiological condition. For example, a normal serum pH value will tend to rule out acidosis or alkalosis. On the other hand, more

TABLE 22-2. Some Serum Electrolyte Patterns

pH	Na^+	K^+	Cl^-	$HCO_3^- + CO_2$	Diagnostic indication
N[a]	High	N	High	N	Probable dehydration
Low	N	Any	Low	Low	Probable metabolic acidosis
Low	N	N or high	Low	High	Probable respiratory acidosis
High	N	N or low	N	Low	Probable respiratory alkalosis

[a] N = normal

than an abnormal pH value is needed to verify whether the acidosis is metabolic or respiratory in nature. Metabolic acidosis is characterized by a low pH, low chloride level, and a low level of carbon dioxide plus bicarbonate ion. In metabolic acidosis a body malfunction causes an increase in the level of acids, such as ketonic acids, and the bicarbonate level falls because the bicarbonate ion reacts with part of the excess acid to attempt to restore a normal pH as follows:

$$HCO_3^- + \text{excess } H^+ \rightarrow H_2CO_3 \text{ (exhaled as } CO_2) \qquad (22\text{-}1)$$

In respiratory acidosis an increase in the carbon dioxide levels releases extra hydrogen ions through ionization as follows:

$$CO_2(g) + H_2O \rightarrow H^+ + HCO_3^- \qquad (22\text{-}2)$$

In contrast, respiratory alkalosis is characterized by a high pH and a low level of carbon dioxide plus bicarbonate ion. In this condition a person is exhaling too much carbon dioxide; this plus a secondary drop in bicarbonate level causes the pH to rise above normal (the reverse of **22-2**).

In the well-known science fiction novel, The Andromeda Strain, *author Michael Crichton narrates a situation in which a deadly strain of bacteria from outer space is able to cause fatal coagulation of the blood within a narrow pH range of 7.39 to 7.43. The story hinges on the scientists finally recognizing that an acidotic old man and an alkalotic, crying baby survived because their blood pH's were outside the pH range in which the strain was fatal.*

Electrolytes in the Urine. The diagnostic significance of chloride, sodium, and potassium ion levels in the urine is a complicated problem since electrolyte levels in the urine are influenced by a large number of factors [3]. The amounts excreted by the body over a 24-hour period are more important than the concentrations of random samples; it is for this reason that Table 22-1 reports electrolyte levels in milligrams per day rather than as concentrations.

Trace Elements

A number of trace elements in body fluids or in the tissues are *essential* to life at low levels [1]; at higher levels some of these elements may be toxic or lethal. There are also a number of elements which may yet be shown to have an essential function at low levels, but at present are only known to be poisonous at higher levels. Finally, there are elements such as iron which perform a useful function at higher than trace levels. The essential elements function by activating or deactivating enzymes, or by controlling vitamin or hormone action. Iron of course is utilized in hemoglobin for binding oxygen. The so-called poisonous elements affect the brain or otherwise impair a vital body function.

Analysis for most trace metals is not done routinely in the clinical laboratory, but can be performed by flame emission or atomic absorption spectrometry.

Analysis for mercury is best done by flameless atomic absorption spectrometry (Sec. 16-3). The analysis for poisonous elements is of course important in situations where a poisoning is suspected. Lead, arsenic, mercury, and thallium are generally considered to be in this class.

Lead poisoning is common among children who have been living in dwellings where lead-based paints are still present, and therefore analysis for lead(II) ion in as small a sample of blood as possible is important. Unfortunately, atomic absorption analysis requires a relatively large sample of blood so that lead(II) ion can be extracted away from interferences before it is measured. A recent fluorometric method appears to be more suitable because only a drop of blood is required for analysis [4]. The drop of blood is placed on a glass slide and inserted in a fluorometer. The sample is excited at 424 nm and fluoresces at 625 nm in the red region. The fluorescence originates from a zinc protoporphyrin compound produced in the blood by the action of the lead(II) ion. Toxic levels of 0.07 mg of lead per 100 ml are readily detected by the method.

22-2 CLINICAL ANALYSIS OF BLOOD

Blood is analyzed for more than just electrolytes in the clinical laboratory; it is also analyzed for the levels of certain compounds, total protein, certain enzymes, and elements such as protein-bound iodine. Before discussing selected analyses, we will discuss sampling.

Sampling of Blood

Samples of blood are frequently taken after a patient has fasted for a certain time, for example, before breakfast. This is particularly important for glucose analysis, but not as important for other species. Since only part of the blood may be analyzed, it is important for you to understand the composition of whole blood. This can be seen easily from the following diagram:

Whole Blood	
Plasma	Cells
Serum + Fibrinogen	Platelets + Erythrocytes + Leukocytes

The liquid portion of blood is the plasma; the majority of analyses are performed on the serum, which is plasma from which the fibrinogen has been removed.

In collecting the blood, the sample is drawn as quickly as possible using syringes that are chemically clean and dry to avoid contaminating the blood as well

[4] *Chem. & Eng. News* **53**, 18 (Feb. 3, 1975).

as coagulating it. The container for the blood sample should also be clean and dry for the same reasons. The sample should then be taken to the laboratory immediately for analysis on the same day the sample is collected.

The above procedure is followed where serum is to be analyzed. If plasma or whole blood is required for analysis, then the blood is put into a container having an anticoagulant in it. Potassium oxalate is one such anticoagulant; the oxalate ion precipitates blood calcium:

$$C_2O_4^{-2} + Ca^{+2} \rightarrow CaC_2O_4(s)$$

preventing the calcium ion from initiating the clotting mechanism. Another anticoagulant is EDTA (Sec. 10-2), which chelates the calcium ion and has the same effect as the oxalate ion.

In the laboratory serum may be obtained from whole blood by allowing the blood to clot. Then the serum is separated from the cells and the fibrinogen by centrifuging the solid matter down, leaving the clear serum in liquid form. Centrifuging is done as quickly as possible to avoid *hemolysis*—the destruction of the red blood cells, which releases hemoglobin and other substances into the serum or plasma. (Since the red blood cells contain larger amounts of potassium, iron, magnesium, urea, etc., the analysis of the serum will give incorrect high results.) The sample of serum is then ready for analysis.

Special techniques are needed if the sample cannot be analyzed on the same day that it is taken [3], but these are beyond the scope of this text.

Substances Determined in the Blood

Of the possible substances in the blood that could be determined, only a dozen or so are routinely determined. These include electrolytes such as sodium, potassium, calcium, magnesium, chloride, and bicarbonate ions, as previously discussed. Other ions that may be determined are phosphate, as inorganic phosphate, iron, and iodine, as protein-bound iodine (PBI). Compounds that are determined include blood urea nitrogen (BUN), glucose, carbon dioxide, and uric acid. Other substances commonly determined include enzymes such as alkaline phosphatase and serum glutamic-oxalacetic transaminase (SGOT), total protein, albumin, and total bilirubin. Table 22-3 lists the normal ranges for some of these substances. The normal ranges listed in the above table cover only the great majority of people, not all people. For example, Figure 22-1 shows that some people's serum chloride nominally falls below 98 mmole/liter, or above 108 mmole/liter [5].

The determination of serum electrolytes has already been discussed in Section 22-1. Of the remaining substances in Table 22-3, most are determined spectrophotometrically (colorimetrically). In some cases, a digestion is necessary to convert organic material to ionic material before the determination can be carried out. For example, in the determination of protein-bound iodine (PBI),

[5] J. C. MacDonald, *Amer. Lab.* **7**, 61 (July, 1975).

TABLE 22-3. Normal Ranges of Substances in the Blood

Substance determined	Normal range in 100 ml serum
Albumin	3.8–5.0 g
Bilirubin	Up to 0.4 mg (direct)
	Up to 1.0 mg (total)
Calcium	2.25–2.75 mmole/liter
Carbon dioxide	25–32 meq/liter
Chloride, serum	98–108 mmole/liter
Cholesterol, total	100–350 mg
Iron, serum	0.050–0.180 mg
Magnesium	0.75–1.25 mmole/liter
Phosphatase, alkaline	Up to 4 Bodansky units
Phosphorus, inorganic	3–4.5 mg
Potassium	3.8–5.0 mmole/liter
Protein, total serum	6.5–8 g
Protein-bound iodine (PBI)	0.0035–0.008 mg
Sodium	138–146 mmole/liter
Sugar (glucose)	65–90 mg
Transaminase (SGO)	Up to 40 units
Urea nitrogen (BUN)	Up to 20 mg
Uric acid	3–6 mg

the organic iodine compounds are decomposed to iodide ion and carbon dioxide first. This is followed by addition of an aqueous solution containing the iodide ion to a cerium(IV)-arsenic(III) mixture (Sec. 14-4). The iodide ion then catalyzes the oxidation-reduction reaction of these two ions.

$$2Ce(IV) + As(III) \xrightarrow{I^-} 2Ce(III) + As(V)$$
(yellow)

The disappearance of the yellow color of cerium(IV) is followed spectrophotometrically and related to the concentration of the iodide.

Determination of Serum Glucose

Since the determination of serum glucose is the most frequently done analysis in the clinical laboratory, we will discuss the various methods for glucose (Table 22-4)

TABLE 22-4. Some Methods for Serum Glucose Determination

Reaction	Reagent, wavelength
Condensation with o-toluidine	$CH_3C_6H_4NH_2$, 630–40 nm
Oxidation by ferricyanide	$Fe(CN)_6^{-3}$, 420 nm
Oxidation with copper(II) neocuproine	Cu(I) neocuproine, 460 nm
Reaction with oxygen using glucose oxidase enzyme	O_2, glucose oxidase (polarographic)

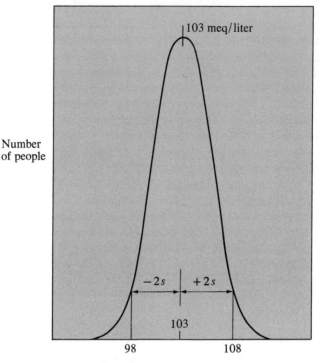

FIGURE 22-1. *Frequency distribution of serum chloride levels of a particular population of people [5]. The normal range of chloride is 98–108 nmole/liter (meq/liter).*

in some detail. Each of the methods in Table 22-4 has its advantages and disadvantages, so there is not always an overwhelming preference for one method in any area. For example, the method used in the largest number of clinical labs in Connecticut is used by less than a majority (43%) of these labs [5]. We will discuss this method first.

o-Toluidine Method for Glucose. This method depends on the fact that glucose possesses a very reactive aldehyde carbonyl group compared to other sugars in body fluids. Addition of *o*-toluidine, an aromatic primary amine, to glucose is followed by a condensation reaction of the primary amino group of *o*-toluidine (R—NH$_2$) with the carbonyl group of glucose.

$$\text{R—NH}_2 + \text{C}_5\text{H}_{11}\text{O}_5\text{—}\underset{\text{H}}{\text{C}}\text{=O} \xrightarrow{\text{heat}} \text{C}_5\text{H}_{11}\text{O}_5\text{—}\underset{\text{H}}{\text{C}}\text{=NR} + \text{H}_2\text{O} \quad (22\text{-}3)$$

(glucose) (Schiff base)

Fructose and sucrose do not interfere significantly because they do not react at the same rate as glucose. Fructose is a ketone sugar; carbonyl groups of ketone sugars (ketoses) are less reactive than carbonyl groups of aldehyde sugars (aldoses).

The reaction produces a Schiff base which has an absorption band at 630–40 nm in the red region. The absorption band covers the yellow, orange, and red regions, so the eye perceives a green color (Table 13-2).

Although glucose is more reactive than other sugars towards *o*-toluidine, the reaction in **22-3** does require a short heating period. Directions for the determination are given in Part IV, Experiment 13.

Ferricyanide Method for Glucose. This method is based on the oxidation of glucose to gluconic acid by ferricyanide.

$$C_5H_{11}O_5-CHO + 2Fe(CN)_6^{-3} + H_2O$$
$$\xrightarrow{95°} C_5H_{11}O_5-CO_2H + 2Fe(CN)_6^{-4} + 2H^+$$

Note that the aldehyde group of glucose is oxidized to a carboxylic acid group in a two-electron oxidation step (Sec. 11-1). This method has been criticized because it is empirical and lacks specificity, but it has been claimed to be the most reliable method for evaluating serum glucose [6].

The ferricyanide ion is an intense yellow color, so the reaction can be followed by measurement of the loss of absorption at 420 nm. Although the ferrocyanide ion also absorbs in that region, its molar absorptivity is only about 1.0 as compared to a molar absorptivity of the order of 10^3 for ferricyanide. Thus the absorption of the ferrocyanide ion is negligible. This reaction is one that has been used in the automated determination of glucose (Fig. 14-10).

Copper(II)-Neocuproine Method for Glucose. This method is also based on the oxidation of glucose to gluconic acid, as in the previous method. The reagent used is a mixture of copper(II) ion and neocuproine, which is the trivial name for 2,9-dimethyl-1,10-phenanthroline.

Once copper(II) has been reduced to copper(I) by glucose, it reacts with neocuproine.

$$Cu(I) + 2C_{14}H_{12}N_2 \rightarrow Cu(C_{14}H_{12}N_2)_2^+$$

In this complex ion the copper(I) is chelated to each neocuproine through the two pairs of electrons on the two nitrogen atoms of each neocuproine, giving it a coordination number of four.

[6] B. R. Sant, *Talanta* **3**, 261 (1960).

The complex ion, or chelate, has a strong absorption band at about 460 nm with a molar absorptivity of over 7.9×10^3 [7]. Because the complex is so stable and so intensely colored, the reaction has also been used in the automated determination of glucose (Fig. 14-10).

Enzyme Kinetics and Determination of Glucose. Since glucose also may be determined with an enzyme (glucose oxidase), we first will review briefly the kinetics of enzyme-catalyzed reactions as background for the determination.

The simplest enzyme reaction involves a reaction between an enzyme E, with a substrate S, to produce a temporary complex ES, which then reacts further to give a product P, and regenerate the original enzyme.

$$E + S \underset{k_3}{\overset{k_1}{\rightleftharpoons}} ES \xrightarrow{k_3} P + E \qquad (22\text{-}4)$$

In the above reaction k_1 is the rate constant for the reaction of E and S to produce ES, k_2 is the rate constant for the reverse reaction in which ES decomposes back to E and S, and k_3 is the rate constant for the decomposition of ES to product.

Let us symbolize the rate of production of P from E and S as rate_p; by a specialized kinetic derivation, it can be shown that when equilibrium is reached (the concentration of ES has reached a constant value at this point also), the rate_p is given by

$$\text{rate}_p = k_3 [E_0] \frac{[S]}{K_m + [S]} \qquad (22\text{-}5)$$

where $[E_0]$ is the enzyme concentration at the start of the reaction, $[S]$ is the concentration of the substrate at any time, and K_m is given by

$$K_m = \frac{k_2 + k_3}{k_1}$$

The K_m term is also known as the Michaelis constant. The $k_3 [E_0]$ product is usually designated as $V_{\max}$, the maximum obtainable rate under specified conditions for a reaction catalyzed by a particular enzyme. Substituting $V_{\max}$ for this product in 22-5 gives

$$\text{rate}_p = V_{\max} \frac{[S]}{K_m + [S]} \qquad (22\text{-}6)$$

Since an enzymatic determination of S is carried out by measuring rate_p, the measurement of this rate must be performed at a concentration of S at which the rate is directly proportional to $[S]$. Since such concentrations of S are quite low, K_m will be much larger than $[S]$, and 22-6 becomes

$$\text{rate}_p = V_{\max} \frac{[S]}{K_m} \qquad (22\text{-}7)$$

[7] H. Diehl and G. F. Smith, *The Copper Reagents: Cuproine, Neocuproine, Bathocuproine,* G. F. Smith Chemical Co., Columbus, Ohio, 1958.

Equation **22-7** now indicates that the rate of the enzyme-catalyzed reaction under these conditions depends directly on only the concentration of the substrate.

The *enzymatic determination of glucose* is in fact performed under conditions where **22-7** is valid. In this method, glucose oxidase enzyme is added to a solution of glucose containing an equilibrium concentration of dissolved oxygen. The glucose oxidase catalyzes the oxidation of the glucose by the dissolved oxygen gas.

$$C_5H_{11}O_5\text{—CHO} + H_2O + O_2(g) \xrightarrow{\text{glucose oxidase}} C_5H_{11}O_5\text{—CO}_2H + H_2O_2 \tag{22-8}$$

In this reaction the rate of formation of product (rate_p) is equivalent to the rate of decrease in concentration of dissolved oxygen ($-\text{rate}_{O_2}$) so **22-7** becomes

$$-\text{rate}_{O_2} = \text{rate}_p = V_{max} \frac{[\text{glucose}]}{K_m} \tag{22-9}$$

If the rate of decrease in dissolved oxygen concentration is measured during the early part of the reaction, it will thus be a measure of [glucose], the initial concentration of glucose.

A commercial *glucose analyzer* instrument has been assembled to measure glucose by using the relationship in **22-9**. It utilizes a polarographic measurement (Sec. 18-4) of the dissolved oxygen by reduction at the cathode.

$$\tfrac{1}{2}O_2 + 2H^+ + 2e^- = H_2O$$

The instrument is calibrated using standard glucose solutions containing, say, 150 mg and 300 mg of glucose per 100 ml. The instrumental readout after 10 s of reduction is set to read exactly 150 mg, and then 300 mg, of glucose. Then the electrodes are rinsed and inserted in the glucose samples, one at a time. After 10 s of reduction of the oxygen in each, the readout from the instrumental scale is recorded directly as milligrams of glucose per 100 ml.

Because the enzymatic determination of glucose is specific for it, the enzymatic method is said to be the only method to give a true measure of glucose. Other methods give slightly high results for glucose because of interferences.

Medical Significance of Glucose Analyses. Some of the dysfunctions associated with low and high serum glucose values are listed in Table 22-5 [5]. The physician may diagnose a high glucose level as *hyperglycemic* and a low glucose level as *hypoglycemic*. Diabetes is the most common condition associated with the hyper-

TABLE 22-5. Dysfunctions Associated with Low and High Glucose Levels

Hyperglycemic dysfunctions (high glucose levels)	Hypoglycemic dysfunctions (low glucose levels)
Endocrine	Postprandial causes
Metabolic	Tumor of the pancreas
Pancreatic	Glucagon deficiency
Ineffective insulin	Liver disease
Stress	

glycemic glucose levels. Lately, it has been recognized that hypoglycemia is fairly common; people who are afflicted with it attempt to raise their blood sugar by eating sweets, which unfortunately enhances body consumption of glucose to the point that blood sugar is lowered rather than raised.

The Glucose Tolerance Test. The glucose tolerance test has already been described in Section 3-1. This test is used to detect both hyperglycemia and hypoglycemia by measurement of the glucose level over a period of time after ingesting a standard amount of glucose. Figure 22-2 shows the tolerance curves for a normal person and for what appears to be a severe diabetic.

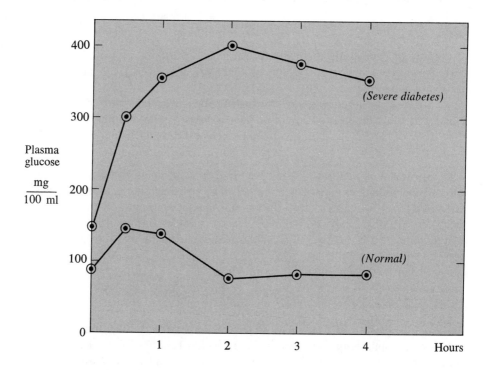

FIGURE 22-2. *Glucose tolerance curves for a normal individual (lower curve) and a diabetic individual (upper curve).*

Determination of Blood Urea Nitrogen (BUN)

Another substance commonly determined in the blood is urea. One method for the determination of urea is also based on an enzyme-catalyzed reaction.

$$NH_2CONH_2 + H_2O \xrightarrow{urease} 2NH_3 + CO_2$$

The ammonia produced in the reaction can be determined with Nessler's reagent; the product is measured spectrophotometrically at 480 nm.

A high BUN value is found in cases of kidney malfunction, carbon tetrachloride or mercuric chloride poisoning, and heart attack. A low BUN value is often found with acute hepatic insufficiency and in normal pregnancy.

QUESTIONS AND PROBLEMS

(Answers to most even-numbered problems are in Appendix 5.)

Definitions and Concepts
1. Define what is meant by clinical analysis.
2. What is the difference between a major electrolyte ion and a trace metal ion?
3. Indicate whether each following group of analyses gives information on the serum electrolyte pattern or on possible toxicity levels.
   a. Determination of serum iron, copper, and lead
   b. Determination of serum chloride, potassium, and sodium
   c. Determination of serum pH, bicarbonate, calcium, and magnesium
4. What is the difference in electrolyte levels between
   a. Respiratory alkalosis and respiratory acidosis?
   b. Respiratory alkalosis and metabolic acidosis?
   c. Dehydration and respiratory alkalosis?
5. List several of the important compounds and elements determined in a clinical blood serum sample.
6. Compare the EDTA titration method with the atomic absorption spectrometric method for the determination of calcium and magnesium in blood serum.

Elements in the Body
7. Analysis gives ppm of the following elements in a blood serum sample. Convert each to molarity.
   a. 3050 ppm sodium ion   b. 200 ppm potassium ion   c. 100 ppm calcium ion
8. Decide whether any of the results in each set below should be checked by further analysis by comparing the sum of the chloride and bicarbonate with the molarity of the sodium ion, and by comparing values with those in Table 22-1.
   a. Na^+: 144 mmole/liter; HCO_3^-: 29 mmole/liter; Cl^-: 112 mmole/liter
   b. Na^+: 140 mmole/liter; HCO_3^-: 30 mmole/liter; Cl^-: 101 mmole/liter
   c. Na^+: 136 mmole/liter; HCO_3^-: 29 mmole/liter; Cl^-: 101 mmole/liter
9. Indicate whether the following blood samples contain toxic amounts of lead(II) ion or not.
   a. 0.05 mg of lead per 100 ml   b. 0.08 mg of lead per liter
   c. 0.8 ppm of lead   d. $1.0 \times 10^{-5} M$ lead

Clinical Analysis of Blood
10. Describe what relationship each following substance has to whole blood.
    a. Plasma   b. Platelets   c. Serum   d. Fibrinogen
11. Indicate whether the following serum chloride values are high or low, and whether they could possibly be diagnosed by a physician as abnormal.
    a. 94 mg/100 ml   b. 99 mg/100 ml   c. 109 mg/100 ml   d. 120 mg/100 ml

12. State whether the following serum glucose values could be diagnosed as hypoglycemia or hyperglycemia, and whether the dysfunction is serious.
    a. 60 mg/100 ml   b. 50 mg/100 ml   c. 400 mg/100 ml   d. 95 mg/100 ml
13. What effect would a high concentration of glucose have on **22-6** in terms of possible enzymatic determination of glucose?

PART IV

Selected Experiments

We have included a number of selected experiments to be used in connection with this text. The theory for essentially all of these experiments has been included in Parts I, II, and III, so there will be little discussion preceding each experiment. You should refer to the relevant material in the front part of the book for background.

To begin with, you should acquire a notebook of sufficient size to record all your laboratory data. General directions for the proper recording of data have been given in Section 2-5. Your instructor will probably wish to supplement this with additional directions.

The first experiment in the laboratory might very well be an experiment involving a simple weighing. We have included such an experiment in the chapter on weighing in Section 4-6; this experiment is listed in the index of experiments.

The first five experiments in this part are concerned with the determination of the chloride ion by five different methods—gravimetric, titrimetric, density determination, index of refraction measurements, and specific ion electrode measurements. These five experiments should provide a useful comparison of the accuracy and precision obtainable by five different experimental approaches. The experiments are also written to accommodate a liquid unknown or a solid unknown.

Prior to the experiment dealing with the analysis of milk of magnesia we have included experiments on the standardization of sodium hydroxide titrant and hydrochloric acid titrant. If other unknowns are to be analyzed with these titrants, you can follow the directions of these experiments to standardize these titrants.

We have also included a few spectrophotometric (colorimetric) experiments because of the importance of this method.

Experiment 1. Five Methods for the Determination of Chloride

This experiment is actually five different experiments; that is, it consists of five different methods for the determination of chloride in the form of sodium chloride. These methods are: 1A, the determination of chloride by density measurement; 1B, the gravimetric determination of chloride; 1C the volumetric determination of chloride; 1D, the determination of chloride by refractive index measurements; and 1E, the determination of chloride by specific ion electrode measurement.

Each experiment can be performed on a liquid sodium chloride unknown containing 0.05–0.17M sodium chloride (see *Instructor's Manual*, p. 43). It will be instructive for you to analyze a liquid unknown by all five methods and compare the accuracy and precision of all five methods. Of course, the gravimetric determination should be the most accurate and precise, but you may not be able to guess which of the other methods is next to the gravimetric method in accuracy and precision.

Each experiment is written so that you can also analyze a chloride sample, such as a solid sample, by using just that one method of analysis.

Experiment 1A. Determination of Chloride by Density Measurement

Introduction: The density of a substance is defined as mass (weight) per unit volume, or stated mathematically $D = \text{weight/volume}$. In metric measurements $D = g/ml$, hence density has the dimensions of grams per milliliter (or cubic centimeter). To determine the density of a substance experimentally, take a known volume of it, weigh it, then compute its density from the above formula. In the following experiment the density of a sodium chloride solution is determined using a volumetric flask to measure the exact volume (Fig. 4-10). First weigh the empty flask and then the filled flask to determine the weight of the solution. In this manner the density may be determined very accurately.

A less accurate but more rapid method frequently used is one which employs a hydrometer. A hydrometer is a hollow glass cylinder which floats in the liquid, and the density is read directly from a scale where the liquid surface meets the hydrometer (Fig. A). The hydrometer is widely used for measuring the density of the sulfuric acid solution in batteries, and in clinical laboratories for measuring the density of urine (a urinometer).
A pycnometer is also used for density measurement (Fig. B). The density of an irregularly shaped solid may be determined by weighing it in air, then weighing it while it is suspended in water. The loss in weight in water is a measure of its volume (Archimedes Principle), hence its density may be computed.

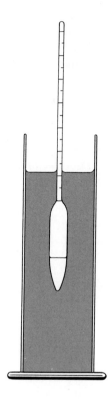

FIGURE A. A hydrometer.

TABLE A-1. DENSITY OF SODIUM CHLORIDE SOLUTIONS, 20°C

Molarity (wt %)	Density, 20° (g/ml)	Refractive index (n_D^{20})
0.000 (0.0%)	0.9982	1.3330
0.0855 (0.5%)	1.0017	1.3339
0.170 (1.0%)	1.0053	1.3347
0.255 (1.5%)	1.0089	1.3356
0.340 (2.0%)	1.0125	1.3365
0.425 (2.5%)	1.0160	1.3374
0.513 (3.0%)	1.0196	1.3382

If you have a pure solution of a single substance, the measurement of its density is a fairly accurate method to determine its concentration. After the density is determined one consults a table or a graph (see Table A-1 and Fig. C) and reads off the concentration corresponding to the density. This technique cannot be used with solutions containing more than one solute. In certain instances, as in the case of urine, one is interested only in a measure of total dissolved solids to determine the concentrating ability of the kidney. In this case the density gives important information, but does not give the individual concentrations of urea, sodium chloride, or other dissolved solids.

Selected Experiments 497

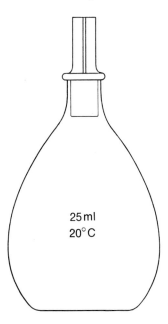

FIGURE B. A pycnometer.

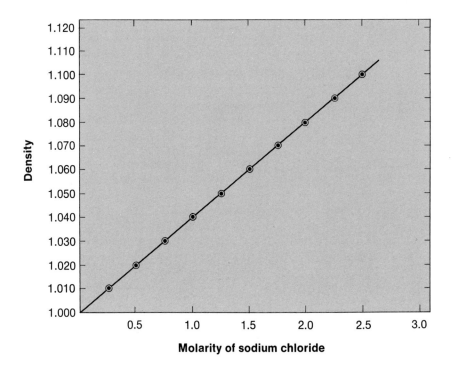

FIGURE C. Density of sodium chloride solutions at 20°C.

LABORATORY PROCEDURE

1. Obtain a small volumetric flask (10 ml, 25 ml, or 50 ml will do), clean it, then dry it in a drying oven, and after cooling weigh it on the analytical balance *along with its stopper*.
2. Fill the flask to the mark with distilled water, stopper it, then weigh again. The weight of the water represents the true accurate volume of the flask. If this number deviates from the given volume of the flask, recheck your measurements and calculations.
3. Pour out the distilled water from the flask, then rinse it three times using 5 ml portions of your unknown solution, discarding each 5 ml portion before adding the next. Using this technique it is unnecessary to redry the flask. Carefully fill the flask to the mark with your unknown solution, stopper it, then weigh.
4. From the weight of the unknown solution and the volume of the flask, calculate the density of your solution.
5. From the table or graph determine the concentration of your solution. Report the concentration of your solution in the following units.
   a. Weight percent NaCl in the solution
   b. Molarity of the NaCl solution
   c. Milligram percent of the chloride ion
   d. Parts per million of the sodium ion

Experiment 1B. Gravimetric Determination of Chloride

The basis for this method is the precipitation of chloride ion by silver (I) nitrate.

$$Cl^- + AgNO_3(excess) \rightarrow AgCl(s) + NO_3^-$$

Either a liquid or a solid sample may be used. Complete directions for the experiment follow. You should read Chapter 5 for a complete background for the experimental details.

Crucibles

1. Sintered-glass crucibles of medium (M) or fine (F) porosity are recommended for filtering silver chloride (do not use crucibles marked C for coarse). Porcelain (Gooch) crucibles with an asbestos mat or glass fiber filter paper may be used but have only the advantage of withstanding temperatures above 500°C, which is usually not necessary. [If your sample is a solid, dry it (Step 1b, *Procedure*) while cleaning the crucibles.]

2. If the crucibles are not new, remove any visible dirt with detergent solution and brush, and rinse. Whether crucibles are new or dirty, assemble filter flask, rubber crucible holder, and crucible. Connect filter flask with rubber suction tubing to aspirator (Fig. D).

Fill the crucible about halfway with concentrated nitric acid and, using gentle suction, draw the acid slowly through the crucible. Fill halfway again, and interrupt the suction briefly to allow a few minutes for the acid to remain in contact with the crucible. Wash the crucible several times with distilled water. Number each crucible. Proceed to Step 3.

If silver chloride is apparently not removed by this procedure, draw $6M$ ammonium hydroxide through the crucible after the acid has been rinsed out with distilled water. Again rinse the ammonium hydroxide out with distilled water.

3. Dry crucibles for 1 hr in an oven at 120°C using a beaker, glass hooks, and a cover glass. Remove crucibles to the desiccator. Cool for 0.5 hr or longer. Weigh. Dry at 120°C for 0.5 hr and cool for 0.5 hr. Reweigh. Constant weight is attained if the second weighings agree with the first weighings within ±0.4 mg. It is advisable to store crucibles in a desiccator until time to use.

PROCEDURE

1a. *Liquid chloride sample.* If the chloride sample is a solution, pipet out exactly 25 ml into each of three 400 ml beakers numbered with numbers corresponding to those on your crucibles. Proceed to Step 2.
  b. *Solid chloride sample.* If the chloride sample is a solid, put it in a weighing bottle, place the open weighing bottle and stopper in a beaker marked with your name, cover the beaker with a cover or watch glass on glass hooks, and dry in the oven at 120°C for 1–2 hr. Let cool in desiccator. While the sample is drying, the crucibles may be cleaned. Next, weigh three samples of from 0.4 to 0.7 g (see the instructor's directions) into clean 400 ml beakers numbered with numbers corresponding to those on your crucibles.
2. Add 150 ml of distilled chloride-free water and acidify with 0.5 ml (10 drops) of concentrated nitric acid. Stir. Use a separate stirring rod for each beaker, and leave it in the beaker throughout the procedure.
3. Consult the instructor's directions for the amount of silver nitrate to add to the liquid sample; it should be a 10 mole % excess. For the solid sample assume the sample is pure sodium chloride, and calculate the millimoles of silver nitrate needed to precipitate it. For instance, 410 mg of sample would be $410/58.5 = 7$ mmoles NaCl = 7 mmoles $AgNO_3$. If a $0.5M$ silver nitrate solution is available, use $7/0.5 = 14$ ml, etc.
4. Add slowly, with stirring, the calculated amount of silver nitrate. Then add a 10% excess of the calculated amount. Heat the beakers almost to boiling, keeping the beakers out of the direct sunlight (a hood is preferred).
5. Stir to aid coagulation. Once the precipitate has coagulated on the bottom of the beaker, test for complete precipitation by adding a few drops of silver nitrate. If a cloud of precipitate appears, add about 5% more silver nitrate and stir. After complete coagulation, test again for complete precipitation, etc.

6. Continue heating, with occasional stirring, until the solution is crystal-clear, indicating that all fine particles of silver chloride have deposited on the coagulated silver chloride on the bottom of the beaker.
7. Remove from the heat, cover with a watch glass, and allow to cool for at least 1 hr in a hood or darkened area. Silver chloride is considerably more soluble in hot water than cold. (The beakers may be left to cool overnight if protected from hydrochloric acid fumes by storing in a desk and covering with watch glasses.)

FILTRATION AND WEIGHING

1. Place weighed filter crucible in the suction filtration appararus (Fig. D) and apply gentle suction. Decant the clear supernate from the corresponding beaker through the crucible, retaining the precipitate in the beaker.

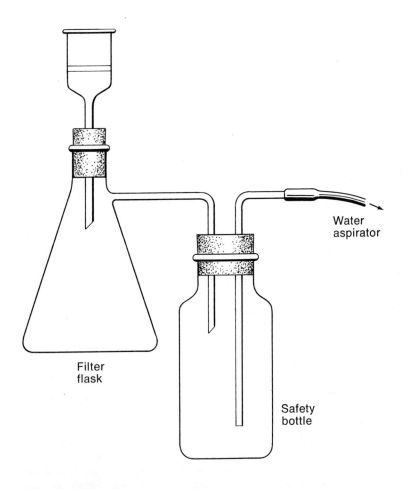

FIGURE D. A suction system for filtering crucibles.

2. Add about 10 ml of wash solution (0.6 ml of concentrated nitric acid per 200 ml of distilled water) to the beaker, and agitate the precipitate. Let it settle, and pour the wash solution through the crucible. Repeat twice. Before proceeding further, examine Figure E for the procedure for decanting and transferring a precipitate.

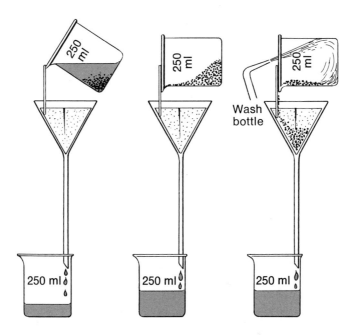

FIGURE E. Decantation and transference of precipitate to filter. (Left) Decantation, keeping most of the precipitate in the beaker. (Center) Transferring the bulk of the precipitate down a stirring rod. (Right) Washing the remainder of the precipitate into the funnel with the wash bottle (a rubber policeman may be needed also in this step).

3. Hold the stirring rod across the lip and top of the beaker with one hand, and direct the stream of wash solution from a wash bottle at the precipitate, washing it into the crucible. If some precipitate sticks to the beaker, loosen it by scrubbing with a rubber policeman. Wash the precipitate completely into the crucible and wash it in the crucible with several 5 ml portions of wash solution. Catch the last washing by itself, and test it for excess silver ion by adding a drop of $12M$ hydrochloric acid. Wash again, if a precipitate is noted.
4. Completely drain the crucible with strong suction, place in a marked beaker, cover with a watch glass on hooks, and dry in an oven at 120°C for 2 hr. Cool in a desiccator for 0.5 hr or longer. Weigh. Dry again for 45 min and cool 0.5 hr or longer. Reweigh. Constant weight is attained if the second weighings agree with the first weighings within ±0.4 mg. Repeat the process, if constant weight is not attained.

5. For the calculation for the liquid chloride sample, use the following setup to calculate the molarity of the chloride.

$$M = \frac{(\text{mg silver chloride})/(143.32 \text{ mg/mmole})}{\text{ml sample}}$$

To calculate the percentage chloride in a solid sample, use 35.45 as the atomic weight of chloride and 143.32 as the formula weight of silver chloride. Report the percentage for each sample and the mean of the three percentages. If one of the results appears questionable, test it, using the Q test (Ch. 3). If the questionable result cannot be rejected, analyze a fourth sample and apply the Q test again. If this does not reject the questionable result, report the median (average of the two middle results) in consultation with the instructor.

Experiment 1C. Volumetric Determination of Chloride by the Mohr and Fajans Methods

Introduction: The reactions involved in the determination of the chloride ion are identical to those in the gravimetric chloride. Silver nitrate is added to the chloride solution, forming a precipitate of silver chloride. The amount of chloride is determined by measuring the amount of a standard solution of silver nitrate required to exactly precipitate all of the chloride ion. In the gravimetric determination the exact concentration of the silver nitrate and the exact amount used were not important except that an excess of silver nitrate was used to completely precipitate the chloride ion.

In the volumetric determination, however, it is necessary to know the exact point at which the silver nitrate has completely precipitated all of the chloride. Two methods are used to detect this point (i.e., the end point of the titration). The first method was developed by C. F. Mohr, a German chemist, over a century ago; hence this very ingenious method is called the *Mohr method* for chloride. A solution of potassium chromate is used as indicator. A known, exact amount of this is added to the unknown chloride solution. When this mixture is titrated with standard silver nitrate, the silver nitrate preferentially reacts with the chloride ion to produce a white insoluble precipitate of silver chloride. This reaction continues until all of the chloride ion is precipitated as silver chloride. At this point the silver nitrate begins to react with the potassium chromate to form a pink-brown precipitate of silver chromate. The formation of this pink-brown color in the solution indicates the complete precipitation of all the chloride, so the buret is then read and the amount of chloride can be computed knowing the exact concentration and volume of the standard silver nitrate used.

The second method was developed by Kasimir Fajans, professor at the University of Michigan. It employs a so-called *adsorption indicator.* Before the end point when an excess of chloride ion is present, the silver chloride particles are negatively charged, owing to the presence of an excess of chloride ions. As soon as precipitation is complete and with the first drop excess of silver nitrate, the silver chloride particles become positively charged, owing to the excess silver ions. The positively charged particles absorb a dye, dichlorofluorescein, on the surface and cause a change in color from yellow to pink. This represents the end point of the titration, and the burette is then read. Dextrin solution is added in this titration to keep the silver chloride particles from coagulating, thus giving a better end point.

Both methods are very sensitive to pH variations and work best in neutral or very slightly acid solution. If the pH of the solution is unknown, it should be measured before the titration is performed and the pH brought in the range of 6–8 by adding dilute acetic acid or ammonium hydroxide as required (pH paper).

Both methods require the use of a standard silver nitrate solution. Solid silver nitrate, $AgNO_3$, is a chemical which can be purchased in a very pure form; hence it can be used as a primary standard. A solution of very exact concentration can be prepared by weighing out an exact known weight of solid silver nitrate and diluting it to an exact known volume in a volumetric flask.

PROCEDURES

I. Preparation of standard silver nitrate solution

1. Clean your weighing bottle. Then dry it in the oven for about 15 min with the top off.

2. Weigh the empty, dry weighing bottle and top and record the weight.

3. Obtain about 8.5 g of pure, primary standard silver nitrate crystals from the instructor. Place it in your weighing bottle, then dry in the oven for 30–60 min. (Be sure to leave the top off the weighing bottle.)

4. Remove bottle and silver nitrate from oven, replace the top and allow to cool, then weigh. Note exact weight of silver nitrate.

5. Quantitatively transfer the solid silver nitrate to a clean 500 ml volumetric flask using a transfer funnel. Wash out weighing bottle and funnel with distilled water and pour washings into the volumetric flask. Remove funnel and dilute to the mark with distilled water. Shake and mix until all solid $AgNO_3$ dissolves.

6. Calculate the exact normality of your silver nitrate solution, using 169.88 as the molecular weight.

II. Preparation of samples

1. Liquid chloride sample. If the chloride sample is a solution, pipet exactly 25 ml into each of three 250 ml erlenmeyer flasks, then proceed to Step III-A or III-B.

2. Solid chloride sample. If the chloride sample is a solid, put it in a weighing bottle, place the open weighing bottle and stopper in a beaker marked with your name, cover the beaker with a cover or watch glass on glass hooks, and dry in the oven at 120°C for 1–2 hr. Let cool in desiccator. Weigh three samples of from 0.4 to 0.7 g (see instructor's directions) into clean 250 ml erlenmeyer flasks and add 25 ml of distilled water. Swirl until the sample is completely dissolved, then proceed to Step III-A or III-B.

III. A. Determination of chloride by Fajan's method
Use the solutions from Step II-1 or II-2.

1. Add 5.0 ml of 2% dextrin and 10 drops of 0.1% dichlorofluorescein in 70% alcohol to each sample.

2. Titrate with your standard silver nitrate from a 50 ml buret, adding the solution fairly rapidly at first while swirling the flask rapidly to insure mixing. As the end point approaches, the local development of a pink color becomes more widespread. The silver nitrate then should be added dropwise until the pink color spreads throughout the flask and does not disappear on swirling. At the end point the entire suspension in the flask changes abruptly from yellow to pink, and the precipitate then settles out as a pink precipitate.

Note the buret reading and calculate the chloride molarity of the sample using the known normality and observed volume of the standard silver nitrate.

B. Determination of chloride by Mohr's method
Use the solution from Step II-1 or II-2.

1. Add exactly 1.0 ml of 0.25M potassium chromate to each sample.

2. Titrate with your standard silver nitrate solution from a 50 ml buret, adding the solution rapidly at first while rotating the flask rapidly to insure mixing. As the end point approaches, the red-brown color of silver chromate becomes more widespread. The silver nitrate then should be added dropwise until a faint red-brownish color of silver chromate persists throughout the yellow solution and suspension. It should not be a dark brown, which indicates the end point is overstepped; the color change should be faint but noticeable. This end point is not as distinct as in Fajan's method, but the observation improves with practice.

Note the buret reading and calculate the chloride molarity of the sample using the known normality and observed volume of the standard silver nitrate.

Experiment 1D. Determination of Chloride by Refractive Index Measurement

When a beam of light passes from one medium to another of different density, its direction is changed slightly (Fig. F). This phenomenon is called *refraction*; the ratio of the sine of the angle of the incident ray, i, with the

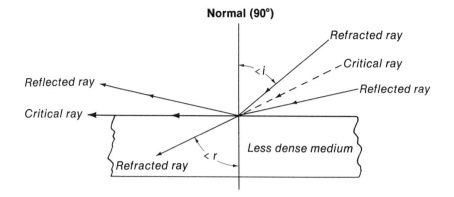

FIGURE F. *Refractive index.*

normal (a perpendicular to the surface of the second medium) to the sine of the angle of the refracted ray, r, is called the *index of refraction*. The index of refraction is usually represented by the Greek letter η (eta). Stated mathematically,

$$\eta = \frac{\sin i}{\sin r}$$

As the angle of incidence is increased and the beam becomes more parallel to the surface, the angle of refraction, r, becomes 90° and the light will no longer travel from one medium to the other, but will pass along the surface and be completely absorbed. The angle at which this occurs is called the *critical angle*. If the critical angle is exceeded, the ray is then reflected from the surface and does not pass into the second medium at all.

An instrument which measures the index of refraction is called a refractometer. The sample, usually a liquid, is placed as a thin layer between two glass prisms. One prism is rotated until the critical angle is reached and a dark and light field is observed at the cross hairs of the eyepiece. The index of refraction is then read from a scale in the instrument. This type of refractometer is called an *Abbe* refractometer; your instructor will explain the details of its operation.

The refractive index varies from one liquid to another and also varies with the temperature and with the concentration of any dissolved solute in the liquid. The index of refraction can be measured to six significant figures. For example, the index of refraction of pure water at 20°C is 1.33299, while a 10% solution of sodium chloride has an index of 1.35021. As with the measurement of density, the index of refraction may be used to determine the concentration of a single solute in a solution. Since the index of refraction varies noticeably with temperature, the temperature must be maintained constant at a certain known value, usually 20°C.

Index of refraction is used frequently by chemists to identify organic liquids and also as a detector in certain instruments, as in liquid chromatography. Transparent solids also have characteristic indexes of refraction. The diamond has one of the highest, and its brilliance arises from the refraction of light from its many surfaces.

PROCEDURE FOR THE ABBE REFRACTOMETER

1. Open up the prism chamber, then clean the chamber with distilled water and wipe the prisms dry using soft tissue paper. Special care should be taken not to scratch the prisms.
2. Start the temperature controlling device and adjust to 20°C ± 1°.
3. Check the instrument by placing a few drops of distilled water in the chamber and taking a reading. It should read 1.3330.
4. Open the chamber and dry with soft tissue paper, then place a few drops of the unknown solution in it.
5. Take the reading and determine the concentration of the solution from Table A-1.

Experiment 1E. Determination of Chloride by Specific Ion Electrode Measurement

A specific ion electrode is constructed of a material sensitive to certain specific ions but not to others. The electrode is immersed in the unknown solution along with a standard reference electrode, and the potential of the system is measured on a potentiometer, as used in pH measurements.

This potential is compared with a standard curve obtained by measuring the potentials of solutions of known concentration, and the unknown concentration is then determined. A variety of specific ion electrodes are commercially available and some of the more common ones are those measuring fluoride, silver, ammonia, and sulfide ions. These electrodes are not completely specific, and foreign ions having similar properties or those which interact with the desired ion may interfere. For example, bromide, iodide, cyanide, and silver ions seriously interfere with the specific chloride ion electrode. The chloride ion electrode consists of a piece of silver metal coated with solid silver chloride (Fig. G). The potential of this electrode is determined by two equilibria:

$$Ag° \leftrightarrow Ag^+ + e^- \qquad (1)$$

$$Ag^+ + Cl^- \leftrightarrow AgCl(s) \qquad (2)$$

The equilibrium between the metal silver electrode and silver ions (1) establishes the potential of the electrode. The silver ion concentration, however, is controlled by the chloride ion concentration (2); hence the potential measured varies with the chloride ion concentration in solution.

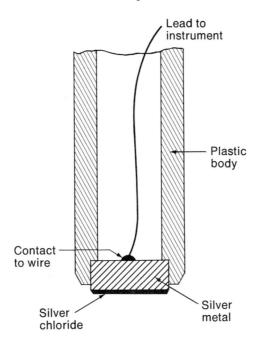

FIGURE G. Chloride ion electrode.

The chloride ion electrode is frequently used in clinical laboratories to determine the chloride in blood plasma. Specific sodium ion electrodes also are available.

PROCEDURE

1. *Preparation of a standard curve.* Prepare 500 ml each of 0.050, 0.100, and 0.150M solutions by weighing out the calculated amounts of chemically pure, dry, solid sodium chloride (mol wt 58.45) and diluting to volume in 500 ml volumetric flasks. Shake until the solid has dissolved completely, then pour about 100 ml of each solution into separate 250 ml beakers. Place the standard reference electrode and the specific ion electrode in each solution and measure its potential. Be sure to dry the electrodes with soft tissue before changing from one solution to the other. Make a plot of potential *vs.* concentration, plotting potential on the ordinate and concentration on the abscissa. If any points deviate from a straight line, recheck those points.
2. *Analysis of unknown solution.* Clean and dry the electrodes, then immerse them in the unknown solution. Stir and let stand for a few minutes until a constant potential is observed. Record this potential, then obtain the concentration of the sodium chloride from the standard curve.

Experiment 2. Preparation and Standardization of NaOH

Sodium hydroxide titrant is best standardized against primary standard potassium acid phthalate (Sec. 9-1). The reaction is

$$NaOH + KHC_8H_4O_4 \rightarrow C_8H_4O_4^{-2} + H_2O + K^+ + Na^+$$
(phthalate ion)

The phthalate anion produced makes the solution slightly basic at the end point, the pH being about 9. This pH is roughly in the middle of the pH transition range of phenolphthalein and thymol blue indicators (Table 9-1). Phenolphthalein is generally used as the indicator; the first perceptible pink is taken as the end point.

Directions are given below for the preparation and standardization of the usual 0.1M sodium hydroxide and the 0.25M sodium hydroxide needed for the analysis of milk of magnesia (Experiment 4).

PREPARATION OF 0.25M AND 0.1M SODIUM HYDROXIDE

1. If it is not available in the laboratory, prepare a saturated (1:1) solution of sodium hydroxide by mixing carefully 50+ g of sodium hydroxide pellets with 50 ml of water in a beaker. When the solution has partly cooled, transfer to a polyethylene bottle, stopper, and let stand for about a week to allow the insoluble sodium carbonate to settle.
2. Obtain about a liter of distilled water and a polyethylene bottle. If the instructor directs, boil the water for 5 min to remove carbon dioxide, and allow to cool to about 40°C before transferring to the polyethylene bottle. If the instructor does not wish the water to be boiled, transfer it directly to the polyethylene bottle.
3a. *0.25M sodium hydroxide for milk of magnesia analysis.* Pipet about 17 ml of a *clear* saturated solution of sodium hydroxide into the polyethylene bottle containing about a liter of distilled water, using a *transfer* pipet and a rubber bulb. Mix thoroughly and keep stoppered.
 b. *0.1M sodium hydroxide.* Pipet about 7 ml of a *clear* saturated solution of sodium hydroxide into the polyethylene bottle containing the liter of distilled water using a *transfer* pipet and a rubber bulb. Mix thoroughly and keep stoppered.

STANDARDIZATION OF SODIUM HYDROXIDE

1. Dry 6–9 g of primary standard potassium acid phthalate in a weighing bottle at 110°C for 2 hr. Allow to cool 30 min in a desiccator before weighing.
2. If your buret is to be used for the first time in the term, fill it with distilled water and allow it to drain to check for dirt. If it does not drain uniformly, it must be cleaned (Sec. 6-3).
3a. To standardize 0.25M sodium hydroxide, weigh out 1.2–1.4 g of potassium acid phthalate into each of three 250 ml flasks. (This is about 6–7 mmoles and should require about 24–28 ml of the sodium

hydroxide.) Dissolve in about 50–75 ml of distilled water, warming if necessary.
- **b.** To standardize 0.1M sodium hydroxide, weigh out 0.8–0.9 g of potassium acid phthalate into each of three 250 ml flasks. Dissolve in about 50 ml of distilled water, warming if necessary.
4. Rinse the buret thoroughly with the titrant (Sec. 6-3), fill, and take an initial reading using the meniscus illuminator (Fig. 6-2).
5. Add three to four drops of phenolphthalein indicator to the first flask and titrate to the first faint pink that persists for 20 seconds. Titrate dropwise near the end point, splitting drops at the end point (Sec. 6-3). Add the same volume of phenolphthalein indicator to the remaining two flasks, and titrate those solutions also. Try to achieve the same shade of pink color for all end points.
6. Calculate the molarity of the sodium hydroxide by using the formula weight of 204.22 for potassium acid phthalate (Sec. 6-2). The molarity is the same as the normality in this case. If one of the values deviates significantly from the others, apply the Q test (Sec. 2-4 and 2-5).

Experiment 3. Preparation and Standardization of HCl

Hydrochloric acid titrant may be standardized against a standard solution of sodium hydroxide or against primary standard sodium carbonate. The equations for both reactions are

$$HCl + NaOH \rightarrow H_2O + Na^+ + Cl^-$$

$$2HCl + Na_2CO_3 \rightarrow H_2CO_3 + 2Na^+ + 2Cl^-$$

When both sodium hydroxide and hydrochloric acid must be used for a method, such as the analysis of milk of magnesia (Experiment 4), it is recommended that the sodium hydroxide be standardized first (Experiment 2), and then used to standardize the hydrochloric acid. If sodium hydroxide is not needed, then it is recommended that the hydrochloric acid be standardized against primary standard sodium carbonate. Since the end point is not sharp, two indicator procedures are given. In the methyl orange procedure, the end point is not sharp, but the solution need not be boiled to remove the carbonic acid product as carbon dioxide. In the methyl red procedure, the end point is sharp, but the solution must be boiled.

PREPARATIOM OF 0.50M AND 0.1M HYDROCHLORIC ACID

1a. *0.50M hydrochloric acid for milk of magnesia analysis.* Fill a 1 liter ground-glass stoppered bottle with almost a liter of distilled water. Add about 45 ml of concentrated hydrochloric acid from a graduated cylinder and mix well. Allow to cool before using.

b. *0.1M hydrochloric acid.* Fill a 1 liter ground-glass stoppered bottle with about a liter of distilled water. Add about 9 ml of concentrated hydrochloric acid from a measuring pipet and mix well.

STANDARDIZATION OF 0.50M HYDROCHLORIC ACID VERSUS SODIUM HYDROXIDE

1. Standardize 0.25M sodium hydroxide as directed in Experiment 2. The sodium hydroxide titrant should be standardized and used within a period of two weeks. If a longer time elapses, it is recommended that you recheck the molarity of the sodium hydroxide by titration.
2. Using a volumetric pipet, transfer exactly 20 ml of 0.50M hydrochloric acid to each of three 250 ml flasks.
3. To the first flask add 2–3 drops of phenolphthalein indicator and titrate to the first faint pink color which will persist for at least 15 s, using 0.25M sodium hydroxide titrant. Use a white paper background to detect the pink color. If necessary, adjust the volume of phenolphthalein indicator for the titration of the remaining two flasks and titrate these also.
4. Calculate the molarity of the hydrochloric acid as follows:

$$M_{HCl} = \frac{(M_{NaOH})(ml_{NaOH})}{ml_{HCl}}$$

STANDARDIZATION OF 0.1M HYDROCHLORIC ACID VERSUS SODIUM CARBONATE

1. Dry 1–1.5 g of primary standard sodium carbonate at 110°C for 2 hr in a weighing bottle.
2. Weigh three 0.2 g samples by difference (to avoid absorption of water from the air) into 250 ml flasks and dissolve in 50–100 ml of water.
3a. *Methyl orange procedure.* Add 2–3 drops of modified methyl orange indicator to each flask. Titrate with 0.1M hydrochloric acid, taking the end point as the point of color change from green to grey. If the color change is difficult to observe, record the color for each increment of titrant added and proceed dropwise past the grey to a purple. By comparison, the change to grey should now be evident. If the color change to purple appears sharper, take this to be the end point, but use the same color change for the unknown analysis so that the errors cancel out.
   b. *Methyl red procedure.* Add 2–3 drops of methyl red indicator to each flask. Titrate with 0.1M hydrochloric acid until the indicator has changed color gradually from a yellow to a definite red. Then boil the solution gently for 2 min (the color should revert to yellow). Cover the flask with a watch glass, cool to room temperature, and continue the titration to a sharp change to red.
4. Calculate the molarity or normality of the acid as follows:

$$N_{HCl} = M_{HCl} = \frac{(\text{mg of } Na_2CO_3/53.0 \text{ mg per meq } Na_2CO_3)}{\text{ml of HCl}}$$

Experiment 4. Analysis of Milk of Magnesia

According to the *United States Pharmacopeia (U.S.P.)*, medicinal preparations of milk of magnesia should contain a minimum of 7% of magnesium hydroxide by weight. Since such solutions contain a suspension of magnesium hydroxide, as well as being saturated with it, an accurate analysis must measure both the suspended and dissolved magnesium hydroxide. To take a representative sample for analysis, it is necessary that the bottle of milk of magnesia be well shaken just before taking the sample.

Since the sample is a white opaque suspension it poses an analytical problem if a *direct* titration is to be used. The suspended magnesium hydroxide particles may cause errors by adhering to the walls of the flask out of contact with the titrant. The opaque sample may also cause difficulties in perceiving an accurate end point color change.

A much better analytical approach is therefore to add a measured excess of a standard hydrochloric acid to neutralize and dissolve all of the suspended particles, giving a clear solution. The unreacted acid is then back titrated with standard base. The equations are

$$Mg(OH)_2(s) + 2HCl(excess) \rightarrow 2H_2O + Mg^{+2} + 2Cl^-$$

$$\text{unreacted HCl} + \text{NaOH} \rightarrow H_2O + Na^+ + Cl^-$$

$$\text{(titrant)}$$

The *U.S.P.* procedure calls for the use of $1M$ acid and $1M$ sodium hydroxide solutions, which can involve large errors for beginning students. The procedure below utilizes $0.5M$ hydrochloric acid and $0.25M$ sodium hydroxide to reduce those errors.

PROCEDURE

1. Prepare and standardize $0.25M$ sodium hydroxide (Experiment 2) and $0.5M$ hydrochloric acid (Experiment 3).
2. Obtain a sample of about 20 g of commercial milk of magnesia (Note 1).
3. Obtain three weighing bottles, each of which should be large enough to contain over 5 ml of solution. Weigh each dry to the nearest mg on the analytical balance.
4. Shake the sample of milk of magnesia thoroughly, and immediately transfer about 5–6 g of the sample to each weighing bottle. Again weigh each bottle to the nearest mg on the analytical balance.
5a. Using a wash bottle, rinse each sample quantitatively out of the weighing bottle into a 250 ml flask. Using a 50 ml volumetric pipet or 50 ml buret, add exactly 50 ml of $0.5M$ standard hydrochloric acid to each of the three flasks. Shake to insure complete reaction. The samples should dissolve completely. If any cloudiness or precipitate remains, this indicates that not enough acid has been added, and a measured additional amount should be added.

b. Add 3–4 drops of methyl red indicator to each flask and titrate the unreacted hydrochloric acid with 0.25M standard sodium hydroxide to the yellow (basic) color. The back titration should require at least 20 ml of sodium hydroxide for accurate analysis; if too large a milk of magnesia sample has been used, less than 20 ml will be required.

6. Calculate the weight per cent of magnesium hydroxide in the sample as follows:

$$\text{mmoles Mg(OH)}_2 = \frac{1}{2}[(M \text{ HCl})(\text{ml HCl}) - (M \text{ NaOH})(\text{ml NaOH})]$$

$$\% \text{ Mg(OH)}_2 = \frac{\text{mmoles Mg(OH)}_2 \times 58.34 \text{ mg/mmole} \times 100}{\text{mg sample}}$$

Note

1. Milk of magnesia preparations from any drug store should make suitable samples. The authors have found that the Phillips' brand assays at significantly above 7% Mg(OH)$_2$.

Experiment 5. Analysis of Impure Sodium Carbonate

Sodium carbonate samples are frequently analyzed by titrating the samples with standard 0.1M hydrochloric acid.

$$Na_2CO_3 + 2HCl \rightarrow 2Na^+ + 2Cl^- + H_2CO_3(\rightarrow CO_2(g) + H_2O)$$

The carbonic acid product slowly decomposes to carbon dioxide and water.

This method involves the same steps as the standardization of hydrochloric acid against pure, primary standard sodium carbonate (Experiment 3). It is recommended that the hydrochloric acid titrant be standardized against primary standard sodium carbonate rather than against standard sodium hydroxide solution. The reason is that the end point is difficult, and the standardization will give you some experience with the end point.

PROCEDURE

1. Dry the impure sodium carbonate sample at 110°C for 2 hr.
2. Weigh out four 0.25–0.35 g samples (check instructor's directions) by difference from a weighing bottle into 250 ml flasks. Dissolve each sample in 50–100 ml of distilled water.
3. Use the first flask to practice the end point by whichever following indicator is specified by the instructor.
4a. *Methyl orange procedure.* Add 2–3 drops of modified methyl orange indicator to each flask. Titrate with 0.1M hydrochloric acid, taking the end point as the point of color change from green to grey. If the

color change is difficult to observe, record the color for each increment of titrant added and proceed dropwise past the grey to a purple. By comparison, the change to grey should now be evident. If the color change to purple appears sharper, take this to be the end point, but be sure that you used the same color change for the standardization of the hydrochloric acid.
   b. *Methyl red procedure.* Add 2–3 drops of methyl red indicator to each flask. Titrate with 0.1M hydrochloric acid until the indicator has changed color from a yellow to a definite red. Then boil the solution gently for 2 min (the color should revert to yellow.) Cover the flask with a watch glass, cool to room temperature, and continue the titration to a sharp change to red.
5. Calculate the per cent sodium carbonate using a formula weight of 106.0 or an equivalent weight of 53.0 for the sodium carbonate.

Experiment 6. Acid-Neutralizing Capacity of an Antacid Tablet

Antacid tablets are made from a variety of substances which react with the hydrochloric acid of the stomach and neutralize it. Most commonly used are solid sodium bicarbonate, magnesium hydroxide, calcium carbonate, aluminum oxide and magnesium trisilicate. Other substances such as flavors are added as well as salicylates and aspirin in small amounts.

The acid-neutralizing capacity is measured by adding an excess of standard hydrochloric to a previously weighed tablet, allowing it to react, then back titrating the excess acid with standard sodium hydroxide solution. The preparation and standardization of these solutions has been discussed previously in Experiments 2 and 3.

The measurement of the pH of a mixture of the tablet with water may give some clue as to its composition. Sodium bicarbonate and magnesium hydroxide form basic solutions of pH 8–9, whereas the other compounds are less basic, forming solutions of pH 6–8.

PROCEDURE

Measurement of pH
1. Grind up a tablet using a mortar and pestle, then mix it with 25 ml of distilled water in a small beaker.
2. Stir for about 5 min, then measure the pH of the solution, first using indicator paper, then using a glass pH electrode.

Acid-neutralizing capacity
1. Weigh a single tablet on the analytical balance, then transfer the tablet to a 250 ml erlenmeyer flask.
2. From a buret add 50.00 ml of standard (about one-half normal) hydrochloric acid to the tablet in the flask.

3. Let the mixture stand for about one-half hour with occasional swirling, then add 4–5 drops of phenolphthalein indicator. If the solution turns pink, add another 50.00 ml of standard hydrochloric acid. Repeat this operation until a colorless solution is obtained.
4. Titrate the mixture with standard sodium hydroxide solution until a faint pink color is obtained. Note the volume of standard sodium hydroxide required.

Calculations
1. Calculate the number of milliequivalents of hydrochloric acid neutralized per tablet.
2. Calculate the number of milliequivalents of hydrochloric acid neutralized per gram of tablet.
3. If normal gastric juices contain 0.4% hydrochloric acid, calculate the number of milliliters of gastric juice neutralized by one tablet.
4. Write balanced equations representing the reaction of hydrochloric acid with $NaHCO_3$, $CaCO_3$, MgO, Al_2O_3, and $MgSi_3O_7$.

Experiment 7. Nonaqueous Titration of Amines and Acid Salts

In glacial acetic acid as solvent, perchloric acid has been found to be the strongest acid; that is, it has a higher degree of ionization than such other acids as sulfuric acid. For this reason it has come to be the titrant of choice for the titration of weak bases in glacial acetic acid. Bases such as aniline and sodium acetate, which are too weakly basic to titrate in water, can be titrated to sharp end points.

The most suitable primary standard for perchloric acid is potassium acid phthalate, which behaves as a Bronsted base in this reaction (**7-1**). Two common types of weak bases that can be analyzed by perchloric acid titration are amines, such as aniline (**7-2**), and salts of carboxylic acids, such as sodium acetate (**7-3**). The reactions are written in molecular rather than ionic form because of the low degree of dissociation in glacial acetic acid; even perchloric acid has an overall K_a of the order of 10^{-5}.

$$HClO_4 + KHP \rightleftharpoons H_2P(s) + KClO_4 \qquad (7\text{-}1)$$

$$HClO_4 + C_6H_5NH_2 \rightleftharpoons C_6H_5NH_3ClO_4 \qquad (7\text{-}2)$$

$$HClO_4 + CH_3CO_2Na \rightleftharpoons CH_3CO_2H + NaClO_4 \qquad (7\text{-}3)$$

PROCEDURE

1. Prepare 0.1M perchloric acid by adding 4.3 ml of 72% perchloric acid ($HClO_4 \cdot 2H_2O$) to 100 ml of glacial acetic acid, mixing well and adding 10 ml of acetic anhydride (Note 1). The mixture will become quite warm; allow it to stand for half an hour, so that the water in the perchloric acid reacts with the acetic anhydride to form acetic acid (Note 2). Then dilute to 500 ml with glacial acetic acid. Allow to cool to room temperature.
2. Prepare the indicator by dissolving 0.2 g of methyl violet or crystal violet in 100 ml of glacial acetic acid or chlorobenzene. This solution may be available in the laboratory.
3. Standardize the perchloric acid by weighing exactly three 0.6–0.7 g samples of primary standard potassium acid phthalate into 250 ml flasks. Dissolve the phthalate by adding 60 ml of glacial acetic acid to each flask and heating to boiling in a hood. If any of the solid creeps up the sides of the flask, wash it down with a little acetic acid.
4. Cool, and add 2–3 drops of methyl violet indicator to each flask. Titrate each, with 0.1M perchloric acid, to the color change of violet to blue (not to blue-green or green or yellow). Calculate the normality of the titrant using an equivalent weight of 204.2 for potassium acid phthalate.
5. Weigh exactly three 0.5 g samples (see the instructor's directions) of solid unknown into 250 ml flasks, or pipet three 25 ml aliquots (see the instructor's directions) of liquid unknown (Note 3) into 250 ml flasks. Dilute to a volume of 60 ml with glacial acetic acid, add 2–3 drops of indicator, and titrate with 0.1M perchloric acid to the color change of violet to blue.
6. For solid unknowns calculate the equivalent weight of the unknown. For liquid samples calculate the milliequivalents of base per a 25 ml aliquot of the liquid unknown.

Notes

1. If the solution is only to be used for the titration of tertiary amines, acetates, or phthalates, add 13 ml of acetic anhydride, instead of 10 ml. This renders the titrant completely anhydrous and permits a slighlty sharper end point.

2. The reaction of water with acetic anhydride is

$$(CH_3CO)_2O + H_2O \rightarrow 2CH_3CO_2H$$

This reaction is actually catalyzed by the perchloric acid and is quite rapid if the solution is not diluted.

3. Liquid unknowns may be potassium acid phthalate dissolved in glacial acetic acid or a solution of an amine in chlorobenzene. Typical amines are n-butylamine, pyridine, or dimethylaniline. Solids unknown may be acetates such as sodium acetate, barium acetate, and strontium acetate. Potassium acetate is not very suitable because it is quite hygroscopic.

Experiment 8. Determination of Water Hardness

The determination of water hardness using EDTA has been discussed in Section 10-2. This analysis involves the determination of the sum of calcium and magnesium carbonate in any type of hard water. If the EDTA is not available in primary standard form, it must be standardized against reagent grade calcium carbonate. Either Eriochrome Black T or Calmagite indicator may be used for the titration; it is recommended that Calmagite be used because it is more stable.

Reagents

1. Prepare $0.01M$ EDTA titrant by dissolving about 1.9 g of the reagent-grade disodium salt ($Na_2H_2Y \cdot 2H_2O$, form wt 372) in 500 ml of distilled water. Add about 0.5 g (6 pellets) of sodium hydroxide and about 0.1 g of magnesium chloride, $MgCl_2 \cdot 6H_2O$. Mix well. Store in a Pyrex bottle. If A.C.S. grade EDTA (99.0% min $Na_2H_2Y \cdot 2H_2O$) is available, the instructor may direct you to use it as a primary standard. Omit the addition of magnesium chloride and follow Note 1.

2. Prepare standard $0.0100M$ calcium(II) solution by weighing 0.500 g ± 0.2 mg of 99+% calcium carbonate into 20 ml of distilled water in a 250 ml beaker. Pipet about 1 ml of concentrated hydrochloric acid down the side of the beaker, and cover with a watch glass placed directly on top of the beaker. When the calcium carbonate has dissolved, rinse the watch glass into the beaker, raise it on glass hooks, and evaporate the solution to a volume of about 2 ml to expel most of the carbon dioxide. Add 50 ml of distilled water, transfer to a 500 ml volumetric flask, and dilute to the mark.

3. Prepare the pH 10 ammonia buffer by diluting 32 g of ammonium chloride and 285 ml of concentrated ammonium hydroxide to 500 ml with distilled water. Store in a polyethylene bottle. The buffer is as concentrated as possible, to avoid contaminating the solutions with interfering metal ions. It is not stored in glass, because metal ions would be leached from the glass.

4. Prepare a dilute, aqueous solution of Calmagite. If Eriochrome Black T indicator is to be used instead, prepare a solid mixture by grinding 50 mg of Eriochrome Black T into 5 g of sodium chloride and 5 g of hydroxylamine hydrochloride.

PROCEDURE

1. Pipet exactly 20 ml of $0.0100M$ calcium(II) solution into each of three clean 250 ml flasks (Note 2), and add about 1 ml of the pH 10 ammonia buffer to each flask. Add enough Calmagite indicator to the first flask to give a wine red color (Note 3). Alternatively add a small scoop of the solid Eriochrome Black T indicator mixture.
2. Titrate with $0.01M$ EDTA from a 25 ml buret until a color change from wine red through purple to clear blue is observed. The color change is somewhat slow, so titrant must be added slowly near the end point (Note 4). Repeat the titration for the remaining two samples. Calculate the molarity of the EDTA, using the $0.0100M$ of the calcium(II) solution.
3. Determine the water hardness of the unknown hard water by pipetting exactly 50 ml of the sample into each of three clean 250 ml flasks. Add

1 ml of ammonia buffer and Calmagite indicator to the first sample and titrate with 0.01M EDTA from a 25 ml buret to a clear blue end point (Notes 4 and 5).

4. Calculate the parts per million of calcium carbonate in the unknowns using the molarity of EDTA and 100.1 mg/mmole as the formula weight of calcium carbonate.

Notes

1. Magnesium chloride is added to the titrant to insure sufficient magnesium(II) for a sharp end point since calcium(II) does not form a strong enough chelate with the indicator. Instead of this, about 0.5 ml of 0.0005M magnesium-EDTA chelate may be added to each flask. Prepare this by mixing equal volumes of 0.010M EDTA and 0.010M magnesium(II) solutions.

2. Remove traces of interfering metal ions on the sides of the titration flasks by rinsing each flask with about 10 ml of 1:1 nitric acid. Tilt and rotate the flask to contact the entire inside. Rinse well with distilled water, and allow to drain upside down.

3. Possible air oxidation makes it advisable to add the indicator just before the titration. Traces of metal ions such as manganese(II) can catalyze this oxidation; this is prevented by adding a bit of ascorbic acid. Avoid adding too much indicator, because the end point color change will be too gradual.

4. To avoid the necessity of titrating slowly near the end point, the solutions may be warmed to about 60°C.

5. If the end point color change is to a violet and not to a clear blue, a high level of iron in the water may be responsible. Avoid this in succeeding samples by adding a few crystals of potassium cyanide (as much as will fit on the tip of a spatula) *after* the buffer has been added. *Caution!* Potassium cyanide is a poison; on contact with acid it forms hydrogen cyanide gas which is especially dangerous.

Experiment 9. Preparation and Standardization of Iodine

A standard solution of iodine may be prepared by weighing it exactly on an analytical balance or by standardizing it against primary standard arsenic(III) oxide. Since it is difficult to handle iodine without losing some of it, it is usually weighed out approximately and standardized. The arsenic(III) oxide is first dissolved in base and then neutralized to arsenious acid.

$$As_2O_3(s) + 2OH^- + H_2O \to 2H_2AsO_3^- + 2H^+ \to 2H_3AsO_3$$

The arsenious acid is then oxidized by the iodine titrant to arsenic acid in a solution buffered at about pH 8 with sodium bicarbonate.

$$I_2 + H_3AsO_3 + H_2O \to 2I^- + H_3AsO_4 + 2H^+$$

The end point is detected by the formation of the deep blue starch-triiodide color. Since iodine exists as the triiodide (I_3^-) ion in aqueous solution, the first excess of iodine titrant added will form the starch-triiodide complex.

PREPARATION OF STARCH AND IODINE SOLUTIONS

1. *Starch solution.* Make a paste of 2 g of soluble starch and 25 ml of water, and pour with stirring into 250 ml of boiling water. Boil the mixture for 2 min, add 1 g of boric acid as preservative, and allow to cool. Store in a glass-stoppered bottle.
2a. *0.03N iodine solution for vitamin C analysis.* Weigh about 3.8 g of reagent grade iodine on a trip scale and transfer it to a 100 ml beaker containing 20 g of potassium iodide dissolved in 25 ml of water. Stir carefully to dissolve all the iodine (it sometimes dissolves slowly). Pour the entire contents of the beaker into a glass-stoppered amber liter bottle. Rinse the beaker with 50 ml of distilled water and pour this into the bottle also. Dilute to a liter using distilled water and shake several times. (If there is any doubt that all of the iodine dissolved, allow to stand until the next lab period.)
b. *0.1N iodine solution.* Weigh about 12.7 g of reagent grade iodine on a trip scale and transfer it to a 250 ml beaker containing 40 g of potassium iodide dissolved in 25 ml of water. Stir carefully to dissolve all the iodine. Pour the solution into a glass-stoppered amber liter bottle. Rinse the beaker into the bottle also. Dilute to a liter using distilled water and shake several times.

STANDARDIZATION OF 0.03N IODINE

1. If a 0.0300N arsenious acid solution is not available in the lab, prepare it by weighing exactly 0.370 g of primary standard arsenic(III) oxide into a 250 ml beaker. Add a solution of 1 g of sodium hydroxide pellets freshly dissolved in 20 ml of distilled water. Swirl until the oxide dissolves completely, warming if necessary. Add 50 ml of water and 2 ml of 12M hydrochloric acid (concentrated). Transfer quantitatively to a 250 ml volumetric flask and dilute to volume.
2. Pipet 25 ml of 0.0300N arsenious acid into each of three flasks. Add 25 ml of water to each. Then add about 3.5 g (trip scale) of sodium bicarbonate to each, and check the pH to be 7–8 with pH paper. If the pH is not in this range, add more sodium bicarbonate.
3. To each flask add about 5 ml of starch solution and titrate with 0.03N iodine to the *first* appearance of the blue starch-triiodide color. (For the first flask it is advisable to check the pH to be in the 7–8 range. If not, add more sodium bicarbonate.) Since the starch-triiodide color is so intense, the depth of the color does not necessarily mean you have overshot the end point. Simply approach the end point more cautiously with the other flasks (all of them should require the same volume of iodine).
4. Calculate the normality of the iodine using the *average* volume consumed as follows:

$$N_{I_2} = \frac{(N_{H_3AsO_3})(ml_{H_3AsO_3})}{ml_{I_2}}$$

STANDARDIZATION OF 0.1 N IODINE

1. Dry about 1.5 g of primary standard arsenic(III) oxide for 1 hr at 110°C. Before weighing out the oxide, put at least 1 g of sodium hydroxide pellets into each of three 250 ml flasks, and add no more than 20 ml of water to one flask at a time. Swirl to dissolve the sodium hydroxide, and then immediately weigh 0.2–0.25 g of arsenic(III) oxide, exactly, into this flask. This utilizes the heat released from dissolution of sodium hydroxide to help the arsenic(III) oxide to dissolve.
2. Swirl to dissolve the arsenic(III) oxide and warm the flasks if necessary to complete dissolution. Inspect each solution for undissolved particles, washing down the sides of the flasks if necessary.
3. To each flask add 50 ml of water and exactly 2.5 ml of concentrated hydrochloric acid from a measuring pipet. The solution should almost be neutral at this point. Add about 3.5 g of sodium bicarbonate to each flask and check the pH to be 7–8.
4. To each flask add 5 ml of starch solution and titrate with 0.1 N iodine to the first appearance of the deep blue starch-triiodide color. (For the first flask, it is advisable to check the pH to be in the 7–8 range. If not, add more sodium bicarbonate.) Since the indicator color is so intense, the depth of the color does not necessarily mean you have overshot the end point. Simply approach it more cautiously with the other samples.
5. Calculate the normality of the iodine of each sample as follows:

$$N = \frac{\text{mg As}_2\text{O}_3/(197.8/4)}{\text{ml}_{\text{I}_2}}$$

where 197.8/4 is the equivalent weight of arsenic(III) oxide in mg/meq.

Experiment 10. Analysis of Vitamin C Tablets Using Iodine

Vitamin C, or ascorbic acid, is oxidized to dehydroascorbic acid.

$$\text{CH}_2\text{OHCHOH}-\overset{\overset{\displaystyle\longleftarrow O \longrightarrow}{|}}{\text{CH}}-(\text{OH})\text{C}=\text{C}(\text{OH})-\text{C}=\text{O}$$

$$\rightarrow \text{CH}_2\text{OHCHOH}-\overset{\overset{\displaystyle\longleftarrow O \longrightarrow}{|}}{\text{CH}}-\underset{\underset{\displaystyle O}{\|}}{\text{C}}-\underset{\underset{\displaystyle O}{\|}}{\text{C}}-\text{C}=\text{O} + 2\text{H}^+ + 2e^-$$

As shown in Self-Test 11-2, the oxidation is a two-electron oxidation process. Iodine is a mild enough oxidizing agent to oxidize the ascorbic acid to dehydroascorbic acid and no further. The titration reaction is

$$\text{I}_2 + \underline{\text{C}_4\text{H}_6\text{O}_4}(\text{OH})\text{C}=\text{COH} \rightarrow 2\text{I}^- + 2\text{H}^+ + \underline{\text{C}_4\text{H}_6\text{O}_4}\text{C}(=\text{O})-\text{C}=\text{O}$$

(Note that we have simplified the above reaction by writing the atoms of both molecules which undergo no electron change as the $C_4H_6O_4$ group and underlining it, as recommended in Sec. 11-1.)

Vitamin C in solution is readily oxidized by dissolved oxygen, so samples should be analyzed as soon as they are dissolved. The titration flask should be covered during the titration to prevent absorption of additional oxygen from the air. A small amount of oxidation from oxygen already dissolved in the solution is usually not significant, but continuous shaking of an open flask will absorb enough oxygen to cause error [1].

PROCEDURE

1. Prepare and standardize 0.03N iodine titrant (Experiment 9).
2. Obtain a sample of 1–4 vitamin C tablets equivalent to 400–500 mg of ascorbic acid. (Anything less will require too small a volume of iodine titrant.)
3. Because a whole tablet may dissolve slowly, it is recommended that you carefully cut the tablet(s) into 5–6 pieces before weighing. (If the instructor wishes you to report the weight of vitamin C per tablet, be careful not to lose any of the tablet in cutting it. If you are required to report only the percentage of vitamin C, then any loss of the tablet(s) is not a problem.)
4. Weigh a sample equivalent to 400–500 mg of ascorbic acid by difference into a clean, dry 100 ml volumetric flask. If you have prepared the iodine titrant and are ready to titrate, dissolve the tablets by adding about 50 ml of water, *stoppering the flask*, and shaking vigorously until the tablets have dissolved. (A small amount of binder in the tablets may not dissolve and may be visible as fine particles but should not cause any error.) Then fill to the mark with distilled water.
5. Fill a buret with 0.03N iodine. Pipet exactly 25 ml of the vitamin C solution into a 250 ml flask and add 5 ml of starch indicator (Experiment 9). Cover the opening of the flask with a piece of cardboard having a small hole for the buret tip.
6. Insert the buret tip through the cardboard covering the flask and titrate with deliberate speed to the first appearance of the blue starch-triiodide color.
7. Repeat Steps 5 and 6 for each of the two remaining 25 ml aliquots of the vitamin C sample. Do not pipet a sample into a flask until you are ready to titrate it.
8. Calculate the results of your titrations. If you are to report the per cent of ascorbic acid in a certain weight, then consult Example 11-3 for the calculation setup. If you are to report the weight of vitamin C in one tablet, use a setup similar to the following.

$$\text{mg/tablet} = \frac{(\text{ml of } I_2)\,(N \text{ of } I_2)\,(88.06 \text{ mg/meq})}{1 \text{ tablet}}$$

[1] C. E. Moore, *J. Chem. Educ.* **25,** 671 (1948).

Experiment 11. Iodometric Determination of Arsenic

Arsenic in the form of arsenic(III) oxide can be determined by oxidation with iodine after dissolving the oxide to form arsenious acid.

$$As_2O_3(s) + 2OH^- + H_2O \rightarrow 2H_2AsO_3^- + 2H^+ \rightarrow 2H_3AsO_3$$

Before the titration, the pH is adjusted to between 5 and 9 because the oxidation is quantitative and rapid only in this range. Sodium bicarbonate is added to buffer the solution at a pH between 7 and 8. Then iodine is used to titrate the arsenious acid to arsenic acid.

$$I_2 + H_3AsO_3 + H_2O \rightarrow 2I^- + H_3AsO_4 + 2H^+$$

Note that acid is released during the titration, which is another reason for adding sodium bicarbonate buffer. The end point is detected by the formation of the deep blue starch-triiodide color.

PROCEDURE

1. Prepare $0.1 N$ iodine and starch solutions (Experiment 9).
2. Dry the unknown arsenic(III) sample for no longer than 1 hr at 110°C. Longer drying may sublime arsenic(III) from some samples.
3. Before weighing out the unknown, put at least 1 g of sodium hydroxide pellets into each of three 250 ml flasks. When you are ready to weigh the first portion of unknown into the flask, add no more than 20 ml of water to the first flask. Then weigh out a 0.3–0.35 g sample (see instructor's directions) immediately into the first flask and swirl to use the heat released from the dissolution of sodium hydroxide. Swirl to dissolve the arsenic(III) oxide and warm the flask if necessary to complete the dissolution.
4. When the first sample has dissolved, repeat the process, adding water to dissolve the sodium hydroxide, and then weighing the arsenic(III) sample into the flask.
5. To each flask add 50 ml of water and not less than 2.5 ml of concentrated hydrochloric acid from a measuring pipet. The solution should almost be neutral at this point. Add about 3.5 g of sodium bicarbonate to each flask and check the pH to be 7–8.
6. To each flask add 5 ml of starch solution and titrate with $0.1 N$ iodine to the first appearance of the deep blue starch-triiodide color. (For the first flask it is advisable to check the pH to be in the 7–8 range. If it is not, add more sodium bicarbonate.)
7. Calculate the percentage of arsenic using an equivalent weight of 74.92/2 mg per meq of arsenic(III). Follow the method of Example 6-8.

Experiment 12. Using a Spectrophotometer to Obtain an Absorption Spectrum

The objectives of this experiment are to become familiar with the operation of a simple spectrophotometer such as the Bausch-Lomb Spectronic 20 and to obtain an absorption spectrum manually. The type of spectrophotometer used will depend on the equipment available in the laboratory. Regardless of the type, you should become familiar with the following controls on the spectrophotometer—the wavelength control, the on-off control, the control to adjust 0%T, the control to adjust 100%T, and the place where cells are inserted.

If you have not already done so, you should read all of the sections in Chapter 13 pertaining to the Beer-Lambert law and the various spectrophotometers.

The absorption spectrum of any species may be measured; we have chosen to have you obtain the spectrum of the $Cu(H_2O)_6^{+2}$ ion as an example of a species that absorbs primarily in the red and near-infrared regions. This ion is formed when copper(II) sulfate is added to water. The copper(II) ion is complexed by six water molecules and forms the light blue-green hexaquocopper(II) complex ion.

Preparation

1. Each student will be assigned one of the following weights of $CuSO_4 \cdot 5H_2O$ to weigh to ±0.001 g—0.499 g, 0.624 g, 0.749 g, 0.874 g, 0.999 g, 1.124 g, 1.249 g, 1.374 g, or 1.498 g. Use the analytical balance.

2. Transfer your assigned amount of $CuSO_4 \cdot 5H_2O$ to a 100 ml volumetric flask. Add distilled water carefully to the mark on the flask and shake until the copper sulfate is completely dissolved.

3. Make sure the Spectronic 20 or other spectrophotometer is equipped with a red measuring phototube and a red filter. (See Fig. H for

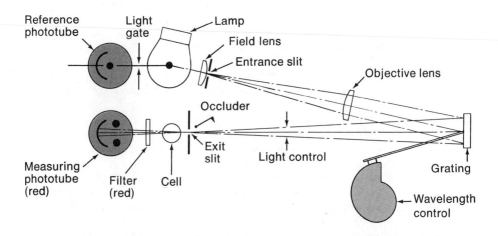

FIGURE H. Optical diagram of the Spectronic 20, top view. (Courtesy of Bausch and Lomb, Inc., Rochester, N.Y.)

the optical diagram of the Spectronic 20.) The following procedure incorporates directions for operating the Spectronic 20, but other spectrophotometers may be used in its place.

Procedure

1. Turn on the Spectronic 20, or other spectrophotometer, and allow to warm up 10 min (Spectronic 20) or longer (instructor's directions) to stabilize. As shown in Figure H, the Spectronic 20 should be equipped with a red measuring phototube and a red filter for measurements above 620 nm.
2. Obtain two test tube cells (Spectronic 20) and clean them before using them—rinse each twice with distilled water. Wipe off the outside of each with a tissue rather than cloth or anything that might scratch the cells.
3. Fill one cell with distilled water for the *blank*. Fill the other with the solution of copper(II) sulfate after rinsing the cell at least once with the copper(II) sulfate solution. Wipe off any liquid on the outside of each cell with a tissue.
4. *Setting 0% transmittance (absorbance = ∞).* The 0% transmittance setting corresponds to no light passing through the solution; it is made without inserting the cell into the instrument on the Spectronic 20. Simply adjust the left-hand knob so that the meter needle on the scale reads 0 on the per cent transmittance scale. Be sure the reading is stable. Check it periodically throughout the experiment and readjust it if necessary.
5. *Setting the wavelength.* Adjust the wavelength to the desired setting, starting at 620 nm. Because the $Cu(H_2O)_6^{+2}$ ion does not absorb appreciably until the start of the red region, it is not necessary to measure below 620 nm. For the Spectronic 20, adjust the wavelength using the knob on the top of the instrument.
6. *Setting 0 absorbance (100% transmittance).* Turn the right-hand knob of the Spectronic 20 counter-clockwise almost to its limit. Insert the test tube cell containing distilled water into the cell holder. Match the line on the test tube cell with the index line on the holder. Close the top of the holder. Adjust the right-hand knob clockwise until the meter needle on the scale reads 0 absorbance. Remove the test tube to avoid instrument fatigue. Check the 0% transmittance setting to be sure it has not changed. If it has, both the 0% transmittance and 0 absorbance settings must be readjusted.
7. *Measurement of absorbance of the $Cu(H_2O)_6^{+2}$ ion.* Insert the test tube cell containing the copper(II) sulfate solution into the holder, again matching the marks. Allow the needle to stabilize, and read the absorbance of the solution to two significant figures; for example, 0.05, 0.10, 0.21. (Recall that for log terms such as absorbance all digits, including zeroes to the right of the decimal point, are significant.) Remove the test tube to avoid instrument fatigue.
8. Repeat Steps 5 through 7 for each wavelength used. It is suggested that the absorbance values be measured every 40 nm, starting at 620

nm. Near the location of the maximum absorbance, absorbance values must be measured every 10 nm since you are required to locate the absorbance maximum within 10 nm.
9. After you have finished your measurements to as far as the instrument can operate, empty your test tube cells and rinse them thoroughly with distilled water. Turn them upside down to drain. Do not turn off the spectrophotometer unless directed to do so.
10. Using graph paper, make a plot of absorbance of $Cu(H_2O)_6^{+2}$ versus wavelength, similar to Figure 13-1. Mark the wavelength of maximum absorbance.
11. Calculate the molarity of the copper(II) sulfate solution. Using $b = 1.1$ cm for the diameter of the cells, calculate the molar absorptivity of $Cu(H_2O)_6^{+2}$ and report it on the graph. Turn in the graph as your report.

Experiment 13. Colorimetric Determination of Glucose

The colorimetric determination of glucose has been discussed in Section 22-2. The method to be used in this experiment is the *o*-toluidine method, which is based on the condensation of glucose, an aldehyde sugar (aldose), and *o*-toluidine, a primary aromatic amine.

$$C_5H_{11}O_5-\overset{H}{\underset{}{C}}=O + C_7H_8-NH_2 \xrightarrow{heat} C_5H_{11}O_5-\overset{H}{\underset{}{C}}=N-C_7H_8 + H_2O$$

The product is a green colored Schiff base which has an absorption band at 630–640 nm in the red region.

Because the absorption maximum is close to the borderline for the blue-sensitive and red-sensitive phototubes of most simple spectrophotometers, measurements may be made at 620 nm using the blue-sensitive phototube, or at 630 (or 640) nm using the red-sensitive phototube. Directions are given below for 620 nm using the blue-sensitive phototube, but the method may be used equally as well with the red-sensitive phototube.

PROCEDURE

1. Prepare fresh each day a solution of *o*-toluidine reagent consisting of 0.28 g of thiourea and 10 ml of pure *o*-toluidine, diluted to 100 ml with glacial acetic acid.
2. Prepare four standard solutions of glucose containing 50, 100, 150, and 200 mg of glucose, each in 100 ml of saturated aqueous benzoic acid.
3. Into five small flasks or test tubes, pipet exactly 5 ml of the freshly prepared *o*-toluidine reagent. Using a 100 microliter (lambda) pipet, add 100 microliters of one of the four standards to 4 of the 5 containers.

(Be sure to blow out the pipet to transfer all of the solution.) Then add 100 microliters of the glucose unknown to the remaining flask or test tube. Mark each container as to what solution it contains.
4. Heat each solution for exactly 8 min in boiling water. Then remove from the heat and cool quickly under tap water for about 30 s. Allow to cool under tap water and in the air a total time of 10 min. During this time, turn on the Spectronic 20 and set the wavelength at 620 nm with the blue-sensitive phototube in the instrument. After it has stabilized, set the 0% transmittance and 0 absorbance settings.
5. After 10 min has elapsed for the first glucose reaction mixture, insert it into the spectrophotometer and read the absorbance quickly. Remove and repeat the measurement for the other reaction mixtures.
6. Plot the absorbance versus concentration for each of the four standard glucose solutions. Enter the absorbance reading of the glucose unknown on the graph and read the concentration of the glucose from it in mg glucose per 100 ml of solution.

Experiment 14. Spectrophotometric Determination of Manganese

Manganese is determined spectrophotometrically by dissolving it in nitric acid as as manganese(II), and then oxidizing it with potassium periodate to the permanganate ion. This ion has a visible absorption band at about 520 nm, so it may be measured with a spectrophotometer or a colorimeter.

A more challenging analytical problem is the spectrophotometric determination of manganese in steel. The manganese steel is first dissolved in nitric acid, yielding manganese(II) and iron(III).

$$Mn° + Fe° + \text{excess } HNO_3 \rightarrow Mn^{+2} + Fe^{+3} + NO_2 + NO$$
$$\text{(yellow)}$$

Since the NO_2 and NO produced may interfere later by reacting with the potassium periodate, they are removed partially by boiling and partially by oxidation with ammonium persulfate.

$$NO + NO_2 + (NH_4)_2S_2O_8 \rightarrow 2NO_3^- + 2SO_4^{-2} + 2NH_4^+$$

Although the persulfate anion has the potential to oxidize manganese(II) to permanganate, the reaction is too slow. Hence potassium periodate is added to oxidize the manganese(II).

$$2Mn^{+2} + 5IO_4^- + 3H_2O \xrightarrow{\text{heat}} 2MnO_4^- + 6H^+ + 5IO_3^-$$

The reaction is carried out at the boiling point to insure a rapid rate of reaction.

Note that at this point, the yellow iron(III) would still be in the solution and would interfere somewhat by absorbing at 520 nm. Therefore phosphoric acid is added to form a colorless phosphate complex ion with the iron(III).

PROCEDURE

This procedure will be performed with a *standard steel* containing a known percentage of manganese and an unknown steel. The final calculation of percentage is very simple if the sample weights of the two are kept the same within ±0.01 g as directed below.

1. Weigh out *equal* weights of the standard steel (known percentage of manganese) and the unknown steel, using 1.00–1.50 g (instructor's directions), to within ±0.01 g. Because of the accuracy of the spectrophotometer, it is necessary to weigh only to the nearest 0.01 g.
2. Place the steels in separate 400 ml beakers, add 50–60 ml of dilute (1:3) nitric acid, and heat both in a hood to dissolve. Finally, cover and boil gently for 2 min to remove most of the nitric oxide. A black residue of carbon may remain at this point.
3. Remove the beakers from the heat and carefully sprinkle 1 g of ammonium persulfate into each beaker. Boil gently for 10–15 min to oxidize the carbon and destroy the excess ammonium persulfate. (Any precipitated manganese dioxide at this point may be removed by adding a few drops of dilute sodium sulfite solution, followed by boiling to expel the excess sulfur dioxide.)
4. Dilute each solution to about 100 ml, add 15 ml of 85% phosphoric acid, or 40 ml of 6M phosphoric acid, and 0.5 g of potassium periodate. Boil gently for about 3 min to effect oxidation to the permanganate ion. Remove the heat and allow to cool below boiling, and then add an additional 0.2 g portion of potassium periodate. Boil for another minute or two.
5. Transfer both solutions to 500 ml volumetric flasks, and dilute to volume when cool to avoid oxidation of any organic matter in the distilled water used for dilution. It is recommended that the solutions be measured spectrophotometrically on the same day they are prepared. (Solutions may last for a few days if carefully prepared.)
6. Turn on the Spectronic 20 spectrophotometer and allow to stabilize. Obtain two test tube cells; fill one with distilled water and the other with a solution of permanganate from the standard steel, after rinsing it once with the solution.
7. Adjust the wavelength to 520 nm and set the 0% transmittance setting. Insert the distilled water cell and set the 0 absorbance setting. Remove the cell containing the distilled water and insert the cell containing the solution from the standard steel. Read its absorbance.
8. Pour the standard steel solution out of the cell and rinse it with the solution containing the unknown steel. Fill the cell with that solution and insert it into the Spectronic 20. Read the absorbance of the solution containing the unknown steel.
9. Rinse both test tube cells with distilled water and allow them to drain upside down. Leave the Spectronic 20 on unless directed otherwise.
10. Calculate the percentage of manganese in the unknown steel by comparing its absorbance with the absorbance of the standard steel. If the weights are the same within ±0.01 g, a simple variation of the following equation from Section 13-2 can be used.

$$c_u = c_s \frac{A_u}{A_s}$$

where c_u and c_s are the concentrations of the unknown and standard, respectively, and A_u and A_s are the absorbances of the unknown and standard, respectively. (*Hint:* Remember that the weights of the steels are the same and that all you know is the percentage of manganese in the standard steel. You do not want the concentration of manganese in the unknown, just the percentage of manganese in the unknown steel.)

Experiment 15. Fluorometric Determination of Quinine in Quinine Water or Bitter Lemon

In this experiment quinine, a fluorescent compound, will be determined in commercial beverages such as quinine water or bitter lemon. These samples will be *real* unknowns since no directions are given how to dilute them or what the solvent composition should be for the sample.

The fluorescence excitation and emission spectra of quinine were given in Figure 15-2. Quinine can be excited at both 254 and 366 nm. If the common Coleman filter fluorometer is used, you will be restricted to exciting at 366 nm. If the Turner or Aminco filter fluorometers are used, you may excite at 254 nm as well as at 366 nm. Be sure to select the proper primary and secondary filters so the secondary filter excludes stray radiation from the source.

We have written the experimental directions for the Coleman filter fluorometer. If another fluorometer is used, the concentrations specified for quinine may have to be changed.

PROCEDURE

1. Obtain five 25 ml volumetric flasks, a 5 or 10 ml microburet, and a 50 ml buret.
2. Prepare a standard quinine sulfate solution containing 10 mg of quinine sulfate in one liter of distilled water.
3. Transfer 1.0, 1.5, 2.0, and 2.5 ml of the standard quinine sulfate solution to 25 ml volumetric flasks marked 0.4, 0.6, 0.8, and 1.0 ppm of quinine, respectively. Use the microburet.
4. Using a 50 ml buret, add sufficient distilled water to each 25 ml flask so that the total volume of the solution in each is 12.5 ml. Then fill each flask to the mark with 0.2N sulfuric acid to give a final concentration of 0.1N acid for optimum fluorescence. Mix the solutions thoroughly.

5. Obtain a solution of quinine water or bitter lemon beverage. Since this is a real unknown, no directions will be given how to dilute or treat it unless your instructor provides some further instructions. Obviously you should make up a dilution in a 25 ml volumetric flask. [If you make a solution that is too concentrated, the fluorometer reading will be off scale (>100 units); if the solution is too dilute, the fluorometer reading will be zero.]
6. Turn on the Coleman filter fluorometer (or other fluorometer) by turning the STD knob clockwise. Allow to warm up 20 min to stabilize the UV source.
7. Insert a primary filter, such as the B-1 or PC-6 Coleman primary filters, into the instrument for 366 nm excitation. Also select an appropriate secondary filter that will not transmit exciting radiation passing through the primary filter.
8. Carefully rinse each of four Coleman test tube cells with the respective quinine solutions (0.4–1 ppm) and fill each with about 15 ml of the respective solutions according to the markings on the tube (0.4, 0.6, etc.). Wipe the outside of each test tube cell with a clean tissue paper. Fill a cell with $0.1N$ sulfuric acid for the blank and another cell with the unknown solution.
9. After 20 min, set the STD knob on the Coleman fluorometer to the full clockwise position and the aperture control to the extreme right position. Set the meter to about zero with the coarse blank control.
10. The procedure below will set the instrument to compensate electrically for the solvent blank. Use the 0.4 ppm standard solution for the "DC solution" mentioned below and set the meter exactly to 20 units with this solution.
    a. Insert the tube containing the DC solution into the fluorometer. Depress the shutter button and move the aperture control toward the left until the meter reads 20 units. Allow the shutter button to return to the normal position.
    b. Remove the tube and insert the blank tube. Depress the shutter button and set the meter to zero with the coarse and then the fine blank controls. Allow the shutter button to return to normal.
    c. Reinsert the tube containing the DC solution and depress the shutter button. Adjust the aperture control to give a meter reading slighlty greater than the recommended setting. Bring the meter exactly to the 20 unit setting with the STD control. Allow shutter button to return to normal.
    d. Recheck the zero setting as directed in *b* and recheck the 20 unit setting as directed in *c*.
    e. Measure the fluorescent intensities of the other quinine solutions and the unknown solution in the same way. Insert the tube and depress the shutter button; record the intensity to the first decimal place. Allow shutter button to return to normal and remove.
11. After you are finished, rinse the test tube cells with $0.1N$ sulfuric acid and again with distilled water. Allow them to drain upside down after wiping off the outside of each tube. Do not turn off the fluorometer unless directed.

12. Plot the fluorescent intensity of each quinine solution against concentration and determine the concentration of the unknown graphically. Turn in the graph.

Experiment 16. Separation of Amino Acids by Paper Chromatography

Amino acids are organic compounds containing the acidic carboxyl group, COOH, and the basic amine group, NH_2. Many amino acids are found in natural products and they compose the fundamental units which combine to form proteins. Organic chemists and biochemists have made extensive studies of proteins and of the amino acids which compose them. Most amino acids are known by their common names rather than by their chemical names. For example, amino acetic acid, H_2N-CH_2-COOH, is commonly called *glycine*, and 1-amino-2-(*p*-hydroxyphenyl) propionic acid is called *tyrosine*. The student should consult a text of organic or biochemistry for a more detailed discussion of the amino acids.

The analytical chemist is frequently called upon to separate and identify amino acids in mixtures. Chromatographic techniques are frequently employed for this purpose. The student should review Chapter 20 before beginning this experiment. The technique specifically employed in this experiment is called *paper chromatography*. Chromatographic paper is similar to filter paper but is usually made in the form of large, rectangular sheets. Whatman Number 1 or a thicker variety, 3 MM, is recommended for this experiment. A small amount (0.2–0.5 microliters) of the unknown solution is placed on the paper about 2 cm from the bottom along with equivalent amounts of a series known solution spaced along the same line at 1 cm intervals. The paper is then placed in an oven to remove the solvent from the spots. Care should be taken to keep each spot small and not allow it to become larger than a millimeter or two. If the initial spot is too large, the final, resolved spots on the chromatogram will be large and diffuse rather than clearly defined. Care also should be taken not to handle the paper excessively since the oils and proteins from the skin may cause smudges and discolorations in the final chromatogram.

After spotting and drying, the paper is rolled into a cylinder, then placed in a beaker containing the developing solvent. Many solvent mixtures have been proposed for the chromatographic separation of amino acids. Commonly used are mixtures of alcohols, acids, and water. The mixture used in this experiment consists of ethyl alcohol, acetic acid, and water. Other mixtures, especially those containing phenol, are more effective, but the above mixture was chosen because the reagents are commonly available and are nontoxic, and the separation can be carried out in a period of one to two hours.

As the developing solvent moves up the paper by capillary action, the amino acids are carried along with the solvent at various rates. Some move almost as rapidly as the solvent, others at an intermediate rate, while the remainder move only slightly. The ratio of the distance the amino acid spot moves to the distance the solvent front moves is called the R_f value for the amino acid. Stated mathematically,

$$R_f = \frac{\text{distance traveled by amino acid}}{\text{distance traveled by solvent front}}$$

Suppose, for example, the solvent front moved 9 cm while a certain spot moved 2 cm. The R_f would be 2/9 or 0.22. Table A-2 gives the R_f values

TABLE A-2. TYPICAL R_f VALUES FOR SOME AMINO ACIDS

Acid	R_f	Color with ninhydrin
Aspartic acid	0.22	Red-violet
Glycine	0.28	Red-violet
D, L-Serine	0.30	Red-violet
L (+)-Glutamic acid	0.36	Red-violet
D, L-Threonine	0.40	Red-violet
D, L-Alanine	0.49	Red-violet
L (−) Proline	0.53	Yellow (faint)
D, L-Methionine	0.62	Red-violet
D, L-Phenylalanine	0.64	Blue-violet
D, L-Isoleucine	0.79	Red-violet

for certain important amino acids using ethanol-acetic acid-water solvent. If some other solvent were used, the R_f values would be different.

When the solvent front approaches the top of the paper, the paper cylinder should be removed from the solvent and the position of the solvent front marked. The paper cylinder is then placed on a watch glass and dried in an oven until all of the solvent has evaporated. The location of the amino acids in the chromatogram is made visible by spraying or immersing the paper in a solution of ninhydrin. Ninhydrin reacts with the amino acids to form a colored compound. The paper is again dried in an oven, then opened up, and R_f values of the unknown spots and the known spots computed and compared. From these data the unknown amino acids may be identified. For a more detailed and comprehensive discussion the student is referred to the literature [1, 2, 3].

PROCEDURE

1. *Preparation of the chromatographic paper.* Obtain a large sheet of Whatman No. 1 or 3 MM paper, and cut a rectangle approximately 12×18 cm. Draw a light line, using a pencil, 2 cm from the bottom along the 18 cm length of the paper. (Do not use a ball-point pen because the ink will smudge the chromatogram.) Mark the line at 1 cm intervals, again using a pencil, and number each mark.
2. *Spotting the paper.* Obtain a small micropipette (or a fine glass capillary) and fill with no more than 0.2 µl of a 0.5% aqueous solution of each known amino acid. Touch this to the chromatographic paper

[1] L. Slaten, M. Willard, and A. A. Green, *J. Chem. Educ.* **33,** 140 (1956).
[2] L. B. Clapp and C. Hansch, *J. Chem. Educ.* **37,** 293 (1960).
[3] E. P. Heimer, *J. Chem. Educ.* **49,** 547 (1972).

at a numbered mark and record the number and name of the amino acid. The final spot should be no larger than 1–2 mm. Larger spots give poor chromatograms.

Repeat the above procedure for each of the known amino acids and the unknown solution, placing each at a new numbered location. Be sure to record the name and number of each.

3. *Drying the spots.* Roll the paper into a cylinder so the spots are at the lower edge of the cylinder. Join the edges of the paper using two hooks made of staples or fine aluminum or stainless steel wire. Stand the cylinder on a watch glass and dry in an oven at 90°–110°C for 10–15 min.

4. *Developing the chromatogram.* In a 400 ml beaker mix 5 ml of 95% ethyl alcohol, 1 ml distilled water, and 0.1 ml of glacial acetic acid. Place this beaker inside a 1000 ml beaker, cover tightly with a watch glass or aluminum foil, and let stand for 10–15 min to saturate the atmosphere in the beaker with the solvent vapors.

Remove the cover and quickly place the dry chromatographic cylinder into the 400 ml beaker so that the lower, spotted edge stands in the solvent mixture. Make certain it stands upright and does not touch the sides of the beaker. Quickly replace the cover and allow the chromatogram to develop until the solvent front reaches about 2 cm from the top of the paper (about $1-1\frac{1}{2}$ h). At this point remove the paper cylinder, place it in a watch glass, and put a pencil mark at the position of the greatest advance of the solvent front. Place the cylinder in an oven (90°–110°C) and dry until no odor of acetic acid remains (10–20 min).

5. *Ninhydrin treatment.* This step is necessary to render the separated spots visible. Use either technique *a* or *b*.

   a. Prepare 0.2% ninhydrin solution by dissolving 0.2 g in 100 ml of 95% ethyl alcohol. Pour enough of the solution into a small shallow tray or dish to fill to a height of 0.5 cm. Rapidly roll the dried chromatographic cylinder through the ninhydrin solution until it is completely wetted. Hold the cylinder upright for a moment to allow it to drain, then place it on a watch glass and return it to the drying oven to dry for 5–10 min.

   b. *Alternate spray technique.* Open up the chromatographic cylinder and spray it with a ninhydrin solution from an atomizer or from an aerosol spray can, if available. After spraying the cylinder thoroughly, dry it in the oven for 5–10 min.

6. *Interpretation of the final chromatogram.* Measure the distance from the original base line to the mark representing the greatest advance of the solvent front. Next measure the distance from the original base line to the center of each colored spot for each known amino acid. Calculate the R_f values for each acid. They should correspond approximately to the values in Table A-2, but some variations may occur owing to variations in experimental conditions such as the nature of the paper, temperature, etc. Finally measure the distance traveled by the spots of the unknown. Calculate the R_f values and compare them with the known standards. The number of individual spots and their R_f values will serve to identify the composition of the unknown amino acid mixture.

Appendix 1. Equilibrium Constants

SOLUBILITY PRODUCT CONSTANTS

(*Note:* The values below are given for *pure* water where the concentration of other ions is zero; hence the ionic strength $\simeq 0$. Values in parentheses are for an ionic strength of 0.1.)

Compound	K_{sp} (Values in parentheses are for $\mu = 0.1$)
Silver salts	
AgBr	$4.9 \times 10^{-13} M^2$ ($8.7 \times 10^{-12} M^2$)
Ag_2CO_3	$8.1 \times 10^{-12} M^3$
Ag_2CrO_4	$1.1 \times 10^{-12} M^3$ ($5.0 \times 10^{-12} M^3$)
AgCl	$1.8 \times 10^{-10} M^2$ ($3.2 \times 10^{-10} M^2$)
AgI	$8.3 \times 10^{-17} M^2$ ($1.5 \times 10^{-16} M^2$)
$AgIO_3$	$3.0 \times 10^{-8} M^2$
AgN_3	$2.9 \times 10^{-9} M^2$
Ag_3PO_4	$1.3 \times 10^{-20} M^4$
AgSCN	$1.1 \times 10^{-12} M^2$ ($2.0 \times 10^{-12} M^2$)
Ag_2HVO_4	$2 \times 10^{-14} M^3$
Aluminum salts	
$AlAsO_4$	$1.6 \times 10^{-16} M^2$
$Al(OH)_3$	$4.6 \times 10^{-33} M^4$
$AlPO_4$	$5.8 \times 10^{-19} M^2$
Gold salts	
$Au(OH)_3$	$5.5 \times 10^{-46} M^4$
$AuK(SCN)_4$	$6 \times 10^{-5} M^6$
Barium salts	
$BaCO_3$	$4.9 \times 10^{-9} M^2$ ($3.2 \times 10^{-8} M^2$)
$Ba(oxinate)_2$	$5.0 \times 10^{-9} M^3$
$BaSO_4$	$1.1 \times 10^{-10} M^2$ ($6.3 \times 10^{-10} M^2$)

Equilibrium Constants

Compound	K_{sp} (Values in parentheses are for $\mu = 0.1$)
Calcium salts	
$CaCO_3$	$4.8 \times 10^{-9} M^2$
$CaMg(CO_3)_2$	$1 \times 10^{-11} M^4$
CaC_2O_4	$2.3 \times 10^{-9} M^2$ ($1.6 \times 10^{-8} M^2$)
$Ca_3(PO_4)_2$	$1 \times 10^{-26} M^5$ ($1 \times 10^{-23} M^5$)
$CaSO_4$	$2.4 \times 10^{-5} M^2$ ($1.6 \times 10^{-4} M^2$)
Iron salts	
$Fe(OH)_3$	$2.5 \times 10^{-39} M^4$ ($1.3 \times 10^{-38} M^4$)
$Fe_4[Fe(CN)_6]_3$	$3.0 \times 10^{-41} M^7$
Magnesium salts	
$MgNH_4PO_4$	$2.5 \times 10^{-13} M^3$
$Mg(OH)_2$	$1.8 \times 10^{-11} M^3$
Mercury salts	
Hg_2Cl_2	$1.3 \times 10^{-18} M^3$
Lead salts	
$PbCl_2$	$1.6 \times 10^{-5} M^3$ ($8.0 \times 10^{-5} M^3$)
$PbSO_4$	$1.66 \times 10^{-8} M^2$ ($1.0 \times 10^{-7} M^2$)
Potassium salts	
$K_2NaCo(NO_2)_6$	$2.2 \times 10^{-11} M^4$
KUO_2AsO_4	$2.5 \times 10^{-33} M^3$
Thorium salts	
$Th(OH)_4$	$4 \times 10^{-45} M^5$
$Th(IO_3)_4$	$2.5 \times 10^{-15} M^5$
Uranium salts	
UO_2NaAsO_4	$1.3 \times 10^{-22} M^3$

IONIZATION CONSTANTS OF ACIDS

Acid	K_a at $\mu = 0$	K_a at $\mu = 0.1$
Acetic acid, CH_3COOH	1.74×10^{-5}	2.24×10^{-5}
Arsenic acid, H_3AsO_4		
K_1	6.5×10^{-3}	8.0×10^{-3}
K_2	1.3×10^{-7}	2.0×10^{-7}
K_3	3.2×10^{-12}	6.3×10^{-12}
Arsenious acid, H_3AsO_3		
K_1	6.0×10^{-10}	8.0×10^{-10}
K_2	—	8.0×10^{-13}
K_3	—	4.0×10^{-14}
Benzoic acid, C_6H_5COOH	6.3×10^{-5}	8.0×10^{-5}
Boric acid, H_3BO_3	5.9×10^{-10}	8.0×10^{-10}

Acid	K_a at $\mu = 0$	K_a at $\mu = 0.1$
Carbonic acid, H_2CO_3		
$\quad K_1$	4.3×10^{-7}	5.0×10^{-7}
$\quad K_2$	4.8×10^{-11}	8.0×10^{-11}
Chloroacetic acid, $CH_2ClCOOH$	1.38×10^{-3}	2.0×10^{-3}
o-Chlorobenzoic acid, $C_6H_4ClCOOH$	1.14×10^{-3}	—
m-Chlorobenzoic acid, $C_6H_4ClCOOH$	1.50×10^{-4}	—
p-Chlorobenzoic acid, $C_6H_4ClCOOH$	1.03×10^{-4}	—
o-Chlorophenol, C_6H_4ClOH	3.33×10^{-9}	—
m-Chlorophenol, C_6H_4ClOH	9.48×10^{-10}	—
p-Chlorophenol, C_6H_4ClOH	4.19×10^{-10}	—
Dichloroacetic acid, $CHCl_2COOH$	5.5×10^{-2}	8.0×10^{-2}
EDTA, H_4Y		
$\quad K_1$	8.5×10^{-3}	—
$\quad K_2$	1.8×10^{-3}	—
$\quad K_3$	5.8×10^{-7}	—
$\quad K_4$	4.6×10^{-11}	—
Formic acid, $HCOOH$	1.70×10^{-4}	2.24×10^{-4}
Fumaric acid, $C_2H_2(COOH)_2$		
$\quad K_1$	9.6×10^{-4}	1.3×10^{-3}
$\quad K_2$	4.1×10^{-5}	7.9×10^{-5}
Hydrazoic acid, HN_3	1.9×10^{-5}	2.5×10^{-5}
Hydrogen cyanide, HCN	4.9×10^{-10}	6.3×10^{-10}
Hydrogen fluoride, HF	6.75×10^{-4}	8.9×10^{-4}
Hydrogen peroxide, H_2O_2	1.8×10^{-12}	2.5×10^{-12}
Iodic acid, HIO_3	1.67×10^{-1}	—
Lactic acid, $CH_3CH(OH)COOH$	1.32×10^{-4}	1.74×10^{-4}
Maleic acid, $C_2H_2(COOH)_2$		
$\quad K_1$	1.2×10^{-2}	1.6×10^{-2}
$\quad K_2$	6.0×10^{-7}	1.2×10^{-6}
Glycine, H_2NCH_2COOH	1.65×10^{-10}	2.0×10^{-10}
Malonic acid, $CH_2(COOH)_2$		
$\quad K_1$	1.4×10^{-3}	2.0×10^{-3}
$\quad K_2$	2.2×10^{-6}	4.0×10^{-6}
o-Nitrophenol, $C_6H_4(NO_2)OH$	6.2×10^{-8}	—
m-Nitrophenol, $C_6H_4(NO_2)OH$	4.0×10^{-9}	—
p-Nitrophenol, $C_6H_4(NO_2)OH$	5.2×10^{-8}	—
Nitrous acid, HNO_2	5.1×10^{-4}	6.3×10^{-4}
Oxalic acid, $(COOH)_2$		
$\quad K_1$	8.8×10^{-2}	8.0×10^{-2}
$\quad K_2$	5.1×10^{-5}	1.0×10^{-4}
Periodic acid, HIO_4	2.3×10^{-2}	—
Phenol, C_6H_5OH	1.4×10^{-10}	1.6×10^{-10}

Equilibrium Constants

Acid	K_a at $\mu = 0$	K_a at $\mu = 0.1$
Phosphoric acid, H_3PO_4		
K_1	7.5×10^{-3}	1.0×10^{-2}
K_2	6.2×10^{-8}	1.26×10^{-7}
K_3	4.8×10^{-13}	2.0×10^{-12}
Phosphorous acid, H_2PHO_3		
K_1	7.1×10^{-3}	1.0×10^{-2}
K_2	2.0×10^{-7}	4.0×10^{-7}
o-Phthalic acid, $C_6H_4(COOH)_2$		
K_1	1.20×10^{-3}	1.6×10^{-3}
K_2	3.9×10^{-6}	8×10^{-6}
Propionic acid, CH_3CH_2COOH	1.34×10^{-5}	—
Pyridinecarboxylic acid, C_5H_4NCOOH	5.0×10^{-6}	—
Salicylic acid, $C_6H_4(OH)COOH$		
K_1	1.05×10^{-3}	1.3×10^{-3}
K_2 (OH)	—	8×10^{-14}
Sulfurous acid, H_2SO_3		
K_1	1.3×10^{-2}	1.6×10^{-2}
K_2	6.3×10^{-8}	1.6×10^{-7}
Tartaric acid, $[CH(OH)COOH]_2$		
K_1	9.1×10^{-4}	1.3×10^{-3}
K_2	4.3×10^{-5}	8.0×10^{-5}
Trichloroacetic acid, CCl_3COOH	2.2×10^{-1}	3.2×10^{-1}

IONIZATION CONSTANTS OF BASES

Base	K_b at $\mu = 0$	K_b at $\mu = 0.1$
Ammonia, NH_3	1.78×10^{-5}	2.34×10^{-5}
Aniline, $C_6H_5NH_2$	4.2×10^{-10}	5.0×10^{-10}
Butylamine, $C_4H_9NH_2$	4.09×10^{-4}	5.1×10^{-4}
Ethanolamine, $HOCH_2CH_2NH_2$	2.8×10^{-5}	—
Ethylamine, $C_2H_5NH_2$	4.7×10^{-4}	—
Ethylenediamine, $H_2NCH_2CH_2NH_2$		
K_1	1.28×10^{-4}	—
K_2	2.0×10^{-7}	—
Glycine, H_2NCH_2COOH	2.2×10^{-12}	3.2×10^{-12}
Hydrazine, NH_2NH_2	1.0×10^{-6}	1.3×10^{-6}
Hydroxylamine, NH_2OH	1.23×10^{-8}	1.6×10^{-8}
Methylamine, CH_3NH_2	4.4×10^{-4}	—
Pyridine, C_5H_5N	1.5×10^{-9}	1.9×10^{-9}

Base	K_b at $\mu = 0$	K_b at $\mu = 0.1$
THAM, tris(hydroxymethyl)aminomethane, $(HOCH_2)_3CNH_2$	1.26×10^{-6}	1.6×10^{-6}
Triethanolamine, $(HOCH_2CH_2)_3N$	6.6×10^{-7}	8×10^{-7}
Trien, $(CH_2NHCH_2CH_2NH_2)_2$		
K_1	1.0×10^{-4}	—
K_2	1.9×10^{-5}	—
K_3	5.6×10^{-8}	—
K_4	2.5×10^{-11}	—
Urea, NH_2CONH_2	1.5×10^{-14}	—

Appendix 2. Standard Electrode Potentials

Oxidizing agent + ne^- → reducing agent	Standard potential, $E°$, in volts ($1M\ H^+$)
$Ce(IV) + e^- = Ce^{+3}$ (in $1M$ $HClO_4$)	1.70[a]
$KMnO_4 + 4H^+ + 3e^- = MnO_2 + 2H_2O + K^+$	1.7 (neutral solution)
$H_5IO_6 + H^+ + 2e^- = IO_3^- + 3H_2O$ (periodic acid)	1.60
$KMnO_4 + 8H^+ + 5e^- = Mn^{+2} + 4H_2O + K^+$	1.51
$Ce(IV) + e^- = Ce^{+3}$ (in $1N$ H_2SO_4)	1.44[a]
$Cl_2 + 2e^- = 2Cl^-$	1.36
$Cr_2O_7^{-2} + 14H^+ + 6e^- = 2Cr^{+3} + 7H_2O$	1.33
$O_2 + 4H^+ + 4e^- = 2H_2O$	1.229 ($= +0.816$ at pH 7)
$IO_3^- + 6H^+ + 5e^- = \frac{1}{2}I_2 + 3H_2O$	1.195
$Br_2 + 2e^- = 2Br^-$	1.065
$Cu^{+2} + I^- + e^- = CuI(s)$	0.86
$Ag^+ + e^- = Ag(s)$	0.800
$Fe^{+3} + e^- = Fe^{+2}$	0.771
$O_2 + 2H^+ + 2e^- = H_2O_2$	0.682
$H_3AsO_4 + 2H^+ + 2e^- = H_3AsO_3 + H_2O$	0.559 ($=0.0$ at pH 7)
$I_2 + 2e^- = 2I^-$	0.535
Dehydroascorbic $+ 2H^+ + 2e^- =$ Ascorbic acid	0.39 ($= -0.023$ at pH 7)
$Fe(CN)_6^{-3} + e^- = Fe(CN)_6^{-4}$	0.36
Cytochrome-a(Fe^{III}) $+ e^- =$ Cytochrome-a(Fe^{II})	0.290
$Hg_2Cl_2 + 2e^- = 2Hg + 2Cl^-$	0.268
$AgCl(s) + e^- = Ag(s) + Cl^-$	0.222
$Sn^{+4} + 2e^- = Sn^{+2}$	0.15
$S_4O_6^{-2} + 2e^- = 2S_2O_3^{-2}$ (thiosulfate)	0.15
$TiO^{+2} + 2H^+ + e^- = Ti^{+3} + H_2O$	0.1
$2H^+ + 2e^- = H_2(g)$	0.0
$Pb^{+2} + 2e^- = Pb(s)$	-0.126
$Cr^{+3} + e^- = Cr^{+2}$	-0.41
$2CO_2(g) + 2H^+ + 2e^- = H_2C_2O_4$	-0.49
$Zn^{+2} + 2e^- = Zn(s)$	-0.763
$Mg^{+2} + 2e^- = Mg(s)$	-2.37

[a] Formal Potentials

Appendix 3. Using Logarithms on the Electronic Calculator to Find Roots

Most of the inexpensive student electronic calculators cannot be used to find roots or they only have a key for finding the square root. By using logarithms on such a calculator, you can quickly find any root from a square root to a ninth or higher root. If you know how to use logarithms, proceed directly to the last section on calculating roots; otherwise read the following brief introduction to logarithms.

USING LOGARITHMS

The logarithm of a number has two parts—the *characteristic*, or whole number part, and the *mantissa*, or decimal part. Once a number is written in semiexponential form, the characteristic is simply the power of ten. The mantissa of the remaining digits must be looked up in a log table, such as the four-place log table given in Appendix 4.

Finding the Logarithm

Before the logarithm (base 10) of a number is found, the number should be written in semiexponential form. For example, suppose that we wish to find the log of the number 10.25. We first write the number as 1.025×10^1. The characteristic part of the log is obviously 1. To find the mantissa, we look at the extreme left-hand column of the four-place log table in Appendix 4. This column has the heading *No.* It is *understood* that the first entry of *10* is really *1.0*; that is, the decimal point is omitted but understood to be present. Since 1.025 has four significant figures, we must report the log to that many digits also. If we look across the column under the columns with the headings *2* and *3*, we see that the mantissa will be between 0.0086 and 0.0128 (each mantissa is *understood* to be preceded by a zero and a decimal point). Since 0.0021 is half (0.5) the distance between 2 and 3, the mantissa is 0.0086+0.0021, or 0.0107. Summarizing the entire calculation, we have

$$\log 1.025 \times 10^1 = \log 1.025 + \log 10^1 = 0.0107 + 1 = 1.0107$$

Finding the Antilogarithm

The reverse process of converting a logarithm into a number is referred to as obtaining the antilogarithm. This is generally the last step in a calculation performed using logs. To do this, the mantissa part of the logarithm must be *positive* so it can be looked up in the log table. If it is not positive, it must be converted to a positive mantissa using a mathematical identity. Suppose, for example, that we wish to find the antilog of -5.740. Since the negative sign implies that the characteristic of 5 and the mantissa of .740 are both negative, we use an identity to obtain a positive mantissa.

$$-5.740 = -6 + 0.260$$

We now look up the antilog of the positive mantissa of 0.260; this is 1.82. The antilog of -6 is of course 10^{-6}. Summarizing, we have

$$\text{antilog of } -5.740 = \text{antilog} -6 + 0.260 = 1.82 \times 10^{-6}$$

CALCULATING ROOTS

We will first discuss the calculation of a square root since that is the most common operation.

Calculating a Square Root

To calculate the square root of a number A, use the following equations.

$$\log \sqrt{A} = \frac{(\log A)}{2}$$

$$\sqrt{A} = \text{antilog of } \frac{(\log A)}{2}$$

For example, suppose we have the calculation

$$\sqrt{2.5 \times 10^{-3}} = ?$$

We recommend that if you have a calculator without scientific notation, you adjust the exponent to be divisible by two and remove the exponent from under the square root by taking the square root of the exponent.

$$\sqrt{2.5 \times 10^{-3}} = \sqrt{25 \times 10^{-4}} = \sqrt{25} \times 10^{-2}$$

Then only the square root of the number remaining under the square root must be found. This is done as follows:

$$\log \sqrt{25} = \frac{(\log 25)}{2} = \frac{1.3979}{2} = 0.6989$$

$$\sqrt{25} = \text{antilog of } 0.6989 = 5.0 \text{ (2 sig. figs.)}$$

$$\text{Final answer} = \sqrt{2.5 \times 10^{-3}} = 5.0 \times 10^{-2}$$

Calculating any root

To find the nth root of a number A, we use the following steps.

$$\log \sqrt[n]{A} = \frac{(\log A)}{n}$$

$$\sqrt[n]{A} = \text{antilog of } \frac{(\log A)}{n}$$

For example, suppose that we have to find the 5th root of 2.00×10^{-29}. There are two methods that can be employed. The less confusing method of the two is to adjust the exponent to be divisible by the root and remove it from under the root sign. Then only the root of the number left under the root sign must be found. This is done as follows:

$$\sqrt[5]{2.00 \times 10^{-29}} = \sqrt[5]{20.0 \times 10^{-30}} = \sqrt[5]{20.0} \times 10^{-6}$$

$$\log \sqrt[5]{20.0} = \frac{\log 20.0}{5} = \frac{1.301}{5} = 0.260$$

$$\sqrt[5]{20.0} = \text{antilog of } 0.260 = 1.82$$

$$\text{Final answer} = 1.82 \times 10^{-6}$$

The other method involves calculating the root without removing the exponent. This is done as follows:

$$\log \sqrt[5]{2.00 \times 10^{-29}} = \frac{\log(2.00 \times 10^{-29})}{5} = \frac{+0.301 - 29}{5} = \frac{-28.699}{5} = -5.7398$$

$$\sqrt{2.00 \times 10^{-29}} = \text{antilog of } -5.7398 = \text{antilog of } (-6 + 0.260)$$

$$= 1.82 \times 10^{-6}$$

$$\text{Final answer} = 1.82 \times 10^{-6}$$

Appendix 4. Four-place Logarithms

No.	0	1	2	3	4	5	6	7	8	9
10	0000	0043	0086	0128	0170	0212	0253	0294	0334	0374
11	0414	0453	0492	0531	0569	0607	0645	0682	0719	0755
12	0792	0828	0864	0899	0934	0969	1004	1038	1072	1106
13	1139	1173	1206	1239	1271	1303	1335	1367	1399	1430
14	1461	1492	1523	1553	1584	1614	1644	1673	1703	1732
15	1761	1790	1818	1847	1875	1903	1931	1959	1987	2014
16	2041	2068	2095	2122	2148	2175	2201	2227	2253	2279
17	2304	2330	2355	2380	2405	2430	2455	2480	2504	2529
18	2553	2577	2601	2625	2648	2672	2695	2718	2742	2765
19	2788	2810	2833	2856	2878	2900	2923	2945	2967	2989
20	3010	3032	3054	3075	3096	3118	3139	3160	3181	3201
21	3222	3243	3263	3284	3304	3324	3345	3365	3385	3404
22	3424	3444	3464	3483	3502	3522	3541	3560	3579	3598
23	3617	3636	3655	3674	3692	3711	3729	3747	3766	3784
24	3802	3820	3838	3856	3874	3892	3909	3927	3945	3962
25	3979	3997	4014	4031	4048	4065	4082	4099	4116	4133
26	4150	4166	4183	4200	4216	4232	4249	4265	4281	4298
27	4314	4330	4346	4362	4378	4393	4409	4425	4440	4456
28	4472	4487	4502	4518	4533	4548	4564	4579	4594	4609
29	4624	4639	4654	4669	4683	4698	4713	4728	4742	4757
30	4771	4786	4800	4814	4829	4843	4857	4871	4886	4900
31	4914	4928	4942	4955	4969	4983	4997	5011	5024	5038
32	5051	5065	5079	5092	5105	5119	5132	5145	5159	5172
33	5185	5198	5211	5224	5237	5250	5263	5276	5289	5302
34	5315	5328	5340	5353	5366	5378	5391	5403	5416	5428
35	5441	5453	5465	5478	5490	5502	5514	5527	5539	5551
36	5563	5575	5587	5599	5611	5623	5635	5647	5658	5670
37	5682	5694	5705	5717	5729	5740	5752	5763	5775	5786
38	5798	5809	5821	5832	5843	5855	5866	5877	5888	5899
39	5911	5922	5933	5944	5955	5966	5977	5988	5999	6010
40	6021	6031	6042	6053	6064	6075	6085	6096	6107	6117
41	6128	6138	6149	6160	6170	6180	6191	6201	6212	6222
42	6232	6243	6253	6263	6274	6284	6294	6304	6314	6325
43	6335	6345	6355	6365	6375	6386	6395	6405	6415	6425
44	6435	6444	6454	6464	6474	6484	6493	6503	6513	6522
45	6532	6542	6551	6561	6571	6580	6590	6599	6609	6618
46	6628	6637	6646	6656	6665	6675	6684	6693	6702	6712
47	6721	6730	6739	6749	6758	6767	6776	6785	6794	6803
48	6812	6821	6830	6839	6848	6857	6866	6875	6884	6893
49	6902	6911	6920	6928	6937	6946	6955	6964	6972	6981
	0	1	2	3	4	5	6	7	8	9

Four-place Logarithms

No.	0	1	2	3	4	5	6	7	8	9
50	6990	6998	7007	7016	7024	7033	7042	7050	7059	7067
51	7076	7084	7093	7101	7110	7118	7126	7135	7143	7152
52	7160	7168	7177	7185	7193	7202	7210	7218	7226	7235
53	7243	7251	7259	7267	7275	7284	7292	7300	7308	7316
54	7324	7332	7340	7348	7356	7364	7372	7380	7388	7396
55	7404	7412	7419	7427	7435	7443	7451	7459	7466	7474
56	7482	7490	7497	7505	7513	7520	7528	7536	7543	7551
57	7559	7566	7574	7582	7589	7597	7604	7612	7619	7627
58	7634	7642	7649	7657	7664	7672	7679	7686	7694	7701
59	7709	7716	7723	7731	7738	7745	7752	7760	7767	7774
60	7782	7789	7796	7803	7810	7818	7825	7832	7839	7846
61	7853	7860	7868	7875	7882	7889	7896	7903	7910	7917
62	7924	7931	7938	7945	7952	7959	7966	7973	7980	7987
63	7992	8000	8007	8014	8021	8028	8035	8041	8048	8055
64	8062	8069	8075	8082	8089	8096	8102	8109	8116	8122
65	8129	8136	8142	8149	8156	8162	8169	8176	8182	8189
66	8195	8202	8209	8215	8222	8228	8235	8241	8248	8254
67	8261	8267	8274	8280	8287	8293	8299	8306	8312	8319
68	8325	8331	8338	8344	8351	8357	8363	8370	8376	8382
69	8388	8395	8401	8407	8414	8420	8426	8432	8439	8445
70	8451	8457	8463	8470	8476	8482	8488	8494	8500	8506
71	8513	8519	9525	8531	8537	8543	8549	8555	8561	8567
72	8573	8579	8585	8591	8597	8603	8609	8615	8621	8627
73	8633	8639	8645	8651	8657	8663	8669	8675	8681	8686
74	8692	8698	8704	8710	8716	8722	8727	8733	8739	8745
75	8751	8756	8762	8768	8774	8779	8785	8791	8797	8802
76	8808	8814	8820	8825	8831	8837	8842	8848	8854	8859
77	8865	8871	8876	8882	8887	8893	8899	8904	8910	8915
78	8921	8927	8932	8938	8943	8949	8954	8960	8965	8971
79	8976	8982	8987	8993	8998	9004	9009	9015	9020	9025
80	9031	9036	9042	9047	9053	9058	9063	9069	9074	9079
81	9085	9090	9096	9101	9106	9112	9117	9122	9128	9133
82	9138	9143	9149	9154	9159	9165	9170	9175	9180	9186
83	9191	9196	9201	9206	9212	9217	9222	9227	9232	9238
84	9243	9248	9253	9258	9263	9269	9274	9279	9284	9289
85	9294	9299	9304	9309	9315	9320	9325	9330	9335	9340
86	9345	9350	9355	9360	9365	9370	9375	9380	9385	9390
87	9395	9400	9405	9410	9415	9420	9425	9430	9435	9440
88	9445	9450	9455	9460	9465	9469	9474	9479	9484	9489
89	9494	9499	9504	9509	9513	9518	9523	9528	9533	9538
90	9542	9547	9552	9557	9562	9566	9571	9576	9581	9586
91	9590	9595	9600	9605	9609	9614	9619	9624	9628	9633
92	9638	9643	9647	9652	9657	9661	9666	0671	9675	9680
93	9685	9689	9694	9699	9703	9708	9713	9717	9722	9727
94	9731	9736	9741	9745	9750	9754	9759	9763	9768	9773
95	9777	9782	9786	9791	9795	9800	9805	9809	9814	9818
96	9823	9827	9832	9836	9841	9845	9850	9854	9859	9863
97	9868	9872	9877	9881	9886	9890	9894	9899	9903	9908
98	9912	9917	9921	9926	9930	9934	9939	9943	9948	9952
99	9956	9961	9965	9969	9974	9978	9983	9987	9991	9996
	0	1	2	3	4	5	6	7	8	9

Appendix 5. Answers to Even-numbered Problems

Where possible, the correct number of significant figures has been used to express the answers. Answers to a few odd-numbered problems are also given.

CHAPTER 1

3. a. Mercury(I) chloride   d. Potassium hexacyanoferrate(II)
4. 0.85_5 mmoles; 85_5 μmoles. $8.5_5 \times 10^{-4} M$
6. 1 mg; 1×10^3 μg. $1.00 \times 10^{-3} M$
8. Using **1-3** for activity coefficient, both = 0.004. Using Table 1-1 for activity coefficient estimation, $a_{Mg} = 0.006$; $a_{SO_4} = 0.005$.

CHAPTER 2

4. a. Two-fourths results used for $n = 4$; only one-third for $n = 3$.   b. Two-fourths results used for $n = 4$; only two-sixths for $n = 6$.
6. More convenient and faster
8. a. 1.0×10^{-7} b. 1.00×10^{-7} c. 1.004×10^{-7}
10. a. 2.5×10^2 b. 2.0×10^2 c. $_2 \times 10^{10}$ (no sig. fig.)
12. a. $M = 9.7\%$ (11.0% differs in 1st digit)   b. $M = \bar{X} = 9.7\%$
14. a. $\bar{X} = 0.120$ ppm; Range $= 0.050$ ppm; $s = 0.021_6$ ppm   b. $\bar{X} = 1.20$ ppm; Range $= 0.50$ ppm; $s = 0.21_6$ ppm
16. Range $= 0.17\%$; $s = 0.065_4\%$
18. a. No   b. Reject 11.00%; $\bar{X} = 10.10\%$ c. Retain 11.00%; $M = 10.20\%$
20. a. $Q_{0.90} = 0.941$ b. $Q_{0.99} = 0.993_9$
22. a. 0.085_6 to 0.154_4 ppm   b. 0.85_6 to 1.54_4 ppm

CHAPTER 3

2. Aspirin capsules near the top of the bottle probably cannot react to a greater degree with water than those elsewhere in the bottle; aspirin tablets will.

4. a. Shake, and sample before homogeneity is lost.   b. Sample at different depths.
6. Where to sample is always one difficulty.
8. Variation in the amount of water causes a variation in sample weight, which then causes the per cent of a constituent to vary.
10. Hydrochloric acid forms insoluble chlorides with these metal ions.
11. a. Most alcohols   b. Water

CHAPTER 4

2. Container weight is subtracted from weight of sample plus container.
4. Weighing is at constant load so that sensitivity is constant at all weights.
6. Use a paper loop, spatula, or scoop.
9. a. 0.06_7 pph   b. 0.1_3 pph
10. a. 10 mg   b. 20 mg
12. Use a balance.
14. 1.594 g/ml
16. 8.0% NaCl
18. Use 1, 2, 4, 8, 16, and 32 g weights.

CHAPTER 5

1. a. Simplicity, usually accurate, usually precise   b. Exacting, lengthy, subject to serious possible errors
4. Crystalline, granular, finely divided, and gelatinous (only 4)
6. a. More convenient   b. Can be heated to higher temperatures   c. Can be used for silver halides   d. Can be heated to higher temperatures
8. The formula weight of $AlCl_3$ contains 3 formula weights of chloride.
9. To insure that a significant portion of the insoluble salt will not dissolve.
12. a. 0.6994_4 b. 1.0000   c. 0.9666   d. 1.056_3
14. 4.0_2 ppm
16. 12.94%
18. $x = 5$; C_6HOCl_5
20. a. $6.5_0 \times 10^{-5} M$ b. $6.8_8 \times 10^{-7} M$ c. $1.2_6 \times 10^{-3} M$ d. $6.3_0 \times 10^{-5} M$
22. a. $1.8 \times 10^{-8} M$ b. Yes
24. 10.35% Fe_2O_3
26. C_4O_3 b. $C_{12}O_9$
28. 61.05% NaCl

CHAPTER 6

2. Known purity, stable, large molecular weight, rapid and stoichiometric reaction
4. TC, to contain; TD, to deliver.
5. Tolerance is the absolute value of the maximum allowable error and it affects accuracy.

8. $M = N = 0.0996_1$
10. a. 20.58% H_3PO_4 b. 50.39% NaH_2PO_4
12. a. 53.17$_5$% Cl b. 71.42$_2$% $MgCl_2$
14. 19.55$_6$% $AlCl_3$
16. a. 22.66$_2$% H_3AsO_3 b. 13.48$_6$% As
18. 1.6$_{03}$ mg% Ca
20. 14.08 ml
22. Sum of $Cl^- + HCO_3^-$ is 137 meq/liter, which is within 10% of 143 meq/liter.

CHAPTER 7

2. K_{rxn} = infinity for all reactions
4. b. K_1 of 3.6×10^{10} is greater than the $K_1 K_2$ of 3.9×10^7; the $K_1 K_2$ product must be used because 3.6×10^{10} represents coordination at two positions.
   c. K_2 of 1×10^9 is greater than the $K_3 K_4$ product of 3.9×10^5
6. pU = 2.0 at start; pU = 6.0 at the equivalence point
8. pH = 3.0 at start; pH = 7.0 at equivalence point
10. a. Not quant.; $K_{rxn} = 3.4 \times 10^8$ b. Not quant.; $K_{rxn} = 2 \times 10^6$ c. Not quant.; $K_{rxn} = 6.7 \times 10^{12}$ d. Quant.; $K_{rxn} = 7.7 \times 10^{19}$
12. a. 0.01: quant.; 0.0001: not quant. b. 0.01: quant.; 0.0001: quant.
14. Min theo $K_{rxn} = \dfrac{[9\% \ NH_4^+][50\% \ A^-]}{[91\% \ NH_3][50\% \ HA]} = 0.1$

    Min theo value of $K_a = \dfrac{(K_{rxn} = 0.1)(K_w = 10^{-14})}{(K_b = 10^{-5})} = 10^{-10}$

CHAPTER 8

2. $[H^+] = [OH^-]$
4. a. 7.80 b. 7.47 c. 6.51
6. a. 1.70 b. 3.82 c. Food has used up some H^+
8. a. $[H^+] = 0.2M$ (same as molarity of acid); pH = 0.7 b. $[H^+] = 0.209$; pH = 0.680
10. $[OH^-] = 4.0 \times 10^{-6} M$; % error = 4.0×10^{-3} %
12. $[OH^-] = 1.8 \times 10^{-4} M$; pH = 10.26
14. a. $\dfrac{[OAc^-]}{[HOAc]} = \dfrac{0.72}{1.0}$ b. $\dfrac{[NH_4^+]}{[NH_4OH]} = \dfrac{1}{1}$ c. $\dfrac{[NH_4^+]}{[NH_4OH]} = \dfrac{1.4_4}{1.0}$
16. pH = 4.33
18. a. pH = 8.34 b. pH = 4.75
20. Slightly less than pH 7.0, the pH of pure water
22. 113.5 mmoles of H_2CO_3; 2270 mmoles of HCO_3^-

CHAPTER 9

2. Easier to find end point
4. Sodium hydroxide reacts with carbon dioxide; hydrochloric acid is of uncertain composition.
6. Equivalence point should fall in the middle half.
8. $H_3BO_3 + NH_4OH = NH_4^+ H_2BO_3^-$
   $NH_4H_2BO_3 + HCl = H_3BO_3 + NH_4Cl$
10. a. Infinity   b. Infinity
12. 1.06; only about 50% of H_3BO_3 is neutralized to $H_2BO_3^-$.
14. $38.01_5\%$
16. $7.6_4\%$ $NaHCO_3$; $16.4_3\%$ Na_2CO_3
18. $2.868_8\%$ N
20. pH = 9.52; thymolphthalein is the best of those listed but isn't very good.
22. a. $n = 2$ b. Formula weight of $R = 82$ c. $R = C_6H_{10}$— and can be —CH=CH—CH$_2$—CH$_2$—CH$_2$—CH$_2$— or cyclohexyl   d. Cyclohexyl (1,2- or 1,4-substituted)   e. Cyclohexane-1,4-dicarboxylic acid

CHAPTER 10

2. Mohr: no; Adsorption indicator: yes if positively charged indicator is used; Volhard: yes
4. a. With nitrobenzene   b. With nitrobenzene   c. Without nitrobenzene   d. With nitrobenzene
6. Acid is a product.
8. a. pAg = 8.31   b. pAg = 6.15   c. pAg = 4.00
10. Mohr %Cl = $12.08_8\%$; Adsorption %Cl = $11.72_5\%$
12. a. Yes; $K_{Ca-EDTA} = 10^{8.41}$ b. No; $K_{Ca-EDTA} = 10^{7.91}$
    b. Yes; $K_{Ca-EDTA} = 10^{8.9+}$
14. 84.0_8 ppm
16. a. pH = 3.0   b. pH = 6.0   c. pH = 5.0
18. %NaCl = 61.05%

CHAPTER 11

2. a. Six-electron change   b. Two-electron change   c. Ten-electron change
   d. One-electron change   e. Five-electron change
4. a. $Cr_2O_7^{-2} + 3H_3AsO_3 + 8H^+ = 2Cr^{+3} + 3H_3AsO_4 + 4H_2O$
   b. $5IO_4^- + 2Mn^{+2} + 3H_2O = 5IO_3^- + 2MnO_4^- + 6H^+$
   c. $IO_3^- + 5I^- + 6H^+ = 3I_2 + 3H_2O$
   d. $2Fe(CN)_6^{-3} + 2NH_2OH = 2Fe(CN)_6^{-4} + N_2(g) + 2H_2O + 2H^+$
   e. $2MnO_4^- + 5H_2O_2 + 6H^+ = 2Mn^{+2} + 5O_2(g) + 8H_2O$

6. a. 1.44 (H_2SO_4) − (+0.771) = +0.66$_9$ (quant.)   b. +0.009 (not quant.)
   c. −0.024 (not quant.)   d. +0.535 (quant.)   e. +0.558 (quant.)
   f. +0.325 (quant.)   g. +0.38$_5$ (quant.)   g. +0.62$_1$ (quant.)
8. $E°_{rxn} = +0.22_9$ V; $\log K = 3.8_8$; $K = 7._6 \times 10^3$
10. 79.43%
12. 7.688$_3$%
14. 88.50$_9$%
16. 37.33$_8$%
18. 0.2002$_0$ N
20. Titrate with iodine.
22. Reduce with iodide, and titrate the resulting iodine with thiosulfate.
24. 10.35% Fe_3O_4
26. $E°_{ox} - E°_{red} = +0.472$ V
28. Use $F = 23{,}062$ cal/(volt)(mole) and $R = 1.9872$ cal/(°K)(mole)

CHAPTER 12

2. A photon is a particle of radiation having energy but no mass (rest mass).
4. a. 185–380 nm   b. 750–2500 nm   c. 380–750 nm   d. 2500–15,000 nm
6. In an electronic ground state all electrons are in their most stable orbitals. In a vibrational ground state all bonds possess the smallest possible vibration energy. In an electronic excited state at least one electron occupies a higher energy orbital than it does in the ground state. In a vibrational excited state one or more of the bonds has a larger amount of vibrational energy than it does in the ground state.
8. The electrical current is too small to give a measurable readout on the spectrophotometer scale.
10. A voltmeter draws enough current from the electrical cell to change the initial concentrations of the ion(s) being measured.
12. The indicating electrode responds to changes in concentration of the ions in the sample solution; the reference electrode's potential remains constant.
14. This is a $Hg°/Hg_2Cl_2$ electrode (Fig. 12-6).

CHAPTER 13

2. a. $A = \log(P_0/P)$ c. $T = P/P_0$ d. $A = \varepsilon bc$
4. 10^{-9} meter. 5.00×10^{-7} meter
6. $[Hb(O_2)_x]_0 + n$ photons$_{560-620 \text{ nm}} \rightarrow [Hb(O_2)_x]_1$ (Repeat, using 520–560 nm photons and 440–460 nm photons)
8. a. Red (weak response)   b. Blue or blue-violet   c. Blue or blue-violet
   d. Red (weak response)
10. a. $A = 0.699$ b. 1.000   c. 1.40   d. 1.80
12. a. 79.$_4$%   b. 62.$_8$%   c. 1%

14. A of 0.18_7 has a negative deviation.
15. a. 4.30   b. 1.04
16. a. $2._0$ b. 1×10^4 c. $_1 \times 10^4$ (no sig. figs.)   d. $2._5 \times 10^3$
18. a. 0.02   b. 0.060   c. 2.0×10^4
20. Two bands. $\varepsilon_{235} = 1.1 \times 10^3 M^{-1}$ cm^{-1}; $\varepsilon_{280} = 1.4 \times 10^3 M^{-1}$ cm^{-1}
22. UV. $c_{203} = 1 \times 10^{-6} M$; $c_{300} = 1._3 \times 10^{-3} M$
24. $\dfrac{\text{(No.)(response) at 766}}{\text{(No.)(response) at 404}} = \dfrac{6._7}{1}$; eye perceives red.
26. a. 5_{55} photons per second

CHAPTER 14

2. a. The electron absorbs a photon and occupies a higher energy orbital.
   b. The bond(s) absorbs a photon and undergoes increased vibrations.
4. a. C—H   b. O—H   c. C=O   d. C≡O   e. Ester C=O
6. a. A d orbital   b. Has no d electrons
8. a. Yes; has π electrons   b. NO and atomic O   c. Alkenes and atomic O
10. a. Will; O—H   b. Will; aldehyde C—H   c. Will; O—H   d. Will not
12. The enol form has an O—H which absorbs at 3.3 μ. Hydrogen bonding to a carbonyl oxygen causes the band to be broad.
14. Tentative indication
16. $7.72_5 \times 10^{-5} M$
18. SCN$^-$
20. The ligand-metal charge transfer transition is twice as probable for Fe(SCN)$_2^+$ as it is for Fe(SCN)$^{+2}$ because it has twice as many SCNs to absorb a photon

CHAPTER 15

2. The spectrofluorometer uses gratings and a xenon arc; the filter fluorometer uses filters and a mercury arc.
4. The 254 nm exciting radiation is directed through the open top of the cell.
6. Turbidimetric analysis involves measurement of the decrease in P_0; nephelometric analysis involves measurement of the scattered portion of P_0.
8. 2.5_2 mg per 100 ml
10. $\dfrac{F_{\text{anthracene}}}{F_{\text{phenanthrene}}} = \dfrac{9}{1}$

CHAPTER 16

2. a. In flame emission it is necessary to optimize atomization and excitation in the flame. In AA it is necessary to optimize atomization and reduce excitation in the flame.

b. In flame emission loss of excitation energy by photon emission is measured, but in AA this is immaterial.

c. Flame emission involves measurement of photons emitted; AA involves measurement of photons absorbed during excitation.

4. a. I_u is compared directly to I_s of the standards.

b. The concentration of the unknown found is multiplied by a correction factor to obtain the actual concentration.

c. Ratios of I_u/I for unknown and internal standard are plotted against the concentration of the unknown.

6. Flame emission is simpler and more sensitive.
8. a. $2.1 \times 10^{-3} M$ b. $1.3_8 \times 10^{-3} M$ c. $1.2_6 \times 10^{-3} M$
10. 0.14% Na_2O

CHAPTER 17

2. The indicating electrode responds to changes in the composition of the solution; the reference electrode's potential is independent of solution composition.
4. It is composed of liquid mercury metal and insoluble mercury(I) chloride (saturated with potassium chloride for a saturated calomel electrode).
6. It is the error caused by the erroneous response of the glass electrode in alkaline medium to [Na^+] as well as to [H^+].
8. The potentiometer measures concentrations without drawing appreciable current; the voltmeter draws appreciable current.
10. Sweating is induced electrically and a flat-headed specific ion electrode is applied directly to the skin for measruement.
11. $1.1 \times 10^{-2} M$ b. $1.15 \times 10^{-1} M$
12. a. Larger   b. Larger   c. Larger   d. Larger   e. Smaller   f. Smaller
13. a. 12.04   b. 13.061
14. a. 1.0 pH   b. 0.2 pH
16. a. $a_{Na^+} = 5.9 \times 10^{-2}$ b. $a_{K^+} = 1.1 \times 10^{-11}$ c. $a_{F^-} = 6.4 \times 10^{-1}$
    c. $a_{Ca^{+2}} = 2.00$

CHAPTER 18

2. a. The measurement step in electrogravimetry is weighing; in direct coulometric analysis it is electrical measurement.

b. The measurement step in electrogravimetry is weighing; in polarography it is done on the polarogram.

c. The measurement step in polarography is done on the polarogram; in an amperometric titration it is done by plotting a titration curve.

d. The measurement step in direct coulometry is a direct electrical measurement of the species determined; in indirect coulometry it is direct electrical measurement of the generated titrant.

4. Oxygen is reduced to H_2O_2 or to H_2O at the cathode.
6. a. Yes b. No c. Yes d. No e. No f. Yes
8. a. A chemical reagent that reacts quantitatively with the species determined
   b. Electrons (direct) or a species generated at the electrode (indirect)
10. a. $0.1888_6 M$ b. $0.0629_5 M$
12. $44.1_{17}\%$ As

CHAPTER 19

2. The separation need not be quantitative.
3. Similarity: both involve formation of an insoluble substance; difference: the electrodeposition uses electrons rather than reagents to produce the insoluble substance.
6. They are (1) distribution between a stationary phase and a carrier phase, (2) a bed of large surface area for a stationary phase, (3) a fluid of liquid or gas for the carrier, and (4) different rates of elution for components of sample.
8. The stationary phase is paper in paper chromatography, whereas it is some type of finely divided adsorbent in thin-layer.
10. The carrier is a gas; the stationary phase is a liquid coating on a solid support.
12. They are immiscible, and one is significantly heavier than the other.
14. The partition coefficient refers to only one form of a species being extracted; the distribution ratio refers to all forms of the species being extracted.
16. Two different equilibria are established.
20. a. $16.6_6\%$ b. $90.9_1\%$ c. $98.3_6\%$ d. $99.75_{06}\%$
22. Ten extractions ($98.0_4\% = \%E$)
24. $D = 0.17_6$

CHAPTER 20

2. R is a measure of the overlap of any two chromatographic zones in general, R_r refers to the fraction of time spent in the mobile phase of a chromatographic column, and R_f refers to the fraction of the distance traveled by the sample compared to solvent on a thin-layer chromatogram.
4. The R_f should reflect the distance traveled by the average sample constituent molecule, not the fastest moving molecules.
6. The time t_r is larger because it includes t_m as well as the time the average constituent molecule spends on the stationary phase.
8. The larger R is, the better the separation between the constituent zones; the longer the column, the better the resolution is, in proportion to the square root of the length.
10. a. 97.73% A and 2.27% B b. 95.46% B, 2.27% A, and 2.27% C
    c. 97.73% C and 2.27% B
12. a. R_r of $A = 0.50$ b. R_r of $B = 0.33$
14. a. 3.2 min   b. 8.0 min

CHAPTER 21

2. a. Silica gel is SiO_2 and alumina is Al_2O_3.
   b. Cellulose is a natural glucose polymer, $(C_6H_{10}O_5)_x$; silica gel is SiO_2.
   c. Kieselguhr is a purified form of diatomaceous earth; alumina is Al_2O_3.
4. a. Faster   b. More accurate, faster
5. Partitioning occurs between the stationary liquid phase held on the solid support and the mobile liquid phase which passes through the support and the column.
8. Blue spot is the more polar salicylic acid.
10. a. *n*-Propanol or ethanol   b. Ethyl acetate or acetone   c. Carbon tetrachloride
12. a. NaCl   b. $MgCl_2$
14. 1-Butanol, 2-butanol, *tert*-butyl alcohol, and 2-methyl-1-propanol

CHAPTER 22

2. A major electrolyte ion is present in major concentrations ($0.01-0.1M$) in body fluids whereas a trace metal is present at concentrations lower than $0.01M$.
4. a. Alkalosis *vs.* acidosis: high *vs.* low pH; normal *vs.* low chloride; low total carbonate *vs.* high total carbonate.
   b. Respiratory alkalosis *vs.* metabolic acidosis: high *vs.* low pH; normal *vs.* low chloride.
   c. Dehydration *vs.* alkalosis: normal *vs.* high pH; high *vs.* normal sodium; high *vs.* normal chloride; normal *vs.* low total carbonate.
6. EDTA titration requires little instrumentation but is slower for many analyses.
8. a. $0.141M$ sum of chloride plus bicarbonate is not $0.01M$ less than sodium; check.
   b. No checking
   c. $0.130M$ sum of chloride plus bicarbonate is not $0.01M$ less than sodium; check.
10. a. Whole blood less cells
    b. Whole blood less plasma; cells less erythrocytes and leukocytes
    c. Whole blood less cells; plasma less fibrinogen
    d. Whole blood less cells; plasma less serum
12. a. Marginal hypoglycemia   b. Hypoglycemia   c. Serious hyperglycemia
    d. Hyperglycemia

Index

(The letter t, followed by a number, indicates the entry is found in a table on that page number.)

Absorbance, 290–92
  self-test, 292
  significant figures, 292
Absorption of radiation:
  by gaseous atoms, 276, 363–68, 378–82
  by gaseous toxic metal atoms, t 380
  by metal ions in solution, 315–18, t 317
  by molecules in solution, 287–89, 310–15, t 313
Absorption spectrum, 287–89, 323–33
Absorptivity, 293
  molar, 293–95
Accuracy:
  of the balance, 54
  of the mean, 24
  of the median, 24
  of a result, 17
  of spectrophotometric measurements, 291, 305
  of volumetric glassware, 112
Acetaldehyde:
  absorption of infrared radiation, 321
  oxidation, 237
Acetic acid, 150, 161
Acetylsalicylic acid (aspirin):
  fluorescence, t 350, 351
  hydrolysis to salicylic acid, 328
  pH of, 150
  sampling, 45
  ultraviolet absorption, 328
Acid-base equilibria, Ch. 8
  of acid salts, 166
  of polyprotic acids, 164
  of strong acids, 147
  of strong bases, 148
  of weak acids, 150

  of weak bases, 152
  self-test, 155
Acid-base titrations, Ch. 9
  calculation of results, 102–09, 186–202
  indicators, 177
  nonaqueous solvents, 203–08
  of boric acid, 197
  of pharmaceuticals, 187, 207
  of polyprotic acids, 193
  of strong acids and bases, 132, 185
  of weak acids and bases, 187
  quantitative reaction, 135
  self-test, 202
Acidosis, t 481, 482
Activity coefficient, 14, t 15
Adsorption chromatography, 462
Adsorption indicator method:
  for chloride, 216
  for sulfate, 219
Air-pollution analysis, 332, 353
Alcohol, t 320, 321
  metabolism (oxidation), 242
Alkalosis, t 481, 482
Alumina, 462
Amines, 208, 514
Amino acids, t 350, 354, 469, 529
Ammonia (ammonium hydroxide), 153, 192
  buffer, 159
Ammonium ion, 151
Amperometric titrations, 409, 417
Amphetamine, 327
Answers to problems, 543
Antacid capacity experiment, 513
Anthracene, t 313, 324, t 350
Aromatic hydrocarbon analysis, 324, 353

553

Arsenic:
  coulometric determination, 409
  in blood, *t* 481
  iodometric determination, 256, 521
Ascorbic acid (see Vitamin C)
Aspirin:
  irritation of stomach, 329
  pain-killing form, 329
  pH of, 150
  sampling of, 45
  tolerance of salicylic acid in, 328
  ultraviolet absorption, 328
Atomic absorption spectrometry, 363–68, 375–82
AutoAnalyzer system, 333–36
Automated analysis, 333–36

Balance, analytical:
  accuracy and precision, 54
  double-pan, 56
  lever principle, 56
  rules for single-pan, 63
  self-test, 63
  single-pan, 58
  sensitivity, 57
Barbiturates, 327
Barium hydroxide, 148
Beer-Lambert law, 289–95
  self-test, 295
Benzene, *t* 313, 349, *t* 350
Benzo[a]pyrene, *t* 350, 353
Bilirubin, 329–31
Blood, 483
  electrolytes, 480
  pH, 147, 149, 480
Bolometer, 301
Boric acid, 197, 351
Bromide, 216–18, 221
Buffers, 159–64, 391
  physiological (blood) buffers, 170
Burets:
  accuracy and precision, 112
  use, 117
  uncertainty of single reading, 112

Calcium, 224, 232, 480, *t* 481
Carbonic acid, 164, 166, 194

Carbon dioxide, 166, 171, *t* 322, 333, 480, 482
Carbon monoxide, *t* 322, 333
Cell:
  electrolytic, 400
  galvanic, 400
  polarographic, 414
Cerium(IV), 259–60
Cholesterol, 73, 88, *t* 485
Chloride:
  coulometric method, 408
  gravimetric method, 73, 84, 86, 498
  in body fluids, 212, 393, 480, *t* 481
  in sweat (cystic fibrosis), 394
  self-test, 219
  specific ion determination, 506
  titration of, 214–17, 220, 502
Chromatography, 426
  applications, Ch. 21
  column, 428, 463
  exclusion, 429, 467
  experiment (paper), 529
  gas, 426, 432, 473
  high-performance (HPLC), 429, 465
  ion-exchange, 431, 470
  liquid, 427, 461
  paper, 427, 430, 469, 529
  theory, Ch. 20
  thin-layer, 427, 429, 467
Chromophores, Ch. 14, *t* 313, *t* 317, *t* 320, *t* 322
Clinical analysis, Ch. 22
  calculations, 110–11
Color, 285–86
Color blindness, 287
Colorimeters, 297, 301
Complexation titrations, 137, 220–33
Complex ions, 126, 137, 220–33
  stability constants, 127, 220, 223
Concentration:
  milligram per cent, 13, 110
  molarity, 13, 102
  normality, 102
  parts per million, 13, 225
Conditional stability constants, 128, 229
Confidence limits, 39
Coordination number, 211, 221
Copper(II) ion:
  absorption band, *t* 317

electrodeposition, 403
  in blood, t 481
  reduction by iodide, 262
Coulometric analysis, 405, 408
Crucibles, 80
Cystic fibrosis, 395

Data treatment, Ch. 2
Debye-Huckel equation, 14
Density measurement, 68, 495
Desiccator, 65
Deviation, 18, 31–32
  standard, 28
Drug screening, 327
Drying samples, 64–66

EDTA (ethylenediaminetetraacetic acid), 222–33
  calculation of results, 224
  ionization constants of, 227
  self-test, 226
  stability constants, 223
  use in lead poisoning, 223
  water hardness, 232
Electromagnetic spectrum, 274, 285
Electrode:
  calomel, 281, 386
  chloride, 264
  dropping mercury, 411
  glass (pH), 281, 387, 390, 395
  indicating, 281, 386
  liquid-membrane, 387
  potentials, 245–52
  potentials (values), t 537
  reference, 281
  solid state, 388, 395
  sweat chloride, 394
Electrodeposition, 400, 402
Electrogravimetry, 402
Electrolysis, 400
End-point detection, 139
Equilibrium constants, 532–36
  acid-base, 122
  complexes, 126
  oxidation-reduction, 128
  solubility product, 90–95, 122
  stability constants, 127

Error, 18, 54
  random, 27
  spectrophotometric, 305
  systematic, 27
Ethyl alcohol metabolism, 242
Exact numbers, 17
Excitation, 275
Extraction, Ch. 19
  countercurrent, 439
  distribution ratio, 435
Eye response to color, 280

Faraday's law, 405
Filter crucibles, 80
Filter fluorometer, 344
Filter paper, 81
Filters (optical), 297, 344
Filtration, 80
Flame emission spectrometry, 363–75
Flame photometry (see Flame emission spectrometry)
Fluoride, 395
Fluorescence, 275, 339–55
  acetylsalicylic acid, t 350, 351
  amino acid analysis, t 350, 354
  instrumentation, 344
  morphine, t 350, 352
  qualitative analysis, 351
  quantitative analysis, 353
  quinine, 341
  self-test, 348, 354

Gas chromatography (see Chromatography)
Glucose:
  colorimetric determination, 524
  determination, 334, 485
  glucose tolerance test, 47, 489
  in blood, t 485
  oxidation, 240
  sampling, 47
Gravimetric analysis, Ch. 5
  calculations, 86–88
  determination of chloride, 73, 84, 86
  determination of cholesterol, 88
  determination of pharmaceuticals, 88, 89

Gravimetric analysis (*continued*)
  gravimetric factor, 83
  report form, 36
  self-test, 85

Hemoglobin, 171, 288, 316, *t* 317, 329
Heroin, 325
High-performance liquid chromatography (HPLC) (see Chromatography)
Hydrochloric acid, 87, 147, 177, 509
  in stomach fluid, 147, 148

Ilkovic equation, 414
Indicators, 140
  acid-base, 177
  complexation (EDTA) titrations, 223, 231, 233
  oxidation-reduction, 254
  precipitation, 215, 218
Infrared radiation, 274, 285, 318–22
Iodide, 335, 262
Iodine:
  oxidation of arsenic(III), 256, 517
  oxidation of pharmaceuticals, 257
  oxidation of Vitamin C, 243, 257
  protein-bound, 335, 484
Ion-electron method, 238, 240
Ionic strength, 14
Ionization:
  polyprotic acids, 164–66
  strong acids and bases, 147–49
  weak acids and bases, 150–52
Iron(II):
  color, 315, *t* 317
  colorimetric determination, 331
  in blood, *t* 481
  oxidation of, 261
Iron(III):
  color, *t* 317
  colorimetric determination, 331
  EDTA titration, 233

Jaundice, 331

Kinetics, enzymatic determination of glucose, 488

Kjeldahl method, 201

Laboratory notebook, 35
Lead(II), 379, 404, 417, *t* 481
  poisoning, 483
Light, 274, 285
  scattering, 275, 340
Logarithms, 539, *t* 541

Manganese(II), 315, *t* 317, 525
Manganese(VII) (see Permanganate)
Mean, 24, 26, 34, 36, 49
Median, 24, 26, 34, 36
Mercury, 25, 31, *t* 380, 381
Mohr method, 215
Molarity, 13, 102, 103–06
Mole, 12
Morphine, *t* 350, 352

Nephelometry, 360
Nickel(II), *t* 317
Nitrate ion, *t* 313, 315
Nomenclature, 9–12
Nonaqueous solvents, titrations in, 203–08
Normal distribution curve, 28
Normality, 102, 106–10, 242
Notebook, 35
Nucleation, 77–79

Oil, crude, 475
Oxidation-reduction methods, Ch. 11
  end-point detection, 253
  equilibrium constants, 249
  iodine methods, 256–62
  quantitative reaction, 137
  potentiometric analysis, 255, 263
  primary standards, 253
Oxidation-reduction reactions:
  equilibrium constants, 249
  equivalent weight, 241
  ion-electron method, 238
  potentials, 247
  stoichiometric calculations, 241

Oxygen, 312, *t* 313, 416
Oxyhemoglobin (see Hemoglobin)
Ozone, 312, *t* 313

Periodate oxidations, 258, 525
Permanganate:
  colorimetric determination, 525
  light absorption, 312, *t* 313
pH, 145
  effect on EDTA titrations, 226
  measurement, 281, 385, 389
  meter, 390
  of aspirin, 150
  of blood, 147, 171
  of stomach, 147, 148
Phenobarbital, 75, 89, 187
Phenol, *t* 350
Phenylalanine, 293, *t* 350, 354
Phenylethylamines, 327
Phenylketonuria, 354
Phosphoric acid, 105, 194
Photomultiplier, 300
Photon, 273, 287
Photosynthesis, 289
Phototube, 277, 300
  diagram, 278
Pipets, 112, 114, 115
Polarography, 409
Potassium, 374, 379
  flame emission determination, 373
  in blood, *t* 481, *t* 485
Potassium acid phthalate, 176, 207
Potential, 246, 385, 393
Potentiometer, 279, 385
Potentiometric analysis, 255, 263, 389, 393
Precipitates, 78–82, 92
Precipitation, Ch. 5
  common-ion effect, 93
  equilibria, 122
  quantitative, 133
  titrations, 211–20
Precision, 19, 27–32
  balance, 54
  buret, 112
Primary standards, 100, 176, 253
Properties, measurement of, 4
Protein-bound iodine (PBI), 335, 484

Q test, 33
  gravimetric analysis, 36, 502
  self-test, 35
  volumetric analysis, 509
Quadratic equation, inside front cover
Qualitative analysis, 4, 323, 351
Quantitative analysis, 4, 329, 353
Quinine, 342, *t* 350, 527

Radiant energy, 273
Range, 30, 33
Refraction, 275, 339, 504
Refractive index, 504
Retardation factor (R_f), 450
Retention ratio, 450

Salicylic acid:
  fluorescence, *t* 350, 351
  irritation of stomach, 329
  pain-killer, 329
  tolerance in aspirin, 328
Salts, pH of, 151
Sample (statistical), 29
Sample handling, 66
Sampling, 43–52
  air, 47
  gases, 48
  liquids, 46
  solids, 44–45
Separations, Ch. 19
  electrodeposition, 425
  extraction, 424, 433
  chromatography, 424–32, 461–73
  precipitation, Ch. 5, 424
  theory, Ch. 20
Significant figures, 19–23
  absorbance, 22, 291
  equilibrium constants, 22, 150, 251
  log terms, 22
  pH, 22, 148
  self-test, 22
Silica gel, 462
Silver(I) chloride, 73, 84, 90, 93
Sodium:
  atomic absorption, 276, 364, 379
  flame emission determination, 364, 373

Sodium (*continued*)
  in blood, *t* 481, *t* 485
  potentiometric determination, 397
Sodium chloride in body fluids, 212, 218
  density determination, 71, 495
  gravimetric determination, 498
  refractive index determination, 504
  specific ion determination, 506
  volumetric determination, 212, 218, 502
Sodium hydroxide, 175, 508
Solubility calculations, 91
Solubility product constant, 90, *t* 532
Specific gravity, 69
Spectrofluorometer, 344, 346
Spectronic 20 spectrophotometer, 302, 522
Spectronic 21 spectrophotometer, 303
Spectrophotometer, 295–306
  detectors, 300
  errors, 305
  monochromator, 297
  schematic diagram, 296, 302
  ultraviolet-visible, 303
  visible, 301
Stability constants, 220, 223
Sulfanilamide, 207
Sulfate, 219
Sulfathiazole, 207
Sulfur dioxide, 332
Sulfuric acid pH, 148
Sweat chloride, 394

Theoretical plates, 457
Thermocouple, 300
Titration:
  acid-base, Ch. 9
  amperometric, 416
  complexation (EDTA), 220–34
  coulometric, 405
  oxidation-reduction, Ch. 11
  precipitation, 211–20
  predicting quantitative reactions, 133–38
Titration curves:
  acid-base, 132, 182, 192
  general, 130
  phosphoric acid, 195
  precipitation, 214
Toluidine method for glucose, 486
Toxic metals, *t* 380
Transducer, 271, 277, 279
Transmittance, 290–92
Tryptophan, *t* 350
Tyrosine, *t* 350

Ultraviolet radiation, 285, 310, 318
Uncertainty, 19
  single buret reading, 112
  single weighing, 55
  weighing by difference, 56
Uranium(VI) fluorescence, 351
Urea (BUN), 490
Urine, 360, *t* 481

Van Deemter equation, 457
Vitamin A, 464, 466
  oxidation, 239
  reduction, 240
Vitamin B separation, 468
Vitamin C (ascorbic acid):
  analysis, 519
  calculation of purity, 243
  oxidation, 243, 245
Vitamin D separation, 466
Vitamin E, 260
Volhard method, 212, 217
Volumetric analysis, 111–18
  calculations, 102–11
Volumetric flask, 112

Water:
  colorimetric analysis, 332
  fluoride analysis, 395
  hardness, 225, 232–33
  ionization, 144
  pH, 146
  values of K_w at different temperatures, 145

/5435324Q>C1/

DATE DUE

DEC 07 1990			
NOV 28 1995			

Demco, Inc. 38-293